# *Mechanics of Fluids* in SI Units

By A. C. WALSHAW
D.Sc., Ph.D., A.C.G.I., C.Eng., F.I.Mech.E.

*Thermodynamics for Engineers*

*SI Units and Worked Examples*

Second Edition

In SI Units

# *Mechanics of Fluids*

## A. C. Walshaw

D.Sc., Ph.D., A.C.G.I., C.Eng.

F.I.Mech.E. *Professor-Emeritus, University of Aston in Birmingham*
*Formerly Head of the*
*Department and Professor*
*of Mechanical Engineering*

## and

## D. A. Jobson

B.Sc., C.Eng., M.I.Mech.E.

A.F.R.Ae.S. *of the Reactor Group, United Kingdom Atomic Energy Authority*
*Both formerly Assistant Professor*
*of Applied Mechanics,*
*Royal Naval College, Greenwich*

Longman

LONGMAN GROUP LIMITED, LONDON
Associated companies, branches and representatives throughout the world

© A. C. WALSHAW, D. A. JOBSON 1962

SECOND EDITION © LONGMAN GROUP LTD 1972

FIRST PUBLISHED 1962
SECOND IMPRESSION 1965
THIRD IMPRESSION 1967

SECOND EDITION 1972

ISBN: 0582 44485 3 Cased
ISBN: 0582 44494 2 Paper

TYPESET BY ETA SERVICES (TYPESETTERS) LTD AND PRINTED IN
GREAT BRITAIN BY HAZELL, WATSON & VINEY LTD.

# Preface

In this edition SI units have been used throughout, and new material has replaced less important sections and problems of the first edition. The book is primarily for those studying for examinations leading to Engineering Degrees and Diplomas, Higher National Diplomas and Certificates, and Examinations of Professional Institutions.

It aims at presenting a straightforward and modern approach to the subject of Fluid Mechanics. Analysis and development of fundamental formulae have been kept as simple as possible. Worked examples are included in the text to illustrate the applications of fundamental principles in design and practical problems. In order to get a thorough grasp of the subject however, it is essential that students should supplement a study of the text and worked examples by working out a selection of the exercises included at the ends of the chapters.

There are few pieces of modern engineering plant or machinery which do not to some extent depend on a measure of understanding of the Mechanics of Fluids for their successful operation or control. In many cases the process of fluid motion is fundamental. Consequently the flow of fluids through prime movers such as water turbines, gas turbines and jet engines has become increasingly important as, also, has the flow of liquids and gases through pumps and compressors. The development of the aeroplane, and the need to develop rockets, missiles, torpedoes, submarines and speedy ships has compelled an intense and thorough study of the motion of fluids generally—gaseous and liquid—relative to solid bodies. Many empirical formulae of the past have now been replaced by rational ones as a result of analysis by the Method of Dimensions, which reveals the non-dimensional groups or parameters by means of which graphs may be plotted in the most condensed and yet most general form.

This book is intended to serve in introducing and assisting students in their early studies of these things.

A. C. W.
D. A. J.

# Contents

## 4  Bernoulli's equation and measurement of flow of incompressible fluids

## 5  Elements of similarity; notches and weirs

## 6  Equations of motion for a fluid element

## 7  Fluid momentum and thrust by reaction

## 8 Behaviour of ideal and viscous fluids

## 9 Viscous films of varying thickness

## 10 Similarity and dimensional analysis

## 11 Steady flow in pipes and channels

## 15  Hydro-kinetic machines

## 16  Positive-displacement machines

# Key to main use of symbols

**Symbols for physical quantities** are usually single letters printed in italics or sloping type. Two-letter symbols (from proper names) are, however, used to represent dimensionless groups of certain physical quantities (e.g. Mach number, $Ma = u/a$; Reynolds' number $Re = ud/v$). Such two-letter symbols can have their indivisibility stressed, if necessary, by placing them within brackets.

| *Symbol* | *Signification* |
|---|---|
| $A$ | Area; constant. |
| $a$ | Area; linear acceleration; local speed of sound; index number; constant. |
| $b$ | Breadth; index number; constant. |
| $C$ | Coefficient; cycles/unit-time; capacitance; heat-capacity. |
| $c$ | Constant; index number; chord length; specific heat-capacity; velocity of sound in water or speed of a wavelet. |
| $D$ | Diameter; drag force or stress; draught of floating body. |
| $d$ | Depth; diameter; index number; relative density ($\rho/\rho_w$). |
| $E$ | Energy; Young's modulus. |
| $F$ | Force; function of. |
| $f$ | Friction coefficient; function of. |
| $Fr$ | Froude number ($u/\sqrt{(gl)}$). |
| $g$ | Acceleration of free fall. |
| $g_n$ | International standard acceleration due to gravity ($9{\cdot}80665$ m/s$^2$). |
| $H, h$ | Head; enthalpy. |
| $h_L$ | Head loss. |
| $I$ | Moment of inertia; second moment of area; specific impulse. |
| $i$ | Slope of hydraulic gradient. |
| $J$ | Non-dimensional group ($V/ND$) for propellers. |
| $K$ | Bulk modulus. |
| $k$ | Radius of gyration; constant or coefficient. |
| $L$ | Length; aero-dynamic force of lift. |
| $l$ | Length. |
| $M$ | Mass; moment or torque. |
| $m$ | Mass; hydraulic mean depth; suffix meaning 'model'. |
| $\dot{m}$ | Rate of mass flow. |

| *Symbol* | *Signification* |
|---|---|
| $Ma$ | Mach number ($u/a$). |
| $N$ | Rotational speed. |
| $n$ | Number; index in gas law ($PV^n$); distance along a normal. |
| $P, p$ | Pressure; wetted perimeter; power. |
| $Q, q$ | Rate of volumetric flow; quantity of heat-transfer. |
| $R$ | Radius; gas constant ($P = \rho RT$); force of reaction. |
| $Re$ | Reynolds' number ($ud/\nu$ or $\rho u l/\eta$). |
| $S$ | Surface area; distance; scale ratio. |
| $s$ | Distance; slip. |
| $T$ | Thrust; torque; absolute temperature; surface tension. |
| $t$ | Time; customary scale-temperature (Celsius, °C). |
| $U$ | Upthrust; linear velocity. |
| $u$ | Linear velocity. |
| $V$ | Volume; linear velocity, |
| $v$ | Specific volume ($V/m$). |
| $W$ | Load or weight; work. |
| $X, x$ | Distance. |
| $Y, y$ | Distance. |
| $Z, z.$ | Height above a datum. |
| $\alpha$ | Angle; angular acceleration. |
| $\beta$ | Angle. |
| $\gamma$ | Ratio of specific heat-capacities $c_p/c_v$ of gases. |
| $\epsilon$ | Eccentricity ratio; efficiency. |
| $\eta$ | Dynamic viscosity; efficiency. |
| $\theta$ | Angle. |
| $\lambda$ | Friction coefficient ($\lambda = 4f$). |
| $\mu$ | Coefficient of friction. |
| $\nu$ | Kinematic viscosity ($\nu = \eta/\rho$). |
| $\pi, \Pi$ | 3·1416; non-dimensional group. |
| $\rho$ | Density ($\rho = m/V = 1/v$). |
| $\Sigma$ | Sum of. |
| $\tau$ | Viscous shear stress; periodic time. |
| $\varphi$ | Angle; function of. |
| $\dot{\varphi}$ | Strain rate. |
| $\psi$ | Angle; function of. |
| $\omega, \Omega$ | Angular velocity. |

The word 'specific' is restricted to the meaning 'divided by mass'—e.g. specific volume $v = V/m = 1/\rho$.

**Symbols for units** are printed in lower case roman (upright) type, except when a unit is derived from a proper name. A capital roman letter is then taken as the unit symbol—e.g. N (newton), K (kelvin), A (ampere), J (joule).

## SI base units

| Quantity | Name of unit | Unit-symbol |
|---|---|---|
| Length | metre | m |
| Mass | kilogramme | kg |
| Time | second | s |
| Thermodynamic temperature | kelvin | K |
| Electric current | ampere | A |
| Luminous intensity | candela | cd |

## Supplementary units

| | | |
|---|---|---|
| Plane angle | radian | rad |
| Solid angle | steradian | sr |
| Quantity of substance | mole | mol |

## Derived units and special names

| Quantity | Name | Symbol | SI units |
|---|---|---|---|
| Force | newton | N | $kg\,m/s^2$ |
| Energy | joule | J | $Nm = kg\,m^2/s^2$ |
| Power | watt | W | $J/s = kg\,m^2/s^3$ |
| Pressure | pascal | Pa | $N/m^2 = kg/s^2 m$ |
| Volume | litre | l | $dm^3 = 10^{-3}\,m^3$ |
| Mass | tonne | t | $Mg = 10^3\,kg$ |
| Time | minute | min | $60\,s$ |
| Dynamic viscosity | poise | P | $10^{-1}\,kg/sm$ |
| Kinematic viscosity | stoke | St | $10^{-4}\,m^2/s$ |
| Frequency | hertz | Hz | $s^{-1}$ |
| Electrical potential | volt | V | $J/sA = kg\,m^2/s^3 A$ |
| Electrical resistance | ohm | $\Omega$ | $kg\,m^2/s^3 A^2$ |

## Multiples and sub-multiples of SI and derived units

Appropriate sizes of units can be provided by use of prefixes signifying multiples and sub-multiples of 10—there being single-letter abbreviations internationally agreed (see table on following page).

*The prefix is part of the unit*—i.e. the combination of a prefix and unit-symbol is considered as *one new unit-symbol*, e.g. $1\,km^2$ is $1\,(km)^2$, *not* $1000\,m^2$.

As a general rule prefixes are used in the numerator only, e.g. $MJ/dm^3 = 10^9\,J/m^3$ is less cumbersome and clearer if written $GJ/m^3$, as is Young's modulus $E = 207\,kN/mm^2$ if written $207\,GN/m^2$.

| Fraction | Prefix | | Multiple | Prefix | |
|---|---|---|---|---|---|
| | Name | Abbreviation | | Name | Abbreviation |
| $10^{-1}$ | deci | d | 10 | deca | da |
| $10^{-2}$ | centi | c | $10^2$ | hecto | h |
| $10^{-3}$ | milli | m | $10^3$ | kilo | k |
| $10^{-6}$ | micro | $\mu$ | $10^6$ | mega | M |
| $10^{-9}$ | nano | n | $10^9$ | giga | G |
| $10^{-12}$ | pico | p | $10^{12}$ | tera | T |

Only one prefix is applied at one time to a unit-symbol, e.g. 1 mega-gramme is written 1 Mg, not 1 kkg. Thus, although the name 'kilo-gramme' has been retained for the SI unit of mass, the *single-prefix rule* is applied in that kg *is regarded as a prefixed unit*. Similarly, although 1 P = 1 dN s/m², one centipoise (1 cP) is not written as 1 cdN s/m² but as 1 mN s/m², in which it will be seen that care is needed with the 'm's'. Hence, *prefixes are printed immediately adjacent to their unit-symbols*, e.g. mN is a prefixed symbol meaning milli-newton. It must not be confused with a metre-Newton which is denoted here either by m × N or by m.N; the use of Nm instead is better in the sense that it eliminates the possible confusion.

In practice, to reduce the number of prefixes which could be used, 'preferred' prefixes have the form $10^{\pm 3n}$, where $n$ is an integer. It may be noted that 'litre' (being $10^{-3}$ m³) and cP (being $10^{-3}$ kg/s m) happen to be preferred sub-multiples of the SI unit. 'Non-preferred' prefixes are not, however, vetoed if the context is appropriate.

**Deviations from the SI**

It has been internationally agreed that certain units which deviate from the coherence of the SI shall be retained, e.g.

(i) the usual larger units of time—minute (min = 60 s), hour (h = 3600 s), etc.

(ii) 360 degrees ($2\pi$ rad) in a circle because of international practice

(iii) the kilo-watt-hour (kWh = 3·6 MJ), which is a deviation from the SI because the hour is not a decimal multiple of the second

(iv) customary Celsius scale-temperature °C, where $t/°C = T/K - 273\cdot15$.

These are seen to be exceptions to the recommendation that only six base units (or their decimal multiples and fractions) should normally be used.

Symbols for *physical quantities* are printed in *italic* (sloping) type, and symbols for *units*, in roman (upright) type. In handwriting and typescript the distinction can be made (when necessary) by under-lining the symbols for physical quantities.

Abbreviations for plurals do not take an 's'—e.g. the abbreviation for kilogrammes is kg, not kgs.

In division of one unit by others, only one solidus is used, e.g. acceleration is $m/s^2$, not m/s/s; and for dynamic viscosity 1 centipoise (1 cP) is $10^{-3}$ kg/sm and 1 centi-stoke ( 1 cSt) is $10^{-6}$ $m^2/s$.

Numbers are printed in upright type, and the decimal sign between digits is a point (·) placed above the line. To facilitate the reading of long numbers the digits are grouped in threes, but no comma is used, and the sign for multiplication of numbers is usually a cross ( × ).

## Chapter 1

# Buoyancy and stability

*'The weight of water displaced by a ship is precisely the same as the weight of that ship.'*
LEONARDO

## 1.1. Introduction and definition of fluid mechanics

The subject of *fluid mechanics* or the *mechanics of fluids* is nowadays generally understood to cover that branch of applied science concerned with substances which cannot preserve a shape of their own. With this definition, the term *fluid* applies equally well to liquids, vapours, and gases, but excludes such semi-solids as fats and waxes. We shall concentrate on those aspects of the mechanical behaviour of fluids which are of direct interest to engineers. Hence we shall be concerned only with those fundamental principles which are essential to a rational understanding of the behaviour of fluids in civil, mechanical, aeronautical, and marine engineering installations. To do this it will be necessary to study those characteristics of a fluid which are described and measured in terms of its mechanical properties, such as density and viscosity. It will be necessary also to know the laws which govern both its equilibrium and its motion, so that we may understand how fluid forces arise. Finally, we shall study how these forces may be usefully employed in the generation, transmission and utilization of power.

## 1.2. Buoyancy and the principle of Archimedes

The earliest physical principle in fluid mechanics which history has preserved for us was established by Archimedes of Syracuse (287–212 B.C.). His ability as a mathematician may be judged from the fact that his estimate of $\pi$ was more accurate than the 22/7 which we so frequently use as a convenient approximation. In addition, he was a brilliant engineer, and designed prodigious war machines to defend Syrácuse from the Roman fleet. His studies in hydrostatics, which had a strictly practical application, led him to the principle which bears his name, i.e.

**Every body experiences an upthrust equal to the weight of fluid it displaces.**

This principle of Archimedes implies that we may imagine replacing the body by fluid of the same kind as that surrounding it, without disturbing the latter. The upward force which the surrounding fluid exerts, is referred to as the *force of buoyancy*, and it is this force which maintains equilibrium against the weight of the imaginary volume of fluid replacing the body. The latter is, however, concentrated at the centroid of the displaced volume (Fig. 1.2(a)) so that, if equilibrium is to be preserved, the

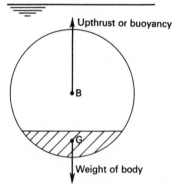

**Fig. 1.2(a).** *The principle of Archimedes*

upthrust must also pass through this point. **The centroid of the displaced volume is therefore referred to as the centre of buoyancy, being the point through which the buoyancy force acts.**

The centre of buoyancy (B) should not be confused with the centre of gravity (G) of the body, see Fig. 1.2(a), which depends on the weight distribution in the latter.

For example, the centre of gravity of a torpedo is determined by the disposition of the weights of the various parts and gear within the hull or casing, whereas the centre of buoyancy is determined solely by the shape of the casing displacing the liquid.

It should be noted that Archimedes' Principle is equally valid for all fluids—gaseous as well as liquid. Thus, when a body is weighed a correction should strictly be made for the buoyancy due to the air. Although this is generally relatively small it is, of course, the reason why, say, a gas-filled balloon rises. Equilibrium is finally established in accordance with the buoyancy principle outlined above, i.e. the upthrust, which decreases with altitude due to the reduced density of the air, eventually becomes equal to the weight of the balloon, its contents, and that of any mooring cable.

## 1.3.  The stability of a submerged body

A submerged body can hover in equilibrium if its weight equals its buoyancy. This condition, which is referred to as neutral buoyancy, can be approached in a submarine, e.g. by flooding tanks with water, or 'blowing' them with air, until the weight is made equal to the upthrust. As the weight may be imagined to be concentrated at G, and the buoyancy force at B (see Fig. 1.2(a)), we need to establish whether the weight will tend to hang from this virtual pivot B, or to rest upon it. Either is nominally an equilibrium position, but only the former configuration is stable. This is a perfectly general conclusion and is demonstrated in Figs. 1.3(a and b) by considering a drum with a weight fastened to its

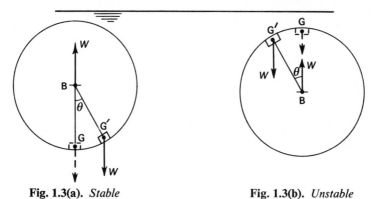

**Fig. 1.3(a).**  *Stable*          **Fig. 1.3(b).**  *Unstable*

*Stability of a submerged body*

side as shown. The weight will naturally set itself at the lowest position, i.e. it is in stable equilibrium when G lies below B.

The test for stability is to imagine the body to be disturbed from the equilibrium position considered, G being moved to G' in each case. In case (a) the body will experience a couple (of magnitude $W . BG' \sin \theta$) tending to return it to its original configuration on being released; it is therefore said to be stable. In case (b) the body will tend to topple on being released, so that the original configuration is said to be an unstable one.

**Thus, if a submerged body initially at rest be slightly displaced so that the force of buoyancy and the force of gravity acting on the body are not in the same vertical line, the body is said to be in stable or unstable equilibrium according as the resulting couple tends to bring the body back to its original position or to give it further displacement.**

## 1.4.  Density and relative density

Just as Newton's Principle of Gravitation is usually said to have been inspired by the fall of an apple from a tree in an orchard, so Archimedes'

Principle is alleged to have resulted from his absent-minded entry into a bath which was full of water. Having realized that his body was displacing a corresponding volume of water, he is said to have rushed out shouting 'Eureka!' ('I have found it!') What he had found was a way of checking the weight per unit volume ($\rho g$) of a crown which King Hiero had ordered to be made of pure gold; he was thus able to confirm the King's suspicions concerning the honesty of the craftsman who had made it.

One most convenient way of estimating the density of a liquid (say, the acid in a battery) relative to pure water is by using a hydrometer. This consists essentially of a highly stable float with a graduated stem

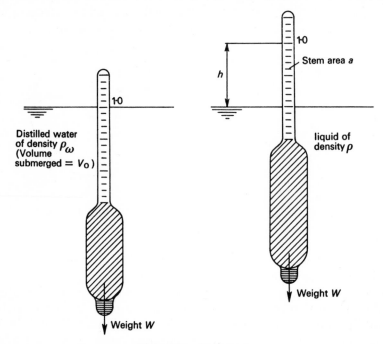

**Fig. 1.4(a).** *Hydrometer*

calibrated in terms of relative density as shown in Fig. 1.4(a). By Archimedes' Principle the hydrometer sets itself so that it displaces its own weight—the depth of immersion of the stem being an inverse measure of the relative density of the fluid. For accurate work a series of sensitive hydrometers is necessary, each covering only a limited range of values. For obvious reasons Customs and Excise Officers carry hydrometers when visiting breweries and distilleries.

Referring to Fig. 1.4(a), the weight $W$ of the hydrometer remains constant. Thus, in distilled water the weight is $\rho_w g V_0$ and the position of the water level on the stem is marked 1·0 to indicate the relative density. When floated in a liquid of density $\rho$ the weight of the hydrometer is $W = \rho g(V_0 - ah)$. Hence, $W = \rho_w g V_0 = \rho g(V_0 - ah)$

and
$$h = \frac{V_0}{a}\left(1 - \frac{\rho_w}{\rho}\right)$$

from which formula the stem of the hydrometer may be graduated to show specific gravity of the liquid in which the hydrometer is floated.

EXAMPLE 1.4 (i)

*What percentage of the total volume of an iceberg of density 912 kg/m³ will extend above the surface of sea water of density 1025 kg/m³.*
   Referring to Fig. 1.4(b):

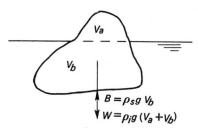

Fig. 1.4(b). *Iceberg*

$$W = B$$

$\therefore$
$$\frac{V_a}{V_b} + 1 = \frac{\rho_s}{\rho_i}$$

*or*
$$\frac{V_a}{V_b} = \frac{1025}{912} - 1 = \frac{113}{912}$$

*Hence,*
$$\frac{V_b}{V_a} + 1 = \frac{912}{113} + 1 = \frac{1025}{113}$$

$$= \frac{V_{total}}{V_{above}}$$

*and*
$$\frac{V_{above}}{V_{total}} = \frac{113}{1025} \text{ or } \mathbf{11} \text{ per cent}$$

EXAMPLE 1.4 (ii)

*A 'ball-cock' type of float valve is required to close when two-thirds of the volume of the spherical float is immersed in water having a density of 1 Mg/m³. The valve has a diameter of 12·5 mm, and the fulcrum of the operating lever is to be 100 mm from the valve and 0·45 m from the centre of the float. Estimate the minimum diameter of the float if it is required to close the valve against a pressure of 138 kN/m² gauge. g = 9·807 m/s².*
   Referring to Fig. 1.4(c), the force required to close the valve is

$$F_A = 138 \frac{kN}{m^2} \times \frac{\pi}{4} \times 156·3 \text{ mm}^2 = 16·94 \text{ N}$$

Hence, the force of buoyancy required to be exerted by the float is

$$F_B = \frac{10}{45} \times 16 \cdot 94 \text{ N} = 3 \cdot 76 \text{ N}$$

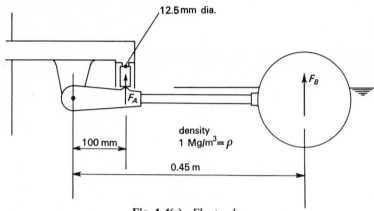

**Fig. 1.4(c).** *Float valve*

Also, if the volume of the sphere is denoted by $V$ and its weight and that of the lever are neglected, then

$$F_B = \tfrac{2}{3}\rho g V = 3 \cdot 76 \text{ N}$$

i.e.

$$V = \frac{3}{2} \times \frac{3 \cdot 76 \text{ N}}{10^3 \dfrac{\text{kg}}{\text{m}^3} \times 9 \cdot 807 \dfrac{\text{m}}{\text{s}^2} \left[ \dfrac{\text{Ns}^2}{\text{kg m}} \right]} = \frac{0 \cdot 577}{10^3} \text{ m}^3$$

The volume $V$ of a sphere is related to its radius $r$ by $V = \tfrac{4}{3}\pi r^3$, hence

$$r = \sqrt[3]{\left( \frac{3}{4} \times \frac{0 \cdot 577}{\pi} \frac{\text{m}^3}{10^3} \right)} = \frac{0 \cdot 516}{10} \text{ m}$$

and the diameter of the float is **103·2** mm.

To allow for the weight of the lever and float, and to ensure that the valve is firmly seated, a larger diameter would be used in practice.

## 1.5. The metacentre and metacentric height of a floating body

Archimedes' statement that a body experiences an upthrust equal to the weight of fluid it displaces is just as valid for bodies floating on the surface as for those which are submerged. It therefore embodies the principle of flotation which states that when at rest **a floating body displaces a volume of fluid equal in weight to its own.** The force of buoyancy acts, as stated in Section 1.2, through the centroid of the displaced volume, but we must note that if a floating body is heeled or pitched, its centre of buoyancy is also moved. The reason for this movement may be seen by referring to

Fig. 1.5(a) in which the heel of a ship is conveniently represented by re-drawing the water surface as RS instead of as PQ. For the ship to displace her own weight in the heeled condition, it follows that she must pivot about O, so that the buoyancy lost, owing to the heeling having rendered the weight of the wedge of fluid OPR inoperative, is balanced by the buoyancy received as a result of the displacement of the corresponding wedge of fluid OQS. If, however, the vessel has a 'flare', i.e. sloping sides, this is not quite true, but the error is generally negligible for small angles of heel.

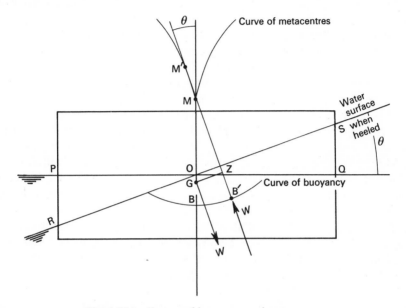

**Fig. 1.5(a).** *Curves of buoyancy and metacentre*

When the vessel is upright the centre of buoyancy is at B, i.e. at the centroid of the volume enclosed by the underwater shape of the vessel and the water surface PQ. When, however, the vessel is heeled, the centre of buoyancy moves to B′, the centroid of the volume enclosed by the underwater shape and the water surface RS.

The upthrust acting through B′, and the weight acting through G constitute a couple which exerts a moment equal to their magnitude, $W$, times the perpendicular distance, GZ, between their lines of action. The distance GZ is known as the *righting lever*, and is positive if the vessel is stable, i.e. the couple tends to restore the ship to the upright position.

The locus of the successive positions of the centre of buoyancy B′ as the angle of heel $\theta$ is increased is known as the *curve of buoyancy*. Its derivation for a particular 'displacement', i.e. weight of ship, constitutes a lengthy exercise in determining centroids and need occupy the attention of naval architects only. The shape of the curve depends exclusively on the lines of the vessel for a particular displacement. Since the centre of buoyancy is the point at which the weight of the vessel is supported, we may imagine that she rests upon that bit of tangent to the curve of buoyancy parallel to the water surface as indicated in Fig. 1.5(a).

Similarly, above the water surface, the curve to which the line of action of the upthrust remains always tangential as the ship (or water surface) is heeled, is known as the *curve of metacentres* for a particular displacement. Mathematically it follows that the curve of metacentres is the evolute of the curve of buoyancy. The cusp (M) in the former curve shown in Fig. 1.5(a) is known as the initial metacentre, and the distance GM is the initial metacentric height. It follows that **the initial metacentre, M, is the point where the line of action of the upthrust intersects the original vertical line through the centre of buoyancy, B, and the centre of gravity, G, for an infinitesimal angle of heel.**

In practice it is found that for *small* angles of heel the line of action of the upthrust passes very nearly through the initial metacentre M. Hence, so long as the angle of heel is less than say 15°, we may assume that the upthrust always acts through the fixed point M, just as the weight always acts vertically downwards through G. Under these conditions the righting moment or couple is $W \times GM \times \sin \theta$, in which expression the distance GM may be considered to be a constant for the vessel. This initial value is generally implied when referring to the *transverse metacentric height* of the vessel, and is a measure of its static stiffness in roll. So long as G lies below M the righting couple will be positive, and this implies stability. In fact, **a floating body is in stable, unstable, or neutral equilibrium according as the metacentre lies above, below, or at the centre of gravity.** Obviously it is important that ships be designed such that M is normally above G for all conditions of loading, and under all circumstances of rolling.

We may note that a negative GM in the upright position is not necessarily catastrophic. Although a vessel in this condition cannot be persuaded to remain upright, she may possibly find a new stable equilibrium position by developing a 'loll' to one side or the other. A vessel with a large metacentric height is said, by naval architects, to be a 'stiff' ship which is found to be correspondingly lively, i.e. the vessel tends to roll with a predominantly greater amplitude and frequency in a rough sea. Merchant ships, especially liners, are therefore designed to have a relatively small metacentric height—say, between 0·3 and 0·6 m, but in warships sea-kindliness is sacrificed so that they have a large reserve of stability—GM varying between, say, 0·6 and 2 m according to displacement. These are the metacentric heights for 'rolling' displacements about a longitudinal axis. The metacentric heights for 'pitching' displacements about a transverse axis are, of course, much larger.

EXAMPLE 1.5 (i)

(a) *Outline briefly how a static stability curve may be obtained, over the whole* range of stability of a ship, for a particular displacement and centre of gravity.

(b) Such a curve of righting levers for a vessel is given by the following ordinates at 10 degree intervals from the upright position:
Righting lever in m: 0, 0·09, 0·50, 0·97, 1·23, 1·16, 0·79, 0·09.

(i) Estimate the influence on the range of stability caused by raising the centre of gravity 0·6 m.

(ii) Discuss the behaviour of the ship with the raised centre of gravity and state what steps should be taken to correct its condition.

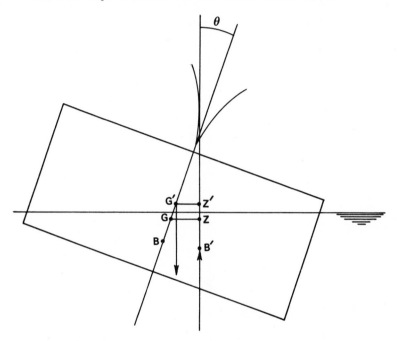

**Fig. 1.5(b).** *Effect of position of centre of gravity on righting lever*

(a) The centroid B′ of the displaced volume must first be determined from the lines of the vessel for a number of angles of heel up to the point of vanishing stability. The water surface is chosen in each case so that the displacement has the correct value.

The lever arm at which the upthrust through B′ acts about the centre of gravity G, gives the righting lever GZ, as indicated in Fig. 1.5(b). A graph showing the variation of the latter with heel is known as a static stability curve; see Fig. 1.5(c).

(b) *Righting Lever* (GZ)

From Fig. 1.5(c) it is seen that the effect of raising the C.G. from G to G′ is to reduce the righting lever from GZ to G′Z′. The reduction in the righting lever is thus:

$$GZ - G'Z' = GG' \sin \theta$$

Hence, if GG′ = 0·6 m we deduce the following values:

| $\theta$ | 0 | 10 | 20 | 30 | 40 | 50 | 60 | 70 | degrees |
|---|---|---|---|---|---|---|---|---|---|
| GG′ sin $\theta$ | 0 | 0·104 | 0·205 | 0·300 | 0·386 | 0·460 | 0·520 | 0·564 | m |

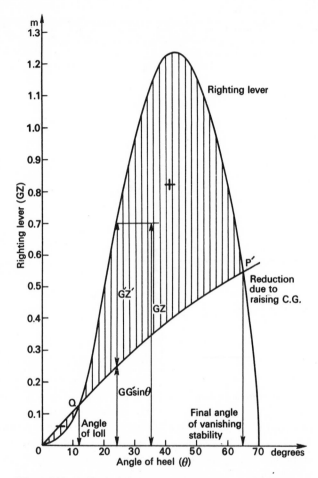

**Fig. 1.5(c).** *Effect of C.G. position on static stability curve*

By plotting these on the original static stability curve, the new righting levers, with the C.G. raised to G′, are represented as differences, namely, the intercept G′Z′ between the two graphs. Thus, the angle of vanishing stability is reduced from that at P (71 degrees) to that at P′ (65 degrees).

Since the righting lever is now negative between O and Q, the vessel will take on a loll of 11 degrees. This condition can be corrected by, say, flooding deep tanks so that G is brought below the curve of metacentres for the new displacement.

## 1.6. The inclining experiment, to determine metacentric height

Although the position of the metacentre M depends only on the geometrical form, that of the centre of gravity G is fixed by the distribution
of the weights in the ship. Hence, although the former can be deduced
with certainty from the lines of the vessel, the latter can only be known
approximately, especially when cargo is carried. Consequently, an inclining experiment may be carried out to deduce the initial metacentric
height, GM, and hence the position of G. The experiment consists
essentially of placing equal masses $\Delta M$ (of weight $\Delta Mg = \Delta W$) on each
side of a deck, as shown in Fig. 1.6(a). As one mass or the other is shifted

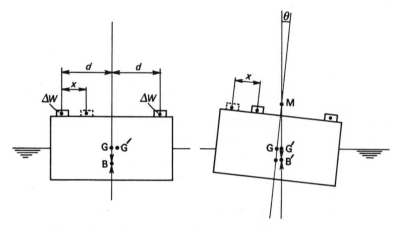

**Fig. 1.6(a).** *The inclining experiment for metacentric height*

transversely through a distance $x$, the ship heels to a new equilibrium
position, which is measured by noting the movement of a plumb bob.

The shift of the centre of gravity due to moving the left-hand mass
through $x$ may be deduced by imagining the ship to remain upright and
taking moments about G for example. Thus, denoting the *total* displacement of mass $M$ (i.e. weight $W = Mg$) we deduce, since the moment of
resultant = sum of moment of parts:

$$W.GG' = \Delta W.d - \Delta W(d-x) = \Delta W.x$$

hence,
$$GG' = \frac{\Delta W}{W}x$$

The ship heels until the upthrust and the weight (which now acts
through G') are in line, i.e. so that:

$$GG' = GM \tan \theta$$

$$\therefore \qquad GM = \frac{\Delta Wx}{W \tan \theta} = \frac{\Delta Mx}{M \tan \theta} \qquad (1)$$

By plotting a curve of the nominal GM from Equation (1), for positive and negative values of $\theta$, and reading off the value of GM for $\theta = 0$, the initial metacentric height is obtained.

EXAMPLE 1.6 (i)

*On moving a mass of 4 Mg or 4 tonne a distance of 9 m across the deck of a vessel of 1 996 tonne displacement the angle of list is changed from +0·015 to −0·015 radians. Find the metacentric height.*

In this example $\Delta M = 4$ tonne, $x = 9$ m, $M = 2\,000$ tonne, $\tan \theta \simeq \theta = 0\cdot03$ radian.

$$\therefore \qquad \text{Metacentric height GM} = \frac{\Delta M . x}{M \tan \theta} = \frac{4}{2\,000} \times \frac{9\text{ m}}{0\cdot03} = \textbf{0·6 m}$$

EXAMPLE 1.6 (ii)

*A ship which has been loaded in an up-river fresh-water harbour proceeds downstream into salt water, consuming 50 tonne of coal and stores on the way. Length of ship 90 m. Displacement 7 000 tonne as originally loaded. Distance of C.G. from bow 49 m as originally loaded. Longitudinal metacentric height 1 000 m. Water-plane area 1 000 m², which may be assumed constant. If 0·97 m³ of salt water and 1 m³ of fresh water each have a mass 1 tonne:*

(a) *Deduce the change of draught from the condition as originally loaded.*
(b) *How much cargo can now be taken on board to bring the ship to its original draught?*
(c) *If the centre of gravity of the 50 tonne of coal and stores consumed is 45 m from the bow of the ship, and if the additional cargo has to be placed at 60 m from the bow, what will be the change of trim (defined as the algebraic difference of changes of draught at bow and stern of ship)?*

(a) If $A$ is the area of the water plane and $M_0$ the original mass then the volume $V_f$ fresh water displaced is given by $M_0 = \rho_f V_f = 7\,000$ tonne. Similarly the volume $V_s$ of sea water displaced is given by $M_0 - 50$ tonne $= \rho_s V_s$.

Hence, the reduction in volume displaced in proceeding from fresh water to sea water and reducing the mass by 50 tonne is:

$$V_f - V_s = \frac{M_0}{\rho_f} - \left(\frac{M_0 - 50 \text{ tonne}}{\rho_s}\right) = A \times \text{decrease of draught}$$

$$\therefore \qquad \text{Decrease of draught} = \frac{1}{1\,000\text{ m}^2}\left\{\frac{7\,000 \text{ tonne}}{\dfrac{1 \text{ tonne}}{1 \text{ m}^3}} - \frac{6\,950 \text{ tonne}}{\dfrac{1 \text{ tonne}}{0\cdot97 \text{ m}^3}}\right\}$$

$$= (7 - 6\cdot74) \text{ m} = 0\cdot26 \text{ m} = \textbf{260 mm}$$

(b) Additional cargo $= \rho_s \times A \times$ increase in draught

$$= \frac{1 \text{ tonne}}{0\cdot97 \text{ m}^3} \times 1\,000 \text{ m}^2 \times 0\cdot26 \text{ m} = \textbf{268 tonne}$$

(c) The new loading in fresh water is as shown in Fig. 1.6(b), and the shift of the centre of gravity GG′ = x is given by:

$$7\,000x = 50(4+x)+268(11-x)$$
$$7\,218x = 3\,148$$

i.e.
$$x = GG' = 0\cdot436 \text{ m}$$

Hence, the change in trim, y, is given by:

$$\varDelta\theta = \frac{GG'}{GM} = \frac{y}{\text{Length of ship}}$$

i.e.     $y = \dfrac{0\cdot436}{122} \times 90 \text{ m} = 0\cdot322 \text{ m} = \textbf{322 mm down by the stern}$

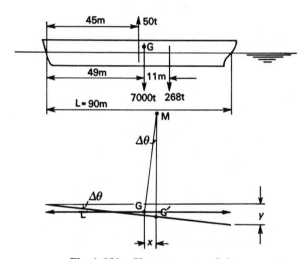

**Fig. 1.6(b).** *Change in trim of ship*

EXAMPLE 1.6 (iii)

*A ship which is* 120 m *long is required to take on additional cargo in a fresh-water dock, but it is necessary that the draught shall not be increased forward. If the centre of flotation (C. of G.) is* 60 m *aft of the bow how far aft must the cargo be placed? The vessel is of* 1·2 MN *per* 100 mm *immersion and requires a moment of* 1·5 MN m *to change the trim by* 25 mm.

$$\frac{M}{Y} = \frac{1\cdot5 \text{ MN m}}{0\cdot025 \text{ m}} = 60 \text{ MN}$$

$$D = \frac{1\cdot2 \text{ MN}}{0\cdot1 \text{ m}} = 12 \text{ MN/m}$$

Referring to Fig. 1.6(c), the additional load w at x is equivalent to w at G and moment w(x−l) about G.

w at G causes uniform sinkage w/D and the moment causes a change of trim

$$\frac{w(x-l)\,Y}{M} = y$$

and a forward lift of

$$l\frac{y}{L} = \frac{l}{L}\frac{w(x-l)\,Y}{M}$$

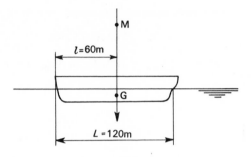

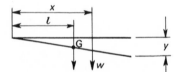

**Fig. 1.6(c).** *Change in trim*

Therefore, for no forward change in draught

$$\frac{w}{D} = \frac{l}{L}\frac{w(x-l)}{M/Y}$$

or
$$x = l + \frac{M}{YD}\frac{L}{l} = 60\ \text{m} + \frac{60\ \text{MN}}{12\ \text{MN}}\times\frac{120}{60}$$
$$\underline{\phantom{12\ \text{MN}}}$$
$$\text{m}$$

$$= (60+10)\ \text{m}$$

i.e. the cargo must be placed 70 m *aft*.


## 1.7. Metacentric radius

The methods outlined in Section 1.5 for determining the curves of buoy-ancy and of metacentres, although simple in theory, are extremely laborious in practice for all but the simplest of underwater shapes. If however a knowledge of the initial stability only is required, a simpler method may be used, by assuming a small angle of heel—exaggerated in Fig. 1.7(a) for clearness sake.

As stated in Section 1.5, the *loss* of buoyancy $\Delta F$ associated with emergence of the wedge OPR, is sensibly equal to the *gain* in buoyancy $\Delta F'$ due to immersion of the wedge OQS. We may thus consider the new buoyant force $F'$ at $B'$ to be the resultant of $F$ at B, together with $-\Delta F$ at C and $+\Delta F'$ at C', i.e. $F' = F - \Delta F + \Delta F'$ (the equivalent system of forces), $\Delta F$ and $\Delta F'$ being equal and opposite, so that the upthrust on the vessel remains unchanged.

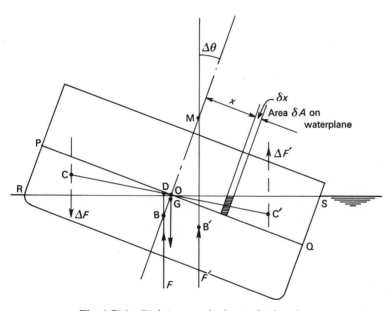

**Fig. 1.7(a).** *Righting couple due to fluid wedges*

By considering the torque, $T$, that the equivalent system of forces exerts about B, we may deduce the shift of the centre of buoyancy. Thus, for a *small* angle of heel, $T = \Delta F.\mathrm{CD} + \Delta F'.\mathrm{C'D} = F'.\mathrm{BB'}$, and since $\Delta F'$ is equal in magnitude to $\Delta F$, they constitute a couple of magnitude $\Delta F.\mathrm{CC'}$ which has the same value about *any* point in their plane. In addition, the force of buoyancy $F'$ is equal to the weight of fluid displaced, i.e. the immersed volume $V$ times the weight per unit volume $w = \rho g$ of the fluid. Hence, for small angles of heel,

$$T = \Delta F(\mathrm{CD} + \mathrm{C'D}) = wV.\mathrm{BB'}$$

or

$$T = \Delta F.\mathrm{CC'} = wV.\mathrm{BB'}$$

and

$$\mathrm{BB'} = \frac{\Delta F.\mathrm{CC'}}{wV} = \frac{T}{wV} = \frac{\text{Moment of wedges}}{\text{Weight of displaced fluid}} \qquad (1)$$

The position of $B'$ establishes the point through which the upthrust acts when a ship is heeled. As this varies with the latter, it is more convenient to work in terms of the metacentre M, as shown in Fig. 1.7(a).

The position of the latter, at least for small angles of heel, remains fixed, i.e. independent of $\Delta\theta$, as the following example will illustrate.

EXAMPLE 1.7 (i)

*Find the metacentric radius of a rectangular pontoon.*

*In certain simple cases Equation (1) may be applied directly. Consider for example a rectangular pontoon of length l, breadth b, and depth of immersion d as indicated in Fig. 1.7(b).*

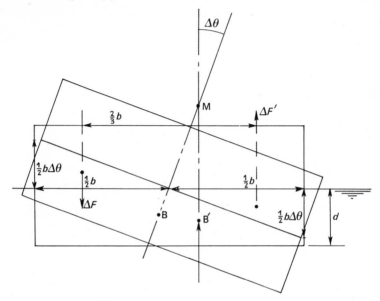

**Fig. 1.7(b).** *Righting couple on rectangular pontoon*

For a small angle of heel, the weight of each wedge is:

$$\Delta F = \Delta F' = w\frac{1}{2}\frac{b}{2}\frac{b\Delta\theta}{2}l = \tfrac{1}{8}wb^2l\Delta\theta,$$

where $w = \rho g$ is the weight of unit volume of fluid of density $\rho$.

Hence the torque due to the wedges is:

$$T = \tfrac{2}{3}b \times \tfrac{1}{8}wb^2l\Delta\theta = \tfrac{1}{12}wb^3l\Delta\theta$$

$$\therefore \qquad \text{BB}' = \frac{\tfrac{1}{12}wb^3l\Delta\theta}{wbdl} = \frac{1}{12}\frac{b^2}{d}\Delta\theta$$

In general, for a small angle of heel:

$$\text{BB}' = \text{BM}.\Delta\theta$$

Hence, the metacentric radius,

$$\text{BM} = \frac{1}{12}\frac{b^2}{d}$$

BM = BB′/$\Delta\theta$, known as the metacentric radius, is shown in Example 1.7 (i) to vary as the square of the beam, $b$, and inversely as the draught, $d$. *For a fixed position of the centre of gravity, an increase in beam, or decrease in draught both increase the metacentric height by raising* M. Lastly, an increase in the top weight or superstructure of a ship will raise G and reduce the metacentric height GM, and thus render the ship less stable. These are the three major factors which influence the stiffness, or initial stability.

In the general case, the moment of the wedges may be related to the shape of an imaginary deck at the water-line (known as the water-plane). First consider the force on a typical element, shown shaded in Fig. 1.7(a) Since the element has a volume $\delta A . x\Delta\theta$ it will contribute a moment:

$$\delta T = xw\delta A x\Delta\theta$$

Since all elements exert a moment in the same sense, no matter whether they be in the immersed or emergent wedge, the total moment (using $\Sigma$ to denote the sum) is:

$$T = w\Delta\theta \sum_{P}^{Q} x^2\delta A = w\Delta\theta I$$

The last symbol, $I$, in this equation depends entirely on the shape of the water-plane and may, more precisely, be expressed in terms of the calculus as:

$$I = \sum_{P}^{Q} x^2\delta A = \int_{P}^{Q} x^2\, dA$$

$I$ is termed the 'second moment of area' of the section and occurs also in other engineering problems as, e.g. the bending of beams. It is analogous to 'moment of inertia', in which mass $m$ replaces area $A$. Sometimes it is convenient to express $I$ in terms of a 'radius of gyration', $k$, defined by the equation: $I = Ak^2$, where $A$ is the area of the water-plane in this case. Hence, the moment exerted by the fluid wedges, due to the heeling of the vessel about a longitudinal axis through O may, for a small angle of heel, be written as:

$$T = wI\Delta\theta = wAk^2\Delta\theta \tag{2}$$

where $I$ and $k$ are respectively the second moment of area and radius of gyration of the water-plane about the longitudinal axis through the centroid of the latter, which is of area $A$.

From Equations (1) and (2) we deduce that the shift of the centre of buoyancy BB′ is:

$$BB' = \frac{T}{wV} = \frac{wI\Delta\theta}{wV} = \frac{I\Delta\theta}{V} \tag{3}$$

From Fig. 1.7(a) it is seen that, for a small angle of heel:

$$BB' = BM.\Delta\theta$$

Hence we deduce that the **metacentric radius:**

$$\text{BM} = \frac{I}{V} = \frac{Ak^2}{V} = \frac{\text{Second moment of area of the water-plane}}{\text{Volume of displaced fluid}} \quad (4)$$

Equation (4) correctly gives the metacentric radius for the *initial* stability, but is generally sufficiently accurate for all small angles of heel. If the position of the centre of gravity is known relative to B, the **initial metacentric height** may be determined from:

$$\text{GM} = \text{BM} \mp \text{BG}$$

the negative sign corresponding to the more usual case, in which G lies above B.

The equations developed in Sections 1.5, 1.6 and 1.7 may be applied to pitching as well as to heeling or rolling. If the displacement is to remain constant, a vessel must pitch as well as heel about an axis through the centre of flotation, i.e. the centroid of the water-plane. Consequently, $I$ or $k$ must be reckoned about a transverse axis through this point when considering pitching. Changes of trim are defined by changes in the difference of draught between bow and stern of a ship, and the moment to change trim may be found from (2). Similarly, fore and aft shift of the centre of buoyancy is given by (3), and the longitudinal BM by (4), provided that the longitudinal values of $I$ or $k$ are used.

As the longitudinal BM is much larger than BG, the difference between BM and GM may frequently be ignored when considering longitudinal stability.

EXAMPLE 1.7 (ii)

*A uniform cube of side L is required to be in stable equilibrium when floating in water. Find the limits between which its relative density $\rho/\rho_w = s$ must lie.*

Weight per unit volume of solid $= \rho g$

Weight of cube $\qquad = \rho g L^3$

Weight of water displaced $\qquad = \rho_w g L^2 d$

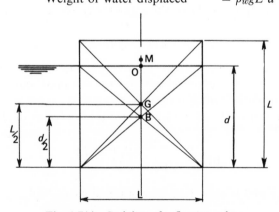

**Fig. 1.7(c).** *Stability of a floating cube*

Therefore, $\rho_w d = \rho L$ or $d = sL$

$$BM = \frac{I}{V} = \frac{\frac{1}{12}L.L^3}{sL^3} = \frac{1}{12}\frac{L}{s}$$

But

$$BG = \frac{L}{2} - \frac{d}{2} = \frac{L}{2}(1-s)$$

Hence, the initial metacentric height:

$$GM = BM - BG = \frac{L}{2}\left(\frac{1}{6s} - 1 + s\right)$$

The equilibrium is stable so long as GM is positive, i.e. if

$$\left(\frac{1}{6s} - 1 + s\right) > 0$$

or

$$(1 - 6s + 6s^2) > 0$$

Denoting the left-hand side of this expression by $y$, we may sketch the parabola:

$$y = 1 - 6s + 6s^2$$

and note that the equilibrium is stable or unstable according as $y$ is positive or negative; see Fig. 1.7(d).

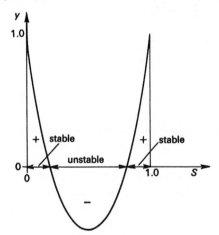

**Fig. 1.7(d).** *Stability diagram*

The borderline case of neutral stability occurs when $y = 0$, i.e. when

$$6s^2 - 6s + 1 = 0$$

or

$$s = \frac{6 \pm \sqrt{(36 - 24)}}{12}$$

$$= \frac{1}{2} \pm \frac{1}{\sqrt{12}}$$

$$= \mathbf{0 \cdot 789}\ or\ \mathbf{0 \cdot 211}$$

It follows that the relative density $s = \rho/\rho_w$ must lie outside these values for stable equilibrium, but must of course be less than unity.

We may conclude that a uniform block of timber will not in general float in stable equilibrium; since the values of the metacentric radii are proportional to the squares of the beam $b$ and length $L$, and vary inversely as the draught $d$. Adequate ratios of $b/d$ and $L/d$ must be allowed if stability is to be assured in any floating body.

EXAMPLE 1.7 (iii)

*A floating dock may be considered as consisting essentially of two side pontoons 110 m long, 3·5 m wide and 12 m deep, attached to a bottom pontoon 150 m long, 25 m wide and 3·5 m deep. The bottom of the side pontoons is 1 m above that of the bottom pontoon.*

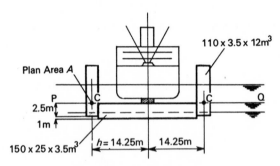

**Fig. 1.7(e).**  *A floating dock*

*Discuss its stability, when a ship is being lifted, and state at what stage the righting moment per unit angle of heel has its minimum value.*

*Determine the height of the metacentre above the centre of buoyancy of the dock for this condition, excluding the free surface effects of the flooded compartments.*

For convenience, consider the dock to be stationary and let the water surface be lowered progressively, as indicated in Fig. 1.7(e).

Initially BM will tend to increase, due to the decreasing draught, but this will be offset to some extent by the raising of the C.G. (due to water being pumped from the pontoons). Thus no large changes in GM need be expected at this stage. The righting moment per unit angle of heel is

$$\frac{T}{\Delta\theta} = W.\text{GM}$$

so that it may be expected to decrease progressively as the displacement of weight $W$ is reduced.

When however the waterline approaches the outer bottom of the ship, the water-plane area of the ship decreases rapidly until, when the bottom pontoon is just awash (as indicated by PQ), only the outer pontoons contribute to

*I* in:

$$\mathrm{BM} = \frac{I}{V}$$

The value of $W.\mathrm{GM}$ may therefore be expected to reach a minimum at this stage. The subsequent surfacing of the bottom pontoon will cause a sudden increase in stability, due to *I* reaching its maximum value.

The *I* of each side pontoon about an axis through its own centroid is:

$$I_c = \frac{bd^3}{12} = \frac{110 \times 3 \cdot 5^3}{12}\ \mathrm{m}^4 = 392\ \mathrm{m}^4$$

The water-plane area of each pontoon is:

$$A = 110\ \mathrm{m} \times 3 \cdot 5\ \mathrm{m} = 385\ \mathrm{m}^2$$

The parallel axes theorem concerning second moments of area enables us to determine *I* about O from *I* about C. It states that:

$$I_o = I_c + AL^2 = 392\ \mathrm{m}^4 + 385\ \mathrm{m}^2 \times 14 \cdot 25^2\ \mathrm{m}^2 = 7 \cdot 85 \times 10^4\ \mathrm{m}^4$$

Hence, when the bottom pontoon is awash, the total *I* is:

$$I = 2I_o = 15 \cdot 70 \times 10^4\ \mathrm{m}^4$$

The corresponding displaced volume is, at this stage:

$$V = 150\ \mathrm{m} \times 25\ \mathrm{m} \times 3 \cdot 5\ \mathrm{m} + 2(110\ \mathrm{m} \times 3 \cdot 5\ \mathrm{m} \times 2 \cdot 5\ \mathrm{m})$$

$$= 13\ 125\ \mathrm{m}^3 + 1\ 925\ \mathrm{m}^3 = 15 \cdot 05 \times 10^3\ \mathrm{m}^3$$

$$\therefore \qquad \mathrm{BM} = \frac{I}{V} = \frac{15 \cdot 70}{15 \cdot 05} \times 10\ \mathrm{m} = \mathbf{10 \cdot 46\ m}$$

## 1.8. The effect of a free surface

If the direction of all the arrows in Fig. 1.7(a) be reversed, we may consider the latter to represent the tilting of a tank of negligible weight containing fluid (see Fig. 1.8(a)). BB′ now represents the shift of the centre of gravity of the fluid and G the pivot point. Since the arrows are all reversed in this case, the equilibrium will be *unstable* if the fulcrum G lies *below.* M.

The moment due to the movement of the fluid wedges, the shift of the centre of gravity, BB′, and the distance between the metacentre M and the centre of gravity of the fluid, B, are given by Equations (2), (3) and (4) in Section 1.7 (p. 14), respectively:

$T = wI\Delta\theta$, where $w = \rho g$ is the weight per unit volume of the displaced fluid of density $\rho$.

$$\mathrm{BB}' = \frac{I\Delta\theta}{V}$$

$$\mathrm{BM} = \frac{I}{V}$$

The presence of free surfaces in a ship, such as those in oil fuel tanks, for example, causes a reduction in the righting moment when the ship is heeled. Denoting the second moment of area of such a free surface by $I'$ and the weight per unit volume of the fluid in the tank by $w'$, the righting moment on a ship is reduced from that given by Equation (2) on page 17 to:

$$T' = wI\Delta\theta - w'I'\,\Delta\theta$$

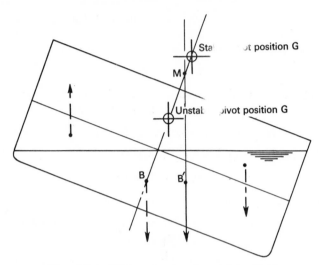

**Fig. 1.8(a).** *Tilting of a vessel containing fluid*

If we imagine this to modify the shift of the centre of buoyancy, the value of the latter is virtually:

$$BB' = \frac{wI\Delta\theta - w'I'\Delta\theta}{wV} = \frac{I - \dfrac{w'}{w}I'}{V}\,\Delta\theta$$

Hence Equation (4) now becomes:

$$BM = \frac{I - \dfrac{w'}{w}I'}{V} = \frac{I - \dfrac{\rho'}{\rho}I'}{V}$$

and it follows that GM is also reduced by $(\rho'/\rho)(I'/V)$ for each tank.

It should be noted that the position of the tank is immaterial, since the movement of the fluid wedges in the latter exerts a couple, and the moment due to this is independent of its point of application.

EXAMPLE 1.8 (i)

*A ship, of displacement mass 4 500 tonne, carries oil fuel in compartments 6 m long and 2·5 m broad. Estimate the loss of metacentric height when three such compartments are in use, if the oil has a relative density of 0·83 and water has a density of* $1t/m^3$.

Reduction in metacentric height caused by each compartment is:

$$\Delta(\text{GM}) = \frac{w'I'}{wV} = \frac{w'I'}{W} = \frac{\rho'I'}{M}$$

$$I' = \frac{6 \text{ m} \times 2 \cdot 5^3 \text{ m}^3}{12} = 7 \cdot 812 \text{ m}^4$$

Reduction caused by three compartments $= 3\Delta(\text{GM}) = 3\dfrac{\rho'I'}{M}$

$$= 3 \times \left(0 \cdot 83 \times \frac{1t}{m^3}\right) \times \frac{7 \cdot 812 \text{ m}^4}{4\,500 \text{ t}}$$

$$= \frac{0 \cdot 404 \text{ m}}{100}\left[\frac{10^3 \text{ mm}}{m}\right] = \mathbf{4 \cdot 04 \text{ mm}}$$

## 1.9. The period of roll of a vessel

The rolling period of a vessel may be determined approximately by making certain simplifying assumptions. The calculation requires a knowledge of the metacentric height and the moment of inertia, $I_G$, of the vessel about a fore and aft axis through G. The restoring moment, $T$, when a vessel is heeled, is assumed to have the same value as that deduced from hydrostatics, the effect of the motion of the surrounding water being neglected. Thus, $T = I_G \alpha$, where $\alpha$ is the instantaneous angular acceleration measured in the same direction as the torque $T$.

Noting that the righting moment is opposite in sense to the angle of heel, $\theta$, and that the moment of inertia may be expressed in terms of the corresponding radius of gyration, $k_G$, we deduce that:

$$- Mg\text{GM}\sin\theta = Mk_G^2 \alpha$$

If the angle of heel $\theta$ is small it is seen that the motion is such that the angular acceleration $\alpha$ is proportional to $\theta$ and opposite in sign to it:

$$\alpha \simeq - \left(\frac{\text{GM}.g}{k_G^2}\right)\theta$$

This however is the condition for simple harmonic motion, for which:

$$\alpha = -p^2\theta$$

where $p$ is related to the periodic time, $\tau$, by:

$$\tau = \frac{2\pi}{p} \quad \text{and} \quad \theta = c\cos(pt+\beta)$$

$$\therefore \qquad \tau = 2\pi\sqrt{\left(\frac{k_G^2}{\text{GM}.g}\right)}$$

$\tau$ is the time taken to roll from one side to the other and back again. The above value is found to give fair agreement with the actual value for rolling, but is less successful when applied to pitching. The major source of error is neglect of movement of the water; this gives rise to an additional virtual inertia, modifies the stiffness and also introduces damping (i.e. it slows the motion and causes the oscillations to die out).

EXAMPLE 1.9 (i)

*Find the periodic time of rolling of the ship in Example* 1.6 (i) *assuming the radius of gyration of the ship about a longitudinal axis through its centre of gravity to be* 3 m.

$$\tau = 2\pi \sqrt{\left(\frac{k_G^2}{g \cdot GM}\right)} = 2\pi 3\text{m} \sqrt{\left(\frac{1}{9 \cdot 806 \frac{\text{m}}{\text{s}^2} \times 0 \cdot 6 \text{ m}}\right)} = \frac{6\pi \text{ s}}{\sqrt{(5 \cdot 884)}} = 7 \cdot 7 \text{ s}$$

## Exercises on Chapter 1

1. A barge, rectangular in cross-section, has a 12 m beam and floats on an even keel with 3 m draught. Estimate how high the centre of gravity of the load could be raised above the water line before the barge would become unstable and likely to capsize in still water.

2. The displacement of a vessel is 50 MN. A weight of 50 kN when moved 13 m across the deck causes the end of a pendulum 6 m long to deflect through 100 mm. Find the metacentric height.

3. A pontoon to carry 2 MN total load is to be of rectangular section in elevation and rectangular in plan. If the length is 15 m and the freeboard to be not less than 1 m, determine suitable dimensions for the breadth and depth. The centre of gravity of the pontoon may be taken as 0·5 m above the centre of the figure, and the metacentre is to be 1·5 m above the centre of gravity when the angle of heel is 10°. Density of water is 1 Mg/m³.

4. State the conditions which govern the stability or instability of a floating vessel.

A buoy carrying a beacon light has the upper portion cylindrical, 2 m diameter and 1·25 m deep. The lower portion, which is curved, displaces a volume of 0·4 m³, and its centre of buoyancy is 1·3 m below the top of the cylinder. The centre of gravity of the whole buoy and beacon is situated 1 m below the top of the cylinder and the total displacement is 25 kN. Find the metacentric height. (Density of sea water = 1·025 Mg/m³.)

5. State the conditions for stable equilibrium of a body floating partially immersed in a liquid.

A cylinder of circular section of diameter $d$ made of uniform material having relative density s floats in a liquid of $s_0$. Find, in terms of $d$, $s$ and $s_0$ the maximum length of the cylinder if equilibrium is to be stable with the axis of the cylinder vertical.

6. Show that if B is the centre of buoyancy and M is the metacentre for rolling of a partially immersed floating body, $BM = I/V$ where $I$ is the second moment of area of the surface of flotation about the longitudinal axis, and $V$ is the immersed volume.

The shifting of a portion of cargo weighing 250 kN through a distance of 6 m at right angles to the vertical plane containing the longitudinal axis of a vessel causes it to heel through an angle of 5°. The displacement of the vessel is 50 MN and the value for $I$ is 6 230 m⁴. The density of sea water is 1·025 Mg/m³. Find (*a*) the metacentric height, and (*b*) the height of the centre of gravity of the vessel above the centre of buoyancy.

7. A pontoon, all transverse and longitudinal cross sections of which are rectangular, has a breadth of 5·5 m and, when floating on an even keel, a draught of 1·25 m. The centre of gravity with symmetrical loading is 2·75 m above the bottom of the pontoon. By calculating the metacentric height show that the pontoon is unstable when floating on an even keel. Calculate the angle through which it will heel to attain a position of stable equilibrium. Work from first principles or prove any formulae used.

8. A hollow cylinder with closed ends is 305 mm diameter and 457·5 mm high, weighs 267 N and has a small hole in the bottom. It is lowered into water so that its axis remains vertical. Calculate the depth to which it will sink, the height to which the water will rise in it, and the air pressure inside it. Disregard the effect of the thickness of the walls but assume that it is uniform and that the compression of the air is isothermal. Atmospheric pressure is 101 kN/m².

Determine also whether the cylinder will be stable in the vertical position when in equilibrium.

9. A hollow wooden cylinder of relative density 0·55 has an outer diameter of 2 m, inner diameter 1 m and has its ends open. It is required to float in oil of relative density 0·84. Calculate the maximum height of the cylinder so that it shall be stable when floating with its axis vertical, and the depth to which it will sink.

10. A ship displaces 1 130 m³ in fresh water when the centre of buoyancy is 1·95 m below the water plane. The second moment of area of the water plane about the fore and aft axis is $3·1 \times 10^3$ m⁴. Calculate the change in transverse metacentric height that will ensue if the vessel passes from fresh to salt water, the relative density of which is 1·025. Assume that the change in immersed depth is small.

11. A cylindrical steel caisson, open at the bottom end, rests with the open end downwards on the sea bed at a depth of $H$. The caisson may be lifted from the bed by forcing a certain quantity of air through the open end, the air expanding according to the law $PV = $ constant, as the caisson rises. Calculate the limiting value of $H$ if the top of the caisson is to reach the surface before the equilibrium becomes unstable. The caisson is 6 m long and 2 m in diameter, the total weight is 60 kN, and the centre of gravity is 3·25 m from the open end. (Weight of sea water = 10 kN/m³; height of barometer = 760 mm Hg.)

12. A rectangular pontoon 7·2 m wide, 15 m long, and 3 m deep weighs 1 MN. A vertical diaphragm divides the pontoon longitudinally into two compartments each 3·6 m wide and 15 m long. The pontoon contains 500 kN of water ballast in equal quantities in the bottom of each of the compartments, the surfaces of the water being free to move. If the centre of gravity of the pontoon with the ballast is vertically above the centre of the plan being 1 m above the bottom, calculate the metacentric height in fresh water of weight 9·81 kN/m³.

13. A cylinder 0·90 m inside and 1·0 m outside diameter having its lower end closed floats in water with its axis vertical.

The cylinder contains a fluid having a relative density of 1·2; the depth of the fluid in the cylinder is 0·9 m. The depth of the bottom of the cylinder below the free surface of the water is 1·7 m. Find the position of the centre of gravity of the cylinder so that it will be just stable for small displacements.

14. A rectangular pontoon is 34 m long by 11 m broad and contains, symmetrically placed, two ballast tanks each 32 m long by 5 m broad, their floors being 0·5 m above the bottom of the pontoon. The weight of the pontoon is 1 MN and its centre of gravity is 1 m above its bottom, while the ballast tanks are filled to a depth of 0·75 m with fresh water. A load of 2 MN, with its centre of gravity 10 m above the bottom of the pontoon, is to be carried on the top of the pontoon. Find the metacentric height of the loaded pontoon in fresh water.

15. (a) Derive a formula for the period of rolling of a ship about the horizontal longitudinal axis through the centre of gravity, and state the assumptions made.

(b) The displacement of a ship is 100 MN. The second moment of the load–water plane about its fore and aft axis is $3 \times 10^4$ m⁴ and the centre of buoyancy is 2·4 m below the C. of G. The period of rolling in sea water, of density 1·025 Mg/m³, is 10 s. Find the metacentric height and relevant radius of gyration of the ship.

16. The pontoon of a floating crane is 18 m long by 6 m wide and has a draught in sea water of 3 m. Estimate the longitudinal metacentric height if the centre of gravity is 4·5 m above the bottom of the pontoon. If the top of the jib is 15 m above the centre of gravity and overhangs the forward end by 3 m, estimate the pitch caused by a 100 kN load on the jib. (Sea water weighs 10·05 kN/m³.)

## Chapter 2

# Hydrostatic forces and centres of pressure

> '*If a drop of water falls into the sea when it is calm it must of necessity follow that the whole surface of the sea is raised imperceptibly seeing that water cannot be compressed within itself like air.*'
>
> LEONARDO

## 2.1. Introduction

It is appropriate and natural that the foundations of modern fluid mechanics should have been laid in a land largely besieged by water. Stevin (1548–1620), who was destined to become a great engineer and one of the founders of modern mechanics, was born in Bruges. Until his time statics had been based on the principle of the lever, but by formulating the principle of the triangle of forces, Stevin led us to the concept of vector quantities, which underlies all mechanics. His achievements in hydro-mechanics were equally far-reaching, for by arguing that each element of a fluid at rest must be held in place by the fluid round it, he established the idea of fluid pressure and determined its variation with depth. We shall in this chapter consider the direct consequences of his results concerning the static pressure distribution in a fluid.

## 2.2. The pressure at a point

When we state that a fluid is readily deformable we imply that it cannot, when in equilibrium, resist any action which tends to change its shape. Such a fluid element at rest can neither resist nor exert a shear force, so that the force acting across any interface in a fluid must act normally to it as indicated in Fig. 2.2(a). We may note that a 'perfect' fluid is defined as one assumed to be non-viscous (i.e. inviscid) under all conditions, and thus offers no resistance to shear even when moving. The intensity of the normal force, i.e. the force per unit area, is termed the pressure on the face considered, and is usually denoted by the symbol $p$; it is considered *positive* if, as is usually the case with fluids, it is *compressive*. It is thus the same quantity as stress, as used in strength of materials, except that the sign convention is reversed. Referring to Fig. 2.2(a) let us examine the equilibrium of the fluid contained within the imaginary interfaces KL, LM and MK in terms of the pressures $p$, $p_x$ and $p_y$ assumed different on each face. The prism of fluid is imagined to be of uniform thickness and to be so small that the pressures may be assumed uniform over each face, and its weight so small as to be negligible.

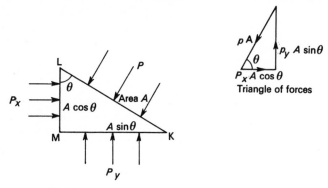

Fig. 2.2(a). *Pressures acting on a fluid element*

Multiplying each pressure by the area on which it acts, and drawing a triangle of forces, we conclude that:

$$pA \cos \theta = p_x A \cos \theta$$
$$pA \sin \theta = p_y A \sin \theta$$
$$\therefore \qquad p_x = p_y = p$$

Noting that the assumption of negligible weight becomes exact when the element is shrunk to a point, and that $\theta$ may have any value, it follows that **the pressure at a point in a fluid at rest has the same value in all directions.** Thus, in the case of stagnant fluids, we may speak of the pressure or stress at a point.

## 2.3. The variation of pressure with depth

To determine the pressure at a depth $d$ below a free surface we may examine the vertical equilibrium of an imaginary cylinder of fluid, as indicated in Fig. 2.3(a). If we denote the atmospheric pressure by $p_0$ then, since the weight $W$ of the fluid column must be supported by the pressure difference across its ends, we deduce that $(p - p_0) A = W$.

If the fluid has a weight per unit volume $\rho g = w$ then:

$$W = wAd$$
$$\therefore \qquad p - p_0 = wd = \rho g d \qquad (1)$$

Hence, so long as *w* may be assumed constant, which is usually justifiable for liquids, the pressure, *p*, increases linearly with the depth. If pressures are measured *above the pressure of the atmosphere*, as is the

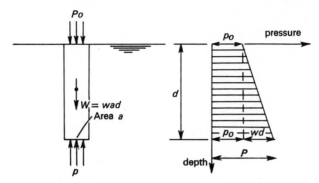

**Fig. 2.3(a).** *Pressure at a depth*

case with ordinary dial pressure gauges, such 'gauge pressures' are represented by the equation:

$$p = wd = \rho gd \qquad (2)$$

By considering the equilibrium of a horizontal cylinder it follows that the pressure is the same on all horizontal planes in a fluid at rest. Hence the pressure at any point is given by Equations (1) or (2), whether or not a vertical fluid column connects it with the free surface, e.g. the resultant force on the bottom of the vessel shown in Fig. 2.3(b) is equal to the weight

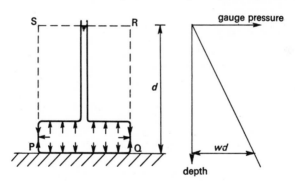

**Fig. 2.3(b).** *The hydrostatic paradox*

of fluid corresponding to PQRS. This is known as '*the hydrostatic paradox*' as it implies that the resultant force on the bottom of a vessel may be many times greater than the weight of fluid contained in it. This will be so, only if the vessel has sides which curve inwards, so that *they* experience an *upthrust*. The *resultant* force on the inside of the vessel is, of course, simply the weight of fluid contained in it.

Although a free surface often serves as a convenient datum, we may denote the pressure at any convenient level by $p_0$. The pressure at any depth $d$ below this—no matter how tortuous the fluid path connecting the two points considered—is:

$$p = p_0 + wd = p_0 + \rho gd$$

Any change in $p_0$ will thus cause an equal change in pressure at all other points. This idea of the transmissibility of fluid pressure implies that if a pressure is applied by means of a force $F$ acting on a piston of area $a$, its effect will be felt throughout the fluid. It may be used to produce a much larger force $(F/a) \times A$ on a piston of area $A$, and is the basic principle of the hydraulic press. There are many variations of this principle in which a hydraulic 'pump' is used to move a hydraulic 'engine' or 'motor'. The flexibility of the connections between them and the ease with which almost any desired mechanical advantage can be obtained results in a versatile energy transformer of high efficiency with many applications. The compressibility of liquids is usually negligible, e.g. the bulk modulus of water is

$$K = -v\frac{dp}{dv} = 2 \cdot 07 \text{ GN/m}^2$$

hence, to compress a one-cm cube of water by $\frac{1}{1000}$ mm or 1 $\mu$m on each side would require an increase of pressure of

$$K\frac{dv}{v} \simeq 3\frac{\delta}{l}K$$

i.e. $\quad \dfrac{3\ \mu\text{m}}{1\ \text{cm}} \times 2 \cdot 07\ \dfrac{\text{GN}}{\text{m}^2} = 6 \cdot 21 \times \dfrac{10^{-6}}{10^{-2}} \times 10^9\ \dfrac{\text{N}}{\text{m}^2} = 0 \cdot 621\ \dfrac{\text{MN}}{\text{m}^2}$

EXAMPLE 2.3 (i)

*If the density of water is 1 Mg/m³ determine:*

(i) *the pressure at a depth of* 100 m,

(ii) *the depth at which the pressure is* 10 kN/m² *above that of the atmosphere.*

(i) $\qquad p = wd = \rho gd = \dfrac{10^3\ \text{kg}}{\text{m}^3} \times 9 \cdot 807\ \dfrac{\text{m}}{\text{s}^2} \times 100\ \text{m} \left[\dfrac{\text{Ns}^2}{\text{kg m}}\right]$

$$= \textbf{980·7 kN/m}^2 \textit{ gauge.}$$

(ii)  since $w = \rho g = 9 \cdot 807$ kN/m³.  $\quad d = \dfrac{p}{w} = 10\dfrac{\text{kN}}{\text{m}^2} \times \dfrac{\text{m}^3}{9 \cdot 807\ \text{kN}}$

$$= \textbf{1·02 m}$$

EXAMPLE 2.3 (ii)

*An air-line exhausts into a water tank 3·5 m below the surface. Find the pressure required to bubble air slowly through the tank.*

Air pressure is required to overcome that due to the head of water, i.e.

$$p = wd = 9 \cdot 807\ \dfrac{\text{kN}}{\text{m}^3} \times 3 \cdot 5\ \text{m} = \textbf{34·6 kN/m}^2 \textit{ gauge.}$$

EXAMPLE 2.3 (iii)

*By how much will a dial pressure gauge read in excess of the pressure $p_1$ of water in a pipe if the dial is placed on an instrument board 10·2 m below the point where it is required to know the pressure?*

The vertical depth between the pipe and the centre of the dial face is $d = 10\cdot2$ m however long the tube connecting the pipe and the gauge. Hence, if $p_2$ is the pressure recorded on the gauge, it is given by the equation: $p_2 = p_1 + wd$ and the excess is

$$p_2 - p_1 = wd = 9\cdot807\,\frac{\text{kN}}{\text{m}^3} \times 10\cdot2\text{ m} = \textbf{100 kN/m}^2$$

It should be noted that this really applies only if the lead to the gauge is full of water. This may not be so unless special precautions are taken to vent the system of any air that may be trapped in it. Bleed points for doing this are particularly desirable at inverted U's etc., where air locks are particularly likely to form.

EXAMPLE 2.3 (iv)

*The cylindrical fuel tank of a motor car is 300 mm diameter with its axis horizontal and contains petrol of density 0·8 Mg/m³.*

*A 37·5 mm diameter filler pipe rises from the top of the tank to a height of 600 mm.*

*Calculate the force on one end of a full tank when the filler pipe is (a) empty, and (b) full.*

Referring to Fig. 2.3(c) if the filler pipe is:

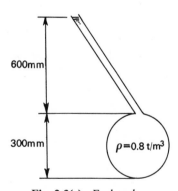

**Fig. 2.3(c).** *Fuel tank*

(*a*) empty, the force on one end of a full tank is

$$F = A(\rho g)\frac{d}{2} = \frac{\pi}{4} \times 0\cdot09\text{ m}^2 \times 800\text{ kg/m}^3 \times 9\cdot807\,\frac{\text{m}}{\text{s}^2} \times 0\cdot15\text{ m}$$

$$= 83\cdot2\text{ kg}\,\frac{\text{m}}{\text{s}^2} = \textbf{83·2 N}$$

(*b*) full, the force on one end of the tank is

$$\frac{\pi}{4} \times 0{\cdot}09 \ \mathrm{m^2} \times 7{\cdot}8456 \ \mathrm{kN/m^3} \times 0{\cdot}75 \ \mathrm{m} = 416 \ \mathrm{N}$$

(Also see Section 2.6).

EXAMPLE 2.3 (v)

*A 1 MN press with a 300 mm stroke is to be operated from a main in which the pressure is 275 kN/m² gauge, via a hydraulic intensifier with a 1·25 m stroke. If the high-pressure cylinders must not be subjected to a pressure greater than 14 MN/m² and the machine is re-charged after each stroke, find the piston and ram areas required, neglecting losses.*

Since the system operates at high pressure, the effect of differences of levels may be neglected.

Referring to Fig. 2.3(d) we see that equilibrium of the press demands that:

$$14 \ \mathrm{MN/m^2} \times A_3 = 1 \ \mathrm{MN}$$

$$\therefore \qquad A_3 = \frac{1 \ \mathrm{m^2}}{14}$$

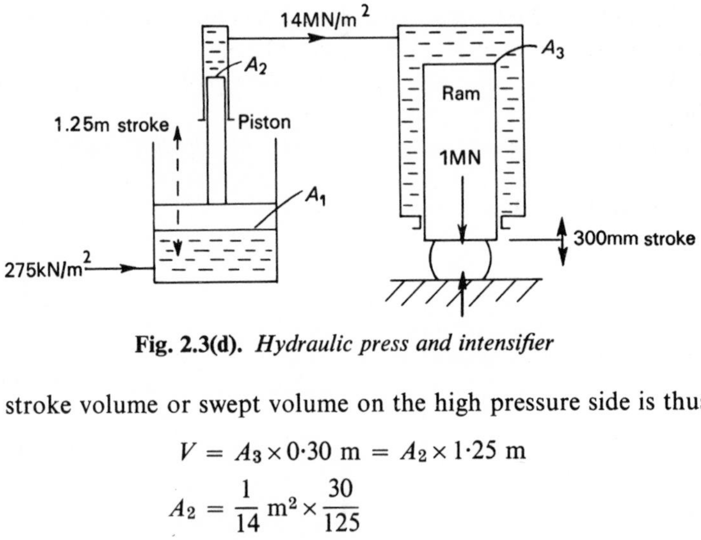

Fig. 2.3(d). *Hydraulic press and intensifier*

The stroke volume or swept volume on the high pressure side is thus:

$$V = A_3 \times 0{\cdot}30 \ \mathrm{m} = A_2 \times 1{\cdot}25 \ \mathrm{m}$$

$$\therefore \qquad A_2 = \frac{1}{14} \ \mathrm{m^2} \times \frac{30}{125}$$

Equilibrium of the intensifier demands that:

$$275 \ \mathrm{kN/m^2} \times A_1 = 14 \ \mathrm{MN/m^2} \times A_2$$

$$\therefore \qquad A_1 = \frac{30}{125} \ \mathrm{MN} \times \frac{\mathrm{m^2}}{275 \ \mathrm{kN}} = 0{\cdot}872 \ \mathrm{m^2}.$$

## 2.4. The mechanism of buoyancy

We are able to explain buoyancy by regarding a body as composed of an infinite number of vertical cylinders, each of which receives an upthrust,

$\delta U$, say. Considering a typical cylinder of cross-sectional area $\delta A$ and length $d$, as indicated in Fig. 2.4(a), then for both an immersed and a floating body:

$$\delta U = wd.\delta A = w\delta V$$

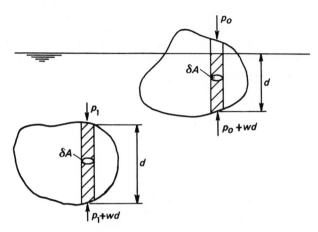

**Fig. 2.4(a).** *Upthrust, caused by pressure difference*

where $\delta V$ is the displaced volume, shown shaded and $w = \rho g$ is the weight of fluid per unit volume. Thus, using $\sum$ to denote 'the sum of', and assuming $w$ to be constant, the total upthrust or force of buoyancy is:

$$U = \sum \delta U = \sum w\delta V = w \sum \delta V = wV$$

This confirms that a body will experience an upthrust equal to the weight of fluid it displaces, in accordance with Archimedes' Principle (Section 1.2), provided that the fluid has free access to the under-side. If, however, as in Fig. 2.4(b), the fluid be excluded over an area $\Delta A$ at a depth $d$ there will be a loss of buoyancy, i.e.

$$U = wV - wd.\Delta A$$

Thus, referring to Fig. 2.4(b), the loss of buoyancy ($wd.\Delta A$) is therefore equal to the weight of the fluid column shown shaded, i.e. the lost buoyancy $= wd.\Delta A = \Delta W = \rho g \, d\Delta A$.

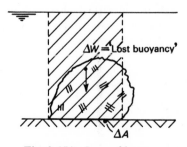

**Fig. 2.4(b).** *Loss of buoyancy*

The grounding of a ship on a sandbank may well cause a serious reduction in buoyancy due to this effect.

### EXAMPLE 2.4 (i)

*A cylindrical diving bell of cross-sectional area 1 m² is 3 m high. When resting on the bed of the river the water is 1 m high in the bell. Calculate the depth of the river and the volume of air needed to be pumped into the bell from the atmosphere in order to expel all the water from the bell. The water barometer may be taken as standing at 10·36 m.*

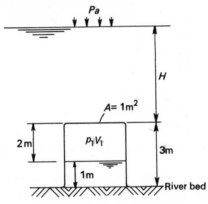

Fig. 2.4(c).  *Diving bell*

Referring to Fig. 2.4(c) we deduce:

$$p_1 = \rho g(H + 2 \text{ m}) + p_a = \frac{RmT}{V_1} = \frac{p_a V_a}{V_1}$$

Hence,

$$H + 2 \text{ m} + \frac{p_a}{\rho g} = \frac{p_a}{\rho g} \cdot \frac{V_a}{V_1}$$

and

$$H = \frac{p_a}{w}\left(\frac{V_a}{V_1} - 1\right) - 2 \text{ m, where } w = \rho g$$

$$= 10\cdot36 \text{ m}\left(\frac{3}{2} - 1\right) - 2 \text{ m} = 3\cdot18 \text{ m}$$

Thus, the depth of the river bed is 6·18 m.

The pressure of the water on the river bed is $p_b = p_a + w \times 6\cdot18$ m and the volume of air in the bell free from water is $V_b = 3$ m³.

Hence, the volume $V_a$ of free atmospheric air to fill the bell under pressure $p_b$ is

$$V_a = \frac{p_b \cdot V_b}{p_a} = \left(\frac{p_a + w \times 6\cdot18 \text{ m}}{p_a}\right) 3 \text{ m}^3$$

$$= \left(1 + \frac{6\cdot18}{10\cdot36}\right) 3 \text{ m}^3 = 4\cdot8 \text{ m}^3$$

The initial volume of free atmospheric air in the bell is 3 m³.

Hence, the volume of air needed to be pumped into the bell from the atmosphere in order to expel all the water is 1·8 m³.

## 2.5. The resultant hydrostatic force on a plane area

The resultant force $F$ on one face of a submerged plane area $A$ is indicated diagrammatically in Fig. 2.5(a) as being caused by the pressure that

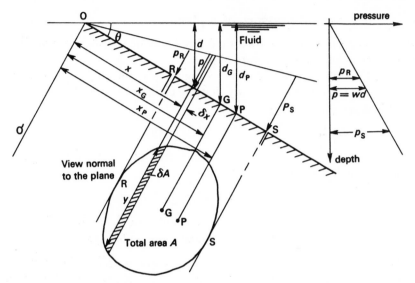

**Fig. 2.5(a).** *Hydrostatic pressure on a plane area*

varies linearly across the plane from R to S. Its magnitude, in the particular case of a plate of *uniform width*, is simply the mean pressure times the area, whatever its inclination:

$$F = \tfrac{1}{2}(p_R + p_S)A = \bar{p}A = wdA$$

where $\bar{p}$ is the average pressure, which occurs at the mean depth $d$, and $w = \rho g$ the weight per unit volume of fluid.

In the general case, however, we must consider the force on each elementary strip of area $\delta A$, namely,

$$\delta F = p.\delta A = wd.\delta A = wx \sin \theta \, \delta A$$

The total force on the plane is thus the sum $\sum \delta F$, namely,

$$F = w \sin \theta \sum x \delta A$$

But $\sum x \delta A$ is the first moment of area of the plane about the line of intersection OO′ of the immersed surface projected to intersect the free surface of the liquid. This first moment is equal to $Ax_G$, where $x_G$ is the 'slant depth' of the centroid G.

$$F = w \sin \theta \, Ax_G$$

Noting that $x_G \sin \theta$ is simply the depth $d_G$ of the centroid, and that $w d_G$ is the pressure $p_G$ at the centroid, we finally deduce that

$$F = w d_G A = p_G A$$

Hence, **the resultant force on one face of a plane area due to hydrostatic pressure is equal to the product of the area, and the pressure at its centroid.**

## 2.6. The centre of pressure of a plane area

**The point at which the resultant fluid force may be considered to act on a plane area is known as its centre of pressure.**

This is denoted by the point P on Fig. 2.5(a), and its *position* may be found by summing the moments that the elementary forces exert about an imaginary axis OO', i.e.

$$M = \sum \delta M = \sum px\delta A = w \sin \theta \sum x^2 \delta A$$

This may be equated to the moment exerted by the resultant force $F$ acting through the centre of pressure $P$.

Thus,    $$M = Fx_P = (w \sin \theta \sum x\delta A)x_P$$

$$\therefore \quad x_P = \frac{\sum x^2 \delta A}{\sum x \delta A} = \frac{\text{second moment of area about OO'}}{\text{first moment of area about OO'}}$$

By using the parallel axes theorem concerning second moments of area, we may relate the value of $\sum x^2 \delta A = I_O$ to the corresponding value through $G$, namely,

$$I_O = I_G + Ax_G^2$$

or, in terms of radii of gyration,

$$k_O = k_G^2 + x_G^2$$

Hence, the expression for $x_P$ may now be written:

$$x_P = \frac{I_O}{Ax_G} = \frac{Ak_O^2}{Ax_G} = \frac{k_G^2 + x_G^2}{x_G} = \frac{k_G^2}{x_G} + x_G$$

**Thus, the centre of pressure P of a plane area lies below the centroid G of the area by a distance PG $= k_G^2/x_G$, measured along the slant of the plane.**

As might be expected the latter expression shows that the distance PG becomes smaller as the plane is taken deeper into the fluid. This is because the pressure distribution becomes more nearly uniform on the face of the plane as the depth increases.

EXAMPLE 2.6 (i)

*Find the total force on one side of a vertical lock gate of breadth b, due to a depth of water d acting on it, and find its point of application.*

**Method 1**

The mean pressure on the gate (Fig. 2.6(a)) due to the wedge shaped loading is simply:

$$\bar{p} = \frac{wd}{2} = \rho g \frac{d}{2}$$

so that the resultant force on the gate is:

$$F = \frac{wd}{2} \times bd = \tfrac{1}{2}wbd^2$$

The line of action of the resultant force will pass through the centroid of the loading triangle, i.e. it will be at a depth:

$$d_P = \tfrac{2}{3}d$$

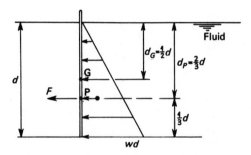

**Fig. 2.6(a).** *Fluid pressure on a lock-gate*

**Method 2**

For a rectangle:

$$d_G = \frac{d}{2}$$

Resultant force = pressure at centroid × area

i.e.  $$F = wd_G \times bd = \tfrac{1}{2}wbd^2$$

For a rectangle:

$$I_G = \frac{bd^3}{12} \quad \text{i.e. } k_G = \frac{d^2}{12}$$

∴  $$PG = \frac{k_G^2}{x_G} = \frac{d^2/12}{d/2} = \tfrac{1}{6}d$$

Hence $P$ lies at a depth:

$$d_p = (\tfrac{1}{2}+\tfrac{1}{6})d = \tfrac{2}{3}d$$

EXAMPLE 2.6 (ii)

*A sluice gate closes a circular port opening 0·30 m diameter and is hinged 1 m below the surface of the water which acts on its face. If the centre of the opening*

*lies at a depth of* 1·25 m, *find the force on the gate due to the fluid pressure.*
*Find, also, the minimum force that must be applied by a clamp, which lies* 0·5 m
*below the hinge, in order to keep the gate water-tight. Density of water is*
1 Mg/m³.

The force on the gate of Fig. 2.6(b) is:

$$F = wd_G A = 9 \cdot 807\,\frac{m}{s^2}\times 1\,\frac{Mg}{m^3}\times\frac{5}{4}\,m\times\frac{\pi}{4}\times 0\cdot 09\ m^2 = 866\ kg\,\frac{m}{s^2} = 866\ N$$

$$PG = \frac{k_G^2}{x_G} = \frac{D^2/16}{x_G} = \frac{9\ m}{16\times 125} = 0\cdot 0045\ m$$

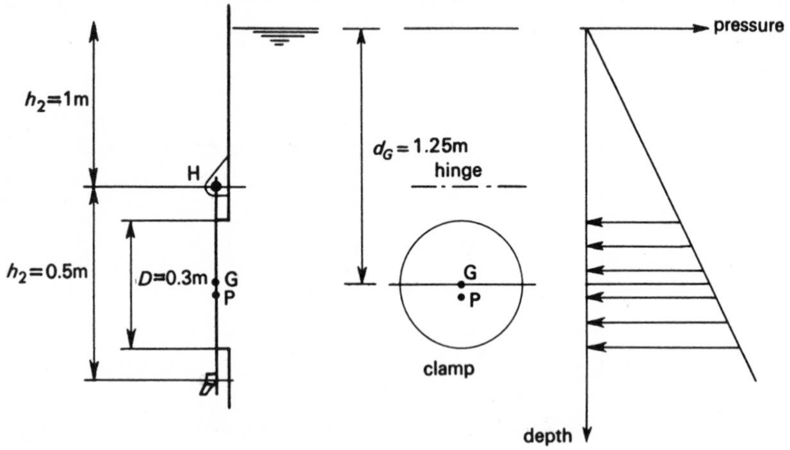

**Fig. 2.6(b).** *Fluid pressure on circular gate*

Taking moments about the hinge, the minimum force required by the clamp
F′ is such that:

$$F' \times 0\cdot 5\ m = F \times HP$$

i.e.          $$F' = \frac{0\cdot 2545\ m}{0\cdot 5\ m}\times 866\ N = \textbf{441 N}$$

EXAMPLE 2.6 (iii)

*Calculate the forces acting at the two hinges and at the clamp of the vertical*
*rectangular door shown in Fig.* 2.6(c) *if it is loaded on one face by sea water of*
*relative density* 1·03. *Density of fresh water is* 1 Mg/m.³

$$\frac{\text{Density of sea water}}{\text{Density of fresh water}} = \text{Relative density of sea water}$$

Hence, density of sea water is $\rho = 1\cdot 03 \times 1$ Mg/m³,
and weight per unit volume of sea water is

$$w = \rho g = 1\cdot 03\,\frac{Mg}{m^3}\times 9\cdot 807\,\frac{m}{s^2}\left[\frac{N\ s^2}{kg\ m}\right]$$

$$= 10\cdot 12\ kN/m^3$$

The static thrust on the door is:

$$F = w \cdot d_G \cdot A$$

$$= 10 \cdot 12 \, \frac{\text{kN}}{\text{m}^3} \times 4 \cdot 5 \text{ m} \times 4 \cdot 5 \text{ m}^2$$

$$= 205 \text{ kN}$$

This thrust acts at the centre of pressure P which is $k_G^2/x_G$ below G. Thus, the distance PG is:

$$\frac{A k_G^2}{A x_G} = \frac{I_G}{A x_G} = \frac{\frac{1}{12} \times 1 \cdot 5 \times 3^3 \text{ m}^4}{4 \cdot 5 \text{ m}^2 \times 4 \cdot 5 \text{ m}} = \frac{1}{6} \text{ m}$$

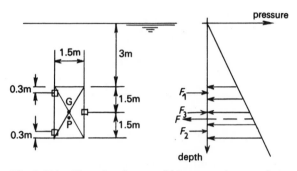

Fig. 2.6(c). *Clamping force and hinge reactions on door*

To find $F_3$, take moments about a line through the two hinges; hence:

$$F_3 \times 1 \cdot 5 \text{ m} = 205 \text{ kN} \times 0 \cdot 75 \text{ m}$$

i.e.

$$F_3 = 102 \cdot 5 \text{ kN} = \frac{F}{2}$$

To find $F_1$, take moments about a horizontal line through $F_2$; hence:

$$F_1 \times 2 \cdot 4 \text{ m} = F \times \left(1 \cdot 2 - \frac{1}{6}\right) \text{ m} - \frac{F}{2}(1 \cdot 2 \text{ m})$$

∴

$$F_1 = \frac{F}{2 \cdot 4}\left(1 \cdot 2 - \frac{1}{6} - \frac{1 \cdot 2}{2}\right) = \frac{F}{2 \cdot 4} \times \frac{26}{60} = \frac{13}{72} \times 205 \text{ kN}$$

$$= 37 \text{ kN}$$

Hence

$$F_2 = F - (F_1 + F_3) = (205 - 139 \cdot 5) \text{ kN}$$

$$= 65 \cdot 5 \text{ kN}$$

EXAMPLE 2.6 (iv)

*Find an expression for the overturning moment caused by the fluid pressure acting on a masonry structure of section shown in Fig. 2.6(d). Also, state the condition for ensuring that tensile stress will not be induced in the masonry at the base of the structure.*

Considering a width $L$ of a dam wall with a plane face inclined at $\theta$ to the water level, the resultant force due to water pressure is:

$$F = wx_G \sin \theta A$$

$$= w \times \frac{D}{2} \times \frac{LD}{\sin \theta}$$

Horizontal component:     $F_H = F \sin \theta = \dfrac{wLD^2}{2}$

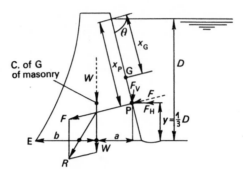

**Fig. 2.6(d).** *Fluid pressure on a dam*

The slant depth of the centre of pressure P is:

$$x_P = \frac{I_0}{Ax_G} = \frac{\frac{1}{3}L\left(\dfrac{D}{\sin \theta}\right)^3}{\left(\dfrac{LD}{\sin \theta}\right)\left(\dfrac{D}{2 \sin \theta}\right)} = \frac{2}{3}\frac{D}{\sin \theta}$$

Hence, the vertical depth of the centre of pressure is $(D-y) = \frac{2}{3}D$.

Thus, as a result of the force $F$ of the fluid and the weight $W$ of the masonry, the resultant moment about the heel E tending to overturn the structure is:

$$F_H \times y - W \times b - F_V \times (a+b)$$

If $R$, the resultant force acting on the masonry structure due to fluid pressure and weight of masonry, intersects the base line within the 'middle third' there will be compressive stress throughout the masonry but if $R$ intersects the base outside the middle third, tensile stress will be induced on the water side of the structure which will crack and allow water to seep in and exert pressure tending to overturn the structure. Also, when the dam is empty $W$ must intersect the base within the middle third or tensile stress will be induced on the heel side E of the structure and cause a crack.

EXAMPLE 2.6 (v)

*A pair of lock gates, each 3 m wide, have their lower hinges at the bottom of the gates, and their upper hinges 5 m from the bottom. The width of the lock is 5·5 m.*

(a) *Find the reaction between the gates when the water level is 4·5 m high on one side and 1·5 m high on the other side.*

(b) *Assuming that this force acts at the same height as the resultant force due to the water pressure, find the reaction forces on the hinges.*

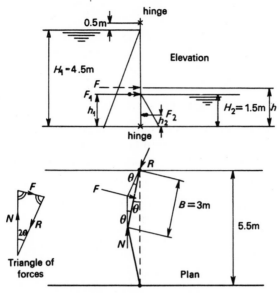

**Fig. 2.6(e).** *Reactions on a lock gate*

Resultant water force on one gate:

$$F = F_1 - F_2$$
$$= \tfrac{1}{2}wH_1 \times BH_1 - \tfrac{1}{2}wH_2 \times BH_2$$
$$= \tfrac{1}{2}wH_1^2 B[1 - \tfrac{1}{9}]$$
$$= \tfrac{1}{2} \times 9{\cdot}807 \, \frac{\text{kN}}{\text{m}^3} \times 20{\cdot}25 \text{ m}^2 \times 3 \text{ m} \times \tfrac{8}{9}$$
$$= 265 \text{ kN}$$

Normal reaction:

$$N = R_t = \frac{F}{2} \times \frac{1}{\sin \theta} = \frac{1}{2} \frac{265 \text{ kN}}{\sqrt{\left(1 - \left(\frac{2{\cdot}75}{3}\right)^2\right)}} = \frac{265 \text{ kN}}{2 \times 0{\cdot}4} = \textbf{331 kN}$$

Height $h$ of $F$ is given by:

$$Fh = F_1 h_1 - F_2 h_2$$
$$(265 \text{ kN}) h = 265 \times 9/8 \text{ kN} \times 1{\cdot}5 \text{ m} (1 - \tfrac{1}{3} \times \tfrac{1}{9})$$
$$\therefore \qquad h = 1{\cdot}5 \times \frac{1}{8} \times \frac{26}{3} \text{ m} = 1{\cdot}625 \text{ m}$$

Assuming $N$ acts at this height, so does $R$, the resultant hinge force; the latter may be split into upper and lower hinge reactions, $R_1$ and $R_2$ respectively.

Thus, by taking moments about the sill, the upper reaction is such that:

$$R_1 \times 5 \text{ m} = 331 \text{ kN} \times 1 \cdot 625 \text{ m}$$
$$\therefore \qquad R_1 = \textbf{106·5} \text{ } kN$$
$$\therefore \qquad R_2 = R - R_1$$
$$= (331 - 106 \cdot 5) \text{ kN} = \textbf{224·5 kN}$$

## 2.7. Fluid pressure on curved surfaces

The resultant force on a curved surface may often be determined by considering the equilibrium of a volume of fluid enclosed by the surface and other conveniently chosen planes. For example, in the subject of

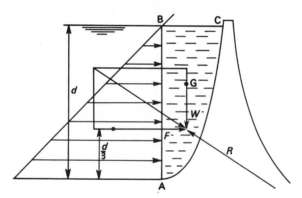

**Fig. 2.7(a).** *Fluid pressure on a curved dam*

Strength of Materials the hoop tension in a horizontal boiler may be found by adding the weight of the fluid below the level considered to the resultant of the pressure acting across this surface. Similarly, the force acting on the face of the dam shown in Fig. 2.7(a) may be calculated by considering the equilibrium of the curved wedge ABC.

It has a weight $W$, which acts through its centre of gravity G, and is held in equilibrium by the forces $F$ and $R$. Since both $F$ and $W$ can easily be determined, both in magnitude and line of action, so may their equilibrant $R$.

EXAMPLE 2.7 (i)

*The figure shows the cross-section of a dam with a parabolic face, the vertex of the parabola being at O. The axis of the parabola is vertical and 13 m from the face at the water level.*

*Estimate the resultant force in mega-newtons per horizontal metre-run due to the water, its inclination to the vertical and how far from O its line of action*

cuts the horizontal OP. *The centroid of the half-parabolic cross-section of water is* 4·75 m *from the vertical through* O.

From the equation of the parabola, namely, $y = ax^2$ we deduce:

$$a = \frac{y}{x^2} = \frac{d}{s^2}$$

$$\therefore \qquad y = \frac{d}{s^2} x^2$$

referring to Fig. 2.7(b).

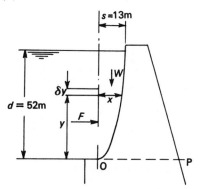

**Fig. 2.7(b).** *Dam with parabolic face*

The weight of the curved wedge of water of breadth $b$ and of weight per unit volume $w$, is:

$$W = w \int_0^d bx\,dy = wb \int_0^d s \sqrt{\left(\frac{y}{d}\right)}.dy = \frac{wbs}{\sqrt{d}} \left(\frac{2}{3} d^{3/2}\right) = \frac{2}{3} wbsd$$

$$= \tfrac{2}{3} \times 9{\cdot}807 \text{ kN/m}^3 \times 1 \text{ m} \times 13 \text{ m} \times 52 \text{ m} = 4{\cdot}415 \text{ MN per m width}$$

The horizontal force on the curved wedge of water is:

$$F = \left(\frac{wd}{2}\right) \times (bd) = \frac{wbd^2}{2}$$

$$= \frac{9{\cdot}807 \text{ kN}}{2 \text{ m}^3} \times 1 \text{ m} \times (52)^2 \text{ m}^2 = 13{\cdot}26 \text{ MN per m width}$$

Hence, the resultant force is $\sqrt{[(13{\cdot}26)^2 + (4{\cdot}415)^2]} = \mathbf{13{\cdot}98}$ MN *per metre width*, inclined at $tan^{-1}$ **13·26/4·415** so its line of action cuts OP at

$$4{\cdot}75 \text{ m} + \frac{52}{3} \times \frac{1326}{442} \text{ m} = \mathbf{56{\cdot}8} \text{ m from O}$$

EXAMPLE 2.7 (ii)

*If the pressure of fresh water is equivalent to a head of* 60 m *find the resultant static thrust on a pipe of* 0·60 m *internal diameter if it is bent through* 30° *in a horizontal plane.*

Referring to Fig. 2.7(c), the static end thrusts are $F = (\pi/4)d^2p$ and the resultant thrust is:

$$T = 2F \sin \frac{\theta}{2} = F\sqrt{[2(1-\cos\theta)]}$$

$$= \frac{\pi}{4} \times 0.36 \text{ m}^2 \left(9.807 \frac{\text{kN}}{\text{m}^3} \times 60 \text{ m}\right) \sqrt{2\left(1-\frac{\sqrt{3}}{2}\right)}$$

$$= 16.65 \sqrt{(0.268)} \text{ kN}$$

$$= 8.62 \text{ kN}$$

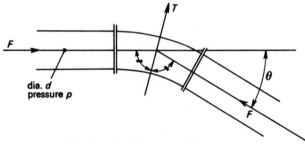

**Fig. 2.7(c).** *Thrust on a bent pipe*

## 2.8. General equilibrium equations for fluids

It is sometimes necessary to consider the equilibrium of a region of fluid in which the specific weight may not be assumed constant. For example, we live on the bottom of a sea of air of which the density $\rho$ decreases considerably with altitude. Solids in suspension, or salts in solution, may similarly cause $\rho$ to vary from point to point in liquids.

In order to derive an equation which is appropriate to such cases, consider the equilibrium of the element of infinitesimally small dimensions in Fig. 2.8(a). If the weight per unit volume of the fluid at the par-

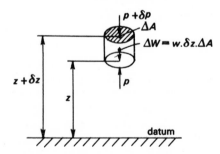

**Fig. 2.8(a).** *Equilibrium of an element of fluid*

ticular point surrounding the element is $w$, the weight $\Delta W$ of the element will be:

$$\Delta W = \rho g \delta z\, \Delta A = w\, \delta z\, \Delta A$$

This weight must be supported by the pressure difference across the ends. We do not presume to guess whether the pressure increases or decreases with height but merely assume that if the pressure is $p$ at a height $z$, it will have some slightly different value $(p + \delta p)$ at $(z + \delta z)$ and the signs will look after themselves. Thus, summing the forces vertically, their resultant must be zero if the element is in equilibrium:

$$p\Delta A - (p + \delta p)\Delta A - w\delta z \Delta A = 0$$

i.e. $$\delta p = -w\delta z$$

and in the limit, as $\delta z$ tends to zero, we deduce that

$$\frac{dp}{dz} = -w = -\rho g \tag{1}$$

the negative sign implying that the pressure decreases with height. **The pressure gradient $dp/dz$ thus depends only on the value of $w$ at the point considered.**

Considering the equilibrium of a horizontal cylinder, we conclude that the pressure must be the same at all points of a given height. Consequently the change in pressure $\delta p$ between any two horizontal layers, $\delta z$ apart, must everywhere be the same, at that height. This in turn implies that $\rho$ must be constant throughout the horizontal layer, so that any region of a fluid at rest must 'stratify' into layers of uniform density.

Additional information is required to integrate Equation (1)

$$\int \frac{dp}{w} = -\int dz$$

in the general case, but for an *incompressible* fluid it yields

$$z + \frac{p}{w} = \text{constant} = z + \frac{p}{\rho g} \tag{2}$$

As $z$ and $p/w$ are respectively termed the height head and the pressure head, this result may be interpreted in words as follows:

**Throughout a continuous region of stagnant incompressible fluid, the sum of the height and pressure heads has the same value at all points, and may be termed the hydrostatic head.**

Applying this result to a point on the free surface of liquid (which is incompressible), such as O in Fig. 2.8(b), the hydrostatic head is seen to

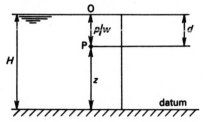

**Fig. 2.8(b).** *Height head and pressure head*

correspond to $H$, since $p/w$ is zero at O if pressures are measured from atmospheric as datum:

Therefore,
$$z + \frac{p}{w} = H$$

This, when interpreted graphically, confirms the previous result of Section 2.3, that the pressure at a depth $d$ is given by:

$$\frac{p}{w} = d$$

The case of a compressible fluid such as the atmosphere is dealt with later in Chapter 13.


## 2.9. Potential energy of a fluid and capacity for doing work

**The potential energy of a body may be defined as its capacity for doing work,** and its magnitude may be determined by considering the work that must be done to bring the body to the state considered. This implies that some datum state has been chosen, from which energy is reckoned, e.g. a mass $M$ will have a potential energy $MgH$ if it is at a height $H$ above the datum level.

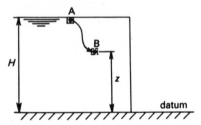

**Fig. 2.9(a).** *Height head and pressure head*

Let us now apply these ideas to the movement of a fluid element from its initial situation on the free surface of a reservoir, as indicated in Fig. 2.9(a). At A, an element of fluid possesses a potential energy per unit mass of magnitude $gH$. No work need, ideally (i.e. without friction, eddies or turbulence), be performed in moving the element to some new position B since, so long as the element is completely immersed, its weight is balanced by its buoyancy at every stage. Hence the total potential energies at A and B must be the same, even though the height head at B is only $z$. This suggests that the balance of the potential energy $gH$ must be in a different form when the element has reached B. Therefore, we infer by reconsidering the equilibrium equation $(p/w + z = H)$ in energy terms that the fluid at B must possess a potential energy $p/w$, in addition to its height energy $z$. The former is therefore the pressure energy per unit weight $(\rho g)$ of fluid or, more simply, the pressure head and the latter is the height energy per unit weight of fluid or simply the height head.

Thus, **a fluid element possesses potential energy because of both the gravitational and the pressure forces which act upon it.** These do work on it as it moves, and give rise to a height energy per unit weight (or height head $z$), and a pressure energy per unit weight (or pressure head $p/w$). Their sum is the total potential head or hydrostatic head per unit weight of fluid at rest. Each element of fluid in the reservoir has this same potential energy or capacity for doing work, namely, an amount of energy $H$ per unit weight of fluid or $wH = \rho g H$ per unit volume.

## Exercises on Chapter 2

1. Calculate the forces and the depths of the centres of pressure on
(*a*) a vertical square plate of 6 m side, and
(*b*) a vertical plate of 6 m diameter, when their centres of area are 4 m below the surface of fresh water of density 1 Mg/m³ or of weight 9·807 kN/m³.

2. A lock gate is 12 m wide and 14 m deep. The fresh water levels are 12 m on one side and 3 m on the other above the lower edge of the gate. Calculate the magnitude and position of the resultant force due to water pressure.

3. A flat ring 600 mm external diameter and 400 mm internal diameter is placed with its centre at a depth of 600 mm in fresh water. The plane of the ring is inclined at 45° to the horizontal. Calculate the force on one side of the ring and the position of the centre of pressure.

4. A symmetrical door in a vertical side of a tank has a top edge 3 m wide and 1 m below the water surface. The lower edge of the door is 5 m wide and 6 m from the top edge.
Calculate the depth of the centre of area, the force due to water pressure on the door and the centre of pressure. $w = \rho g = 9\cdot807$ kN/m³.

5. A rectangular opening, 3 m wide by 2 m deep, in the vertical side of a fresh water tank is closed by a door hinged on a horizontal axis 1 m above the upper edge of the opening. The water level in the tank is 0·5 m below the hinge. Calculate the force of the water on the door and the moment of this force about the hinge.

6. A dam has a vertical inner face 80 m high and the concrete wall is triangular in section and three times the density of water. Find the width of the base to ensure that the resultant thrust intersects the base line within the 'middle-third' when the water level reaches the top of the dam.

7. A circular opening, 1 m diameter, in the vertical side of an open tank, is closed by a door hinged on a horizontal axis 1 m above the centre of the circle. Find the force exerted by the water on the door. Also, find the moment of this force about the axis of the hinge assuming that the water level in the tank is 4 m above the centre of the circle. Density of water is $\rho = 1$ Mg/m³.

8. Develop from first principles a general expression for the position of the centre of pressure of a fluid on one side of a sloping flat plate completely immersed in the fluid.
A rectangular opening in the sloping side of a reservoir containing water is

3 m by 2 m with the 2 m side horizontal. It is closed by means of a gate as shown in the figure. The gate is hinged at the top edge and kept closed against the water pressure partly by its own weight and partly by a weight $W$ on a lever arm.

Assuming the gate to be a uniform flat plate weighing 5 kN and neglecting the weight of the lever arm, calculate the weight $W$ required so that the gate will commence to open when the water level reaches a height of 1 m above the top of the gate.

9. The gates of a lock, which is 8 m wide, make an angle of 120° with each other in plan. Each gate is supported on two hinges, which are situated 1m and 8 m above the bottom of the lock. The depths of water on the two sides of the gates are 12 m and 4 m respectively. Find the force on each hinge and the thrust between the gates. One gate is fitted with a sluice 1·25 m wide and 1 m deep having its upper edge level with the water surface on the low level side of the lock.

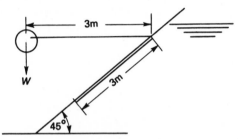

Find the magnitude and point of action of the resultant thrust on the sluice.

10. A tank 1·2 m deep is 3 m square at the top and 2 m square at the bottom. The four sides are plane and have the same trapezoidal shape. The tank is completely full of water. Obtain, working from first principles:

(*a*) the resultant pressure on each side,

(*b*) the position of the point on each side through which the resultant pressure passes.

11. A stream is spanned by a bridge which is a single masonry arch in the form of a parabola, the crown being 8 m above the springings which are 30 m apart. Measured in the direction of the stream the overall width of the bridge is 21 m. During a flood the stream rises to a level of 6 m above the springings. Assuming that the arch remains watertight calculate the force tending to lift the bridge from its foundations.

12. Vertical and horizontal flat plates of a water tank are connected by means of a plate shaped as a quadrant of a circle 1 m radius. The depth of water in the tank is 2·2 m. Find the magnitude and line of action of the horizontal and vertical components of the hydrostatic forces acting on the curved plate per metre of its width.

## Chapter 3

# The measurement of fluid pressure

> 'Every instrument requires to be made by
> experience.'  LEONARDO

## 3.1. The barometer and absolute pressure

Aristotle's dictum that 'Nature abhors a vacuum' passed unqualified in learned circles for more than 1 500 years, although generations of hydraulic engineers had found that even the best of pumps could not 'suck' water through a greater height than 10·4 m. This difficulty which has (and always will) beset pump designers, was brought to the attention of the great Galileo (1546–1642). One of his students, Torricelli, was subsequently to provide the true explanation of this apparent abhorrence, and in doing so pointed out the futility of trying to transcend it. At Torricelli's suggestion, Viviani took a long tube sealed at one end, and filled it with mercury. When this was inverted over a dish containing mercury so that no air could enter the tube, it was found that a vacuous space was left at the top of the tube. The mercury column stood no higher than about 760 mm above the free surface. Torricelli rightly concluded from this device, which we now term a barometer, that the pressure of the atmosphere sustains the column, and that the height of the column of mercury is therefore a measure of the pressure of the atmosphere. The barometric height $h$ may be related to the atmospheric pressure $p_a$ by referring to Fig. 3.1(a).

By equating pressures at A and B, which are at the same level, we deduce that:

$$p_a = \rho_m g h = \frac{\rho_m}{\rho} w h = s w h$$

in which $\rho_m$ is the density of mercury, $s$ is its relative density $(\rho_m/\rho)$ and $w = \rho g$ is the weight of water per unit volume. Thus if $h = 760$ mm

$$p_a = 13\cdot6 \left(\frac{1\ \text{Mg}}{\text{m}^3} \times \frac{9\cdot807\ \text{m}}{\text{s}^2}\right) \times 0\cdot76\ \text{m} \left[\frac{\text{N s}^2}{\text{kg m}}\right] = 101\cdot325\ \text{kN/m}^2$$

In general, therefore, the atmospheric pressure in kN/m² is 101·325/760 (i.e. 0·133 3) of the barometric height in mm of mercury.

Conventional pressure gauges, such as those of the Bourdon type (see Section 3.7, p. 66). detect departures from atmospheric pressure. To convert their readings, i.e. gauge pressures, to absolute values, it is therefore necessary to add the pressure of the atmosphere.

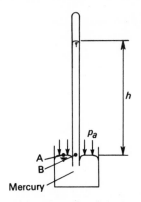

**Fig. 3.1(a).** *Principle of the barometer*

In normal circumstances a liquid will cavitate as the absolute zero of pressure is approached. Cavitation is usually due to dissolved gases, such as air, coming out of a solution but even if dissolved gases are not present the fluid itself will eventually 'boil'. Evaporation or boiling occurs at atmospheric temperature when the pressure is reduced to what is termed the corresponding saturation vapour pressure.

The sealed space above the mercury column of a barometer is, therefore, not a true or perfect vacuum, but contains traces of mercury vapour which exert a slight pressure. The correction for this is very small and can be reliably estimated if necessary. Corrections for surface tension effects are more difficult to make, however, and we shall discuss how they may be overcome in the next Section.

EXAMPLE 3.1 (i)

*The inlet to a pump is 9 m above the bottom of a sump, from which it draws water through a suction pipe. Find the minimum depth of the water in the sump if the pressure at the pump inlet is not to fall below 35 kN/m² absolute.*

Suppose that the free surface lies a distance $h$ below the pump inlet. Then, assuming that the pressure of the atmosphere is 101 kN/m² and the pressure at the pump inlet has its minimum value:

$$\rho g h_{max} = (101 - 35) \text{ kN/m}^2$$

i.e.
$$h_{max} = \frac{66 \text{ kN/m}^2}{9\cdot807 \text{ kN/m}^3} = 6\cdot73 \text{ m}$$

Hence the minimum depth of water in the sump is:

$$9 \text{ m} - 6{\cdot}73 \text{ m} = \mathbf{2{\cdot}27} \text{ m}$$

## 3.2. Surface tension and capillarity

Due to molecular forces there is a natural tendency for a fluid to draw in upon itself. This effect passes unnoticed when other larger forces are present but becomes marked on small volumes of liquid, which conse-quently form themselves into droplets. It is convenient for the purposes of calculation to imagine this phenomenon to be caused by a flexible skin which, like a lightly stretched balloon, exerts a tension round the fluid it contains. Thus it is possible to fill a cup more than brim full, and we shall imagine the extra fluid to be held in place by a tensioned membrane.

The tensile force or tension acting across any imaginary line drawn on the surface is proportional to the length of the line and is perpendicular to it. The magnitude of the *tension per unit width* will be denoted by $T$, and is termed the *surface tension*. Its value decreases as temperature rises, and depends on the qualities and states of the fluids on each side of the surface of separation or 'meniscus'. Its value is, in engineering terms, generally small, and typical values at $20^\circ$ C are as follows:

| Fluids separated | Surface tension $\mu$N/mm |
|---|---|
| water and air | 74 |
| mercury and air | 465 |
| paraffin and air | 27 |
| mercury and water | 427 |
| water and paraffin | 48 |

A meniscus attaches itself to solid boundaries at some angle $\theta$ (as indi-cated in Fig. 3.2(a)) the value of which depends on the relative values of the molecular attractions in this region. If the angle of contact is less

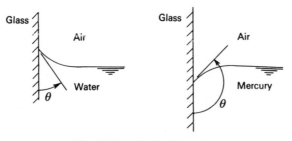

**Fig. 3.2(a).** *Angle of contact*

than 90° the fluid is said to 'wet' the boundary, as water does glass at an air inter-face (Fig. 3.2(a)). The value of $\theta$ in this case is almost zero if the glass is perfectly clean, but in all cases it varies considerably with the state as well as with the nature of the materials. A representative value for mercury on glass in air is about 130°, resulting in an upward curvature of the surface (Fig. 3.2(a)).

Whenever a surface under tension is curved, it follows that there must be a pressure difference across it. Fig. 3.2(b) shows a small element of a

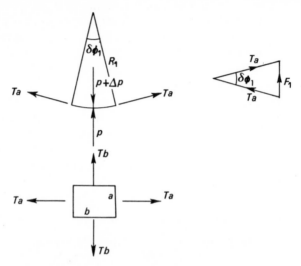

**Fig. 3.2(b).** *Surface tension and pressure difference*

membrane having principal radii of curvature $R_1$ and $R_2$, carrying a tension per unit width $T$ round its sides of length $a$ and $b$. The resultant force due to the tension acting on the sides of length $a$ is, if $\delta\varphi_1$ is very small:

$$F_1 = Ta.\delta\varphi_1 = Ta.\frac{b}{R_1}$$

Similarly for the forces on the faces of length $b$:

$$F_2 = Tb\,\frac{a}{R_2}$$

Both of these forces act vertically upwards as drawn in Fig. 3.2(b), and must be balanced by the pressure difference $\Delta p$ acting on the area $ab$, i.e.

$$Tab\left(\frac{1}{R_1}+\frac{1}{R_2}\right) = \Delta p.ab$$

$\therefore$
$$\Delta p = T\left(\frac{1}{R_1}+\frac{1}{R_2}\right) \tag{1}$$

Thus the pressure within a spherical droplet ($R_1 = R_2 = R$) is greater than that outside by an amount:

$$\Delta p = \frac{2T}{R}$$

This may easily be checked by considering the equilibrium of one half, for which:

$$\Delta p . \pi R^2 = T . 2\pi R$$

or
$$\Delta p = \frac{2T}{R}$$

In the case of a soap bubble there are two liquid surfaces, an inner and an outer, so that the excess pressure inside is, in this case:

$$\Delta p = \frac{4T}{R}$$

Thus, we deduce that the pressure difference across a fluid surface is negligible so long as its radii of curvature are large, but *may be appreciable*

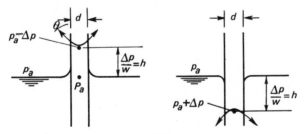

**Fig. 3.2(c).** *Capillary action*

for liquids in, for example, *tubes of narrow bore. If the liquid wets the tube, the pressure below the surface must be less than that immediately above it according to Equation* 1. Hence if the tube tips into a reservoir of fluid as indicated in Fig. 3.2(c), the level will rise within the tube. If the angle of contact is greater than 90°, as with mercury for example, the pressure in the capillary tube immediately below the surface of separation, or meniscus, must be greater than atmospheric so that the level will drop to a depth corresponding to this pressure.

A *simple* interpretation of this capillary action, as it is called, is to say that the meniscus attaches itself to the bore of the tube, and the vertical force which it exerts lifts or depresses the fluid in the tube a distance $h$. If we consider this column to be sensibly cylindrical in shape its weight will be:

$$W = w \frac{\pi}{4} d^2 h \quad \text{or} \quad \rho g \frac{\pi}{4} d^2 h$$

where $\rho$ is the density of the fluid ($\rho g = w$) and the tube is of diameter $d$.

Assuming the angle of contact is known, this weight is carried by a force:

$$\pi dT \cos \theta = W$$

$$\therefore \qquad h = \frac{4T \cos \theta}{wd}$$

which is positive or negative according as $\theta$ is less or greater than 90°. This action may cause serious errors when measuring pressures in terms of head and can be *minimised by using gauge glasses of large diameter.* If the bore of the tube is of inadequate diameter it may prove difficult to correct for capillary effects, owing to uncertainty concerning the value of $\theta$. Fig. 3.2(d) shows a Fortin-type barometer which, among other

**Fig. 3.2(d).** *Fortin-type barometer*

refinements, has an enlarged bore at its upper end to reduce capillary action. In cases where a difference of two pressures is measured by means of a difference of height of fluid in a uniform-bore U-tube (as in Section 3.3) *the effect of capillary action is the same in the two legs of the tube and, therefore, does not affect the difference in height of the fluid in the two legs provided that the angle of contact is the same in each.* The best way of ensuring the latter is to keep each limb scrupulously clean.

EXAMPLE 3.2 (i)

*What will be the capillary rise at 20°C in a clean glass tube of 1 mm bore containing (a) water, (b) mercury and (c) what will be the corresponding values in a tube 10 mm bore? Density of water is 1 Mg/m³ and of mercury 13·6 Mg/m³.*

(a) Average rise:

$$h = \frac{4T \cos \theta}{\rho g d} = \frac{4 \times 74 \, \frac{\mu N}{mm} \times 1}{1 \, \frac{Mg}{m^3} \times 9{\cdot}807 \, \frac{m}{s^2} \times 1 \, mm} \left[ \frac{Mg \, s^2}{kN \, m} \right] = 30 \, mm$$

(b) $$h = \frac{4 \times 465 \, \mu N/mm \times \cos 130°}{13{\cdot}6 \times 9{\cdot}807 \, kN/m^3 \times 1 \, mm} = -8{\cdot}5 \, mm$$

(c) In a tube 10 mm bore, which is about the size used in gauges in many engineering recorders, the figures would be **3·0** mm and **−0·85** mm.

## 3.3. Manometers and measurement of pressure difference

A simple manometer consists essentially of a U-tube containing fluid of known density $\rho$. Fig. 3.3(a) shows such a gauge, used to measure

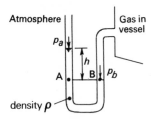

**Fig. 3.3(a).** *Simple manometer*

the difference of pressure $(p_b - p_a)$ between that of a gas in a vessel and that of the atmosphere outside, the density of the gas being negligible relative to that of the manometric liquid.

Since the pressures at corresponding levels in each limb must be the same, then

$$p_a + \rho g h = p_b$$

or $$(p_b - p_a) = \rho g h$$

i.e. the pressure difference is proportional to the difference in level, and the device needs no calibration.

If a manometer is used to measure pressure differences in a fluid *whose relative density (s) may not be neglected*, it is necessary to allow for the difference in head of both fluids in the legs of the U-tube. A typical example is a differential manometer where a U-tube containing mercury is used to measure a pressure difference in, say, a hydraulic main, as shown in Fig. 3.3(b). From Fig. 3.3(b) we may deduce the following equation based on the fact that the pressures in the two legs on the level AB must be equal, namely,

$$p_a + \rho g z_a = p_b + \rho g z_b + s \rho g h$$

i.e. $$(p_a - p_b) = s \rho g h - \rho g(z_a - z_b) = (s-1)\rho g h$$

or expressed as a head

$$\frac{p_a - p_b}{\rho g} = (s - 1)h = \frac{p_a - p_b}{w}$$

Thus, in the case of a mercury–water differential manometer for which

$$s = 13{\cdot}6 \text{ and } w = \rho g = 1 \frac{Mg}{m^3} \times 9{\cdot}807 \frac{m}{s^2} \left[ \frac{N\,s^2}{kg\,m} \right] = 9{\cdot}807 \text{ kN/m}^3$$

we deduce:

$$\frac{p_a - p_b}{w} = 12{\cdot}6\,h$$

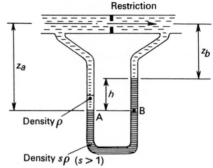

**Fig. 3.3(b).** *Differential manometer*

i.e. the pressure difference measured as head of water is 12·6 times the difference in height of the mercury levels. Such a manometer with limbs 1 m long may therefore be used to measure water head differences

$$\frac{p_a - p_b}{w} = h(s - 1)$$

up to $1 \text{ m} \times 12{\cdot}6 = 12{\cdot}6$ m, which corresponds to a pressure difference

$$(p_a - p_b) = wh(s - 1) = 9{\cdot}807 \frac{kN}{m^3} \times 12{\cdot}6 \text{ m} = 123{\cdot}6 \text{ kN/m}^2$$

The sensitivity of such a gauge may be defined as the ratio of the observed difference of levels, $h$, to the difference of pressure head

$$\frac{(p_a - p_b)}{w}$$

of water being measured, i.e.

$$\frac{h}{\dfrac{p_a - p_b}{w}} = \frac{1}{s - 1} = \frac{\rho g h}{p_a - p_b}$$

Thus, in the case of mercury the sensitivity is 1/12·6, whereas in the case of paraffin of relative density 0·85 the sensitivity is −1/0·15, the negative sign indicating that the gauge must be inverted as in Fig. 3.3(c). The sensitivity of a water–paraffin gauge is, therefore, 12·6/0·15 = 84 times that of a water–mercury differential manometer gauge.

Thus, less or more dense liquids are used in differential manometers according as the pressure difference in the fluid under measurement is small or large, respectively. In the case of a water-paraffin gauge the U-tube must be inverted, otherwise the paraffin would rise up one of the legs and flow into and away down the water main. Fig. 3.3(c) shows such

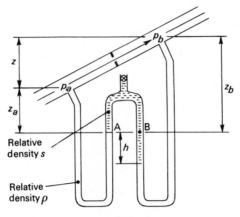

**Fig. 3.3(c).** *Inverted U-tube manometer*

an inverted manometer connected between two points at different levels. The pressures at A and B are equal, hence the equation in which $w = \rho g$:

$$(p_a + wz_a) = (p_b + wz_b) + wh - swh$$

$$\therefore \qquad h = \frac{1}{w(1-s)} [(p_a + wz_a) - (p_b + wz_b)]$$

or
$$\frac{p_a - p_b}{w} = z + h(1 - s) = \frac{p_a - p_b}{\rho g}$$

If the pipe is horizontal $z = 0$, and the latter equation reduces to that previously obtained, except that there is a reversal of sign. This implies that with an inverted U-tube the manometric fluid rises, instead of being depressed, in the limb subjected to the greater pressure.

It should be noted that the gauge reading $h$ is proportional to the differences between the values of $(p + wz)$ at A and B. Thus even though the gauge be connected between two points which are at different levels, and hence at different pressures, there will be no reading on the manometer if the system is at rest. This follows since the value of $(p + wz)$ is the same at all points in such cases (see Section 2.8, p. 44).

Hence the gauge responds to departures from the hydrostatic pressure distribution, and this is an important characteristic of all differential pressure gauges. The correction for differences of level is, in effect, automatically applied by the leads to the gauge, so long as they are filled with fluid. This assumes that there is no air in the system, and care should be taken to minimize potential traps for air. If the latter cannot be avoided, suitable vents must be provided, from which the system may be bled until all the air is removed. If the pressure at these points is below atmospheric, a suitable aspirator or suction pump must be applied to remove air which may, e.g., have come out of solution due to the reduction in pressure at this point. Fig. 3.3(d) shows a typical industrial manometer

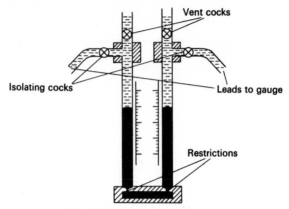

**Fig. 3.3(d).** *Industrial manometer*

fitted with cocks for venting and for isolating the gauge. The latter will minimize the risk of sweeping the manometric fluid into the system under heavy surge conditions. The effect of the latter, and other dynamic effects, may also be limited by the incorporation of restrictors of small bore; these damp the response of the gauge as the levels change and thus prevent excessive oscillations.

### 3.4. Two-fluid differential manometers

In the single-fluid uniform U-tube manometer of Fig. 3.3(a) the level of the liquid in each leg moves the same distance in opposite directions—i.e. each leg shows only half the total motion.

It is, however, possible to increase the sensitiveness of a gauge (i.e. increase the motion of the level of fluid measuring a given pressure difference) by means of a two-fluid manometer with enlarged ends, as shown in Fig. 3.4(a).

Assume the fluids settle to positions represented by $x$, $y$, $h$. When the gas pressure is $p$ and the atmospheric pressure on the free surface is $p_a$, then the hydrostatic equilibrium equation (in which $w_1 = \rho_1 g$ and $w_2 = \rho_2 g$) is:

$$p + w_2(x+h) = p_a + w_1(y+h) \tag{1}$$

Similarly, if the gas pressure increases to $p'$ and the fluids move to positions represented by $x'$, $y'$, $h'$ the corresponding equilibrium equation is

$$p' + w_2(x'+h') = p_a + w_1(y'+h') \tag{2}$$

Hence, by subtraction of Equation (1) from (2), we deduce that the increase in pressure of the gas $(p' - p)$ is given by

$$(p' - p) + w_2\{(x' - x) + (h' - h)\} = w_1\{(y' - y) + (h' - h)\} \qquad (3)$$

Also, since the volume displaced is $A(x - x') = a(h' - h) = A(y' - y)$,

i.e. $$-(x' - x) = \frac{a}{A}(h' - h) = (y' - y) \qquad (4)$$

we deduce by substituting (4) in (3) that:

$$(p' - p) + w_2(h' - h)\left\{1 - \frac{a}{A}\right\} = w_1(h' - h)\left\{1 + \frac{a}{A}\right\} \qquad (5)$$

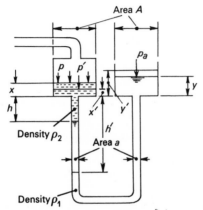

**Fig. 3.4(a).** *Two-fluid differential manometer*

Hence, defining the sensitiveness of a manometer gauge as the ratio

$$\frac{\text{movement of the indicating level}}{\text{change in pressure being measured}}$$

the sensitiveness of this gauge may be deduced from Equation (5) rearranged as $(p' - p)A = (h' - h)\{w_1(A + a) - w_2(A - a)\}$, i.e.

$$\frac{h' - h}{p' - p} = \frac{A}{A(w_1 - w_2) + a(w_1 + w_2)} \qquad (6)$$

Thus, using Equation (6) in the two extreme cases, (i) a plain U-tube manometer with a single fluid, i.e. $A = a$ and $w_1 = w_2$, we may deduce

$$\frac{h' - h}{p' - p} = \frac{1}{2w_1} = \frac{1}{2\rho_1 g}$$

and (ii) of a two-fluid differential manometer of Fig. 3.4(a) in which area $a$ is negligible relative to $A$ we deduce that

$$\frac{h' - h}{p' - p} = \frac{1}{w_1 - w_2} = \frac{1}{g(\rho_1 - \rho_2)}$$

From these two results we may deduce that the limiting ratio of the sensitiveness of the two-liquid differential manometer compared with the plain U-tube single-liquid manometer is

$$\frac{2w_1}{w_1 - w_2} = \frac{2\rho_1}{\rho_1 - \rho_2}$$

## EXAMPLE 3.4 (i)

*What is the limiting ratio and the actual ratio of the sensitivity of a two-liquid differential manometer gauge with enlarged ends to that of a single-liquid uniform U-tube gauge if in the actual gauge A is 40a and*

$$\frac{\rho_1}{\rho_2} = \frac{w_1}{w_2} = \frac{0.81}{0.80}$$

Limiting ratio of sensitivities $= \dfrac{2\rho_1}{\rho_1 - \rho_2} = \dfrac{2 \times 0.81}{0.01} = 162$

Actual ratio of sensitivities $= \dfrac{2\rho_1 A}{A(\rho_1 - \rho_2) + a(\rho_1 + \rho_2)}$

$$= \frac{2 \times 0.81}{0.01 + \frac{1}{40}(0.81 + 0.80)}$$

$$= \frac{1.62}{0.01 + 0.040\,25} = \frac{1.62}{0.050\,25} = \mathbf{32.3}$$

Thus, showing that this differential gauge with ends enlarged to areas 40 times that of the tube is 32·3 times as sensitive as the plain U-tube manometer.

## EXAMPLE 3.4 (ii)

*A pressure gauge for use in measuring the pressure of flue gases at the base of a chimney is made of a glass U-tube with enlarged ends. One end is connected to the gases in the chimney and the other end is exposed to the atmosphere. The gauge is filled with water of density 1 Mg/m³ or weight per unit volume 9·807 kN/m³ on the gas side and with oil of relative density 0·90 on the open side such that the surface of separation is in the vertical part of the U-tube below the enlarged ends.*

*If the areas of the enlarged ends are 40 times the cross-sectional area of the tube what is the sensitivity of this gauge? How many mm of water pressure in the chimney correspond to a displacement of 20 mm in the surface of separation and what is the increase in pressure of the gas in kN/m²?*

Referring to Fig. 3.4(a) and using $w = \rho g$, the weight of water per unit volume, we deduce:

$$p + w(x+h) = p_a + sw(y+h) \tag{1}$$
$$p' + w(x'+h') = p_a + sw(y'+h') \tag{2}$$
$$A(x-x') = (y'-y)A = a(h-h') \tag{3}$$

From (1) and (2)

$$p' - p + w\{(x' - x) + (h' - h)\} = sw\{(y' - y) + (h' - h)\}$$

From (3)

$$-(x' - x) = \frac{a}{A}(h - h') = +(y' - y)$$

$$\therefore \qquad p' - p + w(h - h')\left\{-\frac{a}{A} - 1\right\} = sw(h - h')\left\{\frac{a}{A} - 1\right\}$$

$$p' - p = (h - h')\left\{w\left(1 + \frac{a}{A}\right) - sw\left(1 - \frac{a}{A}\right)\right\}$$

$$\therefore \qquad \text{Sensitivity} = \frac{h - h'}{p' - p} = \frac{A}{w\{(A + a) - s(A - a)\}}$$

$$= \frac{1}{9 \cdot 807 \text{ kN/m}^3 \{(1 + \frac{1}{40}) - 0 \cdot 90(1 - \frac{1}{40})\}}$$

$$= \frac{1}{9 \cdot 807 \left(\dfrac{41}{40} - \dfrac{0 \cdot 90 \times 39}{40}\right)} \frac{\text{m}^3}{\text{kN}} \left[\frac{10^3 \text{ mm}}{\text{m}}\right]$$

$$= \frac{40 \times 10^3}{9 \cdot 807(41 - 35 \cdot 1)} \frac{\text{m}^2}{\text{kN}} \text{ mm}$$

$$= \mathbf{692} \frac{\text{mm}}{\text{kN/m}^2}$$

Since

$$\frac{p' - p}{w} = (h - h')\left\{\left(1 + \frac{a}{A}\right) - s\left(1 - \frac{a}{A}\right)\right\}$$

then if $(h - h') = 20$ mm

$$\frac{p' - p}{w} = 20 \text{ mm}\left\{\left(1 + \frac{1}{40}\right) - 0 \cdot 9\left(1 - \frac{1}{40}\right)\right\} = 20 \text{ mm}\left(\frac{41 - 35 \cdot 1}{40}\right) = \mathbf{2 \cdot 95} \text{ mm}$$

and the 20 mm displacement of the surface of separation is caused by a change in pressure of the gas of

$$p' - p = w \times 2 \cdot 95 \text{ mm} = 9 \cdot 807 \frac{\text{kN}}{\text{m}^3} \times 2 \cdot 95 \frac{\text{m}}{10^3} = 28 \cdot 9 \text{ N/m}^2$$

## 3.5. Inclined-tube manometers

The sensitivity of a U-tube manometer may be increased by inclining the tube as indicated in Fig. 3.5(a).

As before, $\qquad (p_b - p_a) = wh = \rho g h$

Also, referring to Fig. 3.5(a),

$$d = \frac{h}{\sin \alpha} = \frac{1}{\sin \alpha} \frac{(p_b - p_a)}{\rho g}$$

Hence the magnification $d/h$ is equal to $1/\sin \alpha$, but surface tension effects limit the increase in sensitivity that can be obtained in this way. The uncertainty concerning the lowest point on the meniscus can be minimized only by reducing the bore of the tube but this will increase the

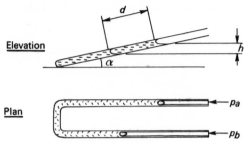

**Fig. 3.5(a).** *Inclined U-tube*

capillary effect. For the latter to be the same in each leg, the bores must be both uniform and equal. Some uncertainty will still remain, as the angle of contact in each leg is critically dependent on their cleanliness.

The need for matching each leg may be avoided by inclining only one leg and making the other a large reservoir as indicated in Fig. 3.5(b).

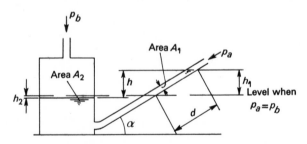

**Fig. 3.5(b).** *Inclined-tube manometer*

Such an instrument may be used for measuring small pressure differences, e.g. such as exist between the flue gases in a chimney and the outside air.

If the pressure difference $(p_b - p_a)$ causes changes in level $h_1$ and $h_2$ as indicated in Fig. 3.5(b), then the difference in level $h$ is given by:

$$h = h_1 + h_2$$

Again equating pressures at the lowest level:

$$p_b = p_a + wh$$

$$\therefore \qquad (p_b - p_a) = w(h_1 + h_2)$$

If the shape of the meniscus in the inclined tube remains unchanged as the level rises, the total volume of fluid displaced into this tube is:

$$A_2 h_2 = A_1 d$$

In addition, if we note that:

$$h_1 = d \sin \alpha$$

both $h_1$ and $h_2$ may be expressed in terms of the deflection $d$, giving:

$$(p_b - p_a) = wd\left(\sin \alpha + \frac{A_1}{A_2}\right)$$

or

$$\frac{p_b - p_a}{w} = \frac{p_b - p_a}{\rho g} = h = d\left(\sin \alpha + \frac{A_1}{A_2}\right)$$

### EXAMPLE 3.5 (i)

*Find the magnification factor of the gauge shown in Fig. 3.5(b) if the slope of the inclined tube is* 1 *in* 10, *i.e.* $\sin \alpha = 0 \cdot 1$ *and the ratio of the areas is* 40 *to* 1.
   The difference in levels is:

$$h = \frac{p_b - p_a}{w}$$

and the calibration of the gauge is:

$$d = \frac{p_b - p_a}{w} \frac{1}{\sin \alpha + A_1/A_2}$$

Hence the magnification of the difference in levels is:

$$\frac{d}{h} = \frac{1}{\sin \alpha + A_1/A_2} = \frac{1}{\frac{1}{10} + \frac{1}{40}} = \frac{40}{5} = 8$$

   Very small inclinations can be employed only if the bore of the capillary tube is made very small. The consequent possible inaccuracies due to this require that the gauge be calibrated. The latter may conveniently be carried out by introducing additional fluid into the reservoir, as indicated by the following example.

### EXAMPLE 3.5 (ii)

*An inclined-tube micro-manometer of the type shown in Fig. 3.5(b) is required to measure gauge pressures up to* 35 N/m².
   *Find the approximate inclination required if alcohol is used as the gauging fluid and the full scale deflection is to be about* 120 mm.
   If the inclination and level are so arranged that the meniscus travels the 120 mm length of the inclined tube as the pressure is increased from 0 to 35 N/m², then:

$$p_{max} = wh_{max} = wd \sin \alpha$$

in which $d$ is 120 mm and $p_{max}$ is 35 N/m².

$$\therefore \qquad \sin \alpha = \frac{35 \text{ N/m}^2}{0 \cdot 8 \times 9 \cdot 807 \text{ kN/m}^3 \times 0 \cdot 12 \text{ m}} = 0 \cdot 0372$$

$$\therefore \qquad \alpha \simeq 2°$$

## 3.6. Tilting manometers

In low speed wind-tunnels it is necessary to measure small pressure differences very accurately, and developments from the inclined tube micro-manometer are frequently used for this purpose. Such instruments are generally made 'null-reading' (that is, the meniscus is returned to its original position) usually by tilting the gauge. For example an inclined tube may be mounted on a cradle which can be rotated to bring the meniscus back to the same mark on the tube. This principle is indicated diagrammatically in Fig. 3.6(a) where, for simplicity, the tube is imagined

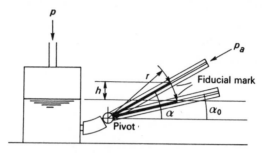

**Fig. 3.6(a).** *Null reading micromanometer*

to pivot about a point on its bore. If the area of the reservoir is much greater than that of the capillary tube, and $\alpha_0$ is the inclination corresponding to no pressure difference:

$$h = r(\sin \alpha - \sin \alpha_0)$$

$$\therefore \quad (p_b - p_a) = \rho g r (\sin \alpha - \sin \alpha_0)$$

Instruments operating on this principle have been constructed to detect pressure differences of the order of 0·001 mm of water. Water is not recommended as the manometric fluid in such instruments however. Alcohol is generally preferred, since it acts as a solvent for grease (which otherwise may affect the angle of contact), has a lower relative density (approximately 0·8), and has a much lower surface tension (about 22 $\mu$N/mm).

An even more sensitive gauge, which is widely used in this country, is that due to Chattock. It consists essentially of two reservoirs each containing water, connected by what is, in effect, an inverted U-tube containing oil. Its principle of operation is indicated diagrammatically in Fig. 3.6(b). When a pressure difference is applied to the gauge, the water–oil meniscus is displaced. The gauge is then tilted until the meniscus is brought back to its original position. The inclination of the gauge, which is represented on the figure by rotating the horizontal datum, is thus a measure of the pressure difference.

In an actual gauge the limbs of the inverted U-tube are concentric, being formed by an inner tube with a surrounding vessel, as shown in

Fig. 3.6(c). The instrument is adjusted so that one meniscus is formed at the top of the inner tube.

This may be observed through a telescope with crossed wires and the tilt adjusted by means of a screw of fine pitch. It is the distortion (rather

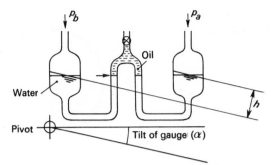

**Fig. 3.6(b).** *Principle of Chattock gauge*

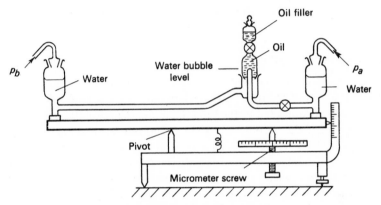

**Fig. 3.6(c).** *Chattock gauge*

than the displacement) of the observed meniscus which makes the instrument so sensitive. Pressure differences up to about 50 mm of water can be measured and such gauges are sensitive to differences of less than one micro-metre of water. With further refinements, both of these limits may be considerably exceeded.

## 3.7. Industrial pressure gauges

The most common form of pressure gauge found in industry is probably that which works on the principle of Bourdon's pressure gauge. This type of gauge is illustrated in Fig. 3.7(a). It consists of a curved tube BB, oval in cross section as shown at C. One end of the tube is closed and attached to a sector D geared so as to move the pointer EF when the tube tends to straighten due to the pressure of the fluid which enters the tube at the end connected to cock A. Gauges of this type indicate pressures above that of the surrounding atmosphere; hence 'super-atmospheric' pressures are usually referred to as gauge pressures.

Such gauges are usually calibrated by means of a dead-weight gauge tester.

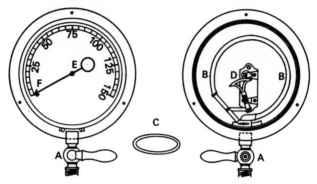

**Fig. 3.7(a).** *Bourdon's pressure gauge*

## Exercises on Chapter 3

1. A vertical U-tube is used to measure the pressure of a gas in a pipe. Water in the arm of the tube open to the atmosphere stands 300 mm higher than in the arm connected to the pipe. Estimate the absolute pressure of the gas in the pipe if the barometric height is 760 mm of mercury.

2. If the difference of levels in a vertical U-tube containing paraffin of relative density 0·798 is 76 mm calculate the pressure difference in kPa, mm of water and mm of mercury taking the density of water as 1 Mg/m³ and the relative density of mercury 13.6.

3. Two tanks A and B contain water. The water surface level in A is 254 mm higher than that at B. A hole in A is connected by tubing to one limb of a vertical U-tube manometer containing mercury. The other limb of the manometer is connected to a hole in tank B. Find the difference in level of the mercury surfaces in the two limbs of the manometer.

4. Estimate the absolute pressure on the surface of water in a condenser if the vacuum gauge placed 1 m below the level of the water reads 600 mm of mercury when the mercury barometer stands at 760 mm.

5. A U-tube gauge with enlarged ends has each leg connected to points in a gas main in order to measure the change in difference of pressure of the gas at the two points when the rate of flow is changed.

Water of density 1 Mg/m³ and oil of relative density 0·83 are the two fluids used in a tube 10 mm inside diameter which enlarges to 50 mm diameter at each end, and in which the gas is in contact with the two fluids. If the surface of separation of the oil and water in a leg of the U-tube moves one inch, calculate the change in pressure difference in mm of water.

6. A U-tube has limbs which are of cross-sectional area *A* and its lower

part is filled with mercury. Each limb is reduced in area at the top, well above the mercury levels. One limb is connected by a lead containing water to the base of an overhead water-tank. The other limb is surmounted by a gauge-glass of area $A/100$ containing indicating fluid. Show that if the specific gravity of the latter is 0·745, changes of level in the gauge-glass show to full scale any corresponding change of water level in the water-tank overhead, provided that no air is trapped in the system.

7. A differential gauge consists essentially of a hollow toroidal (i.e. tyre shaped) tube of mean radius $r$ and cross-sectional area $a$. It is pivoted horizontally about its axis and contains fluid in its lower part. The latter, in conjunction with a sealing diaphragm inside, at the top of the ring, divides it into two compartments. Differential gas pressure, when applied through flexible leads to each space tends to tilt the ring against the resisting moment exerted by a weight $W$ attached to the ring at a radius $R$. Deduce an expression to relate the tilt of the ring to the differential pressure.

# Bernoulli's equation and measurement of flow of incompressible fluids

'*In order to write with the true science of the flight of birds through the air it is necessary first to give the science of winds, which we shall prove by the motion of water itself; and the understanding of this science will lead to the required knowledge.*'  LEONARDO

## 4.1. Bernoulli's equation and total head

Just as J. S. Bach is generally considered to have fathered much of the classical music since his day, so Daniel Bernoulli (1700–82) may be thought of as the father of modern fluid mechanics. The two men were contemporaries, and they came of families which were brilliant in their respective fields—music in the one case, and mathematics in the other. In hydraulics Bernoulli's name is coupled with the equation which he derived in 1738 for steady, frictionless flow of an incompressible fluid, namely,

$$gz + \frac{p}{\rho} + \frac{u^2}{2} = \text{constant for a particular } g$$

Bernoulli's equation may be compared with that for the motion of a particle which, if of mass $M$ and velocity $u$, possesses a potential energy $Mgz$ and a kinetic energy $Mu^2/2$. If gravity is the only force which acts on it, any loss of potential energy increases its kinetic energy, and vice versa, their sum remaining constant at a particular place:

$$Mgz + M\frac{u^2}{2} = \text{constant}$$

i.e.
$$gz + \frac{u^2}{2} = \text{constant}$$

In Section 2.9 it is indicated that when an incompressible fluid element is moved, the work done by the pressure $p$ gives rise to an additional potential energy, of magnitude $p/\rho$ per unit mass. Hence Bernoulli's equation states that the sum of the height, pressure and kinetic energies per unit mass remains constant, i.e.

$$gz + \frac{p}{\rho} + \frac{u^2}{2} = gH = \text{constant at a particular location,}$$

there being no friction (by assumption) to raise the temperature. Each term in this equation has the dimensions of energy per unit mass.

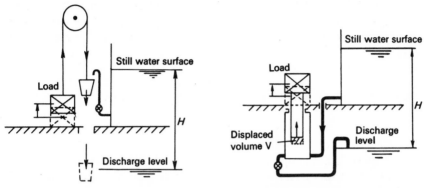

**Fig. 4.1(a).** *Gravity hoist*          **Fig. 4.1(b).** *Hydraulic ram*

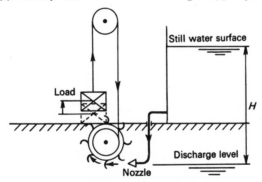

**Fig. 4.1(c).** *Pelton wheel*

Just as $z$ and $p/w$ are respectively termed the height and pressure heads, so $u^2/2g$ may be referred to as the kinetic (or velocity) head, and $H$ is termed total head. The term 'head' is appropriate as each quantity has the units of energy per unit weight, such as Nm/N, or, more simply, m—hence the term 'head'. The term 'total' merely implies that all forms of energy likely to undergo significant changes have been included.

The total head $H$ is the quantity of mechanical energy which would be obtained if the whole of the height, pressure and kinetic heads per unit weight of fluid could be transformed into work. That each separately represents a capacity for doing work is demonstrated by the three hydraulic machines shown in Figs. 4.1(a, b and c).

Each is imagined to draw fluid from a reservoir in which the level of the still water surface is maintained at a height $H$ above that of the datum (or tail race) at which fluid is discharged from the machine. The machines shown in (a), (b) and (c) respectively operate by converting height, pressure and kinetic energy into useful work. It should be noted that machines shown in Figs. 4.1(a and b) are intermittent in operation; the bucket in the former and the ram in the latter need re-setting after each stroke. If we imagine each of these operations to be preceded by discharge of the fluid at the datum level, no work is theoretically required to reset the machine. Hence, as a mass $M$ of weight $W = Mg$ passes through the gravity hoist the machine ideally performs an amount of work:

$$Wz = WH$$

Similarly the work done by the hydraulic ram is, in theory:

$$pV = \rho g H \frac{M}{\rho} = MgH = WH$$

Unit mass of fluid supplied to the Pelton-wheel theoretically acquires (in the frictionless case) a kinetic energy $U^2/2g$ in falling through a height $H$, so that the energy available for conversion into work from mass $M$ of weight $W = Mg$ is in this case also:

$$\frac{MU^2}{2} = MgH = WH$$

Power is rate of doing work, so that the power potentially available depends on the rate at which fluid is drawn from the reservoir. This power, $P$ say, may be obtained by multiplying the total head $H$ by the weight of fluid, $wQ$, drawn off in unit time, where $Q$ is the volumetric flow rate, i.e.

$$P = wQH = \rho QgH$$

This is the maximum power that a perfect machine (100 per cent efficient) would deliver. Conversely it represents the minimum power required to pump fluid from the datum level into the reservoir.

EXAMPLE 4.1 (i)

*Find, neglecting losses, the power required to transfer 28 m³/min of oil of relative density 0·8 from one tank to another against a difference in levels of 10 m.*

Choosing a datum through the still water surface in the supply tank, the total head in this is zero. Since that in the delivery tank is 10 m, the work done in transferring unit weight of fluid is ideally:

$$H = 10 \text{ Nm/N}.$$

The rate of working or power required when transferring a volume $Q$ in unit time is, for fluid of weight per unit volume $w$,

$$P = wQH = sw_wQH$$
$$= 0{\cdot}8 \times 9{\cdot}807 \text{ kN/m}^3 \times 28 \frac{\text{m}^3}{\text{min}} \times 10 \text{ m} \left[\frac{\text{W s}}{\text{Nm}}\right]\left[\frac{\text{min}}{60 \text{ s}}\right]$$
$$= 36{\cdot}7 \text{ kW}$$

EXAMPLE 4.1 (ii)

*Tests on a hydraulic ram supplied with liquid at constant pressure show that, when the control valve is fully opened, a thrust of 200 kN can be exerted at 3 m/min and a thrust of 80 kN at 15 m/min. Assuming that the force of resistance is proportional to the square of the speed, estimate the maximum power of the ram and the corresponding thrust and speed.*

Force or thrust at any speed $u$ is given by the law

$$f = F - ku^2,$$

where $F$ is the maximum thrust and the constant

$$k = \frac{F - 200 \text{ kN}}{9(\text{m/min})^2} = \frac{F - 80 \text{ kN}}{225(\text{m/min})^2}$$

Hence, $F = 205$ kN and $k = \frac{5}{9} \frac{\text{kN}}{(\text{m/min})^2}$.

Power $P = fu = Fu - ku^3$ and is a maximum when $0 = F - 3ku^2$, i.e. when

$$u^2 = \frac{F}{3k} = 123 \left(\frac{\text{m}}{\text{min}}\right)^2 \text{ or } u = 11{\cdot}1 \text{ m/min}.$$

Thus, when operating at maximum power, the thrust is

$$f = 205 - \tfrac{5}{9} \times 123 = 136{\cdot}7 \text{ kN}$$

and the maximum power

$$P = 136{\cdot}7 \text{ kN} \times 11{\cdot}1 \frac{\text{m}}{\text{min}} \left[\frac{\text{min}}{60\text{s}}\right]\left[\frac{\text{kW s}}{\text{kN m}}\right] = 25{\cdot}2 \text{ kW}.$$

## 4.2. Bernoulli's equation in terms of work and energy

In describing flow patterns, frequent use is made of the terms 'stream-lines' and 'stream-tubes'.

A *stream-line* is a continuous line drawn in a fluid so that it has the direction of the fluid velocity at every point. As a particle always moves tangential to a stream-line, there can be no flow across a stream-line.

A *stream-tube* is the surface or tube formed by the stream-lines passing through a small closed curve. There is no flow through the walls of a stream-tube.

Bernoulli's equation strictly applies only to the steady motion of an ideal fluid along a fixed stream-tube, across which conditions may be assumed constant. Such flows are said to be *steady*, i.e. the variables at any section do not change with time. If the fluid is assumed incompressible and inviscid, no energy or fluid can be added or subtracted from a fixed stream-tube.

This may be demonstrated by considering Fig. 4.2(a) in which fluid is shown contained between two imaginary pistons which move forward

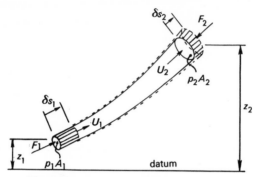

**Fig. 4.2(a).** *Steady flow along a stream-tube*

with the fluid. If the fluid is incompressible each piston will sweep out the same volume $\delta V$ during a short interval of time such that:

$$A_1 \, \delta s_1 = A_2 \, \delta s_2 = \delta V$$

Hence, on balance, the net work done by the pistons (the one pushing, and the other being pushed) on the fluid between them is the pumping work:

$$F_1 \, \delta s_1 - F_2 \, \delta s_2 = p_1 A_1 \, \delta s_1 - p_2 A_2 \, \delta s_2$$
$$= (p_1 - p_2) \, \delta V$$

As fluid enclosed between the pistons advances, a portion $\delta V$ of it vacates the lower shaded region at 1, and $\delta V$ enters the shaded territory at 2. Consequently potential energy of the fluid between the pistons is increased by an amount:

$$\delta(\text{P.E.}) = \rho \, \delta V g (z_2 - z_1)$$

A similar argument concerning the difference between the kinetic energies of the fluid occupying these regions shows that the latter has increased by:

$$\delta(\text{K.E.}) = \rho \, \delta V \left( \frac{u_2^2}{2} - \frac{u_1^2}{2} \right)$$

This assumes that the kinetic energy of the unshaded portion of the cylinder is unchanged. This is true only for *steady flows*. For such a flow, the velocity at each point always has the same value, i.e. it is independent of time. Under these conditions, as each element of fluid moves forward to the position occupied by its forerunner, it acquires the velocity that is appropriate to that point. These conditions are obviously not satisfied by flows in which surges are taking place.

An examination of the above expressions indicates that Bernoulli's equation assumes that the pumping work (i.e. the net work done on the fluid) goes to increase the potential and kinetic energies:

$$\text{Pumping work} = \delta(\text{P.E.}) + \delta(\text{K.E.})$$

i.e.
$$(p_1 - p_2)\,\delta V = \rho \delta V g(z_2 - z_1) + \rho\,\delta V \left(\frac{u_2^2}{2} - \frac{u_1^2}{2}\right)$$

i.e.
$$g z_1 + \frac{p_1}{\rho} + \frac{u_1^2}{2} = g z_2 + \frac{p_2}{\rho} + \frac{u_2^2}{2}$$

This equation is probably more widely used in hydraulics than any other and is capable of explaining, at least qualitatively, many of the

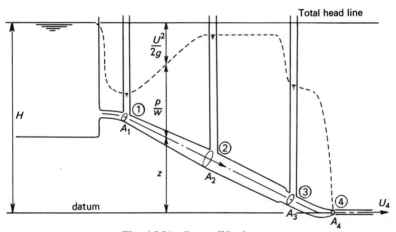

Fig. 4.2(b). *Bernoulli's theorem*

phenomena that are encountered in fluid mechanics. It suggests that the height, pressure and velocity cannot all increase simultaneously along a stream-tube, nor can any two of them, if the third remains constant.

If, for example, changes in height may be neglected, the pressure must decrease if the velocity increases, and vice versa. Contraction of a stream-tube implies an increase in velocity, in order that the same quantity of fluid $Q$ may be discharged in unit time through the reduced area $A$, since

$$Q = uA$$

By Bernoulli's equation, therefore, contraction of a stream is associated with a decrease in pressure and it is this principle which is used in carburettors, aspirators and metering devices of many kinds.

Fig. 4.2(b) represents a hypothetical (frictionless) hydraulic system in which $\rho g = w$ for the liquid and steady conditions are maintained by keeping the level in the supply tank constant, namely

$$H = z + \frac{p}{w} + \frac{u^2}{2g}$$

Each gauge glass records the pressure head at the section of the pipe to which it is attached so that, according to Bernoulli's equation, the difference in levels between each gauge glass and the reservoir must represent the kinetic head at that section. Since the same volume of fluid must be discharged through each section of the pipe in unit time, the volumetric flow rate:

$$Q = u_1 A_1 = u_2 A_2 = u_3 A_3 \ldots \text{etc.}$$

It follows that the velocity varies inversely as the pipe area, so that whenever the pipe is constricted, the velocity, and hence the kinetic head, will be large, and vice versa.

At exit the pressure head must fall to zero, as the fluid is discharged into the atmosphere, the pressure of which is usually taken as the datum when incompressible fluids are being considered. Thus, denoting the exit velocity by $u_4$, it follows that

$$\frac{u_4^2}{2g} = H$$

i.e.
$$u_4 = \sqrt{(2gH)}$$

since both $p/w$ and $z$ are zero at the exit section.

If the exit area is $A_4$ the ideal rate of volumetric flow may be determined from

$$Q = u_4 A_4 = A_4 \sqrt{(2gH)}$$

and this enables the velocity at any other section of the pipe-line to be deduced. In practice, however, there is friction in pipes and losses at sudden discontinuities of section the effect of which can be seen on Fig. 4.3(b).

EXAMPLE 4.2 (i)

*Apply Bernoulli's equation to show the principle of action of injectors and ejectors.*

The principle of action of an injector and of an ejector or aspirator can be demonstrated by applying Bernoulli's equation to Fig. 4.2(c), neglecting change in height. Thus,

$$\frac{p_2}{\rho} + \frac{u_2^2}{2} = \frac{p_3}{\rho} + \frac{u_3^2}{2}$$

and hence $\dfrac{p_3 - p_2}{\rho} = \dfrac{u_2^2 - u_3^2}{2} = k$ (always positive).

By choice of suitable throat and pipe areas $A_2$ and $A_3$, the pressure difference $p_3 - p_4$ can be made large enough to inject, say, water at atmospheric pressure from a feed tank connected to pipe 4 into a boiler at pressure $p_3$.

Alternatively, in the case of an ejector or simple aspirator, water may be forced to flow down pipe 1, through the throat 2 and to waste at 3 where the

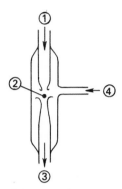

**Fig. 4.2(c).** *Injector and ejector aspirator*

pressure is atmospheric. Thus, in this case $0 - (p_2/\rho) = k$ and hence a partial vacuum can be created in a vessel connected to pipe 4.

## EXAMPLE 4.2 (ii)

*A low-speed wind-tunnel of the open-circuit type draws in air from the laboratory through a contraction. A U-tube manometer or draught gauge measures the static pressure at the working section in mm of water. Assuming Bernoulli's equation, find the air-speed through the tunnel at a section where the gauge reads 25 mm vacuum, if the density of air is 1·228 kg/m³, and that of water is 1 Mg/m³.*

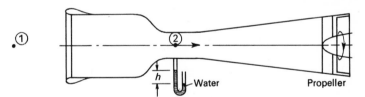

**Fig. 4.2(d).** *Open circuit wind-tunnel*

Assuming that the air in the laboratory is at rest, and applying Bernoulli's equation between points 1 and 2 in Fig. 4.2(d) we have:

$$gz_1 + \frac{p_1}{\rho} + \frac{u^2}{2} = gz_2 + \frac{p_2}{\rho} + \frac{u^2}{2}$$

$$0 + 0 + 0 = 0 + \frac{p_2}{\rho} + \frac{u_2^2}{2}$$

$$\therefore \qquad u_2 = \sqrt{\left(\frac{2(-p_2)}{\rho}\right)}$$

If the specific weight of water is $w_w$ the pressure in the working section is $p_2 = -\rho_w gh$ (i.e. is below atmospheric).

$$\therefore \quad u_2 = \sqrt{\left(2g\frac{\rho_w}{\rho}h\right)} = \sqrt{\left(2\times 9{\cdot}807\frac{m}{s^2}\times\left(\frac{1\times 10^3}{1{\cdot}288}\right)\times 0{\cdot}025\ m\right)} = \mathbf{19{\cdot}5\ m/s}$$

## EXAMPLE 4.2 (iii)

*A circular flat plate is placed horizontally parallel to a corresponding flat plate in which there is a central hole through which water is forced.*

*Find an expression showing how the pressure head of the water between the plates varies with radius. Hence, prove that a suction effect is created between the plates. Assume the momentum of fluid supplied may be neglected.*

Referring to Fig. 4.2(e),

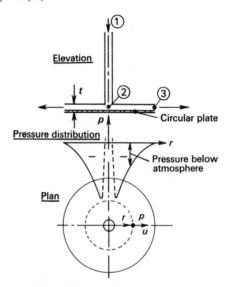

**Fig. 4.2(e).** *Pressure on a plate*

(i) By law of conservation of matter,

$$\text{rate of mass flow} = \rho_1 a_1 u_1 = \rho_2 a_2 u_2 = \rho_3 a_3 u_3$$

Hence, if the fluid is assumed incompressible, the rate of volumetric flow is

$$Q = \frac{\pi}{4}d_1^2 u_1 = 2\pi r t u = 2\pi r_3 t u_3 \tag{1}$$

Therefore,
$$ur = u_3 r_3 \tag{1a}$$

(ii) By the law of conservation of energy, the energy per unit mass of fluid remains constant, namely:

$$gH = \frac{p}{\rho}+\frac{u^2}{2}+gz+\text{friction heat} \tag{2}$$

Hence, neglecting friction,

$$g(H-z) = C \text{ (constant)} = \frac{p}{\rho} + \frac{u^2}{2} = 0 + \frac{u_3^2}{2} \qquad (2a)$$

since $p_3/\rho = 0$ reckoned from atmospheric pressure as datum.

Thus, from (1a) and (2a) we deduce

$$\frac{p}{\rho} = \frac{u_3^2 - u^2}{2} = \frac{u_3^2}{2}\left[1 - \left(\frac{r_3}{r}\right)^2\right] = -\frac{u_3^2}{2}\left[\left(\frac{r_3}{r}\right)^2 - 1\right]$$

Therefore, for any particular rate of flow the pressure head at radius $r$ is

$$\frac{p}{\rho} = -\text{constant}\left[\left(\frac{r_3}{r}\right)^2 - 1\right]$$

which is always negative (i.e. below atmosphere) as shown plotted in Fig. 4.2(e).

Hence the suction effect, assuming the momentum of the fluid supplied from pipe 1 may be neglected (see Chapter 6).

## 4.3. Effect of losses

*The justification for applying Bernoulli's equation to the motion of real fluids along stream-tubes ultimately depends on comparisons with the results of experiments.* These confirm that good agreement is obtained only if the effect of friction is small, and conditions at each section are reasonably steady and uniform. Fig. 4.3(a) is typical of the behaviour

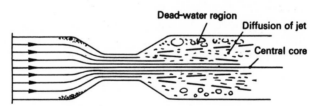

Fig. 4.3(a). *Separation at an expansion*

of a real fluid (water in this case) during its passage through a convergent–divergent channel. So long as the fluid is accelerating the channel tends to 'flow full', but beyond the throat the main stream separates from the boundaries forming a central core of moving fluid, surrounded by a 'dead water region'. The latter is filled with eddying fluid which is constantly feeding on the energy of the jet. Eventually these two regions become indistinguishable and conditions are once again reasonably uniform across the section.

The energy which is carried into the eddying motions is lost in the mechanical sense, being finally converted into heat. Consequently the total head $H$, which has been assumed to remain constant in Bernoulli's equation, is decreasing continuously along the stream-tube in a real fluid.

Hence if we modify Bernoulli's equation to take account of the loss of total head $H_L$ between any two sections 1 and 2, it becomes:

$$z_1 + \frac{p_1}{w} + \frac{u_1^2}{2g} = z_2 + \frac{p_2}{w} + \frac{u_2^2}{2g} + H_L$$

The loss of head $H_L$ is small so long as the stream is contracting, but may be large if the walls of the channel diverge too quickly as in Fig. 4.3(a). In addition, there is a cumulative friction loss down the length of the pipe which increases with the velocity of the fluid. Fig. 4.3(b) shows the effect

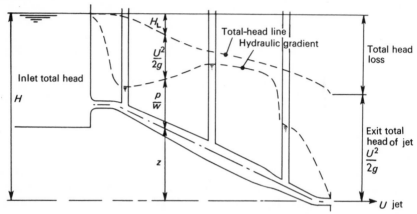

**Fig. 4.3(b).** *Effect of losses on hydraulic gradient*

of these losses, which reduce the total head progressively and conse- quently modify the levels in the gauge glasses. The imaginary profile formed by the latter is known as the *hydraulic gradient*, and the vertical intercept between this and the centre line of the pipe gives the gauge pressure in terms of head. If the hydraulic gradient falls below the centre- line of the pipe it follows that the pressure in the pipe at this point is below atmospheric.

EXAMPLE 4.3 (i)

*A long pipe-line carries water from a point A where the pressure is 5·15 MN/m²* *gauge, to a point B, 3·6 m above A, where the pressure is 4·7 MN/m² gauge.* *If the velocity of the fluid is 1 m/s at point A, where the pipe is of 50 mm* *diameter, and the pipe is of 25 mm diameter at point B, find the power expended* *in overcoming the losses.*

From energy considerations we may write:

Total head at A = total head at B + head loss between A and B

i.e.                    $H_A = H_B + H_L$

$$z_A + \frac{p_A}{w} + \frac{u_A^2}{2g} = z_B + \frac{p_B}{w} + \frac{u_B^2}{2g} + H_L$$

Since water is almost incompressible, the volumetric flow rates at $A$ and $B$ are the same:

$$Q_A = Q_B$$

$$u_A \frac{\pi}{4} d_A^2 = u_B \frac{\pi}{4} d_B^2$$

i.e.

$$u_B = \left(\frac{d_A}{d_B}\right)^2 u_A = 4u_A$$

Hence the head loss:

$$H_L = (z_A - z_B) + \left(\frac{p_A - p_B}{w}\right) + \left(\frac{u_A^2 - u_B^2}{2g}\right)$$

$$= -3 \cdot 6 \text{ m} + \frac{0 \cdot 45 \text{ MN/m}^2}{9 \cdot 807 \text{ kN/m}^3} + \frac{1 \text{ m}^2/\text{s}^2}{19 \cdot 614 \text{ m/s}^2} (1 - 4^2)$$

$$= -3 \cdot 6 \text{ m} + \frac{450}{9 \cdot 807} \text{ m} - \frac{1 \times 15}{19 \cdot 614} \text{ m}$$

$$= (-3 \cdot 6 + 45 \cdot 9 - 0 \cdot 76) \text{ m} = 41 \cdot 5 \text{ m}$$

The corresponding power loss is:

$$P = wQH_L = wu_A \frac{\pi}{4} d_A^2 H_L$$

$$= 9 \cdot 807 \frac{\text{kN}}{\text{m}^3} \times \frac{1\text{m}}{\text{s}} \times \frac{\pi}{4} \frac{25}{\times 10^4} \text{m}^2 \times 41 \cdot 5 \text{ m}$$

$$= 0 \cdot 798 \frac{\text{kNm}}{\text{s}} = 0 \cdot 798 \text{ kW}.$$

## 4.4. Venturi meter

If direct measurement of rate of discharge by weighing or volumetric measurement is impossible or inconvenient, then a secondary method employing an orifice, nozzle, weir or Venturi meter may be used.

The Venturi meter, shown diagrammatically in Fig. 4.4(a), is one device for measuring the rate of flow of liquids through pipes. In effect it is a convergent–divergent nozzle, smoothly bored with a 'quick' contraction and a 'slow' expansion. The divergence of the expanding portion, or 'diffuser', is made gradual to reduce the tendency of the fluid to separate from the boundaries and form eddies. The difference of pressure between the inlet section 1 and the throat may be used to deduce the flow rate. Assume, in the first instance, that there is no loss of head due to friction as water flows from section 1 to section 2. Water fills the leads to the U-tube, which contains mercury, of relative density $d = 13 \cdot 6$.

Thus, applying Bernoulli's equation, we have:

$$z_1 + \frac{p_1}{w} + \frac{u_1^2}{2g} = z_2 + \frac{p_2}{w} + \frac{u_2^2}{2g}$$

in which

$$Q = u_1 A_1 = u_2 A_2, \quad \text{and} \quad w = \rho g.$$

Hence, expressing the velocity of approach in terms of the throat velocity,

$$\frac{u_2^2}{2g}\left[1 - \left(\frac{A_2}{A_1}\right)^2\right] = \left(z_1 + \frac{p_1}{w}\right) - \left(z_2 + \frac{p_2}{w}\right)$$

The right-hand side is the departure from the hydrostatic distribution and it may be recalled that a U-tube responds to this (see Section 3.3).

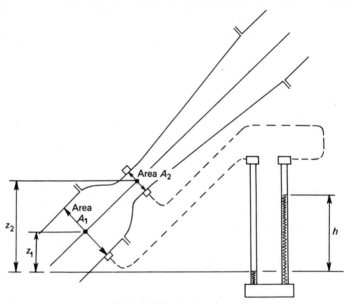

**Fig. 4.4(a).** *Venturi meter*

Its calibration may be deduced by equating pressures at the datum level:

$$p_1 + wz_1 = p_2 + w(z_2 - h) + dwh$$

i.e.

$$\left(z_1 + \frac{p_1}{w}\right) - \left(z_2 + \frac{p_2}{w}\right) = (d - 1)h$$

Hence, the throat velocity $u_2$ may be deduced from $h$ by means of:

$$\frac{u_2^2}{2g}\left[1 - \left(\frac{A_2}{A_1}\right)^2\right] = (d - 1)h$$

i.e.

$$u_2 = \frac{1}{\sqrt{[1 - (A_2/A_1)^2]}} \sqrt{[2g(d - 1)h]}$$

and the volumetric flow rate is:

$$Q = u_2 A_2 = \frac{1}{\sqrt{[1 - (A_2/A_1)^2]}} A_2 \sqrt{[2g(d - 1)h]}$$

It should be noted from this formula that the discharge depends on the gauge reading *h* irrespective of the orientation of the Venturi meter which may be inclined, vertical or horizontal.

If the fluid being metered is not water, then *d* is the ratio of the density of the manometric fluid to that of the liquid flowing through the meter. This should be proved as an exercise.

The first term in this expression for *Q* depends only on the contraction ratio $A_2/A_1$. This determines the magnitude of the *velocity of approach* $u_1$, for as $A_2/A_1$ tends to zero $u_1$ also tends to zero. In this extreme case, the first term simply becomes unity. Thus, if the flow issued from a large reservoir the volumetric rate of flow would be:

$$Q = A_2 \sqrt{[2g(d-1)h]}$$

If the ratio $A_1/A_2$ is not very large we see that the latter formula gives an underestimate of the flow rate and that the factor which corrects for the velocity of approach is of magnitude:

$$K_a = \frac{1}{\sqrt{[1-(A_2/A_1)^2]}}$$

This factor is always greater than 1, and really allows for the additional kinetic head of the approaching fluid.

Thus, for a given installation, theory suggests that:

$$Q \propto \sqrt{h}$$

and the theoretical (i.e. frictionless) 'constant' for the meter is:

$$C = K_a A_2 \sqrt{[2g(d-1)]}$$

It should be noted that the numerical value of the meter constant depends on the units used, and these should always be quoted as in Example 4.4 (i).

For accurate work the meter should be calibrated ($Q = C' \sqrt{h}$) after installation, but given favourable approach conditions, the measured value of the constant $C'$ is usually about 97 per cent of *C*, the calculated or theoretical value. The ratio of actual to theoretical *Q* is often known as the coefficient of discharge, $C_d$, of the meter and is equal to the ratio $C'$ to *C*, i.e.

$$C_d = \frac{C'\sqrt{h}}{C\sqrt{h}} \quad \text{or} \quad \text{Actual } Q = C'\sqrt{h} = C_d C \sqrt{h}$$

EXAMPLE 4.4 (i)

*If the constant of a Venturi meter is 3 per cent less than that calculated theoretically, estimate the flow rate in m³/s through it if the U-tube records a difference in the mercury levels of 200 mm. The meter has a throat diameter of 25 mm and is installed in a 75 mm diameter water main.*

The velocity of approach correction is:

$$K_a = \frac{1}{\sqrt{[1-(A_2/A_1)^2]}} = \frac{1}{\sqrt{[1-(2{\cdot}5/7{\cdot}5)^4]}}$$

$$= \frac{1}{\sqrt{(1-\frac{1}{81})}} = 1{\cdot}006$$

The theoretical meter constant is:

$$C = K_a A_2 \sqrt{[2g(d-1)]}$$
$$= 1{\cdot}006 \times \pi \times 12{\cdot}5^2 \text{ mm}^2 \sqrt{(2 \times 9{\cdot}807 \text{ m/s}^2 \times 12{\cdot}6)}$$
$$= 7\,750 \text{ mm}^2/\text{s} \times \sqrt{\text{m}}$$

The actual meter constant is therefore:

$$C' = 0{\cdot}97 \times 7{\cdot}75 \times 10^{-3} \text{ m}^{5/2}/\text{s}$$

and the discharge, for 200 mm of head on the gauge, is:

$$Q = C' \sqrt{h} = 7{\cdot}53 \times 10^{-3} \times \sqrt{0{\cdot}2} \text{ m}^3/\text{s}$$
$$= 3{\cdot}36 \times 10^{-3} \text{ m}^3/\text{s or } 3{\cdot}36 \text{ litre/s.}$$

EXAMPLE 4.4 (ii)

*A Venturi meter has its axis vertical, the inlet and throat diameters being* 150 *mm and* 75 *mm respectively. The throat is* 200 *mm above the inlet and the coefficient of discharge is* 0·96. *Petrol of relative density* 0·78 *flows through the meter at the rate of* 40 *litre/s. By application of Bernoulli's principle, find:*

(a) *the pressure difference between inlet and throat;*

(b) *the difference of level which would be registered by a vertical mercury manometer, the tubes above the mercury being full of petrol. (Relative density of mercury* 13·6.)

Referring to Fig. 4.4(a) (which can be imagined vertical for this example and metering petrol) and the subsequent analysis, we conclude that the actual rate of volumetric flow of liquid through the meter is

$$Q = C' \sqrt{h} = C_d C \sqrt{h} = C_d \frac{A_2 \sqrt{[2g(d-1)h]}}{\sqrt{[1-(A_2/A_1)^2]}}$$

where *d* is the ratio of the density of the manometric fluid (mercury) and the density of the fluid (petrol) flowing through the meter, i.e.

$$40 \times 10^6 \text{ mm}^3/\text{s} = 0{\cdot}96 \times \frac{\pi}{4} \times 75^2 \text{ mm}^2 \sqrt{\left\{ \frac{2 \times 9807 \dfrac{\text{mm}}{\text{s}^2} \left(\dfrac{13{\cdot}6}{0{\cdot}78} - 1\right)h}{(1-\frac{1}{16})} \right\}}$$

$$= 2{\cdot}49 \times 10^6 \text{ mm}^{5/2}/\text{s} \sqrt{h}$$

hence, $\sqrt{h} = 16{\cdot}05 \text{ mm}^{1/2}$ and $h = 258$ mm.

Equating pressures at the datum level in the U-tube of Fig. 4.4(a) (the

Venturi being imagined vertical in this example and metering petrol):

$$p_1 + w_p z_1 = p_2 + w_p(z_2 - h) + w_m h$$

$$\therefore \quad \frac{p_1 - p_2}{w_p} = (z_2 - z_1) + \left(\frac{w_m}{w_p} - 1\right)h$$

$$= 200 \text{ mm} + \left(\frac{13 \cdot 6}{0 \cdot 78} - 1\right) 258 \text{ mm}$$

$$= (200 + 4\ 245) \text{ mm} = 4 \cdot 445 \text{m}$$

Hence, $\quad p_1 - p_2 = 4 \cdot 445 \text{ m} \times \left(0 \cdot 78 \times 9 \cdot 807 \dfrac{\text{kN}}{\text{m}^3}\right)$

$$= \textbf{34 kN/m}^2$$

## 4.5. Jet contraction and flow through an orifice

When a fluid is discharged through a sharp-edged orifice, as indicated in Fig. 4.5(a), the issuing jet contracts downstream, forming a throat.

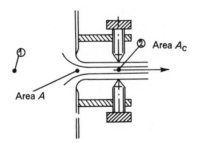

**Fig. 4.5(a).** *Contraction at an orifice*

By fitting such an orifice in the side of a tank and allowing water to discharge from it into the atmosphere, it is possible to measure the size of the 'vena contracta' or throat that is formed. It is desirable to do this fairly close to the orifice, to minimize the effect of gravity. The ratio of the contracted area of the jet to that of the orifice is known as the *contraction coefficient* of the orifice, i.e.

$$C_c = \frac{\text{Contracted area of jet}}{\text{Area of orifice}} = \frac{A_c}{A}$$

The value of $C_c$ is critically dependent on the sharpness of the upstream edge of the orifice. For a perfectly sharp lip $C_c \simeq 0 \cdot 6$, and this increases progressively with the lip radius until finally it approaches the value $1 \cdot 0$ for a 'bell-mouthed' opening, which flows full.

If we choose as our datum the centre-line through the orifice, which is assumed small in diameter compared with its depth $H$ below the still-water surface in the tank, then Bernoulli's equation simplifies to:

$$H = z_2 + \frac{p_2}{w} + \frac{u_2^2}{2g} = 0 + 0 + \frac{u_2^2}{2g}$$

Hence, the theoretical velocity of the jet is $u_2 = \sqrt{(2gH)}$, a result originally due to Torricelli.

In practice there is a slight energy loss in the contraction, and conditions are not quite uniform across the throat. These effects reduce the effective mean velocity to some fraction, $C_v$, of the theoretical value, i.e. the actual velocity of the jet is

$$u_2 = C_v\sqrt{(2gH)}$$

The factor $C_v$ is termed the *coefficient of velocity* and generally has a value which varies between 0·95 and 0·99.

The volumetric flow rate $Q$, which is equal to $u_2A_2$, is thus theoretically:

$$Q = C_cA\sqrt{(2gH)}$$

and in practice:

$$Q = C_vC_cA\sqrt{(2gH)}$$

This may be written:

$$Q = C_dA\sqrt{(2gH)}$$

where $C_d$ is the *coefficient of discharge* as determined directly from the measured discharge for a constant head $H$. The value of the discharge coefficient, $C_d$, may be compared with independent measurements of $C_c$ and $C_v$, to which it is related by:

$$C_d = C_vC_c$$

*Provided that the pressure drop across an orifice is small compared with the absolute pressure, the above equations are equally appropriate when the working fluid is a gas.* The pressure difference in such cases may be measured by an inclined gauge or U-tube, as described in Chapter 3.

EXAMPLE 4.5 (i)

*If the pressure difference across a sharp-edged circular orifice of 50 mm diameter is 18 mm of water when air flows through it, calculate the volumetric flow rate assuming a discharge coefficient of 0·6. The density of the air is 1·228 kg/m³ and the barometer stands at 760 mm Hg.*

Since 760 mm Hg corresponds to 13·6 × 760 mm of water the ratio of the pressure drop to the absolute pressure is:

$$\frac{\Delta p}{p} = \frac{1\cdot8}{13\cdot6 \times 76} = \frac{1}{575}$$

and the fractional change in density $\Delta \rho/\rho$ is of the same order; hence changes in $\rho$ may be assumed negligible.

The volumetric rate of flow of air is given by

$$Q = C_d A \sqrt{(2gH_a)}$$

where $H_a$ is the total head of air in the reservoir relative to the pressure downstream. The difference in the water-gauge reading may be expressed as a head of air by noting that:

$$H_a = \frac{\Delta p}{w_a} = \frac{\Delta p}{w_w} \times \frac{\rho_w}{\rho_a} = H_w \times \frac{\rho_w}{\rho_a}$$

$$\therefore \qquad H_a = 18 \text{ mm} \times \frac{1 \text{ Mg/m}^3}{1 \cdot 228 \text{ kg/m}^3}$$

$$= 14 \cdot 65 \text{ m}$$

Thus: $\qquad Q = 0 \cdot 6 \times \frac{\pi}{4} \; 25 \times 10^{-4} \text{ m}^2 \sqrt{\left( 2 \times 9 \cdot 807 \, \frac{\text{m}}{\text{s}^2} \times 14 \cdot 65 \text{ m} \right)}$

$$= 3 \cdot 75\pi \times 10^{-4} \text{ m}^2 \sqrt{(2 \cdot 873 \times 10^2 \text{ m}^2/\text{s}^2)}$$

$$= \mathbf{19 \cdot 95 \times 10^{-3}} \text{ m}^3/\text{s or } \mathbf{19 \cdot 95} \text{ litre/s}$$

EXAMPLE 4.5 (ii)

*Indicate how $C_v$, $C_d$, and $C_c$ may be estimated in the laboratory.*

The coefficient of velocity, $C_v = u/\sqrt{(2gh)}$, can be estimated experimentally as a deduction from measurements of length $x$, $y$ and $h$ indicated on Fig.

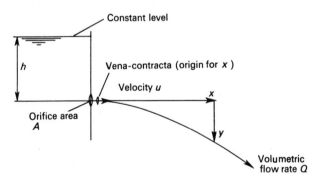

Fig. 4.5(b). *Estimation of orifice coefficients*

4.5(b). For, at any time $t$ after a particle of fluid has left the vena contracta, we have $x = ut$ and $y = \frac{1}{2}gt^2$. Hence,

$$t = \frac{x}{u} = \sqrt{\frac{2y}{g}}$$

$$\therefore \qquad u = x \sqrt{\frac{g}{2y}} \quad \text{and} \quad C_v = \frac{x}{2\sqrt{(yh)}}$$

The coefficient of discharge, $C_d = Q/A \sqrt{(2gh)}$, can be determined by collecting the liquid in a calibrated tank when the head $h$ is kept constant. The coefficient of contraction, $C_c$, may be obtained as a deduction $C_d/C_v$, or an attempt made to measure the diameter of the jet at the vena contracta by adjusting three pointed screws in a ring as indicated in Fig. 4.5(a).

### EXAMPLE 4.5 (iii)

*A vertical cylindrical tank 2 m in diameter has, at the bottom, a 50 mm diameter sharp-edged orifice for which the discharge coefficient is 0·6.*

   (a) *If water enters the tank at a constant rate of 10 litre/s, find the depth of water above the orifice when the level in the tank becomes steady.*
   (b) *Find the time taken for the level to fall from 2·5 m to 0·5 m above the orifice, if there is no inflow.*
   (c) *If water is now run into the tank at a constant rate of 18 litre/s, the orifice remaining open, find the rate of rise in water when this level has reached 2 m above the orifice.*

If $Q_i$ is the rate of inflow to the tank of surface area $A$ and $Q_o$ is the rate of outflow from the tank through the orifice of area $a$ then the rate at which the head, $h$, of water above the orifice rises is:

$$\frac{dh}{dt} = \frac{Q_i - Q_o}{A} \quad \text{and} \quad Q_o = C_d a \sqrt{(2gh)}$$

Hence, (a) if $Q_i = 10 \text{ l/s} = 10^{-2} \text{ m}^3/\text{s}$ and $dh/dt = 0$

then
$$Q_i = Q_o = C_d a \sqrt{(2gh)}$$

and
$$h = \frac{Q_i^2}{C_d^2 a^2 2g} = \left(\frac{10^{-2} \text{ m}^3/\text{s}}{0\cdot6 \times \pi 6\cdot25 \times 10^{-4} \text{ m}^2}\right)^2 \times \frac{1 \text{ s}^2}{19\cdot614 \text{ m}} = \textbf{3·68 m}$$

(b) if $Q_i = 0$

then
$$\frac{dh}{dt} = -\frac{C_d a}{A} \sqrt{(2gh)}$$

$$\int_1^2 dt = -\frac{A}{a C_d \sqrt{(2g)}} \int_1^2 h^{-1/2} dh = \frac{A}{a C_d \sqrt{(2g)}} 2(\sqrt{h_1} - \sqrt{h_2})$$

$$= \frac{10^4}{6\cdot25} \frac{2}{0\cdot6 \times 4\cdot43} \frac{\text{s}}{\sqrt{\text{m}}} (\sqrt{(2\cdot5 \text{ m})} - \sqrt{(0\cdot5 \text{ m})}) \left[\frac{\text{min}}{60 \text{ s}}\right] = \textbf{17·5 min}$$

(c) if $Q_i = 18 \text{ l/s}$ and $h = 2 \text{ m}$

then
$$\frac{dh}{dt} = \frac{Q_i - Q_o}{A} = \frac{Q_i - C_d a \sqrt{(2gh)}}{A}$$

$$= \frac{18 \times 10^{-3} \text{ m}^3/\text{s} - 0\cdot6\pi \times 625 \times 10^{-6} \text{ m}^2 \sqrt{(19\cdot614 \times 2 \text{ m}^2/\text{s}^2)}}{\pi \text{m}^2}$$

$$= \frac{18 \text{ mm}}{\pi \quad \text{s}} - \frac{2\cdot35 \text{ mm}}{\pi \quad \text{s}}$$

$$= \frac{15\cdot65 \text{ mm}}{\pi \quad \text{s}} \left[\frac{60 \text{ s}}{\text{min}}\right] = \textbf{299 mm/min}$$

## 4.6. Effect of velocity of approach

An orifice or nozzle may be used for metering purposes in a pipe-line as an alternative to a Venturi meter. The major advantage of the latter is that the diffuser (i.e. the diverging cone) reduces the energy loss (i.e. the amount dissipated into heat). In fact a metering nozzle fitted in a pipe-line may be considered as equivalent to a Venturi meter without a diverging portion. The velocity of approach correction is, therefore, the same as for the Venturi, namely:

$$K_a = \frac{1}{\sqrt{[1-(A_2/A_1)^2]}}$$

see page 81.

For an orifice (Fig. 4.5($a$)) $A_2 = C_c A$ so that in theory:

$$K_a = \frac{1}{\sqrt{[1-C_c^2(A/A_1)^2]}}$$

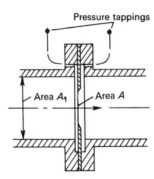

**Fig. 4.6(a).** *Metering orifice*

In practice $C_c$ increases as $A/A_1$ (Fig. 4.6(a)) is increased in accordance with the following table:

| $A/A_1$ | 0·1 | 0·2 | 0·3 | 0·4 | 0·5 | 0·6 | 0·7 | 0·8 | 0·9 | 1·0 |
|---------|-----|-----|-----|-----|-----|-----|-----|-----|-----|-----|
| $C_c$ | 0·624 | 0·632 | 0·643 | 0·659 | 0·681 | 0·712 | 0·755 | 0·813 | 0·892 | 1·00 |

In addition, the upstream pressure tapping is usually located for convenience at the flanges, as indicated in Fig. 4.6(a), so that the pressure at this point is not the true static value, but also depends on the dynamic head ($u_1^2/2g$). Consequently it is more usual, and generally more accurate, to make a purely nominal correction for the velocity of approach, i.e. to write:

$$Q = K_a C_d A \sqrt{(2gH)} = \frac{1}{\sqrt{[1-(A/A_1)^2]}} C_d A \sqrt{(2gH)}$$

$H$ is the measured head difference and $C_d$ (the coefficient of discharge) is, at least for relatively small velocities of approach, very nearly 0·6 for a sharp-edged orifice under most practical conditions. If very great accuracy is required it is necessary to calibrate the installation or alternatively to reproduce a standard design recommended by, say, the British Standards Institution. Similar remarks apply to all such metering devices, it being particularly important to achieve as uniform a flow as possible in the approach channel.

### EXAMPLE 4.6 (i)

*A pair of hydraulic buffers is used to arrest the motion of railway trucks. Each buffer consists of a piston working in a cylinder of 200 mm diameter. The working fluid is water which is discharged to atmosphere through an orifice in each cylinder of 40 mm diameter when the piston is forced into the cylinder.*

*If a truck of mass 25 tonne strikes the buffers when travelling at 32 km/h, determine the distance each piston moves in reducing the truck speed to 8 km/h and the time taken to do it. Assume the coefficient of discharge of the orifice to be 0·65, and neglect the friction of the piston in the cylinder.*

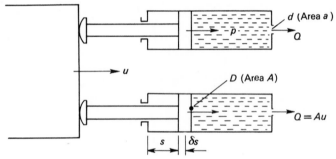

**Fig. 4.6(b).** *Hydraulic buffers*

At the instant shown in Fig. 4.6(b) the rate of discharge from each cylinder

$$Q = C_d a \sqrt{\left(2g\frac{p}{w}\right)} = Au \tag{1}$$

Hence, the pressure
$$p = \left(\frac{A}{aC_d}\right)^2 \rho\frac{u^2}{2} \tag{1a}$$

Also as the rate of reduction in kinetic energy of truck = the rate at which work is done on water in two cylinders.

$$-\frac{d}{dt}\left(\frac{1}{2}Mu^2\right) = 2pAu = 2pA\frac{dS}{dt} \tag{2}$$

i.e.
$$-Mu\delta u = 2pA\delta S \quad \text{or} \quad p = -\frac{M}{2}\frac{u}{A}\frac{du}{dS} \tag{2a}$$

Hence, equating (1a) and (2a):

$$p = -\frac{M}{2}\frac{u}{A}\frac{du}{dS} = \left(\frac{A}{aC_d}\right)^2 \rho\frac{u^2}{2}$$

or
$$\int_1^2 dS = -\frac{Ma^2 C_d^2}{A^3 \rho} \int_1^2 \frac{du}{u}$$
(3)

$$S_2 - S_1 = \frac{M}{\rho}\left(\frac{a}{A}\right)^2 \frac{C_d^2}{A}\log_e\left(\frac{u_1}{u^2}\right) = \frac{25\ t}{1\ \text{t/m}^3}\left(\frac{4}{20}\right)^4 \times \frac{0\cdot65^2}{\pi \times 0\cdot01\ \text{m}^2}\times\log_e 4$$

$$= 0\cdot538\ \text{m} \times \log_e 4$$

i.e.   *Distance moved* = **0·746 m**

Dividing Equation (3) by $u$ we have

$$\int_1^2 \delta t = -\frac{M}{A}\left(\frac{a}{A}\right)^2\frac{C_d^2}{\rho}\int_1^2\frac{du}{u^2}$$

i.e.   $$t_2 - t_1 = \frac{M}{\rho}\left(\frac{a}{A}\right)^2\frac{C_d^2}{A}\left(\frac{1}{u_2}-\frac{1}{u_1}\right) = 0\cdot538\ \text{m}\left(\frac{9}{20}-\frac{9}{80}\right)\frac{\text{s}}{\text{m}}$$

$$= 0\cdot538 \times 9\left(\frac{3}{80}\right)\text{s}$$

$$= \mathbf{0\cdot1816\ s}$$

## 4.7. Pitot tube or total-head tube

When a stream divides to pass round an obstacle, as indicated by Fig. 4.7(a), the area of each dividing stream tube increases considerably, and this effect becomes more marked as the tubes approach closer to P.

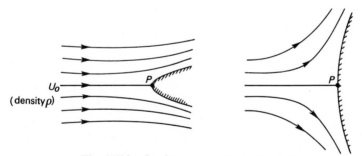

**Fig. 4.7(a).** *Conditions near stagnation point*

Consequently P is termed a stagnation point, and a pressure tapping located there will record a pressure given by:

$$gz_0 + \frac{p_0}{\rho} + \frac{u_0^2}{2} = gz_P + \frac{p_P}{\rho} + \frac{u_P^2}{2}$$

i.e.   $$p_P = p_0 + \tfrac{1}{2}\rho u_0^2$$

since $z_0 = z_P$ and $u_P = 0$. Alternatively, $(p_P - p_0)/w = u_0^2/2g$, where $w = \rho g$.

Hence, for an incompressible fluid, the pressure at a stagnation point is the sum of the static and dynamic pressures of the free stream. Any device which consists essentially of a forward-facing pressure tapping, which records the stagnation pressure, as indicated by Fig. 4.7(b), is termed a Pitot tube, or total-head tube.

An open-ended circular tube, together with the stationary fluid contained within it, forms an obstacle which records satisfactorily without calibration. In addition, such a tube is insensitive to small angles of yaw, i.e. inclination of the tube to the direction of the on-coming stream.

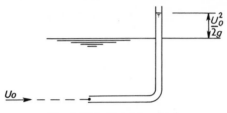

**Fig. 4.7(b).** *Total-head tube*

EXAMPLE 4.7 (i)

*Calculate the dynamic pressure at the stagnation point on the wing of an aeroplane travelling at 720 km/h assuming the density of air (at sea level) to remain constant at 1·228 kg/m³.*

Dynamic pressure is:

$$\tfrac{1}{2}\rho u^2 = \tfrac{1}{2} \times 1\cdot228 \, \frac{\text{kg}}{\text{m}^3} \times 200^2 \, \frac{\text{m}^2}{\text{s}^2} \left[ \frac{\text{N s}^2}{\text{kg m}} \right]$$

$$= \textbf{24·56 kN/m}^2$$

## 4.8. Pitot-static tube

When a measurement of the mean static pressure at a section of a pipe or duct is required, wall tappings are normally adequate if free from burrs. If, however, the pressure varies across the flow, e.g. owing to the presence of an obstacle (such as a model in a wind tunnel), local readings may be required. A simple *static tube* consists essentially of a streamlined tube with side tappings as indicated in Fig. 4.8(a). The presence of the body constricts the streamlines in its close proximity, causing an increase in velocity along its surface, whereas the presence of the stem however produces a damming-up effect which tends to produce an error of opposite sign. Hence, by careful placing of the side tappings relative to the nose and the stem a satisfactory static reading may be obtained. The side tappings are placed at regular intervals round the tube, but the reading obtained is much more sensitive to yaw than that of the total-head tube. By running a lead through the inside of a static tube to an additional pressure tapping at the nose, an instrument, which simultaneously records the stagnation and static pressures, is obtained. This is termed a *Pitot-static tube* and Fig. 4.8(b) shows a design due to Prandtl.

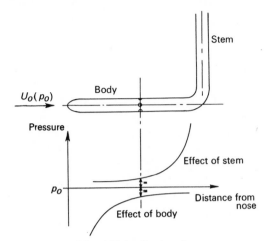

**Fig. 4.8(a).** *Static tube*

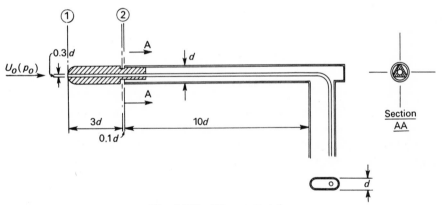

**Fig. 4.8(b).** *Pitot-static tube*

Since the 'total-head' tapping records:

$$\frac{p_1}{\rho} = \frac{p_0}{\rho} + \frac{u_0^2}{2}$$

and the 'static' tapping records:

$$\frac{p_2}{\rho} = \frac{p_0}{\rho}$$

it follows that a differential manometer connected between them records the kinetic, or dynamic, head:

$$\frac{p_1 - p_2}{\rho} = \frac{u_0^2}{2}$$

The kinetic head, when expressed as a pressure (i.e. the dynamic pressure) is

$$p_1 - p_2 = \tfrac{1}{2}\rho u_0^2$$

$$\therefore \qquad u_0 = \sqrt{\left(\frac{2(p_1 - p_2)}{\rho}\right)}$$

A Pitot-static tube can be fitted to a pressure gauge of the aneroid barometer type whose scale is graduated in knots (1 international knot being 0·514 m/s). Such an instrument gives what is termed the indicated air-speed (I.A.S.) of an aircraft, the instrument being calibrated on the basis of a nominal value of $\rho$ at sea level, $\rho_0$ say. The indicated air-speed at altitude thus requires correcting by the factor $\sqrt{(\rho_0/\rho)}$ due to the increasing tenuity of the air. This is apart from the effect of compressibility, which further affects the reading obtained—causing the above expression for the velocity to be about 3 per cent high at one-half the speed of sound, i.e. at a Mach Number of 0·5.

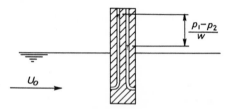

**Fig. 4.8(c).** *Current meter*

A number of commercial instruments are of simpler construction, generally employing forward- and backward-facing pressure tappings (see Fig. 4.8(c)). Such current meters require calibration to find the value of $C$ in the following formula which applies over limited ranges:

$$u_0 = C \sqrt{\left[\frac{2(p_1 - p_2)}{\rho}\right]}$$

The variation in the value of $C$ is primarily due to the fact that the downstream tapping is recording in the wake, where conditions tend to be, not only unsteady, but also sensitive to scale effects.

EXAMPLE 4.8 (i)

*A Pitot-static tube in an air-stream of density $\rho = 1\cdot228$ kg/m³ is connected to a manometer gauge containing water. If the velocity u of the air is such as to cause a vertical difference of head h of water in the manometer tube find a formula relating h in mm and u in m/s.*

$$p_1 - p_0 = \tfrac{1}{2}\rho_{air} u_0^2 = \rho_{water} g h$$

$$u_0^2 = 2 \frac{\rho_w}{\rho_a} g h = 2 \times \frac{1 \text{ t/m}^3}{1\cdot228 \text{ kg/m}^3} \times 9\cdot807 \frac{\text{m}}{\text{s}^2} h \left[\frac{\text{m}}{10^3 \text{ mm}}\right]$$

i.e. 
$$\left(\frac{u_0}{\text{m/s}}\right)^2 = \frac{2 \times 10^3 \times 9\cdot807}{1\cdot228 \times 10^3}\left(\frac{h}{\text{mm}}\right)$$

Hence, 
$$\frac{u_0}{\text{m/s}} = 4\cdot0\sqrt{\left(\frac{h}{\text{mm}}\right)}$$

## Exercises on Chapter 4

1. A Venturi meter has a diameter of 400 mm in the large part, and 125 mm in the throat. With water flowing through it the pressure head is 10 m in the large part and 9·7 m in the throat. Find the velocity in the throat and the discharge through the meter. Take the meter coefficient as 1.

2. A Venturi meter has an entrance diameter of 60 mm and a throat diameter of 20 mm. Pipes from the entrance and the throat lead water to the limbs of a U-tube containing mercury and the difference in pressure at these two places is thus recorded by a difference in the level of the mercury.

If the coefficient of discharge of the meter is 0·96, find the constant $D$ in the law of the meter.

$Q = D \sqrt{h}$ where $Q$ is in litre/min and $h$ is in mm.

3. A Venturi meter placed in a 90 mm diameter pipe has a throat diameter of 30 mm. The coefficient of discharge of the meter is 0·97.

(a) Determine the litres of water passing per minute when the Venturi head is 486 mm of water.

(b) If the frictional loss in the diverging cone is double that in the converging cone, find the total head lost in the meter due to friction when the water is passing at the above rate.

4. A Venturi meter in a horizontal 300-mm diameter water pipe has a throat diameter of 100 mm and a discharge coefficient of 0·98. A U-tube mercury gauge was used to measure the difference of head between the pipe inlet and the throat at points 600 mm apart. If the mercury gauge reading was 600 mm, find the flow through the pipe in $m^3$/min. If the meter had been placed vertically with the flow upwards, what would have been the gauge reading for the same rate of flow?

5. A sharp-edged orifice of 50 mm diameter has a coefficient of velocity 0·97 and a coefficient of contraction 0·64 and discharges into a pond the surface of which is 1 m below the centre of the orifice. The jet strikes the surface at a horizontal distance of 3·3 m from the orifice.

Estimate the rate of discharge and the head over the orifice.

6. A tank of square cross-section, each side measuring 300 mm is open at the top, and is fixed in an upright position. A 6 mm diameter circular orifice is situated in one of the vertical sides near the bottom. Water flows into the tank at the top at a constant rate of 47 l/min. At a particular instant, the jet strikes the floor at a point 640 mm from the vena contracta, measured horizontally and 520 mm below the centre-line of the orifice measured vertically. Determine whether the water surface in the tank is rising or falling at the instant under consideration. Also find the height of the surface above the

centre-line of the orifice and the rate of change of height. (Take $C_v$ as 0·97 and $C_d$ as 0·64.)

7. Describe, with the help of neat diagrams, the construction and operation of (*a*) a Pitot-static tube, and (*b*) a manometer suitable for use with such a tube when differences of head are very small.

A Pitot-static tube used to measure air velocity along a wind channel is coupled to a manometer which shows a difference of head of 3·8 mm water. The density of the air is 1·22 kg/m³ and that of water 1 Mg/m³.

Obtain the air velocity assuming that the coefficient for the Pitot tube is unity.

8. A Venturi meter (coefficient of discharge 0·97) is incorporated in a 300 mm diameter pipe. The diameter at the throat is 100 mm and the rate of flow of water along the pipe is 8·4 m³/min. What is the pressure difference between the entrance and the throat of the meter?

What reading will be registered in a manometer in the form of an inverted U-tube if the Venturi meter is vertical?

There is air on top of the water in the two limbs of the manometer.

9. A Venturi meter is tested with its axis horizontal and the flow measured by means of a tank. The pipe diameter is 80 mm, the throat diameter 40 mm, and the pressure difference is measured by a U-tube containing mercury, the connections being full of water. If the difference of levels in the U-tube remains steady at 270 mm of mercury while 2·4 Mg of water are collected in 4 min, what is the coefficient of discharge? (Relative density of mercury is 13·6.)

10. A tank measures 6 m long by 3 m wide in plan view. It is divided into two unequal compartments by a thin vertical partition 2 m from one end. This partition contains a sharp-edged orifice 80 mm square through which water may pass from one compartment to the other. The coefficient of discharge of the orifice is 0·61. Let *h* be the difference of level in the two compartments at a certain instant. Obtain an expression for the rate of change of *h* with time. Also find the time taken for *h* to decrease from 3 m to 1 m.

11. Water is to be discharged from a pit through a 230 mm diameter pipe 60 m long, at the outlet end of which is fitted a horizontal Venturi meter having a throat diameter of 150 mm and a coefficient of discharge of 0·96. A constant head of 21 m of water (gauge) is maintained at the pipe inlet, which is at a level 18 m below that of the Venturi meter. If the barometric pressure corresponds to 10·3 m of water and the absolute pressure at the Venturi throat is not to fall below 3 m of water, find the maximum discharge that may be permitted. (For the pipe $f = 0·007$ in the formula $H_L = (4fl/d)(u^2/2g)$.)

Under these conditions what would be the difference of level between the columns of a U-tube mercury manometer connected between inlet and throat of the Venturi, the connecting tubes above the mercury being full of water? (Relative density of mercury = 13·6.)

12. The discharge coefficient for a Venturi meter was found to be constant for rates of flow exceeding a certain value. Show that, for this condition, the

loss of head due to friction in the convergent part of the Venturi can be expressed by $kQ^2$ where $k$ is a constant and $Q$ is the rate of flow. Obtain the value of $k$ if the inlet and throat diameters of the Venturi are 100 mm and 50 mm, respectively, and the discharge coefficient is 0·96.

13. Sketch a modern form of Pitot tube and show how its readings may be applied to determine the velocity of the stream. State also how the result may be affected when the orientation of the tube does not coincide accurately with that of the fluid current.

In measuring the air flow in a 250 mm diameter pipe by a Pitot tube placed in the centre of the pipe, the difference of levels in the gauge was 50 mm of water. If the coefficient of the instrument was 0·98, and taking the mean velocity as 0·82 of the maximum, find the discharge through the pipe in kg/s and m³/s if the mean static pressure of the air in the pipe is 138 kPa and temperature 15°C. Assume, for air, $\rho = 1·25$ kg/m³, $R = 288$ J/kgK, and for water $\rho = 1$ Mg/m³.

## Chapter 5

# Elements of similarity; notches and weirs

*'Observe the motion of the surface of the water which resembles that of hair which has two motions, of which one depends on the weight of the hair, the other on the direction of the curls.'*

LEONARDO

## 5.1. Geometric similarity

Two systems are said to be geometrically similar if they can be made to appear photographically alike, i.e. they differ only in their absolute size, see Fig. 5.1.(a). A series of geometrically similar systems has the following properties:

If one length is arbitrarily chosen as a reference length to denote the scale ($r$ in Fig. 5.1(a)), all other lengths are proportional in magnitude to $r$.

Thus:

$$d_1 \propto l_1 \propto r_1$$
$$d_2 \propto l_2 \propto r_2$$

The corresponding constants of proportionality ($d_1/r_1$) and ($d_2/r_2$) are the same, as are ($l_1/r_1$) and ($l_2/r_2$). These are *shape coefficients*, which define the geometry of the system, independent of its size, and are not to be confused with the *linear scale ratio* of geometrically similar bodies ($r_1/r_2$), which has the *same* value for all pairs of corresponding lengths:

$$\frac{d_1}{d_2} = \frac{l_1}{l_2} = \frac{r_1}{r_2}$$

Since all corresponding areas ($A$) are the product of two lengths which are each proportional to $r$:

$$A \propto r^2$$

Similarly all volumes (and hence weights, if the materials correspond) are the product of three lengths each proportional to $r$, i.e.

$$V \propto r^3$$

Thus if we double the linear size as typified by $r$, we increase all areas by 4 times, and all volumes by 8 times.

It should be noted that all corresponding angles (such as $\theta_1$ and $\theta_2$) on geometrically similar bodies are equal. This follows since, by definition, an angle is the ratio of two lengths, that of an arc ($s$) to a radius ($a$). If each of these is respectively in the same ratio, then their quotients will be equal:

$$\frac{s_1}{s_2} = \frac{a_1}{a_2} \left( = \frac{r_1}{r_2} \right)$$

i.e.
$$\frac{s_1}{a_1} = \frac{s_2}{a_2} \quad \text{or} \quad \theta_1 = \theta_2$$

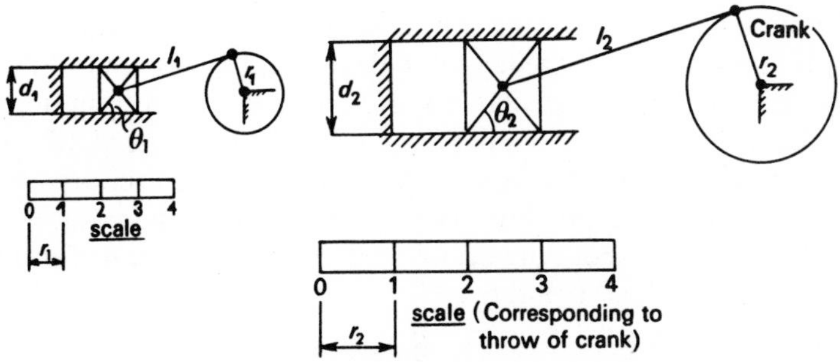

**Fig. 5.1(a).** *Geometrically similar reciprocating pumps*

EXAMPLE 5.1 (i)

*A reciprocating pump delivers 25 litre/min of oil when running at 1 100 rev/min. What will be the delivery of a geometrically similar machine of twice the original size and at what speed should it be run if the maximum piston velocity is limited to the same value on each machine, and their volumetric efficiencies are the same?*

The swept volume per rev is proportional to the cube of the size, i.e.

$$V \propto r^3$$

Hence the volumetric discharge:

$$Q = \epsilon V N \propto r^3 N$$

Since the strokes are proportional to $r$ the piston speeds:

$$u \propto rN$$

Eliminating $N$ between the two latter expressions;

$$Q \propto r^3 \frac{u}{r}$$

Hence if $u$ is the same for each machine

$$Q \propto r^2$$

i.e. the discharge is proportional to the square of the size, so that:

$$Q_2 = 2^2 \times 25 \text{ l/min} = \mathbf{100} \text{ litre/min}$$

and the corresponding allowable speed:

$$N \propto \frac{u}{r} \propto \frac{1}{r}$$

i.e.
$$N_2 = \tfrac{1}{2} \times 1\ 100 \text{ rev/min} = \mathbf{550} \text{ rev/min}$$

## 5.2. Kinematic similarity

If two systems are such that corresponding elements traverse correspond-ing distances in corresponding times, their motions are said to be kine-matically similar. Thus if the length scales ($r_1$ and $r_2$) shown in Fig. 5.1(a) are supplemented by clocks marked off in intervals corresponding to the time of 1 revolution of the crank ($T_1$ and $T_2$), the motions of all elements in the machines will appear kinematically similar when referred to these scales of length and time. For such kinematically similar motions the times taken for corresponding events (e.g. the length of time for which a valve is open), $t$ say, is such that:

$$t_1 \propto T_1$$
$$t_2 \propto T_2$$

Since, for kinematically similar motions the corresponding distances and times are respectively proportional to $r$ and $T$, the corresponding velocities, such as the mean piston speed, for example:

$$u \propto \frac{r}{T}$$

and the corresponding accelerations are:

$$a \propto \frac{r}{T^2}$$

If we eliminate $T$ by squaring the expression for $u$ and dividing by that for $a$, we obtain:

$$\frac{u^2}{a} \propto r$$

$$a \propto \frac{u^2}{r}$$

EXAMPLE 5.2 (i)

*Compare the maximum allowable running speeds of a series of geometrically*

*similar pumps if the valve accelerations are limited to a constant maximum
value.*

Choosing a typical length such as the stroke $l$, and a typical time $T$ such as
that to complete 1 revolution, all corresponding accelerations:

$$a \propto \frac{l}{T^2} \propto lN^2$$

where $N$ is the number of revs per unit time. If $a$ is limited to a constant
maximum value by the values:

$$lN^2 = \text{constant}$$

i.e.
$$N \propto \frac{1}{\sqrt{l}}$$

i.e. the angular speed must vary inversely as the square root of the linear size
of the machine.

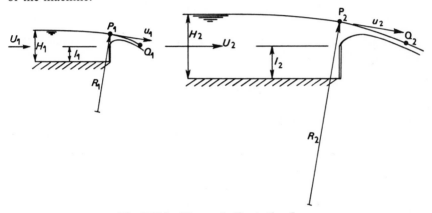

**Fig. 5.2(a).** *Kinematically similar flows*

The concepts of corresponding velocities and accelerations in kinematically
similar flows are so fundamental to later work that it is worth while to
reconsider them from a slightly different point of view.

Fig. 5.2(a) shows two streams with approach velocities $U_1$ and $U_2$ moving
with similar flow patterns (i.e. with velocities in the same directions at corresponding points) flowing over two geometrically similar obstacles of height
$l_1$ and $l_2$ respectively.

*Geometric similarity is a necessary condition but not a sufficient condition* (as
is the case with machinery to ensure kinematically similar motions) *for kinematic similarity of flow patterns past two bodies. The velocities* (which in flow
patterns are not wholly dependent on geometrical similarity of bodies) *must
be in the same ratio at corresponding points in the flows.*

If all velocities are proportional we may select one, such as the velocity $U$
of the free stream, and express all the others in terms of this as: $u = k_u U$.

In particular $u_1 = k_u U_1$ and $u_2 = k_u U_2$ the value of $k_u$ being the same in
each system if the motions are kinematically similar. In this case we have

selected a reference *velocity* rather than a reference *time*. The latter, however, may be derived from $U$ and $l$ as the time taken by the free stream to traverse the length $l$, i.e. $T = l/U$. If this time is doubled then the time for any other event, such as the time which elapses while an element is moving from P to Q, is also doubled.

Referring again to Fig. 5.2(a), the particles at the points P are moving on paths having radii of curvature $R$ which are proportional to $l$. Hence the centripetal accelerations ($u^2/R$) are, for kinematically similar motions:

$$\frac{u^2}{R} \propto \frac{U^2}{l}$$

This is a general result for all the accelerations, centripetal or otherwise, in kinematically similar motions. For such motions, since acceleration is change of velocity (proportional to $U$) per unit time (proportional to $l/U$), i.e.:

$$a \propto \frac{U}{l/U} \propto \frac{U^2}{l} \text{ in general}$$

In the particular problem considered one of the accelerations acting is that of gravity so that, if the flows are similar then:

$$g \propto \frac{U^2}{l}$$

i.e. the value of $U^2/lg$ must be the same for each. Extensive use will be made of this result at a later stage—see Chapter 14. Further, for similar flows of this kind the value of $H$ must be proportional to the corresponding value of $l$ so that equally well:

$$U \propto \sqrt{(gl)} \propto \sqrt{(gH)}$$

Thus the discharge per unit width down each channel $Q'(= UH)$ is such that

$$Q' \propto \sqrt{(gH)} . H \propto g^{1/2} H^{3/2}$$

provided that the two flows are similar. Further the actual volumetric discharge rate $Q$ for two similar installations is, if the flows are similar:

$$Q \propto Q'l \propto Q'H \propto g^{1/2} H^{5/2}$$

This expression applies to geometrically-similar systems and thus requires that the breadth of each channel is proportional to its height.

EXAMPLE 5.2 (ii)

(a) *Show that, for kinematically similar wave motions created on the surface of still water, geometrically similar ships must move at speeds proportional to the square root of their linear size.*

(b) *Hence deduce the speed at which a 4 m model should be towed in order that it may have a similar wave pattern to that round the full-scale ship which travels at 15 knot and has a length of 144 m.*

(a) The motion of the waves ('gravity waves') is governed primarily by the speed of the ship, $U$ say, and the acceleration due to gravity, $g$. For kinematic similarity:

$$g \propto \frac{U^2}{L} \quad \text{or} \quad \frac{U^2}{Lg} = \text{constant} \; (Fr^2 \text{ say})$$

i.e. the speed must be such that $U/\sqrt{(Lg)}$ is the same on both model and full scales, and this is known as the Froude number $Fr$ (see Chapter 14). From it we deduce (taking $g$ constant) that the law of corresponding speeds is:

$$U \propto \sqrt{L}$$

a result first established by William Froude.

(b) As the length of the model is 4/144 (i.e. 1/36) that of the ship, the speed of the model should be:

$$U_m = \frac{1}{\sqrt{(36)}} \times 15 \text{ knot} = \mathbf{2 \cdot 5} \text{ knot}$$

## 5.3. Scaling of lengths and times

The magnitude of a quantity is expressed as being so many times a unit of the same *kind* as itself. Thus, a particular length is 1 000 mm or 1 m, and to distinguish this kind of quantity from, for example, 1 second, we say that any length ($l$) is of the dimension [L] and any time ($t$) is of the dimension [T]. This is conventionally written here as:

$$l = [L]$$
$$t = [T]$$

the square brackets implying 'has the dimensions of'. Dimensional concepts are particularly useful in similarity; for example if two systems are geometrically similar, then all quantities having the same dimensions, in terms of [L], are in the same ratio.

As an area possesses both length and breadth, we may quite simply derive the dimensions of the form as:

$$A = l_1 \times l_2 = [L^2]$$

Similarly the dimensions of a volume are:

$$V = [L^3]$$

The dimensions of a velocity ($u$) follow from those of distance per unit time:

$$u = \frac{l}{t} = \left[\frac{L}{T}\right]$$

which distinguishes this kind of quantity from, say, acceleration ($a$) for which

$$a = \frac{u}{t} = \left[\frac{L}{T^2}\right]$$

For kinematically similar systems all quantities of the same dimensions in terms of [L] and [T] are in the same ratio.

That all lengths will be in a certain ratio presupposes geometric similarity, and the fact that corresponding events take place in a certain time ratio implies kinematically similar motions.

EXAMPLE 5.3 (i)

*A model lock installation built to $\frac{1}{64}$ full scale is found to have a filling time of 75 s. Assuming that the flow is kinematically similar to that on full scale, compare the volumes of water transferred and the corresponding filling times.*

All lengths ($l$) are in the ratio $\frac{1}{64}$. As each installation fills under the action of gravity, which is the same for each, all accelerations ($a$) are in the ratio 1/1.

Using suffixes $m$ and $f.s.$ respectively to denote model and full scale, i.e.:

$$\frac{l_m}{l_{f.s.}} = \frac{1}{64} \quad \text{and} \quad \frac{a_m}{a_{f.s.}} = 1$$

The corresponding volumes $V$ (of dimensions L³) are thus in the ratio:

$$\frac{V_m}{V_{f.s.}} = \frac{1}{64^3}$$

and corresponding times follow from:

$$\frac{a_m}{a_{f.s.}} = \frac{l_m/t_m^2}{l_{f.s.}/t_{f.s.}^2} = 1$$

i.e.

$$\frac{t_{f.s.}}{t_m} = \sqrt{\left(\frac{l_{f.s.}}{l_m}\right)} = 8$$

∴

$$t_{f.s.} = 8 \times 75 \text{ s} = \mathbf{10 \text{ min}}$$

Hence it takes 8 times as long, i.e. 10 min, to transfer a volume of water 64³ times as great as on the model.

## 5.4. Dimensional checking of equations

It is possible to check the form of most physical equations by testing the dimensions of the quantities involved. This follows since any formula for a physical quantity (such as a volumetric flow rate $Q$) must be an expression having the correct dimensions (L³/T in the case considered).

Consider, for example, an expression developed in the previous chapter for the discharge through an orifice:

$$Q = C_d A \sqrt{(2gH)}$$

see Section 4.5.

The dimensions of the left-hand side are those of volume per unit time:

$$Q = \left[\frac{L^3}{T}\right]$$

Those of the right-hand side are, since $C_d$ is a numerical coefficient:

$$C_d A \sqrt{(2gH)} = \left[ L^2 \left( \frac{L}{T^2} \right)^{1/2} [L]^{1/2} \right]$$

for $A$ is an area (of dimensions $[L^2]$), $g$ is an acceleration (of dimensions $[L/T^2]$), and $H$ is a length (of dimensions $[L]$).

By equating the indices of L and T for the left-hand side to those of the right-hand side, we confirm that the equation is, at least, a possible one:

L:    $3 = 2 + \frac{1}{2} + \frac{1}{2}$

T:    $-1 = -2 \times \frac{1}{2}$

In order to check equations in this way (and, later, to extend the method by applying it in reverse to forecast the form of physical equations), it is necessary to be able to relate the dimensions of any physical quantities.

Although L and T may conveniently be chosen as primary dimensions from which those of any other *kinematic* quantity may be derived, it is necessary to introduce a third dimension to express the relationships between *kinetic* quantities. For example kinetic energy ($E$) is defined as $\frac{1}{2}mu^2$, which involves mass ($m$) as well as velocity ($u$). If we denote the dimensions of mass by M, i.e.

$$m = [M]$$

then those of energy are, in terms of $[M]$, $[L]$ and $[T]$:

$$E = \tfrac{1}{2}mu^2 = \left[ \frac{ML^2}{T^2} \right]$$

It is found that these three $[M]$, $[L]$ and $[T]$ are sufficient for all mechanical, i.e. kinetic, problems, since, by using such mechanical laws as:

$$\text{Force} = \text{Mass} \times \text{acceleration}$$
$$\text{Impulse} = \text{Change of momentum}$$
$$\text{Work done} = \text{Change of kinetic energy}$$

the commonly occurring quantities may easily be linked:

| *Quantity:* | Mass | Force | Impulse (Momentum) | Work (Energy) |
|---|---|---|---|---|
| *Dimensions:* | $[M]$ | $[MLT^{-2}]$ | $[MLT^{-1}]$ | $[ML^2T^{-2}]$ |

The choice of primary dimensions is to some extent governed by convenience. It is possible for example (and sometimes preferable) to work in terms of $[F]$, $[L]$ and $[T]$ rather than $[M]$, $[L]$ and $[T]$. In terms of force, length and time, the above table becomes:

| Quantity: | Mass | Force | Impulse (Momentum) | Work (Energy) |
|-----------|------|-------|--------------------|---------------|
| Dimensions: | $[FT^2L^{-1}]$ | $[F]$ | $[FT]$ | $[FL]$ |

The subject of dimensional analysis forces a clear distinction to be made between mass and weight (which is force due to gravity). They are quantities which are completely different in *kind*. The mass has the dimensions:

$$[M] = [FT^2L^{-1}]$$

and the weight (being a force of attraction) has the dimensions:

$$[F] = [MLT^{-2}]$$

Similarly the density of a fluid $\rho$ is mass per unit volume:

$$\rho = [ML^{-3}] = [FT^2L^{-4}]$$

and the weight per unit volume ($w$), being a force which the earth exerts on unit volume, has the dimensions:

$$w = [FL^{-3}] = [MT^{-2}L^{-2}]$$

EXAMPLE 5.4 (i)

*The velocity u of an airstream is alleged to be related to the pressure difference ($\Delta p$) across a Pitot-static tube by the expression:*

$$u = \sqrt{\left(\frac{2g\Delta p}{\rho}\right)}$$

*where $\rho$ is the air density and g is the acceleration due to gravity.*

*Carry out a dimensional check on the validity of this expression and discuss its implications.*

After squaring the expression for convenience the left-hand side has the dimensions:

$$u^2 = \left[\frac{L^2}{T^2}\right]$$

For the right-hand side:

$$\frac{2g\Delta p}{\rho} = \left[\frac{L}{T^2}\frac{F}{L^2}\frac{L^3}{M}\right]$$

Working in terms of $[M]$, $[L]$ and $[T]$ by noting that:

$$[F] = \left[\frac{ML}{T^2}\right]$$

$$\frac{2g\Delta p}{\rho} = \left[\frac{L}{T^2}\frac{ML}{L^2T^2}\frac{L^3}{M}\right] = \left[\frac{L^3}{T^4}\right]$$

which is the dimension of the left-hand side, multiplied by the factor $L/T^2$.

This suggests that the '$g$' has no place in the equation and that the formula should be $u = \sqrt{(2\Delta p/\rho)}$, which is dimensionally correct.

## 5.5. Dimensional analysis of a vee notch

A method which is particularly convenient for measuring the rate of flow from one tank to another is illustrated diagrammatically in Fig. 5.5(a). The height of the still-water surface measured from the base of

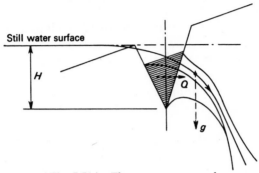

**Fig. 5.5(a).** *Flow over a vee notch*

the notch determines the rate of discharge. Although calibration is desirable, the method of dimensional analysis quite simply suggests the form of the equation relating the volumetric rate of discharge ($Q$) to the head over the notch ($H$).

As the fluid falls under the action of gravity, any equation for $Q$ is likely to involve both $H$ and $g$. Let us assume for the moment that $Q$ varies as $H^a g^b$ (we shall elaborate this argument at a later stage, see Chapter 10), i.e. suppose:

$$Q = kH^a g^b$$

where $k$ is a coefficient which has no dimensions and $a$ and $b$ are undetermined numerical indices. Since the dimensions of the left-hand side are:

$$Q = \left[\frac{L^3}{T}\right]$$

and those of the right-hand side are

$$kH^a g^b = [L^a L^b / T^{2b}]$$

by equating the indices of L and T we infer:

L:                3 $= a+b$
T:              $-1 = -2b$

Hence $$b = \tfrac{1}{2}$$
and $$a = 3 - \tfrac{1}{2} = \tfrac{5}{2}$$

Thus if $Q$ depends on $g$ and $H$ only:

$$Q = kg^{1/2}H^{5/2}$$

A series of calibrating experiments to determine $k$ will establish whether our initial assumption is correct. It will generally be found that for a wide range of heads $k$ is constant. If it is not, then some other factor we have ignored is likely to be influencing the problem. The effect of viscosity and surface tension will be considered in Chapter 10. Both of these effects, and the condition of the inner face of the weir plate may modify the flow in practice, by retarding the upward flow. This tends to increase the sectional area of the sheet of falling liquid or 'nappe', and consequently the rate of flow. At very low heads the nappe may even cling to the downstream face. If the sides and the bottom of the approach channel are not sufficiently far away they too will tend to increase the rate of flow, as will any rounding of the sharp edges of the notch. The theoretical expression tends to correspond to the case for which the rate of flow is a minimum for a given head, and if this ideal is approached, consistent and accurate measurements may be obtained in practice.

The fundamental importance of the vee notch lies in the fact that no matter what the head, it always presents a geometrically similar shape to the oncoming flow. Since this is a prerequisite for kinematically similar flow patterns, it is the only shape likely to have a simple calibration over a wide range of heads. For a rectangular notch for example, of width $L$, the shape of the falling sheet of fluid will depend on $H/L$.

Great care is needed in observing the head over notches, particularly to find the correct zero reading of the head gauge. Surface tension may prevent flow even when the water is more than 2 mm above the base of the notch. Consequently it is advisable to insert a template of known height into the vee, and to observe when its upper point breaks the still-water surface. The head reading should then correspond to the known value.

## 5.6. Dimensional analysis of a rectangular notch

Fig. 5.6(a) shows a rectangular notch and dimensional arguments can take us no further than the result that *for a given H/L, Q is likely to depend only on H and g* so that, if we write for any one liquid:

$$Q = kH^a g^b$$

we again obtain the result:

$$Q = kg^{1/2}H^{5/2}$$

but this time it applies, not to a weir over which the head may vary, but only to a number of different weirs working at the same $H/L$. It implies that if one notch or weir is $N$ times the size of the other then the corresponding discharges will be in the ratio $N^{5/2}$ (see Section 5.2). The stumbling block to further progress in this direction is the fact that dimensional analysis cannot distinguish between two qualities of the same kind, such as $H$ and $L$. With two lengths involved in the problem it is only for similar systems, where one is proportional to the other, that simple results are obtained.

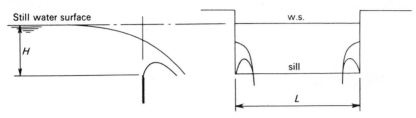

**Fig. 5.6(a).** *Flow over a rectangular notch*

For weirs of considerable width, however, the end effects may be neglected and the discharge per unit width $Q'$ ($= Q/L$) may simply be related to the height $H$ of the still-water surface over the sill. Here only one length is involved, and we may justifiably suppose that:

$$Q' = kH^a g^b$$

The dimensions of the left-hand side are:

$$Q' = Q/L = \left[\frac{L^2}{T}\right]$$

and those of the right-hand side are:

$$kH^a g^b = \left[\frac{L^a L^b}{T^{2b}}\right]$$

as before. Equating indices:

L:                    $2 = a + b$
T:                    $-1 = -2b$
Hence                 $b = \frac{1}{2}$
                      $a = 2 - \frac{1}{2} = \frac{3}{2}$

Thus if the flow per unit width depends on $g$ and $H$ only:

$$Q' = kg^{1/2}H^{3/2}$$

i.e.                  $$Q = kLg^{1/2}H^{3/2}$$

where $k$ must be determined by calibrating experiments, which will also test the validity of the above expression.

EXAMPLE 5.6 (i)

*A reservoir has an area of 50 000 m², and a weir 4 m long has its sill 0·60 m below the surface. Find the time required to reduce the level of the water in the reservoir by 0·5 m.*

If the discharge rate is $Q$, a volume $Q\delta t$ leaves the reservoir in a time $\delta t$. This causes a *decrease* in head (i.e. $-\delta H$) which, for a plan area $A$ is given by

$$-A\delta H = Q\delta t$$

in which

$$Q = kg^{1/2}LH^{3/2}$$

Hence, assuming that $k = 0{\cdot}57$

$$\delta t = -\frac{A}{0{\cdot}57\, g^{1/2}L}\frac{\delta H}{H^{3/2}}$$

The time taken for the head to change from $H_1\,(= 0{\cdot}60$ m$)$ to $H_2\,(= 0{\cdot}10$ m$)$ is thus given by:

$$\int_0^t dt = -\frac{A}{0{\cdot}57\, g^{1/2}L}\int_{H_1}^{H_2} H^{-3/2}dH$$

i.e.

$$t = \frac{A}{0{\cdot}57\, g^{1/2}L}\times 2[H_2^{-1/2}-H_1^{-1/2}]$$

$\therefore$

$$t = \frac{50\,000\ \text{m}^2\times 2}{0{\cdot}57\times 4\ \text{m}\ \sqrt{(9{\cdot}81\ \text{m/s}^2)}}\left[\frac{1}{\sqrt{0{\cdot}1}}-\frac{1}{\sqrt{0{\cdot}6}}\right]\frac{1}{\sqrt{\text{m}}}$$

$$= 0{\cdot}263\times 10^5\ \text{s or 7 h 18 min.}$$

If the plan area of the free surface of the reservoir varies with the depth of water in it, and hence with $H$, the above integral expression must be written:

$$\int_0^t dt = -\frac{1}{0{\cdot}57\, g^{1/2}L}\int_{H_1}^{H_2} AH^{-3/2}dH$$

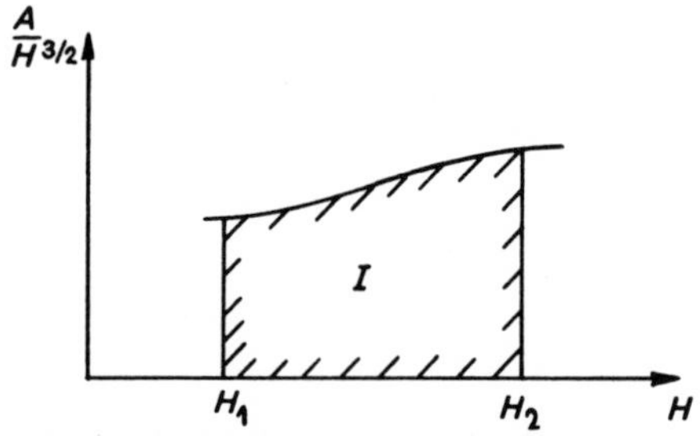

Fig. 5.6(b). *Evaluation of* $I = \int_{H_1}^{H_2} AH^{-3/2}\,dH$

The integral on the right-hand side, $I$ say, is shown graphically on Fig. 5.6(b), and may in general be evaluated in this way.

## 5.7. End effects and Francis' formula

It is observed in practice that if a weir or notch is of sufficient width, the overshooting action at the sides appears to influence only a limited region of fluid, namely, near the sharp edges (see Fig. 5.7(a)).

It remains to determine what width of fluid, $b$ say, is influenced by the end contraction. The value of the width, $b$, affected will certainly vary with $H$, and if this is the only factor on which it depends, dimensional analysis tells us that the expression connecting $b$ and $H$ is of the type:

$$b = kH$$

in which $k$ is a number (i.e. a non-dimensional coefficient).

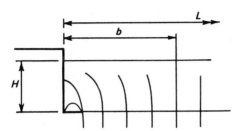

**Fig. 5.7(a).** *End effect at a notch or weir*

Thus, on the above assumption, the end effect is restricted to a rectangle of width $b$, which is proportional to the depth $H$. This is analogous to the vee notch (see Section 5.5), as in each case the discharge depends only on one length dimension, viz. $H$. Thus the flow through the end portions ($Q_1$ say) is such that:

$$Q_1 \propto H^{5/2} \quad \text{or} \quad Q_1 = K_1 H^{5/2}$$

If the stream contracts at both ends of the notch or weir (which is of breadth $L$) the central region which is free from end effects, and over which the stream-tubes or sections may be assumed parallel, has a breadth $(L-2b)$.

For this central region the law deduced in Section 5.6 applies, i.e.

$$Q_2 \propto (L-2b)H^{3/2} \quad \text{or} \quad Q_2 = K_2(L-2b)H^{3/2}$$

The total flow is thus:

$$Q = Q_1 + Q_2$$

or

$$Q = K_1 H^{5/2} + K_2(L-2b)H^{3/2}$$

Thus, since $b = kH$, the expression for the total discharge rate consists of two terms, one proportional to $LH^{3/2}$ and the other to $H^{5/2}$. The latter will, in addition, be proportional to the number of end contractions, $n$, say, i.e.

$$Q = K' LH^{3/2} + K'' nH^{5/2}$$
$$= K'\{L - K''' nH\}H^{3/2}$$

Francis found from experiments that over a wide range of conditions:

$$Q = 1{\cdot}85 \,\frac{\text{m}^3/\text{s}}{\text{m}^{5/2}} \left\{ L - \frac{nH}{10} \right\} H^{3/2}$$

Thus, the working rule

$$Q = 1\cdot85\left\{L - \frac{n}{10}H\right\}H^{3/2}$$

gives the flow directly in m³/s, if $L$ and $H$ are in metres. That this formula is rational in form may be seen from the above argument concerning end effects. If either, or both, of the end plates are removed, the corresponding suppression of the end effect makes $n = 1$ and $n = 0$, respectively.

A trapezoidal notch whose sides are sloped outwards at 1 in 4 to the vertical, is referred to as a Cippoletti weir. The discharge over it is, according to the above argument, independent of end effect, so that

$$Q \propto LH^{3/2}$$

According to Cippoletti, experiments indicate that:

$$Q = 1\cdot875 \frac{\text{m}^3/\text{s}}{\text{m}^{5/2}} LH^{3/2}$$

The working rule **$Q = 1\cdot875LH^{3/2}$** thus gives the discharge in m³/s directly when the values of $L$ and $H$ are in metres.

## 5.8. Effect of velocity of approach

If the fluid in the approach channel to a notch or a weir has a mean velocity $u$, it possesses, in addition to its pressure and height heads, a kinetic head $u^2/2g$. Had the fluid issued from a large reservoir, the still-water surface in the latter would be $u^2/2g$ above that in the approach channel. In energy terms, the head over the weir should be measured to the still-water surface, the level in the approach channel merely being a first approximation to this. In practice, the second approximation is usually sufficiently accurate, as indicated by the following example.

EXAMPLE 5.8 (i)

*Water flows over a 90° vee notch under a measured head of 150 mm in the approach channel, where the cross-sectional area of the stream is 0·25 m². Find the discharge, assuming that the coefficient k has a value of 0·45 in the formula:*

$$Q = kg^{1/2}H^{5/2}$$

**First approximation**

$$Q = 0\cdot45 \,\sqrt{(9\cdot81 \text{ m/s}^2)} \times (0\cdot15 \text{ m})^{5/2}$$

$$= 0\cdot01228 \text{ m}^3/\text{s} \left[\frac{60\text{s}}{\text{min}}\right] = 0\cdot7368 \text{ m}^3/\text{min}.$$

The mean approach velocity is thus:

$$u = \frac{0 \cdot 01228 \ \text{m}^3/\text{s}}{0 \cdot 25 \ \text{m}^2} = 0 \cdot 04912 \ \text{m/s}$$

and the corresponding kinetic head is:

$$\frac{u^2}{2g} = \frac{(0 \cdot 04912 \ \text{m/s})^2}{2 \times 9 \cdot 807 \ \text{m/s}^2}$$

$$= 0 \cdot 000123 \ \text{m}$$

**Second approximation**

Adding the kinetic head to the measured head in the approach channel a closer approximation to the true $H$ is $0 \cdot 1501$ m,

$$\therefore \qquad Q = 0 \cdot 45 \ \sqrt{(9 \cdot 81 \ \text{m/s}^2)} \times (0 \cdot 1501 \ \text{m})^{5/2}$$

$$= 0 \cdot 01231 \ \text{m}^3/\text{s} \left[\frac{60 \text{s}}{\text{min}}\right] = 0 \cdot 7386 \ \text{m}^3/\text{min}$$

The above argument ignores the fact that the upstream channel causes the oncoming fluid to approach the weir less obliquely than it would if it issued from a large reservoir. Fig. 5.8(a) shows a plan view of the

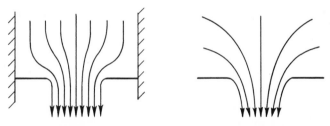

**Fig. 5.8(a).** *Streamlines in approach area to a weir*

streamlines in the approach area for the two cases, assuming that both weirs have end contractions. A similar situation arises in the approach to orifices, the straightening of the streamlines increasing the component of the momentum normal to the opening. This tends to reduce the overshooting action, and thus increases the discharge (see Sections 4.6 and 11.2).

More elaborate methods have consequently been proposed to make a more satisfactory correction for approach effects. As its effect is, however, generally small, the simple method outlined above is sufficiently accurate for most purposes.

## Exercises on Chapter 5

1. Find the time required for the head of liquid over a rectangular notch to fall to one-quarter of its initial value if the surface area of the tank is $A$.

2. The area of the water surface of a reservoir is 20 000 m². Estimate the time required for the surface to fall 0·36 m when water discharges over a rectangular weir 1·5 m long with no end contractions, the initial head of water over the sill being 1 m.

3. 148·8 litre/min of water flows over a right-angled vee notch when the head is 100 mm and the same rate of flow over a rectangular notch occurs when the head is 53 mm. If the heads in each case are liable to errors of ±0·5 mm owing to the human factor find the percentage errors in rates of flow in each case.

4. Use Francis' formula to estimate the rate at which water leaves a turbine house in which the width of a rectangular notch, with two end contractions, is 370 mm and the head of water above the sill is 250 mm. If the total energy of the water supplied to the turbine is 15 m head, find the efficiency of the turbine when developing 7·5 kW.

5. Develop an expression for the quantity of liquid flowing over a sharp-edged vee notch of total angle $2\theta$, in terms of the head $H$ above the bottom of the notch, the angle $\theta$, and the coefficient of discharge $C_d$, assuming the velocity of the approach to be small. State carefully what other assumptions are made.

If the rate of flow of water over a vee notch, having $\theta = 35°$, is 2·55 m³/min, calculate the head. (Take $C_d$ as 0·62.)

6. Derive a formula for discharge across a sharp-edged rectangular notch without allowance for end contractions. Also, give the formula suggested by Francis to allow for the effect of end contractions and explain why this formula must not be used when the head exceeds about one-quarter of the width of the notch. The rate of flow across a rectangular notch is 2 m³/min and the undisturbed level upstream of the notch is not to rise more than 150 mm when the discharge is increased to 8 m³/min. Using the formula without end contraction allowance estimate the minimum allowable width of the notch.

## Chapter 6

# Equations of motion for a fluid element

'*The book on the science of mechanics must precede the book of useful inventions.*'

LEONARDO

## 6.1. Newton's second law of motion

There can be little doubt that the most far-reaching physical principle in the mechanical sciences is that which is embodied in Newton's second law of motion:

**The rate of change of momentum (mass × velocity) is proportional to the impressed force, and takes place in the direction of that force.**

The time rate of change of *mu* is, for a particle:

$$\frac{d}{dt}(mu) = m\frac{du}{dt} = ma$$

i.e. it equals mass times acceleration, and as no other factors are known to influence the problem we write:

$$F = \frac{d}{dt}(mu)$$

i.e.
$$F = ma \tag{1}$$

Thus a unit of force is obtained through multiplying unit mass by unit acceleration. The product of 1 kg and 1 m/s², i.e. 1 kg m/s², is the SI standard example, and is called the newton (N). Straightforward application of the above equation states that the force required to accelerate 1 000 kg at 10 m/s² is 10 000 kg m/s², i.e. 10 000 N. Conversely, a force of 10 000 N, when acting on a mass of 1 000 kg produces an acceleration:

$$a = \frac{10\,000\ N}{1\,000\ kg}\left[\frac{kg\ m}{N\ s^2}\right] = 10\ m/s^2$$

The kg, the m/s² and the newton (N) are referred to as a 'consistent' set of units. This implies that if we use two of them when inserting values into Equation (1) (p. 113), the answer will always be in units of the third. This is one of the advantages of the Système International d'Unités or the SI.

## 6.2. Pressure distribution in a moving fluid

Bernoulli's equation indicates that, under certain conditions (see Section 4.2), if the hydrostatic head $(p/w + z)$ decreases in the direction of motion, the fluid must speed up, and vice versa:

$$\left(\frac{p_1}{w} + z_1\right) - \left(\frac{p_2}{w} + z_2\right) = \frac{u_2^2 - u_1^2}{2g}$$

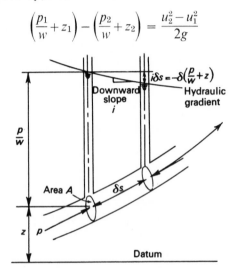

Fig. 6.2(a). *Hydraulic gradient causing acceleration*

If sections 1 and 2 are close enough (Fig. 6.2(a)) for the acceleration $a$ to be considered constant over the short distance $\delta s$ between them, then:

$$u_2^2 = u_1^2 + 2a\delta s$$

$$\therefore \quad \left(\frac{p_1}{w} + z_1\right) - \left(\frac{p_2}{w} + z_2\right) = \frac{2a.\delta s}{2g}$$

Recalling that $(p/w + z)$ is the height of the levels in imaginary gauge-glasses as shown in Fig. 6.2(a), the left-hand side is the drop in the hydraulic gradient over the length $\delta s$. If its *downward* slope *with respect to s* be denoted by $i$, then:

$$i.\delta s = \frac{2a.\delta s}{2g}$$

i.e. $$ig = a$$

It should be noted that the figures 6.2(a) and 6.2(b) are diagrammatic only.

Symbol *i denotes the downward (i.e. negative) gradient of* $(p/w+z)$ *with respect to distance s along the pipe* or stream-tube and *not*, as shown, with respect to horizontal distance *x*. *Only if the pipe is horizontal does the figure give to scale the value of i.*

According to Newton's second law of motion acceleration *a* equals force per unit mass:

$$\frac{F}{m} = a$$

Hence *ig* must be the force acting on a particle as it moves along the stream-tube.

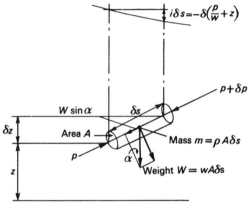

**Fig. 6.2(b).** *Forces on fluid element*

This may be confirmed by reference to Fig. 6.2(b) neglecting friction on the surface of the cylinder (i.e. for an inviscid fluid) the resultant force in the direction of motion is:

$$F = -\delta p A - W \sin \alpha$$

$$= -\left(\delta p A + mg \frac{\delta z}{\delta s}\right)$$

The corresponding force per unit mass ($m = \rho A \delta s$) is thus:

$$\frac{F}{m} = -\left(\frac{1}{\rho}\frac{\delta p}{\delta s} + g\frac{\delta z}{\delta s}\right) = -g\frac{\delta}{\delta s}\left(\frac{p}{w}+z\right) = ig$$

This important result may also be established by noting that so long as a fluid remains at rest, the level in each gauge-glass is the same and, referring again to Fig. 6.2(a), that a resultant force is produced by a hydraulic gradient. The latter represents a *departure from the static pressure distribution, and it is this which produces the unbalanced force, causing acceleration.* A difference in levels $i\delta s$ causes a force $wi\delta s \times A$ on the mass $m = \rho A \delta s$. Equating the unbalanced force per unit mass to the acceleration:

$$\frac{F}{m} = \frac{wi\delta s A}{\rho A \delta s} = a$$

i.e.                          $$ig = a$$

This derivation of the above equation emphasizes that *the absolute pressure or hydrostatic load on the system is of no account in determining the accelerations and, hence, the flow pattern.* For an incompressible fluid the absolute height of the fluid in the imaginary gauge-glasses is immaterial; it is the differences in level (i.e. changes in hydrostatic head) which govern the motion.

Although Fig. 6.2(b) shows a fluid element with its axis lying along the direction of motion the above argument is equally valid whatever the orientation of the element. For an incompressible fluid, a falling hydraulic gradient $i$ in *any* direction produces a corresponding force per unit mass in that direction equal to $ig$. If this is the only force which acts (i.e. if friction is negligible) it may be equated to the corresponding acceleration in that direction.

In the notation of partial differentials, i.e.

$$\frac{F_x}{m} = -\left(\frac{1}{\rho}\frac{\partial p}{\partial x} + g\frac{\partial z}{\partial x}\right) = -g\frac{\partial}{\partial x}\left(z + \frac{p}{w}\right) = a_x$$

where $a_x$ is the *total* acceleration in any direction $x$. The negative sign follows since $i$ represents the *rate of decrease* of $(p/w + z)$ in the direction considered.

### EXAMPLE 6.2 (i)

*A fire nozzle which is 0·3 m long is supplied at a steady pressure of 300 kN/m².
It is so profiled that the pressure drops uniformly along its axis. Neglecting friction, estimate the acceleration of each fluid element as it is discharged into the atmosphere.*

Neglecting changes in $z$ it follows from Fig. 6.2(c) that

$$i = \frac{p}{wl} = \frac{p}{\rho g l}$$

Hence the corresponding acceleration follows from the equation of motion:

$$a = ig = \frac{p}{\rho l} = 3 \times 10^5 \,\frac{N}{m^2} \times \frac{m^3}{10^3 \text{ kg}} \times \frac{1}{0\cdot3 \text{ m}}\left[\frac{\text{kg m}}{\text{N s}^2}\right] = 1 \text{ km/s}^2$$

### EXAMPLE 6.2 (ii)

*A circular tank of 0·3 m diameter is set spinning about its axis, which is vertical, at 60 rev/min. Find the slope of the free surface of the fluid contained within it at the outermost radius.*

Since the fluid has no acceleration in the vertical direction the free surface represents the transverse hydraulic gradient as shown in Fig. 6.2(d). The

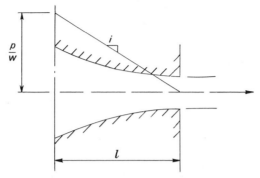

**Fig. 6.2(c).** *Hydraulic gradient through nozzle*

outermost layer of fluid will be carried round with the drum and thus has a centripetal acceleration which acts *horizontally*:

$$a = \omega^2 R$$

$$= 60^2 \frac{\text{rev}^2}{\text{min}^2} \times 0 \cdot 15 \text{ m} \left[ \frac{\text{min}^2}{60^2 \text{s}^2} \right] \left[ \frac{(2\pi)^2}{\text{rev}^2} \right]$$

$$= 0 \cdot 15 \times 4\pi^2 \text{ m/s}^2 = 5 \cdot 92 \text{ m/s}^2$$

$$\therefore \qquad i = \frac{a}{g} = \frac{5 \cdot 92}{9 \cdot 81} = 0 \cdot 603$$

Noting that $i = \tan \theta$ for each *horizontal* element it follows that $\theta = \mathbf{31 \cdot 1°}$.

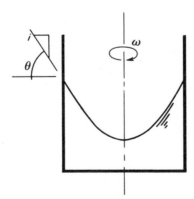

**Fig. 6.2(d).** *Transverse hydraulic gradient across spinning fluid*

EXAMPLE 6.2 (iii)

*A pipe-line which is 60 m long supplies water from a tank in which the level is 15 m above that of the stop-valve which terminates the other end of the pipe. Find the initial acceleration of the fluid in the pipe when the valve is suddenly opened.*

If the fluid is incompressible the 'plug' of fluid in the pipe of Fig. 6.2(e) must move forward with a constant acceleration throughout its length. Hence, from the equation of motion:

$$ig = a$$

the hydraulic gradient must be of uniform slope, and its initial value is:

$$i = \frac{H}{l} = \frac{15}{60} = 0\cdot25$$

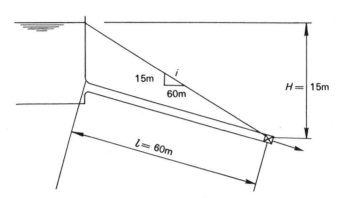

**Fig. 6.2(e).** *Acceleration of fluid in pipe-line*

The corresponding initial acceleration is thus:

$$a = ig = 0\cdot25 \times 9\cdot81 \text{ m/s}^2 = 2\cdot45 \text{ m/s}^2$$

From the above examples it may be seen that there are a number of different ways in which a fluid element can be induced to accelerate. In Example 6.2(i) each element is moving along a contracting stream and must therefore speed up, i.e. accelerate, as it approaches the throat. This requires that the hydraulic gradient shall fall in accordance with the equation of motion:

$$ig = a$$

That the sum of $z$ and $p/w$ must decrease towards a throat has been previously noted from Bernoulli's equation (see Fig. 4.2(b), p. 73).

In Example 6.2(ii) the fluid moves on a curved path, so that there must be a transverse hydraulic gradient to produce the corresponding centripetal acceleration.

Example 6.2(iii) is a simple case of an unsteady flow, i.e. one in which conditions vary at each point from one instant to the next—i.e. conditions vary with time.

In the following sections we shall consider each of the above types of acceleration in turn and so deduce corresponding equations of motion.

## 6.3. Steady flow along a stream-line; compressibility

*In a steady flow, conditions at each point do not change with time.* As each element passes from one point to the next however it must, in general, accelerate so that it may arrive with the velocity appropriate to that point. Referring to Fig. 6.3(a), each element must change its speed by an amount $\delta u$ whilst it is travelling the distance $\delta s$.

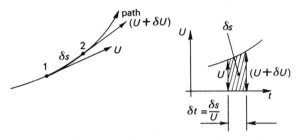

**Fig. 6.3(a).** *Convective acceleration of a particle*

The distance $\delta s$ is the shaded area under the speed–time curve, and if this distance is sufficiently small we may write:

$$\delta s = u \delta t$$

i.e. $$\delta t = \frac{\delta s}{u}$$

Hence, the corresponding acceleration of the fluid element along the path in the $s$ direction is, for a steady flow:

$$a_s = \frac{\delta u}{\delta t} = u \frac{\delta u}{\delta s} \tag{i}$$

Equating this to the corresponding acceleration deduced from force per unit mass, namely, *ig*, and noting (as in Fig. 6.2(a)), that *iδs* is the decrease in $(z+p/w)$ in the direction of motion, then

$$ig = u \frac{\delta u}{\delta s} \quad \text{or} \quad i\delta s = \frac{u\delta u}{g},$$

and $$i\delta s = -\delta \left( z + \frac{p}{w} \right) = \frac{u \delta u}{g}$$

Thus, when a number of such changes are summed, i.e. integrated along a stream-line, we rediscover Bernoulli's equation:

$$\delta z + \frac{\delta p}{w} + \frac{u \delta u}{g} = 0$$

or $$z + \frac{p}{w} + \frac{u^2}{2g} = \text{constant} \tag{ii}$$

The equation has, in this case, been derived for an incompressible fluid by assuming that the acceleration of an element in the direction of motion is entirely due to the departures from the static pressure distribution, i.e. the hydraulic gradient which acts upon it.

If the fluid were at rest the static pressure distribution would be given by:

$$z + \frac{p_s}{w} = \text{constant}$$

Thus the *departure* from the static pressure distribution, $p_d$ say, which is such that:

$$p_d = p - p_s$$

is obtained by subtraction of

$$z + \frac{p}{w} + \frac{u^2}{2g} = \text{constant}$$

and

$$z + \frac{p_s}{w} = \text{constant}$$

and results in

$$\frac{p - p_s}{w} + \frac{u^2}{2g} = \text{constant}$$

i.e.

$$\frac{p_d}{\rho} + \frac{u^2}{2} = \text{constant} \qquad \text{(iii)}$$

Such an equation is the same as that which would be obtained by neglecting gravity (i.e. height head $z$) except that $p_d$ replaces $p$. This important result, implies that, for incompressible fluids the *deviations* of pressure caused by fluid motion are independent of gravity. Gravity thus

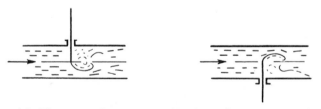

(A) *Flow past valve*          (B) *Same flow past inverted valve*
Fig. 6.3(b). *Flow past valves*

has no intrinsic effect on the flow pattern, except in so far as it may, by causing free surfaces, change the boundary conditions, i.e. the geometry of the problem. Consequently the flow past the gate valve shown diagrammatically in Fig. 6.3(b) is just the same when it is inverted, or set at any other angle, provided only that the driving pressures, i.e. the departures from the hydrostatic pressure distribution correspond. These conclusions obviously cannot apply if the pipe is not flowing full since inversion of the valve completely changes the geometry of the free surfaces, as shown in (A) and (B) of Fig. 6.3.(c). **Hence, for problems not**

**involving a free surface, the orientation of the system relative to the earth is of no account in determining the flow pattern.** Strictly this is not true for gases, as the static pressure distribution modifies the density but, in such cases, the weight of the fluid elements is so small that the pressure differences caused by gravity may be neglected. Thus, the effect of gravity

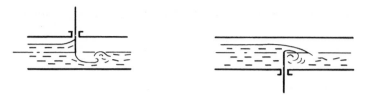

*(A) Flow with free surface     (B) Corresponding flow, valve inverted*
**Fig. 6.3(c).** *Free surface flow past valves*

may usually be allowed for subsequently in liquids, and its effect may usually be ignored completely in gases. We shall, therefore, establish an alternative expression to *ig* for force per unit mass acting on an element, this time *ignoring gravity* (and friction as previously), but allowing for compressibility. These conditions are generally appropriate to gas flows.

Referring to Fig. 6.2(b), if the pressure increases from $p$ to $(p+\delta p)$ over the length $\delta s$, the corresponding force in the direction of motion is $-\delta p A$. This force, acting on the mass $\rho A \delta s$, produces the acceleration $(a = F/m)$ in the direction of motion, i.e.:

$$a_s = -\frac{\delta p A}{\rho A \delta s} = -\frac{1}{\rho}\frac{\delta p}{\delta s}$$

In general, the force per unit mass in *any* direction $x$, say, is $-(1/\rho)(\partial p/\partial x)$, and if this is the only force acting, it may be equated to the total acceleration in that direction.

In the present case, according to Equation (i) (p. 119), the acceleration for steady flow is $u(\delta u)/(\delta s)$ in the direction of motion, i.e.

$$-\frac{1}{\rho}\frac{\delta p}{\delta s} = u\frac{\delta u}{\delta s}$$

Hence,
$$\frac{\delta p}{\rho} + u\delta u = 0 \qquad\qquad \text{(iv)}$$

If we sum (i.e. integrate) the changes along a stream-line between any two sections 1 and 2:

$$\int_{p_1}^{p_2}\frac{dp}{\rho} + \int_{u_1}^{u_2} u\,du = 0$$

If changes in $\rho$ may be neglected, we again rediscover Bernoulli's equation—this time ignoring gravity (i.e. assuming changes in datum head $z$ are negligible)

$$\frac{1}{\rho}(p_2 - p_1) + \tfrac{1}{2}(u_2^2 - u_1^2) = 0$$

which, when multiplied through by $\rho$ gives energies per unit volume. Thus,

$$\frac{p_1}{\rho} + \frac{u_1^2}{2} = \frac{p_2}{\rho} + \frac{u_2^2}{2}$$

and $$p_1 + \tfrac{1}{2}\rho u_1^2 = p_2 + \tfrac{1}{2}\rho u_2^2$$

Both expressions are statements of the equation of motion along a stream-tube of an incompressible fluid when friction and gravitational forces are neglected.

If compressibility effects may *not* be neglected the expression for the increase in kinetic energy per unit mass depends on the nature of the variation of $\rho$ with $p$. In such cases, Equation (iv) for frictionless flow (ignoring any changes of height) integrates to:

$$\frac{u_2^2}{2} - \frac{u_1^2}{2} = -\int_{p_1}^{p_2} \frac{dp}{\rho} = \int_{p_2}^{p_1} \frac{dp}{\rho} \tag{v}$$

The quantity $1/\rho$ is often referred to as the specific volume, $v$. Evaluation of the integral in Equation (v) for particular pressure–density relationships will occur in Chapter 13.

## 6.4. Curved stream-lines; transverse hydraulic gradient

If a fluid element is moving along a curved path its velocity must change in direction and may also change in magnitude. Dealing with the former, it gives rise to a centripetal acceleration, the magnitude of which may be determined from Fig. 6.4(a). If $\delta s$, and hence $\delta\theta$, are very small the following expression may be deduced:

$$\delta\theta = \frac{\delta s}{R} = \frac{\delta u_n}{u}$$

Hence, an expression for the change in velocity normal to the path results, namely,

$$\delta u_n = \frac{u}{R}\delta s$$

The corresponding normal acceleration is:

$$a_n = \frac{\delta u_n}{\delta t} = \frac{u\,\delta s}{R\,\delta t} = \frac{u^2}{R}$$

The approximations made in developing this result become increasingly valid as $\delta s$ is decreased indefinitely, so that the final expression is exact at the point 1.

This acceleration, which acts inwards, may be equated to the corresponding specific force or force per unit mass. For an incompressible

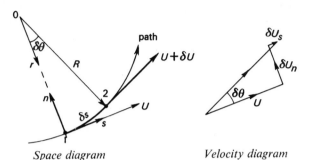

Space diagram       *Velocity diagram*
**Fig. 6.4(a).** *Centripetal acceleration*

fluid the latter is equal to the downward slope of the hydraulic gradient towards the centre of rotation, times $g$ (see Section 6.2).

i.e. $$i_n g = a_n$$

Hence, **the slope of the hydraulic gradient across a curved streamline:**

$$i_n = \frac{u^2}{Rg}$$

If we ignore the force (weight) due to the effect of gravity, the force per unit mass in the $n$ direction is:

$$\frac{F_n}{m} = -\frac{1}{\rho}\frac{\partial p}{\partial n}$$

(as shown in Section 6.2)

The equation of motion is then:

$$-\frac{1}{\rho}\frac{\partial p}{\partial n} = \frac{u^2}{R}$$

an expression which is true for both compressible and incompressible fluids.

If, for example, a region of liquid is *circulating steadily about a fixed common axis* (as in Fig. 6.2(d)), the value of $i_n$ at any radius $r$ is:

$$i_n = \frac{a_n}{g} = -\frac{d}{dn}\left(z+\frac{p}{w}\right) = \frac{d}{dr}\left(z+\frac{p}{w}\right)$$

since (Fig. 6.4(a)) the positive direction of the normal is opposite to that of the radius from the centre. Thus, the slope of the transverse hydraulic gradient is:

$$i_n = \frac{dz}{dr} + \frac{1}{w}\frac{dp}{dr} = \frac{u^2}{rg}$$

Hence, from Bernoulli's equation [$H = z + (p/w) + (u^2/2g)$], the corresponding variations in the total head from one stream-line to the next may be deduced. Its gradient, in the radial direction measuring from the centre of rotation, is:

$$\frac{dH}{dr} = \frac{d}{dr}\left(z + \frac{p}{w} + \frac{u^2}{2g}\right)$$

Therefore, substituting $u^2/rg$ for the transverse hydraulic gradient $d[z + (p/w)]/dr$, it follows that:

$$\frac{dH}{dr} = \frac{u^2}{rg} + \frac{1}{2g}\frac{du^2}{dr}$$

i.e.

$$\frac{dH}{dr} = \frac{u}{g}\left(\frac{u}{r} + \frac{du}{dr}\right)$$

We may deduce an expression of this type which is *valid across any curved stream-lines*. In general, however, $u$ may vary both along and normal to the path, so that the basic equation of motion should be written:

$$i_n = -\frac{\partial}{\partial n}\left(z + \frac{p}{w}\right) = \frac{u^2}{Rg}$$

The negative sign follows since $i_n$ is the *downward* gradient of $(z + p/w)$ in the $n$ direction, see Fig. 6.4(a).

The corresponding expression for the variation of total head is:

$$\frac{\partial H}{\partial n} = \frac{\partial}{\partial n}\left(z + \frac{p}{w} + \frac{u^2}{2g}\right)$$

so that this becomes:

$$\frac{\partial H}{\partial n} = -\frac{u}{g}\left(\frac{u}{R} - \frac{\partial u}{\partial n}\right)$$

## 6.5. Free, or potential, vortex

Circulatory motions are frequently observed in practice to be created in regions which are sensibly of constant energy throughout. This suggests that we should examine both along and normal to the stream-lines the type of circulating motion corresponding to $H = $ constant in all places at all times. It is termed a *free, or potential, vortex* and must be such that $dH/dr = 0$. Hence, using the final equation of Section 6.4 we may state that:

$$\frac{u}{r}+\frac{du}{dr}=0$$

i.e.

$$\frac{du}{u}+\frac{dr}{r}=0$$

Integration of this expression between any two radii results in:

$$\log\frac{u_2}{u_1}+\log\frac{r_2}{r_1}=0$$

i.e.       $u_1r_1 = u_2r_2$    or    **$ur = $ constant**

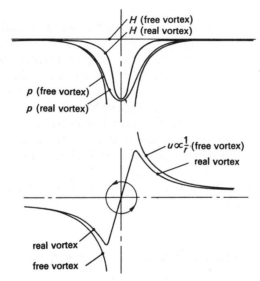

**Fig. 6.5(a).** *Comparison of real and potential (free) vortex*

A motion of this type, in which the velocity varies inversely as the radius, is often established in practice round anything which tends to set up a rotating core of fluid.

The corresponding pressure distribution across such a free vortex may be obtained from Bernoulli's equation since $dH/dr = 0$ for this motion, i.e. $H$ has the same value along each stream-line:

i.e.

$$z+\frac{p}{w}+\frac{u^2}{2g}=H$$

in which       $ur = c$

$c$ being a constant which depends on the strength of the vortex.

$\therefore$

$$p = w(H-z)-\frac{w}{2g}\frac{c^2}{r^2}$$

If the axis of spin is vertical, so that $z$ is constant along any radius,

$w(H-z)$ is simply the value of the pressure, $p_\infty$ say, at infinite radius:

$$p = p_\infty - \frac{\rho}{2}\frac{c^2}{r^2}$$

i.e.    $(p-p_\infty) = -\frac{\rho}{2}\frac{c^2}{r^2}$    or    $\frac{p-p_\infty}{w} = -\frac{1}{2g}\frac{c^2}{r^2}$

Under these conditions the pressure at each point is $w$ times the head of fluid above it, so that the surface profile is given by:

$$(z-z_\infty) = -\frac{1}{2g}\frac{c^2}{r^2}$$

This may alternatively be obtained from Bernoulli's equation, by setting the gauge pressure $p$ to zero along the free surface.

Fig. 6.5(a) compares a free vortex with that which may be shed from the tips of a wing, or form in the draft tube of a turbine.

The theory, which predicts an infinite velocity and suction at the axis, must break down as the core is approached. Within the core, the velocity is seen to vary with the radius, rather than its reciprocal, so that conditions here must be very different from those just considered (see Section 6.6). Outside this narrow central region (Fig. 6.5(a)) the agreement of theory and practice is often extremely good, i.e. conditions may closely approach those for a free, or potential, vortex in which $H$ is constant everywhere.

## 6.6. Forced vortex and Rankine combined vortex

The fluid entrained in the impeller of a centrifugal pump before the discharge valve is opened undergoes what is virtually forced vortex motion. Such a vortex is created whenever a region of fluid is whirled bodily about an axis with a constant angular speed. In such circumstances the linear velocity is proportional to the radius, the fluid being rotated as if it were a solid body, i.e.

$$u = r\omega$$

Since the transverse hydraulic gradient maintains each element on a circular path; and $i_n g = a_n$ (Section 6.4) then,

$$i_n = \frac{u^2}{rg} = \frac{\omega^2 r}{g}$$

If, for example, the fluid is being rotated about a vertical axis, and has a free surface, the radial slope of the latter ($i_n$) increases linearly with $r$. This implies that the cross-section is a parabola so that the free surface forms a paraboloid, see Fig. 6.6(a).

In symbols:

$$\frac{a_n}{g} = i_n = -\frac{d}{dn}\left(z+\frac{p}{w}\right) = +\frac{d}{dr}\left(z+\frac{p}{w}\right)$$

the positive direction of the radius being outwards from the centre of rotation and, therefore, opposite to that of the normal which is inwards towards the centre of rotation (see Fig. 6.4(a)).

Hence,
$$d\left(z+\frac{p}{w}\right) = \frac{\omega^2}{g}r\,dr$$

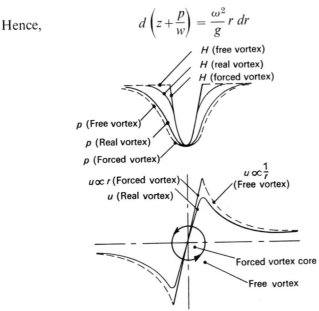

Fig. 6.6(a). *Combined vortex*

and integration of this expression between any two radii $r_1$ and $r_2$ results in:

$$z_1+\frac{p_1}{w}-\frac{\omega^2 r_1^2}{2g} = z_2+\frac{p_2}{w}-\frac{\omega^2 r_2^2}{2g}$$

i.e.
$$z_1+\frac{p_1}{w}-\frac{u_1^2}{2g} = z_2+\frac{p_2}{w}-\frac{u_2^2}{2g} \tag{i}$$

Comparison of this expression with Bernoulli's equation confirms that $H = z+(p/w)+(u^2/2g)$ must vary from one stream-line to the next. The above equation may be rewritten as:

$$H_1 - \frac{u_1^2}{g} = H_2 - \frac{u_2^2}{g}$$

Thus, if $H$ has the value $H_0$ at the axis (i.e. whe $r = 0$ and $u = 0$), its value at any other radius is given by:

$$H - H_0 = \frac{u^2}{g} = \frac{\omega^2 r^2}{g} \tag{ii}$$

This may be compared with the corresponding expression for the surface profile along which $p$ is zero.

Thus, if $p_1 = p_2 = 0$, Equation (i) becomes:

$$z_1 - \frac{u_1^2}{2g} = z_2 - \frac{u_2^2}{2g}$$

and $z$ at any radius $r$ is given by:

$$z - z_0 = \frac{u^2}{2g} = \frac{\omega^2 r^2}{2g}$$

where $z_0$ is the value at the axis where $r = 0$.

Similarly, since the pressure variation across any *horizontal* plane is simply $w$ times the head of fluid above it:

$$p - p_0 = \frac{\rho u^2}{2} = \frac{\rho \omega^2 r^2}{2}$$

Hence, for a forced vortex, both the pressure and the total head increase parabolically with the radius as indicated in Fig. 6.6.(a).

A forced vortex at the core surrounded by a free vortex, forms a very close approximation to most circulatory motions which occur in practice. This is termed a Rankine combined vortex and is suggested by the dotted lines in Fig. 6.6(a) which may be compared with the corresponding profiles for a real vortex, see Fig. 6.5(a).

### EXAMPLE 6.6 (i)

*A tube ABCD comprises a straight vertical part ABC and a curved part CD whose shape is a quarter-circle of radius 250 mm with its centre at B. The end A is open to atmosphere but the end D is closed. The tube, which is arranged so that it can be rotated about a vertical axis coincident with ABC, is completely filled with water up to a height in the vertical part 300 mm above C.*

*Find (a) the speed of rotation which will make the pressure head at D equal to the pressure head at C, and (b) when running at this speed, the value and position of the maximum pressure head in the curved part CD. The formula for pressure head in a forced vortex may be assumed.*

(a) Using the forced vortex (Equation (i)) for points C and D of Fig. 6.6(b)

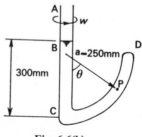

**Fig. 6.6(b)**

we have:

$$z_C + \frac{p_C}{w} - \frac{u_C^2}{2g} = z_D + \frac{p_D}{w} - \frac{u_D^2}{2g}$$

or

$$0 + 0 \cdot 3 \text{ m} - 0 = 0 \cdot 25 \text{ m} + \frac{p_D}{w} - \frac{r_D^2 \omega^2}{2g}$$

Hence, when $p_D/w = p_C/w = 0 \cdot 3$ m

$$\frac{r_D^2 \omega^2}{2g} = 0 \cdot 25 \text{ m} \quad \text{and} \quad \omega^2 = 19 \cdot 61 \, \frac{\text{m}}{\text{s}^2} \times \frac{1}{4} \text{ m} \left( \frac{4}{1 \text{ m}} \right)^2$$

$$= 78 \cdot 44 \text{ /s}^2$$

$$\therefore \quad \omega = \frac{8 \cdot 85}{\text{s}} \left[ \frac{\text{rev}}{2\pi} \right] \left[ \frac{60 \text{ s}}{\text{min}} \right] = 84 \cdot 5 \frac{\text{rev}}{\text{min}}$$

(*b*) The pressure at any point P varies radially due to the vortex action and vertically due to gravity. As P lies on a circular arc the most convenient co-ordinates are $a$ and $\theta$, see Fig. 6.6(c).

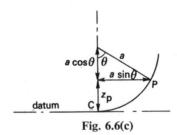

**Fig. 6.6(c)**

Choosing a datum through C:

$$z_C + \frac{p_C}{w} - \frac{u_C^2}{2g} = z_P + \frac{p_P}{w} - \frac{u_P^2}{2g}$$

and noting that:

$$u_C = 0 \qquad\qquad u_P = r_P \omega = \omega a \sin \theta$$

$$z_C = 0 \qquad\qquad z_P = a - a \cos \theta = a(1 - \cos \theta)$$

$$\frac{p_C}{w} = a(1 - \cos \theta) + \frac{p_P}{w} - \frac{\omega^2 a^2}{2g} \sin^2 \theta$$

P is required such that $p_P$ is a maximum, i.e. $(dp_P/d\theta) = 0$; hence, differentiating:

$$0 = a \sin \theta + \frac{1}{w} \frac{dp_P}{d\theta} - \frac{\omega^2 a^2}{2g} 2 \sin \theta \cos \theta$$

i.e. the maximum pressure head occurs where:

$$a \sin \theta = \frac{\omega^2 a^2}{2g} 2 \sin \theta \cos \theta \qquad\qquad \text{(i)}$$

i.e. either when:

$$\cos \theta = \frac{g}{w^2 a}$$

∴    $$\cos \theta = \frac{9 \cdot 807 \text{ m/s}^2}{19 \cdot 614 \text{ m/s}^2} = \tfrac{1}{2}, \text{ i.e. } \theta = 60°$$

or when:        $\sin \theta = 0$            i.e. $\theta = 0$

as either of these conditions satisfies the Equation (i).
    When $\theta = 60°$: $\cos \theta = g/w^2 a = \tfrac{1}{2}$

$$\frac{p_P - p_C}{w} = \frac{w^2 a^2}{2g} \sin^2 60° - a(1 - \cos 60°)$$

$$= a(\sin^2 60° - 1 + \cos 60°) = +\frac{a}{4}$$

which is greater than the pressure at C, so that $\sin \theta = 0$ corresponds to a minimum. Thus the maximum pressure head occurs where **θ = 60°**, and its value is $(p_C/w) + (a/4) = (0 \cdot 3 + 0 \cdot 25/4)$ m = **0·363** m, water.

## 6.7. Unsteady or accelerating flows

The third type of acceleration which we referred to at the end of Section 6.2 was that which occurs in *non-steady flows*. In such problems the velocity at each point is continuously changing with time, thus causing local accelerations.

    As a simple example we may consider the oscillation of fluid in a U-tube, see Fig. 6.7(a).

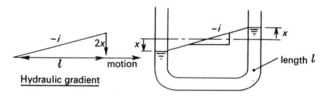

**Fig. 6.7(a).** *Oscillation in a U-tube*

    If the tube is of constant cross-section each element must have the same acceleration. From the equation of motion it follows that the instantaneous hydraulic gradient in the direction of the acceleration, i.e. along the axis of the tube, is also uniform. If at any instant the fluid column is displaced a distance $x$ away from its equilibrium position the *downward* slope of the hydraulic gradient is thus:

$$i = -\frac{2x}{l}$$

The negative sign arises since when $x$ is *positive* the slope of $i$ is *upwards* and vice versa. Hence, since $ig = a$, the equation of motion is:

$$-\frac{2g}{l}x = a$$

This is the equation of a frictionless or undamped simple harmonic oscillation for which the acceleration is proportional to the displacement and opposite in sign to it, i.e.:

$$a = -\omega_n^2 x$$

Hence, the general solution for the period of the motion is

$$T = \frac{2\pi}{\omega_n}$$

and which, in this problem becomes:

$$T = 2\pi \sqrt{\left(\frac{l}{2g}\right)}$$

Thus, the periodic time corresponds to that of a simple pendulum having a length which is one-half of that of the fluid column.

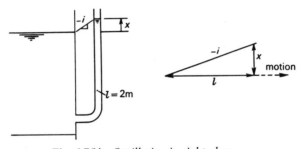

**Fig. 6.7(b).** *Oscillation in sight glass*

EXAMPLE 6.7 (i)

*In order that the liquid level may be recorded outside a large storage tank, a vertical sight glass is attached to the side of it. If the length of the fluid column is 2 m, estimate, neglecting losses, the natural period of the liquid level for small oscillations.*

Changes of level in the storage tank may be neglected if the area of the latter is large. Therefore, referring to Fig. 6.7(b), the hydraulic gradient down the length of the side tube has a slope:

$$i = -\frac{x}{l}$$

and the corresponding equation of motion is:

$$ig = a$$

i.e.

$$-\frac{g}{l}x = a$$

Hence, the undampened natural period of the system is:

$$T = 2\pi \sqrt{\left(\frac{l}{g}\right)} = 2\pi \sqrt{\left(\frac{2}{9\cdot807}\, \text{s}^2\right)} = 2\cdot84 \text{ s}$$

This example is similar in nature to the system which results when a surge tank is fitted at the delivery end of a pipe-line leading from a reservoir to a turbine, see Fig. 6.7(c). The surge tank acts as an accumulator to and from which water may be delivered during sudden changes in demand at the turbine. This reduces the accelerations and hence the inertia effects of the column of fluid in the supply pipe.

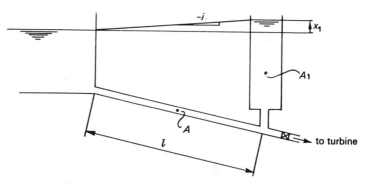

**Fig. 6.7(c).** *Surge tank at delivery end of pipe-line*

We may estimate the undamped natural frequency of such a system quite simply by neglecting losses and assuming that the cross-sectional area of the surge tank $A_1$ is very much larger than that of the pipe $A$. Under these conditions, only the fluid in the pipe undergoes appreciable acceleration. If the level in the surge tank lies a distance $x_1$ above its equilibrium position the corresponding adverse hydraulic gradient is:

$$i = -\frac{x_1}{l}$$

If the turbine valve is shut, the velocity and hence the displacement, of each element in the pipe, $x$ say, is $A_1/A$ times $x_1$. The equation of motion at each element in the pipe-line is thus:

$$ig = a$$

$$-\frac{g}{l}x_1 = a$$

i.e.

$$-\frac{g}{l}\frac{A}{A_1}x = a$$

Therefore, the undamped natural period of oscillation is:

$$T = 2\pi \sqrt{\left(\frac{A_1}{A}\frac{l}{g}\right)}$$

In practice, friction and other losses damp the motion, i.e. they slow the period and cause the oscillations to die out, as indicated by Fig. 6.7(d).

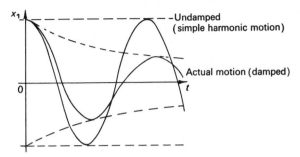

**Fig. 6.7(d).** *Effect of damping on natural oscillation*

Such oscillations occur whenever there is a sudden change in demand, the fluid oscillating not about the static position but about the final equilibrium running position.

If a surge tank were not fitted, sudden closure of the valve would require the moving column of fluid in the supply pipe to be arrested instantaneously. The only factors limiting this deceleration, and hence the pressure gradient, would be the elasticity of the fluid and of the pipe-line. As both of the latter are extremely stiff a severe pressure wave would be propagated up the pipe, giving rise to what is termed water hammer.

If a surge tank cannot conveniently be installed at a sufficient height, a closed stand-pipe may be used with compressed air above it. This is

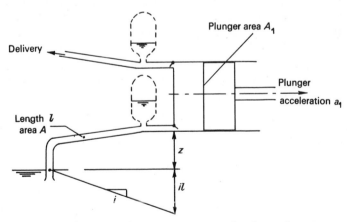

**Fig. 6.7(e).** *Hydraulic gradient in suction pipe of reciprocating pump*

similar in nature to the use of air vessels on a reciprocating pump as indicated by Fig. 6.7(e). If no air vessel were fitted on the suction side the fluid would be drawn through the pipe with an acceleration $a$ which is $A_1/A$ times that, $a_1$ say, of the plunger. The corresponding downward slope of the hydraulic gradient due to this acceleration, $a$, of fluid in the pipe would therefore be:

$$i = \frac{a}{g} = \frac{A_1}{A}\frac{a_1}{g} = \frac{\text{head required to accelerate water}}{\text{length of suction pipe}} = \frac{h_a}{l}$$

If for example we imagine the piston to be driven with simple harmonic motion by a crank having a radius $r$ and an angular speed $\omega$, then $a_1 = \omega^2 r \cos \omega t$, and the head required to accelerate the water would be

$$h_a = \frac{A_1}{A}\frac{l}{g}\,\omega^2 r \cos \omega t$$

The maximum value of $a_1$ would be:

$$a_1 = \omega^2 r$$

when $\omega t = 0$ and this would occur at the beginning of the suction stroke. Hence at this instant:

$$i = \frac{A_1}{A}\frac{\omega^2 r}{g} = \frac{h_a}{l}$$

Referring to Fig. 6.7(e), the hydraulic gradient would then lie below the pump inlet by an amount:

$$z + il = z + \frac{A_1}{A}\frac{\omega^2 r}{g}\,l$$

If suction is to be maintained this cannot possibly exceed 10·4 m of water, for the pressure at the pump inlet would then have fallen to absolute zero. In practice, however, cavitation would occur at a higher pressure than this owing to the release of air or other dissolved gases. Even if the latter were absent the fluid would eventually 'boil' when its vapour pressure was reached.

Thus the maximum height to which a column of liquid could rise (no matter how good the suction) would be the height equivalent to the barometric pressure less the head equivalent to the vapour pressure of the liquid. The latter depends on temperature—e.g. the maximum height of a column of water at 60°C with corresponding vapour pressure 0·01922 MPa and relative density $d = 0·98$ would be

$$\frac{p - p_v}{\rho g} = \frac{p - p_v}{d\rho_w g} = \frac{(0·10132 - 0·01992)\ \text{MN/m}^2}{0·98 \times 10^3\ \text{kg/m}^3 \times 9·807\ \text{m/s}^2}\left[\frac{\text{kg m}}{\text{N s}^2}\right] = 8·46\ \text{m}$$

for a vacuum pump pulling on fluid at standard atmospheric pressure. There is, therefore, a limit to the acceleration which the fluid in the suction column can acquire. If the piston accelerates faster than this the fluid will separate from the piston face and reach it again subsequently. This gives rise to water hammer or 'knocking' when a pump is run under these conditions.

As with a surge tank, the air vessel acts as an accumulator. The pump makes a varying demand on the suction side which is met largely by the air vessel; the cyclic fluctuations of speed in the suction pipe are consequently considerably reduced, and if the air vessel is increased in size the fluid velocity in the suction pipe becomes more uniform (see Section 16.3). Similar considerations apply on the delivery side in which an air vessel filters out the cyclic fluctuations in the pump delivery and thus enables an almost uniform rate of flow to be maintained in the delivery pipe beyond the air vessel.

EXAMPLE 6.8 (ii)

*The vertical suction pipe of a single-ram pump is 2 m long and is of 50 mm diameter. The stroke of the plunger is 300 mm and its diameter is 150 mm. Assuming that it moves with simple harmonic motion, estimate the maximum speed at which it may run in order to avoid breaking the suction column given that separation is likely if the absolute pressure head falls below 1·4 m water at the beginning of the stroke.*

So long as continuity is maintained, simple harmonic motion of the plunger will imply simple harmonic motion of the suction column with a velocity ratio depending on the area ratio, in this case $(15/5)^2$, i.e. 9 to 1. Since acceleration is simply change of velocity in a given time, the former is also 9 times that of the plunger, and the maximum value for the fluid in the pipe is thus:

$$a = 9 \times \left( \omega^2 \times \frac{L}{2} \right) = 9 \times \omega^2 \times 0·15 \text{ m} = 1·35 \text{ m} \times \omega^2$$

This acceleration is related to the hydraulic gradient, and the latter is, under limiting conditions:

$$i = -\frac{\delta(z+p/w)}{\delta l} = -\frac{\{(2+1·4)-10·4\} \text{ m}}{2 \text{ m}} = 3.5$$

neglecting the immersion of the suction pipe and taking the barometer to stand at 10·4 m water.

$$a = 3·5 \times 9·807 \text{ m/s}^2 = 34·3 \text{ m/s}^2$$

Thus:

$$\omega = \sqrt{\left( \frac{34·3}{1·35 \text{ s}^2} \right)} = 5·04 \frac{1}{\text{s}} \left[ \frac{\text{cycle}}{2\pi} \right] \left[ \frac{60 \text{ s}}{\text{min}} \right] = \mathbf{48·1} \frac{\text{cycles}}{\text{min}}$$

## Exercises on Chapter 6

1. Water flows with a velocity of 9 m/s along a 75 mm diameter fire hose to a nozzle 300 mm long and which is profiled so that the pressure drops uniformly along its length. The pressure of the water at the entrance to the nozzle is 294 kN/m². Neglecting friction, estimate the acceleration of the water in passing through the nozzle, the velocity and diameter of the jet and the vertical height to which the jet would rise.

2. An open vertical cylinder 160 mm diameter contains water to a depth of 80 mm when stationary. What will be the actual height of the water at the walls when the cylinder is rotated at a speed of 120 rev/min? Also, estimate the speed at which the centre of the base becomes dry.

3. A closed vertical cylinder 900 mm diameter is full of water and rotates by paddles 300 mm diameter which revolve at 120 rev/min concentric with the axis of the cylinder. Calculate the velocity of the water at radii of 150 mm and 300 mm and the excess pressure head at these radii over that at the centre.

4. A hollow cylindrical vessel 600 mm diameter open at the top spins about its axis which is vertical, thus producing a forced vortex motion of the liquid contained in it. Calculate the height of the vessel so that the liquid first reaches the top and just begins to uncover the base when rotating at 120 rev/min.

If the speed is now increased to 150 rev/min what area of the base will be uncovered and if the speed is now reduced until the base is just covered again, how high will the liquid stand at the sides?

5. (a) Plot a curve showing the relation between radius and pressure of water inside the impeller of a centrifugal pump on no flow. The inner (i.e. eye) radius is 300 mm and the outer (i.e. tip) radius is 600 mm. Assume that the pressure of water at the eye of the impeller is 70 kPa absolute and the speed is 336 rev/min.

(b) If the same impeller rotates at the same speed concentrically in water contained in a closed cylinder 900 mm radius, extend the curve of part (a) to show the change of pressure as the radius increases from 600 to 900 mm.

6. A reciprocating pump has a plunger 300 mm diameter, stroke 450 mm and is placed with its centre-line 3 m vertically above the suction pump. The suction pipe is 4·5 m long and 150 mm diameter and the delivery pipe 36 m long rising uniformly to the delivery tank, whose water level is 18 m above the suction level. If separation occurs when the absolute pressure head is 2·5 m of water, find the speed of the pump in rev/min at which separation commences (a) in the suction pipe, and (b) in the delivery pipe, assuming normal barometric pressure and simple harmonic motion of the plunger.

7. Show that, if the liquid in a U-tube is suddenly disturbed from its condition of equilibrium, the resulting motion will be simple harmonic of periodic time $2\pi \sqrt{(L/2g)}$, where $L$ is the total length occupied by the liquid.

A vertical U-tube of uniform bore, 6 mm diameter, is rounded at the lower end to a radius of 75 mm and is charged with a liquid of relative density 0·95, which, for zero pressure difference rises to a height of 600 mm above the centre of curvature. If, when in this condition, the column is momentarily disturbed so that the maximum difference of levels is 150 mm, calculate the frequency and the maximum energy of the ensuing oscillation. Weight of water is 9·807 $kN/m^3$.

8. Deduce an expression for the head required to accelerate the water in the suction pipe of a reciprocating pump assuming simple harmonic motion of the plunger.

A single-acting pump having a plunger diameter 200 mm and stroke 300 mm is placed with its centre line 3·6 m above the level of the water in the suction tank. The suction pipe is 75 mm diameter and 4·5 m long. If separation occurs when the absolute pressure head is 2·5 m of water, find the maximum speed of the pump to avoid separation at the commencement of the suction stroke. Assume a normal barometric height of 10·3 m of water and simple harmonic motion of the plunger.

## Chapter 7

# Fluid momentum and thrust by reaction

> '*A substance offers as much resistance to the air as the air does to the substance.*'   LEONARDO

## 7.1. Impact of jets

In the previous chapter Newton's second law, in the form $F = ma$, was applied to the motion of a fluid element. An alternative approach, which is particularly useful when studying jets, and the motion of whole regions of fluid, is to consider the fluid *momentum* directly. The time rate of change of the latter is, according to the second law of motion, proportional to the impressed force. Thus the force due to the impact of a jet is equal to the momentum change of the latter per unit time. As a simple example, we may wish to find the force required to arrest the motion of a uniform stream, as indicated by Fig. 7.1. First consider a block of mass $m$, which has an initial velocity $U$ and brought to rest; its loss of momentum is $mU$. If this takes place in a time $t$, the rate of change of momentum is $mU/t$ so that the arresting force, which is equal and opposite to the force exerted by the jet is:

$$F = \frac{m}{t} U$$

Applying this idea to the jet, $m/t$ is the rate ($\dot{m}$) of mass flow, i.e. $\rho$ times the volumetric flow rate $Q$. Hence:

$$F = \frac{m}{t} U = \dot{m}U = \rho QU$$

EXAMPLE 7.1 (i)

*A jet of water*, 50 mm *in diameter, issues with a velocity of* 15 m/s, *and impinges on a flat plate which destroys its forward motion.*

*Find the force exerted on the vane* (a) *if the latter is stationary,* (b) *if it is moving away from the jet at the rate of* 3 m/s.

(*a*) If we denote the area of the jet by $A$ and its velocity by $U$, the volumetric flow rate is:

$$Q = AU$$

The corresponding mass flow rate is:

$$\rho Q = \rho A U = \dot{m}$$

so that the force exerted is:

$$F_a = \dot{m}U = \rho A U^2$$

i.e.

$$F_a = 10^3 \frac{\text{kg}}{\text{m}^3} \times \frac{\pi}{4} \frac{25\,\text{m}^2}{10^4} \times 225 \frac{\text{m}^2}{\text{s}^2}$$

$$= \mathbf{441}\ \text{N}$$

(*b*) The initial velocity of the fluid relative to the plate, $U_r$ say, equals $(15-3)$ m/s and this is reduced to zero after impact.

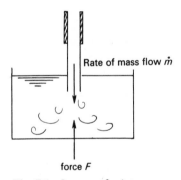

Rate of mass flow $\dot{m}$

force $F$

**Fig. 7.1.** *Impact of a jet*

Therefore, the rate at which the fluid strikes the plate is 12 m/s, and its change of velocity is also 12 m/s. Thus the force on the plate is, in this case:

$$F_b = \rho A U_r^2$$

Comparison with part (*a*) indicates that:

$$\frac{F_b}{F_a} = \left(\frac{12}{15}\right)^2 = 0.64$$

i.e.

$$F_b = 0.64 \times 441\ \text{N} = \mathbf{282}\ \text{N}$$

Let us suppose that we observe the changes in the motion of the fluid relative to the earth in part (*b*) of the previous problem. We conclude that *the force exerted is equal to the product of the mass per unit time which suffers a change of velocity and the magnitude of that change.* It will be proved in Section 7.3 that this is a very general conclusion, i.e. **the force associated with the changing momentum of a fluid is equal to the rate of mass flow times its change in velocity.** Thus force may be induced by changing the *direction* of flow without necessarily changing the speed of fluid; also, forces may be induced by changing the speed of flow without changing direction as, for instance by changes of area in pipe-lines.

It should be noted that if a jet strikes a *series* of moving vanes, the mass which suffers a change of momentum in unit time is different from that caused by a *single* vane.

The former case is more common in hydraulic turbines, where a series of buckets or blades experience a force by changing the momentum of a stream of fluid issuing from nozzles or guide blades.

EXAMPLE 7.1 (ii)

*Find the force exerted by the jet described in Example 7.1(i), when it impinges on a series of vanes moving away from it at 3 m/s. Assume, as before, that the fluid moves with the velocity of the vanes after impact.*

In this case the mass flow striking the series of vanes is:

$$\dot{m} = \rho A U$$

$$= 10^3 \frac{\text{kg}}{\text{m}^3} \times \frac{\pi}{4} \frac{25\,\text{m}^2}{10^4} \times 15 \frac{\text{m}}{\text{s}}$$

$$= 29\cdot45 \text{ kg/s}$$

As the change in a quantity is the difference between its final and initial values:

$$\Delta U = \text{(final velocity)} - \text{(initial velocity)}$$

i.e.                $$\Delta U = U_2 - U_1 = (3-15) \text{ m/s} = -12 \text{ m/s}$$

The minus sign indicates that the velocity is *decreased*. The force *on the fluid* is thus:

$$F = \dot{m}(U_2 - U_1) = 29\cdot45 \frac{\text{kg}}{\text{s}} \times \left(-12 \frac{\text{m}}{\text{s}}\right) = -353\cdot4 \frac{\text{kg m}}{\text{s}^2}$$

This is the force exerted *by* the blades so that the force *on* the blades is equal and opposite, namely:

$$F = +353\cdot4 \text{ N}$$

in the direction of motion of the incident jet.

## 7.2. Pelton wheels

Energy considerations suggest that a turbine should be so operated that the fluid leaving the rotor should have little or no residual kinetic energy. For example, if a jet impinges on a number of buckets spaced round the circumference of the rotor, as in a Pelton wheel, and the fluid simply falls away from the buckets after impact, all the energy of the incident fluid would (neglecting losses) be transferred to the wheel. Such ideal conditions can be approached by shaping the buckets as shown in Fig. 7.2 (a), and running the wheel at such a speed that the relative velocity of the fluid leaving the buckets is equal and opposite to that of the buckets themselves. In theory the fluid would then have zero velocity relative to the earth. The wheel operates at atmospheric pressure and therefore must be installed above the flood-water level of the river into which it discharges. As the area of the jets is relatively small compared with the size of the machine, high powers can be obtained only if the jet velocities are high. This fact combined with the fall in efficiency at other than the design head and the inability to use the available head above the tail race causes Pelton wheels to be employed for high heads (say upwards of 150 m). Under these conditions they achieve efficiencies of over 80 per cent (see Chapter 15).

If the fluid is at constant pressure (a condition which is achieved by allowing air to enter the casing) it moves round the blade at constant

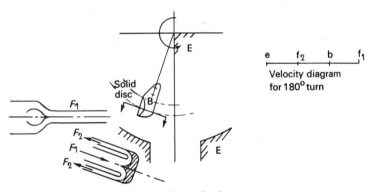

Fig. 7.2(a). *Pelton wheel*

speed, neglecting friction. This follows since there can be no acceleration round the surface of the blade unless there is a pressure gradient to produce it, see Section 6.3. Hence, in the ideal case, the fluid velocity changes in direction (turning through 180°) but not in magnitude, relative to the buckets (B). If we denote the earth by E and the fluid before and after impact by $F_1$ and $F_2$ respectively, the ideal (180° turn) velocity diagram is as shown. The notation for this is that commonly used when studying mechanisms, i.e. the velocity of the incident jet ($F_1$) relative to the buckets (B) is denoted by $bf_1$, etc. In this frictionless case $bf_1 = f_2b$, and if the outgoing fluid ($F_2$) is to have no final kinetic energy, then $f_2e$ should be zero. This will be achieved provided that $eb = f_2b$. If the latter equals $bf_1$, it follows from the diagram that $eb = \frac{1}{2}ef_1$, i.e. the efficiency will be a maximum if the speed of the buckets is one half that of the jet.

In practice the buckets come into action before they reach the vertical position and in addition they can seldom be allowed to turn the fluid through a greater angle than about 165°, otherwise the used water would be in the way of the oncoming buckets. These factors, together with friction, modify this simple analysis, and tests indicate that the best blade speed is about 0·46 of the jet speed, and the ratio of jet diameter to wheel diameter is about 1 to 18 for high efficiency.

The force impressed on the buckets is caused by the changing momentum of the fluid and is equal to the product of the mass flow and its change in velocity. The latter is, in general, the vector difference between the final and initial velocities, and its component in any direction gives the magnitude of the component force in that direction per unit mass flow. As a Pelton wheel bucket divides the jet symmetrically, so that the resultant force acts in the direction of whirl, we may in general consider the velocity diagram in Fig. 7.2(b) to represent the whirl components of the velocities, i.e. those tangential to the bucket circle. If the diagram be interpreted in this way, i.e. if $f_1 f_2'$ represents the change in the velocity of whirl of the

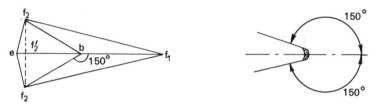

**Fig. 7.2(b).**  *Velocity diagram for Pelton wheel*

fluid, the force exerted on the wheel is $\dot{m}.f_2'f_1$, where $\dot{m}$ represents the rate of mass flow issuing from the nozzle. The product of this force and the blade speed gives the rate at which work is being done, i.e. the output of power. The ratio of the power obtained from the velocity diagram in this way to the power theoretically available from the jet is termed the diagram efficiency.

Example 7.2 (i)

*Deduce a numerical formula for the power delivered by a Pelton wheel in terms of the kg/s flowing, the change in the velocity of whirl, and the blade speed, the velocities being in m/s.*

In general, the force exerted on the water, which is equal and opposite to that exerted on the wheel is:

$$F = \dot{m}.f_1 f_2'$$

If $D$ kg/s change their velocity of whirl by $U_w$ m/s the force is thus:

$$F = \left(D\,\frac{\text{kg}}{\text{s}}\right)\left(U_w\,\frac{\text{m}}{\text{s}}\right)\left[\frac{\text{N s}^2}{\text{kg m}}\right]$$

i.e.

$$F = DU_w \text{ N}$$

For a blade speed of $V$ m/s, the corresponding power is thus:

$$P = DU_w \text{ N} \times V\,\frac{\text{m}}{\text{s}}\left[\frac{\text{W s}}{\text{N m}}\right]$$

$$= U_w VD \text{ W}$$

It should be noted that in this expression the symbols represent numbers of particular units, i.e. it is a numerical formula, which is correct only for the particular units stated here.

EXAMPLE 7.2 (ii)

*A Pelton wheel is driven by a jet having a velocity of 36 m/s. The buckets turn the water through an angle of 150°, friction causing a 25 per cent reduction in speed over their surfaces.*

*Calculate the diagram efficiency of the wheel when the speed is 240 rev/min, if the blades have a pitch circle diameter of 1·2 m.*

The speed of the blades (B) relative to the earth (E) is:

$$eb = R\omega$$

$$= 0\cdot 6 \text{ m} \times 240 \frac{\text{rev}}{\text{min}} \left[\frac{2\pi}{\text{rev}}\right] \left[\frac{\text{min}}{60 \text{ s}}\right]$$

$$= 15\cdot 1 \text{ m/s}$$

Hence the oncoming fluid ($F_1$) strikes the buckets (B) with a speed:

$$bf_1 = (36 - 15\cdot 1) \text{ m/s} = 20\cdot 9 \text{ m/s}$$

If there is a 25 per cent loss in speed of fluid relative to the buckets, the fluid leaves the latter with a relative velocity:

$$bf_2 = 0\cdot 75 \times bf_1$$

$$= 0\cdot 75 \times 20\cdot 9 \text{ m/s} = 15\cdot 7 \text{ m/s}$$

The resultant change in the velocity of the fluid is $f_1 f_2$ and this has a component in the direction of motion equal to $f_1 f_2'$. The corresponding force exerted per kg/s of fluid is thus:

$$F = 1 \text{ kg/s} \times (20\cdot 9 + 15\cdot 7 \cos 30°) \text{ m/s} = 35\cdot 4 \text{ N}$$

so that the power developed is $34\cdot 5 \text{ N} \times 15\cdot 1 \text{ m/s}$, i.e. 521 W.

The kinetic energy of the jet is:

$$\frac{U^2}{2} = \tfrac{1}{2}(36 \text{ m/s})^2 = 648 \text{ Nm/kg}$$

so that the power available from a mass flow of 1 kg/s is 648 Nm/s or 648 W.

Hence the efficiency of the wheel is:

$$\epsilon = \frac{521}{648} = \mathbf{0\cdot 802} \quad or \quad \mathbf{80\cdot 2} \text{ } per \text{ } cent$$

EXAMPLE 7.2 (iii)

*(a) A jet of water 50 mm diameter impinges on a single bucket of a Pelton wheel held stationary and is deflected through a total angle of 165°. The measured jet force was 2 120 N in the direction of the jet when the total head at the nozzle was 30·5 m. Assuming the coefficient of velocity of the nozzle 0·97*

*find the ratio k of the velocity of the water leaving the bucket to that of the jet.*

(b) *Assuming the same value of k and the angle of deflexion find the force which the same jet would exert on a series of these buckets revolving at velocity 0·46 that of the jet. Hence find the power developed by the buckets and, assuming windage and mechanical friction losses amount to 4 per cent of the water power supplied, find the brake power and the overall efficiency of the turbine.*

(a) Velocity of jet is $v_j = C_v \sqrt{(2gH)} = 0\cdot97 \sqrt{(19\cdot61 \text{ m/s}^2 \times 30\cdot5 \text{ m})}$ $= 23\cdot7$ m/s.

For one side of a stationary bucket the velocity triangle is shown in Fig. 7.2(c) in which length OB is $k$ times length OA.

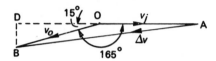

**Fig. 7.2(c).** *Velocity diagram for stationary bucket*

The change in velocity of the water is $\varDelta v$ with components $\overline{\text{AD}}$ and $\overline{\text{DB}}$. The force exerted by the water in the direction of the jet is $F = \dot{m} \times \overline{\text{DA}}$ $= 2\ 120$ N,

i.e.                          $F = \rho a v_j (v_j + k v_j \cos 15°)$

or

$$k = \frac{1}{\cos 15°}\left(\frac{F}{\rho a v_j^2}-1\right) = \frac{1}{0\cdot966}\left(\frac{2\ 120 \text{ N}}{10^3 \dfrac{\text{kg}}{\text{m}^3} \times \dfrac{\pi}{4}\dfrac{25 \text{ m}^2}{10^4} \times (23\cdot7)^2 \dfrac{\text{m}^2}{\text{s}^2}}-1\right)$$

$$= \frac{1}{0\cdot966}\left(\frac{2\ 120 \text{ N}}{1\ 103 \text{ N}}-1\right) = \frac{(1\cdot923-1)}{0\cdot966} = \mathbf{0\cdot955}$$

(b) When the jet impinges on a series of moving buckets the velocity diagram is similar to that of Fig. 7.2(b) but on which $\text{ef}_1 = v_j = 23\cdot7$ m/s and $\text{eb} = v_B = 0\cdot46 \times 23\cdot7 = 10\cdot9$ m/s. Hence, $\text{bf}_1 = 12\cdot8$ m/s and $\text{bf}_2 = k.\text{bf}_1$ $= 12\cdot2$ m/s. Change of velocity of water in direction of jet is $\text{f}'_2\text{b} + \text{bf}_1$ or $\text{f}'_2\text{f}_1 = 12\cdot2 \cos 15° + 12\cdot8 = 24\cdot6$ m/s.

Therefore, force exerted on buckets in $\dot{m} \times \text{f}'_2\text{f}_1$ and the power developed by the water is $P_i = \rho a v_j \times \text{f}'_2\text{f}_1 \times \text{eb}$

$$= 10^3 \frac{\text{kg}}{\text{m}^3} \times \frac{\pi}{4}\frac{25 \text{ m}^2}{10^4} \times 23\cdot7 \frac{\text{m}}{\text{s}} \times 24\cdot6 \frac{\text{m}}{\text{s}} \times 10\cdot9 \frac{\text{m}}{\text{s}}\left[\frac{\text{N s}^2}{\text{kg m}}\right]$$

$= 12\cdot48$ kN m/s   or   12·48 kW.

Water power supplied in jet $= \frac{1}{2}\dot{m}v_j^2 = \rho a v_j^3/2 = 13\cdot06$ kW; losses in windage and mechanical friction $= 0\cdot04 \times 13\cdot06 = 0\cdot52$ kW. Hence, shaft power $= 12\cdot48 - 0\cdot52 = 11\cdot96$ kW, and overall efficiency is $11\cdot96/13\cdot06$ $= 0\cdot915$ or **91·5** *per cent.*

## 7.3. Proof of expression for momentum force

In Section 4.2 we studied the energy changes taking place within an imaginary control surface (stream-tube and pistons) moving with the flow. We shall now consider the corresponding changes in its momentum as it moves. Referring to Fig. 4.2a (p. 72) and again assuming steady flow, the volume enclosed between the imaginary pistons has, after a time $\delta t$, vacated the lower shaded region at 1 but has occupied the shaded volume at 2. If the fluid is incompressible, these two volumes are equal, but for compressible fluids only their masses, $\delta m$, would be the same.

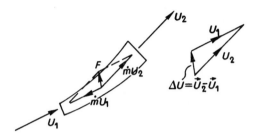

**Fig. 7.3(a).** *Force acting on moving fluid to change its momentum*

In a time $\delta t$, the control volume has acquired additional momentum at 2, equal to $\delta m U_2$ but has lost the momentum $\delta m U_1$. All the other fluid elements have, as it were, moved up one, but in doing so have acquired the velocity appropriate to their new positions. Hence, there is no change in the momentum of the unshaded portion of the enclosed fluid. Using the symbol $\dot{m}$ to denote the rate of mass flow $\delta m / \delta t$, the control volume is acquiring momentum at the rate $\dot{m} U_2$ and losing it at the rate $\dot{m} U_1$. The former requires a force of magnitude $\dot{m} U_2$ in the direction of $U_2$ and the latter a force $\dot{m} U_1$ against the direction of $U_1$.

Their resultant is indicated in Fig. 7.3(a) and is equal to the mass flow rate times the change in velocity,

i.e.
$$\vec{F} = \dot{m} \Delta U = \dot{m}(\vec{U_2} - \vec{U_1})$$

the overhead arrows implying the vector difference. This equation may be re-expressed in terms of the volumetric flow rate $Q$ as:

$$\vec{F} = \rho Q(\vec{U_2} - \vec{U_1})$$

It should be noted that the above expression gives the force which must act *on* the enclosed fluid in order to change its momentum.

EXAMPLE 7.3 (i)

*A cascade of turning vanes is used at a 90° bend in the return circuit of a wind-tunnel which is 3 m square at this point. Find the force on the bend if the velocity of the air-stream is 12 m/s and its density is 1·28 kg/m³.*

Neglecting compressibility: $U_1$ and $U_2$ are equal in magnitude, namely 12 m/s. Hence the change in velocity as deduced from Fig. 7.3(b) is:

$$\Delta U = (\overrightarrow{U_2} - \overrightarrow{U_1}) = 12 \sqrt{2} \text{ m/s}$$

The rate of mass flow may be written as:

$$\dot{m} = \rho Q = \rho A_1 U_1$$

i.e.
$$\dot{m} = 1 \cdot 28 \text{ kg/m}^3 \times 9 \text{ m}^2 \times 12 \text{ m/s}$$
$$= 138 \cdot 2 \text{ kg/s}$$

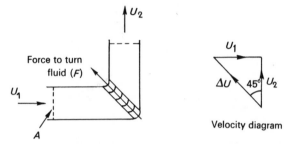

**Fig. 7.3(b).** *Force exerted by a cascade of turning vanes*

Hence, the resultant force on the fluid, which acts in the direction of its change in velocity, is:

$$F = \dot{m}(\overrightarrow{U_2} - \overrightarrow{U_1}) = 138 \cdot 2 \text{ kg/s} \times 12 \sqrt{2} \text{ m/s}$$
$$= 2\ 345 \text{ N}$$

This is the force which acts *on* the fluid (see Fig. 7.3(b)). The bend experiences its reaction, i.e. 2·345 kN *acting outwards* at 45°.

As there is no change in the velocity of the fluid normal to the cascade the transverse force on it is, theoretically, zero.

This is because we have neglected the effect of friction and its associated pressure drop; for a well-designed cascade these effects are however relatively small.

EXAMPLE 7.3 (ii)

*Deduce an expression for the force experienced by a fixed plate flat inclined at an angle θ to a jet of fluid issuing with a velocity U and cross-sectional area A as shown in Fig. 7.3(c).*

The oncoming fluid has a velocity $U \sin \theta$ normal to the plate, and if it leaves the plate tangentially the final velocity in the normal direction is zero:

$$\overrightarrow{F} = \dot{m}(\overrightarrow{U_2} - \overrightarrow{U_1})$$

∴
$$\overrightarrow{F} = \dot{m}(0 - U \sin \theta)$$

This is the force required to change the direction of motion of the fluid. The corresponding normal *reaction* on the plate is, therefore:

$$F_n = \dot{m}U \sin \theta$$

and, if friction may be neglected, the force along the plate is zero.

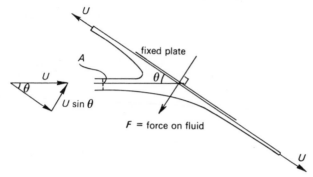

**Fig. 7.3(c).** *Force due to an inclined flat plate*

If the fluid is of density $\rho$, the rate of mass flow is:

$$\dot{m} = \rho A U$$

$\therefore$
$$F_n = \rho A U^2 \sin \theta$$

The component of this in the direction of the jet is:

$$F_n \sin \theta = \rho A U^2 \sin^2 \theta$$

and this has a maximum value when the plate is normal to the jet:

$$F_{max} = \rho A U^2$$

as is also proved in Example 7.1 (i).


## 7.4. Force and power exerted on blading

In turbo-machinery, forces are exerted on moving blading primarily by a change in the momentum of the fluid traversing it. The inlet blade angle should be so adjusted that, as far as possible, the fluid enters without shock, and the blading should ideally be of sufficiently close pitch for the fluid to leave it at an angle dictated by the trailing edges, see Fig. 7.4(a).

If the blading is of the so-called impulse type there is no change of pressure across the moving blades. Under these conditions the fluid, viewed relative to the blades, can change its speed only through friction which causes a reduction. Neglecting the latter, the magnitudes of the velocities of fluid relative to blade and inlet and outlet are equal, i.e. (bi) = (bo) on the velocity diagram.

If, however, the machine is so arranged that all the passages are completely drowned, it is possible to design for a pressure drop across the moving blades ($\Delta p$ say). This is called reaction blading, and will cause a corresponding change in kinetic energy. Viewed relative to the blades, if friction and density changes may both be neglected, Bernoulli's equation applies:

$$p_i - p_o = \tfrac{1}{2}\rho(bo)^2 - \tfrac{1}{2}\rho(bi)^2$$

or

$$\frac{\Delta p}{\rho} = \frac{(bo)^2 - (bi)^2}{2}$$

On reaction blading $\Delta p$ causes an end thrust but this has no component in the direction of whirl.

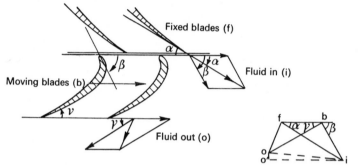

**Fig. 7.4(a).** *Blades of a turbine*          **Fig. 7.4(b).** *Velocity diagram*

The force on the blading due to the momentum changes is equal to the mass flow times the change in velocity across it (oi), no matter whether the blading is of the impulse or reaction type. The component of this velocity change in the direction of whirl, i.e. (o′ i), gives the force per unit mass flow performing work and as the velocity of the blading is (fb), the rate of working, i.e. the power, again per unit mass flow is (o′ i) × (fb).

$$F = o'\,i \times \dot{m} = o'\,i \times \rho Q$$
$$P = o'\,i \times fb \times \dot{m} = o'\,i \times fb \times \rho Q$$

in which $Q$ is the volumetric discharge rate.

Example 7.4 (i)

*Using Fig. 7.4(b) deduce an expression for diagram efficiency for impulse blading.*

In an impulse turbine the energy supplied to the moving blade ring is in the form of kinetic energy, namely $\tfrac{1}{2}(fi)^2$ per unit mass or $\tfrac{1}{2}\dot{m}(fi)^2$ per unit time which is the power of the fluid entering the moving blade ring. The force on the blading in direction of motion (or whirl) is $\dot{m}(o'i)$ which is the rate of change of momentum in direction of whirl, and the rate at which work is done or power developed is $P = \dot{m}(o'i)(fb)$. Hence, the efficiency as deduced from the diagram is the ratio of the mechanical power developed to the power of the fluid supplied, namely

$$\frac{\dot{m}(o'i)(fb)}{\frac{1}{2}\dot{m}(fi)^2}$$

i.e. the diagram efficiency is

$$\frac{2(o'i)(fb)}{(fi)^2}$$

EXAMPLE 7.4 (ii)

*When working at its best point, the blading at the mean radius (0·5 m) of a propeller type pump deflects a stream approaching it at a relative angle of 60° to the axis, through 15°, so that the water leaves it at a relative angle of 45°. Assuming the water approaches it axially and that the axial velocity remains constant, draw the inlet and outlet velocity triangles under these conditions for a rotational speed of 600 rev/min.*

*Estimate the corresponding ideal pressure rise across the moving blades.*

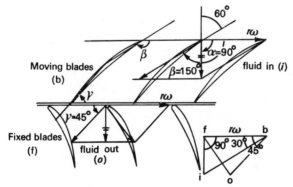

Fig. 7.4(c). *Blades of an axial flow pump*  Fig. 7.4(d). *Velocity triangles*

The blade velocity of whirl $= r\omega = 0\cdot5$ m $\times (10 \times 2\pi)/s = 10\pi$ m/s. Viewed relative to the moving blades, the inlet fluid velocity

$$\text{bi} = (2/\sqrt{3} \times 10\pi) \text{ m/s} = (20\pi/\sqrt{3}) \text{ m/s}$$

the corresponding axial velocity being

$$\text{fi} = \tfrac{1}{2}\text{bi} = (10\pi/\sqrt{3}) \text{ m/s}$$

Hence the relative fluid velocity at outlet $= \text{bo} = \text{fi } \sqrt{2} = 10\pi \sqrt{2/3}$ m/s.

The corresponding kinetic head change across the moving blades is, ideally:

$$\frac{p_i - p_o}{w} = \frac{(\text{bo})^2 - (\text{bi})^2}{2g} = \frac{\pi^2}{2g}\left(\frac{200}{3} - \frac{400}{3}\right) \text{ m}^2/\text{s}^2$$

i.e.

$$\frac{p_o - p_i}{w} = \frac{200\pi^2}{6 \times 9\cdot807} \text{ m} = 33\cdot6 \text{ m}$$

This is the ideal rise in head across the moving blades, but it should be noted that the water does not leave them axially, and this swirl must be taken out by the fixed blades indicated downstream of the propeller.

## 7.5. Thrust by reaction

The above examples indicate that in practice we usually require the reaction which the fluid exerts on its surroundings, rather than the force on the enclosed fluid. From Fig. 7.5(a) it may be seen that no matter

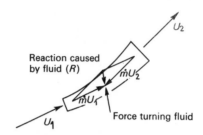

**Fig. 7.5(a).** *Force of reaction due to changing momentum*

what the sense of the velocity, the force of reaction ($R$) may be found by summing the $\dot{m}U$ vectors so drawn that they act inwards with respect to the control volume, e.g. a jet engine experiences a momentum drag from the intake air and a forward thrust from the gas issuing from the exit nozzle as in the following Example 7.5(i).

EXAMPLE 7.5 (i)

*A jet engine travelling at 240 m/s, discharges the gases through the exit nozzle at atmospheric pressure with a velocity of 760 m/s. If the air:fuel ratio is 40:1 and the fuel consumption is 0·9 kg/s, find the thrust and the thrust power developed under these conditions.*

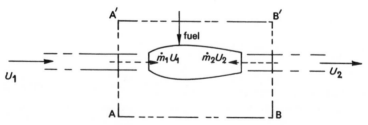

**Fig. 7.5(b).** *Momentum forces on a jet engine*

From Fig. 7.5(b) we deduce that the resultant thrust on the engine is:

$$F = \dot{m}_2 U_2 - \dot{m}_1 U_1$$

If the fuel consumption is 0·9 kg/s:

$$\dot{m}_1 = 40 \times 0\cdot9 \ \text{kg/s} = 36 \ \text{kg/s}$$
$$\dot{m}_2 = (36+0\cdot9) \ \text{kg/s} = 36\cdot9 \ \text{kg/s}$$

∴
$$F = (36\cdot9 \times 760 - 36 \times 240) \ \text{kg} \, \frac{\text{m}}{\text{s}^2}$$

$$= \textbf{19·41 kN}$$

The corresponding thrust power is equal to the force times the velocity with which it moves, i.e.:

$$P = FU_1$$

$$= 19\cdot41 \ \text{kN} \times 240 \, \frac{\text{m}}{\text{s}} \left[ \frac{\text{W s}}{\text{N m}} \right]$$

$$= \textbf{4·66 MW.}$$

With a supersonic stream the gases are not necessarily expelled through the exit nozzle at atmospheric pressure. If the gas pressure is greater or less than this value, the nozzle is said to be either under- or over-expanded, respectively. Slight under-expansion causes a negligible reduction in thrust.

## 7.6. Pressure and momentum forces

It is often convenient to separate the force which is exerted by the walls of the stream-tube from that caused by the pressures acting across the flow areas. Referring to Fig. 7.6(a) the forces *on* the enclosed fluid

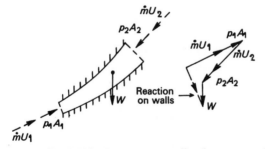

**Fig. 7.6(a).** *Reaction on walls of a stream-tube*

consist of $W$ due to gravity, $p_1A_1$ on the inlet area, $p_2A_2$ on the exit area, together with the force exerted by the walls on the fluid. Their resultant is equal and opposite to the total reaction $R$ which, from the previous section, may be obtained by summing the momentum vectors $\dot{m}U_1$ and $\dot{m}U_2$ pointing inwards.

If the above forces are not equal and opposite to $R$, the flow is *unsteady* in the sense that there will be a time rate of change of velocity, i.e. a local acceleration within the control volume. Such conditions exist during surges for example.

The above result may be stated alternatively as follows:

*For steady flow through a stream-tube the reaction on the walls is equal to the sum of the inward acting momentum and pressure forces across the flow areas, together with the force of gravity.*

An expression for the thrust exerted by a rocket motor, which expels the gases relative to the casing with a velocity $U$ and density $\rho$ at a pressure $p$ through the exit area $A$, may be deduced by reference to Fig. 7.6(b).

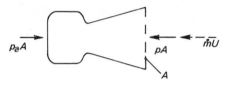

**Fig. 7.6(b).** *Forces on rocket motor*

As the initial velocity of the fuels relative to the casing is either negligible or zero, we may neglect their initial momentum.

Hence, the fluid reaction on the inside walls is the sum of the inward acting momentum and pressure forces across the exit area, i.e.

$$F_i = pA + \dot{m}U$$

If the area is $A$ and the velocity is $U$, the volumetric flow rate is $UA$; the mass flow is thus:

$$\dot{m} = \rho UA$$

Hence the expression for $F_i$, the resultant of the internal forces, may be written as:

$$F_i = (p + \rho U^2)\, A$$

This suggests that it is sometimes convenient to consider the resultant pressure across the exit section to be the sum of the static pressure $p$, together with an additional momentum pressure $\rho U^2$.

It is of interest to note that, according to kinetic theory, the static pressure is also due to momentum flux. The random motions of the molecules back and forth across any interface, with normal root mean square velocity $U'$, gives rise to what we term a static pressure:

$$p = \rho(U')^2$$

As the temperature is proportional to the kinetic energy of the molecular motions, $T$ is proportional to $(U')^2$ so that:

$$p \propto \rho T$$

the constant of proportionality varying inversely as mass of molecules. Hence we may write:

$$p = \rho RT$$

in which $R$ is the 'specific gas constant' which is inversely proportional to the relative molecular mass (previously known as 'molecular weight') of the gas.

If atmospheric pressure $(p_a)$ acts round the outside of the casing, this has a resultant $F_o = -p_a A$, the negative sign indicating that this force is a drag. Hence the resultant force on the motor is given by the vectorial equation:

$$F = F_i + F_o = [(p + \rho U^2) - p_a] A$$

That is, if the gauge pressure at exit is $p_G$, the thrust is:

$$F = (p_G + \rho U^2) A$$

With an under-expanded nozzle, the gauge pressure is positive. Its thrust can theoretically be increased by extending the nozzle, so that the gases are accelerated to a higher velocity. If however the gases are over-accelerated, so that the gauge pressure at exit becomes negative, the thrust again falls. The thrust at the optimum expansion ratio is thus:

$$F = \rho U^2 A = \dot{m} U$$

A measure of merit or performance is the thrust per unit mass flow, and under optimum conditions:

$$\frac{F}{\dot{m}} = U$$

This quantity is termed the *effective gas velocity*.

## 7.7. Momentum pressure and shear

Generalising the result obtained above with particular reference to a rocket motor, we may state that: *wherever fluid crosses an imaginary boundary normally we may imagine an additional momentum pressure of magnitude $\rho U^2$ to act inwards on the volume it encloses*, see Fig. 7.7(a).

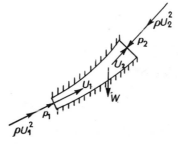

**Fig. 7.7(a).** *Static and momentum pressures*

This idea of a momentum pressure may be extended to cases where fluid is moving obliquely relative to the fixed imaginary inter-face, as shown in Fig. 7.7(b).

Fluid crosses the latter at the rate $\rho A u$, the $v$ component merely causing a simultaneous drift along the area $A$. Hence the rate of change of momentum vector for this surface is:

$$\dot{m}U = \rho A u . U$$

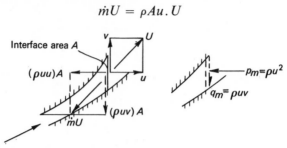

Fig. 7.7(b). *Momentum flux across oblique interface*

and this has normal and tangential components which are respectively:

$$\dot{m}U \times \frac{u}{U} = \rho A u . u = \rho A u^2$$

$$\dot{m}U \times \frac{v}{U} = \rho A u . v$$

These may be considered to be due to a momentum pressure $p_m$ of magnitude $\rho u^2$ and a momentum shear $q_m$ of magnitude $\rho u v$, each acting over the area $A$, as shown in Fig. 7.7(b).

EXAMPLE 7.7 (i)

*A horizontal pipe-line, 30 m long, terminates with a 45° bend in the downward direction. It is 100 mm in diameter and conveys water at a speed of 4·5 m/s. Due to losses, the pressure drops from 345 kN/m² to 205 kN/m² over its length. Find the resultant fluid force on the pipe under steady conditions.*

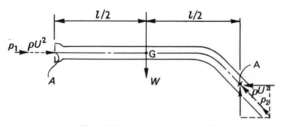

Fig. 7.7(c). *Forces on pipeline*

The mass of water in the pipe as shown in Fig. 7.7(c) is:

$$M = \rho A l = 1 \frac{Mg}{m^3} \times \frac{\pi}{400} \ m^2 \times 30 \ m$$

$$= 235\!\cdot\!5 \text{ kg and its weight is}$$

$$W = Mg = 235\!\cdot\!5 \text{ kg} \times 9\!\cdot\!807 \text{ m/s}^2 = 2\!\cdot\!31 \text{ kN}$$

Assuming a uniform velocity distribution at inlet and outlet the 'momentum pressure' is:

$$\rho U^2 = 1\,\frac{Mg}{m^3} \times 4\cdot 5^2\,\frac{m^2}{s^2}\left[\frac{N\,s^2}{kg\,m}\right]$$

$$= 20\ kN/m^2$$

Hence the effective pressures at inlet and outlet are, respectively, 365 kN/m² and 225 kN/m². The corresponding fluid forces are thus:

$$F_1 = 365\,\frac{kN}{m^2} \times \frac{\pi}{400}\,m^2 = 2\cdot 866\ kN$$

$$F_2 = 225\,\frac{kN}{m^2} \times \frac{\pi}{400}\,m^2 = 1\cdot 767\ kN$$

The latter may be split up into horizontal and vertical components of $1\cdot 767/\sqrt{2}$ kN, i.e. $1\cdot 250$ kN. There is thus a resultant force down the length of the pipe of $(2\cdot 866-1\cdot 250)$ kN, i.e. $1\cdot 616$ kN. Resolving vertically, there is a resultant downward force of $(2\cdot 310-1\cdot 250)$ kN, i.e. $1\cdot 060$ kN.

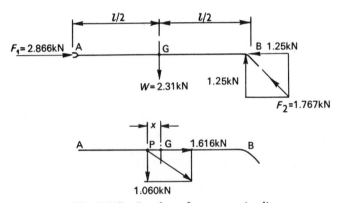

**Fig. 7.7(d).** *Resultant force on a pipe-line*

If we neglect the small offset caused by the bend, the only force which exerts a moment about G is the vertical component of $1\cdot 25$ kN, which acts at a lever arm of approximately 15 m and thus exerts a moment of $18\cdot 75$ kNm. As the resultant force down the length of the pipe exerts a negligible moment, the vertical component of $1\cdot 06$ kN must act at a point $x$, *upstream* of G, such that:

$$1\cdot 06\ kN \times x = 18\cdot 75\ kNm$$

i.e.     $x = 17\cdot 68\ m$

Hence the resultant force ($F$) on the pipe has horizontal and vertical components of $1\cdot 616$ kN and $1\cdot 060$ kN respectively, as shown on Fig. 7.7(d). It acts at the point P which is $17\cdot 68$ m from G.

## 7.8. Efficiency of propulsion

All vehicles may be thought of as deriving their thrust by reaction from momentum changes—even the forward motion of a car imparts an (imperceptible) backward motion to the earth. Ships, aircraft and rockets all overcome their resistance by using some mechanism to produce a more or less continuous increase in the backward component of the fluid momentum. The paddle, the sail, the propeller or the jet all achieve this result with varying degrees of success. The figure of merit for this conversion of available work into jet energy is termed the *conversion efficiency*, and this is virtually 100 per cent for a pure jet. A screw propeller however does not generate a uniform jet, and in particular it imparts a rotation to the screw race. Consequently some of the work delivered to the propeller does not appear as useful jet energy; the conversion efficiency seldom exceeds 85 per cent in this case, and is generally considerably less.

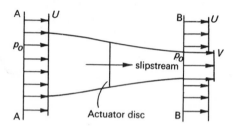

**Fig. 7.8(a).** *Ideal thrust actuator*

The basic difference between the propeller and the jet, apart from their conversion efficiencies, is that the former exerts a thrust by imparting a small increase in velocity to a relatively large mass of fluid, whereas a jet generally discharges a much smaller mass flow at a much higher speed. The question arises as to which of these is fundamentally more efficient in any given circumstances. The answer to this was established by Rankine in 1865, and his theory was elaborated by R. E. Froude who, with his father W. Froude, laid the foundations of resistance and propulsion—and, indeed, of much of naval architecture. According to the Rankine–Froude momentum theory the propulsive element is assumed to be replaced by a hypothetical actuator disc as shown in Fig. 7.8(a) and which is assumed to accelerate the fluid axially without rotation.

Although this may appear to be a very abstract concept it comes much nearer to describing how a propeller acts than the more popular but erroneous notion that a propeller 'screws' its way through a fluid. A chordwise section through a propeller blade encounters an oncoming stream which it deflects in a fore and aft direction, so that the axial component of the fluid velocity is increased. The various helical motions which are also imparted to the fluid are merely parasitic in so far as they represent a useless waste of energy.

Considering the motion relative to the actuator, fluid approaches it with a velocity $U$. If we concentrate on the fluid which is destined to pass through it, the stream will contract as the fluid accelerates to its final velocity, $V$, say.

If the rate of mass flow is $\dot{m}$, the force or thrust exerted is:

$$T = \dot{m}(V - U)$$

and the thrust power is:

$$P = TU$$

Hence the work output or useful work done in a time $t$ is:

$$W_0 = Pt = \dot{m}tU(V - U)$$

During this time, the actuator has increased the velocity of a mass ($\dot{m}t$) from $U$ to $V$, so that the work input $W_i$, which has been used to increase the kinetic energy of the fluid is:

$$W_i = \tfrac{1}{2}(\dot{m}t)(V^2 - U^2)$$

Hence the ideal, or Froude, efficiency:

$$\eta_F = \frac{W_o}{W_i} = \frac{U(V-U)}{\tfrac{1}{2}(V^2 - U^2)} = \frac{2}{1 + \left(\dfrac{V}{U}\right)}$$

For the Froude efficiency to be high, $V$ should be only slightly greater than $U$, i.e. the increase of velocity, necessary to produce a forward thrust, should be a minimum. The cause of inefficiency is the jet energy of the wake; the useful thrust $T$ is proportional to the increase of velocity $(V - U)$, whereas the energy imparted to the fluid increases as its square, $\tfrac{1}{2}u_{\text{jet}}^2$, the energy rejected being $\tfrac{1}{2}(V - U)^2$ per unit mass. Thus the efficiency is a maximum when the increase of velocity is a minimum, so that it is fundamentally more efficient to obtain thrust by imparting a relatively small increase in velocity to a large mass flow.

It should be borne in mind that it is the ratio of $V/U$ which is important and that in practice $U$ is the forward speed. Thus the allowable jet speed $V$ (relative to the vehicle) for reasonably efficient propulsion increases with the speed of the vehicle itself, so that although an aircraft jet engine may deliver sufficient thrust to drive a ship, the efficiency would be extremely low due to its relatively small forward speed $U$. For example, an exit nozzle velocity of 600 m/s and a ship speed of 6 m/s (i.e. 11·7 knot) would result in an efficiency of only $2/(1 + 100)$, i.e. less than 2 per cent. Put in other terms, if the variation of engine performance with forward speed be neglected, the engine would be delivering a thrust power which is 40 times greater when driving an aircraft at $40 \times 6$, i.e. 240 m/s, for the same fuel consumption. Thus it is only on fast-moving vehicles that high jet velocities can efficiently be used. For high efficiency, not only should the average kinetic energy of the jet along the direction of motion be reduced to a minimum, but the jet should also be uniform and free from eddying and rotational motions. It is the energy wasted in these sub-sidiary motions which gives rise to a conversion efficiency $\eta_C$. The overall *propulsion efficiency* $\eta$, which relates the power available from the engine or shaft (i.e. brake power $P_B$) to the useful thrust power (or power output of propeller $P_T$), is thus:

$$\eta = \frac{P_T}{P_B} = \eta_F\eta_C$$

A propeller has a high Froude efficiency because it acts on a large mass flow, but its conversion efficiency drops sharply at high sub-sonic speeds. The pure jet has a small Froude efficiency at low speeds, but this continues to rise as the disparity between the jet and forward speeds is reduced. As its conversion efficiency is virtually 100 per cent, the jet

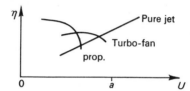

Fig. 7.8(b). *Aircraft propulsive efficiencies*

eventually tends to be more efficient than the propeller. The turbo-fan in which the jet velocity is reduced by using some of the energy to drive the fan from a turbine, tends to be more efficient (and hence more economical) than either, over a limited range of speeds, see Fig. 7.8(b).

EXAMPLE 7.8 (i)

*Examine the mechanism by which an ideal actuator experiences a thrust, and show that half the added velocity in the slipstream occurs ahead of the actuator. Hence deduce a simple expression for the ideal or Froude efficiency.*

Referring to Fig. 7.8(c), suppose that the velocity of the oncoming stream ($U$) is increased by some amount $aU$ on reaching the actuator, and is eventually increased by $bU$ well downstream, where the pressure has again returned to the undisturbed value $p_0$. Continuity requires that the velocity immediately ahead of the actuator must be the same as that immediately behind it. Hence, the increase in total head, which corresponds to the energy input from the actuator must, between these sections, cause a corresponding pressure jump, $\Delta p$ say. This pressure difference acting over the area of the actuator $A$ is the mechanism by which thrust is exerted.

Applying Bernoulli's equation firstly to the region ahead of the disc where the total pressure head is $H$ say, and then to the region behind, where the total head has been increased to $(H+\Delta p)$:

$$H = p_0+\tfrac{1}{2}\rho U^2$$
$$H+\Delta p = p_0+\tfrac{1}{2}\rho U^2(1+b)^2$$

we obtain an expression for $\Delta p$ from energy considerations:

$$\Delta p = \tfrac{1}{2}\rho U^2[(1+b)^2-1] = \rho U^2 b[1+(b/2)]$$

Recalling that the thrust, which may be written as:

$$T = \Delta p . A$$

is, from the momentum law, equal to the mass flow $\rho AU(1+a)$ times its change of velocity $bU$:

$$T = \rho AU^2 b(1+a)$$

we obtain a second expression for $\Delta p$:

$$\Delta p = \rho U^2 b(1+a)$$

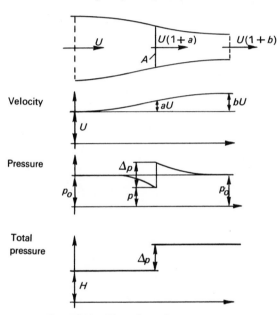

**Fig. 7.8(c).** *Flow through an actuator*

Comparison of the two expressions for the pressure rise shows that:

$$a = \frac{b}{2}$$

Thus one half of the overall increase in velocity ($bU$) is already achieved by the time that the fluid has reached the actuator, i.e. the axial velocity of flow of fluid through the disc is the arithmetic mean of the forward speed of the propeller and the axial velocity of the slipstream.

As no energy is added to the fluid until it has reached the actuator, there is a consequent reduction in pressure ahead of it; this corresponds to the *suction* generated over the forward faces of a screw propeller, from which it derives much of its thrust. The quantity $a$, which may be thought of as a measure of the stream contraction, is referred to as the axial inflow factor; the Froude efficiency:

$$\eta_F = \frac{2}{1+\dfrac{U(1+b)}{U}} = \frac{1}{1+\dfrac{b}{2}}$$

may be most simply expressed in terms of it as $\eta_F = 1/(1+a)$.

### EXAMPLE 7.8 (ii)

*An airscrew 3 m diameter absorbs 1 350 kW brake power when moving with a forward speed of 360 km/h in air of density 1·23 kg/m³. Estimate the ideal efficiency of the airscrew and state the probable efficiency of the actual propeller.*

From work in Example 7.8 (i) on the ideal actuator:

$$\text{Thrust force } T = 2\rho A U^2 a(1+a)$$

and its ideal (Froude) efficiency is $\eta_F = 1/(1+a)$. The thrust power or power expended in flight is $P = TU = \eta_F P_B$ where $P_B$ is the brake power of the engine. Hence $T = \eta_F P_B/U = 2\rho A U^2 a(1+a)$

or

$$\frac{\eta_F P_B}{2\rho A U^3} = a(1+a) = \left(\frac{1}{\eta_F}-1\right)\frac{1}{\eta_F} = \frac{1-\eta_F}{\eta_F^2}$$

i.e.

$$\frac{1-\eta_F}{\eta_F^3} = \frac{P_B}{2\rho A U^3} = \frac{1\cdot35\times10^6\,\text{W}\left[\dfrac{\text{N m}}{\text{W s}}\right]\left[\dfrac{\text{kg m}}{\text{N s}^2}\right]}{2\times1\cdot23\dfrac{\text{kg}}{\text{m}^3}\times\pi2\cdot25\ \text{m}^2\times10^6\ \text{m}^3/\text{s}^3}$$

$$= 0\cdot077\ 6$$

Therefore, the ideal efficiency $\eta_F = 0\cdot94$ or **94** *per cent*.

The actual efficiency would probably be about **80** *per cent* owing to kinetic energy in the rotation of the slipstream, drag of propeller blades and the fact that the thrust force $T$ is not uniformly spread over the disc.

### EXAMPLE 7.8 (iii)

*An airscrew, 3 m diameter absorbs a brake power of 670 kW in level flight at 360 km/h at an altitude of 4·5 km where the relative density of air is $\sigma = 0\cdot629$. Applying the simple momentum theory and assuming that the efficiency of the airscrew is 0·84 of the ideal value, calculate: (i) the axial velocity of the air in the slipstream and (ii) the pressure increase across the airscrew disc.*

*At sea level the density of air $\rho_0$ is 1·23 kg/m³.*

Propulsive power = Efficiency of propeller × Shaft power of engine

$$P = TU = \eta P_B$$

where $\eta = 0\cdot84\eta_F$. Thus, $2\rho A a(1+a)U^3 = 0\cdot84[1/(1+a)]P_B$

i.e. $\quad a(1+a)^2 = \dfrac{0\cdot42P_B}{(\sigma p_0)AU^3} = \dfrac{0\cdot42 \times (670 \times 10^3 \text{ W}) \left[\dfrac{\text{N m}}{\text{W s}}\right]\left[\dfrac{\text{kg m}}{\text{N s}^2}\right]}{\left(0\cdot629 \times 1\cdot23 \dfrac{\text{kg}}{\text{m}^3}\right)\left(\dfrac{\pi}{4} 9 \text{ m}^2\right)\left(10^6 \dfrac{\text{m}^3}{\text{s}^3}\right)}$

$$= 0\cdot0514$$

Hence, $\qquad a = 0\cdot047 \quad \text{and} \quad b = 2a = 0\cdot094$

and (i) the axial velocity of the slipstream is:

$$U(1+b) = 100 \text{ m/s} \times 1\cdot094 = 109\cdot4 \text{ m/s}$$

and (ii) the pressure increase across the airscrew disc is:

$\Delta p = \rho U^2 b(1+a)$

$\quad = \left(0\cdot629 \times 1\cdot23 \dfrac{\text{kg}}{\text{m}^3}\right)\left(10^4 \dfrac{\text{m}^2}{\text{s}^2}\right) \times 0\cdot094(1\cdot047) = \mathbf{761} \text{ N/m}^2$

## 7.9. Thrust coefficient and propulsive efficiency

The ratio of the thrust of a propeller to the dynamic pressure $p_0$ times the disc area $A$ is termed the thrust coefficient $C_T$. Since $p_0 = \frac{1}{2}\rho U^2$, $C_T$ is given by:

$$C_T = \frac{T}{p_0 A} = \frac{T}{\frac{1}{2}\rho A U^2} = 2b(1+a)$$

Hence, since $b = 2a$:

$$C_T = 4a(1+a)$$

Thus, as the axial inflow factor $a$ is increased, the thrust coefficient (and hence the thrust) increases but the efficiency falls, as shown in Fig. 7.9(a) (see Example 7.8(i) $\eta_F = 1/(1+a)$).

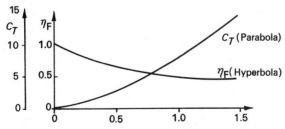

**Fig. 7.9(a).** *Effect of axial inflow factor on efficiency and thrust coefficient*

Example 7.9 (i)

(a) *Show that the ideal efficiency of a propeller is related to its thrust coefficient by:*

$$\eta_F = \frac{2}{1+\sqrt{(1+C_T)}}$$

*Hence, draw conclusions as to how the efficiency of propellers may be increased and mention limitations.*

(b) *A ship's propeller is required to deliver a thrust of 50 kN at 15 knot and its diameter is limited to 2 m. Estimate the ideal efficiency and the theoretical power required to drive the propeller. What would the probable actual power be?*

(a) $$C_T = 4a(1+a)$$

in which $$1+a = \frac{1}{\eta_F}$$

i.e. $$a = \frac{1}{\eta_F} - 1$$

∴ $$C_T = 4\left(\frac{1}{\eta_F} - 1\right)\frac{1}{\eta_F}$$

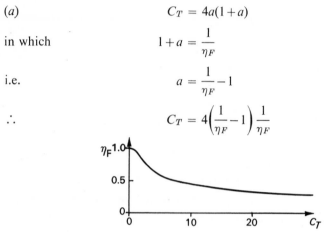

Fig. 7.9(b). *Variation of ideal (Froude) efficiency with thrust coefficient*

Denoting $1/\eta_F$ by $x$, then $4x^2 - 4x - C_T = 0$

i.e. $$x = \frac{4 \pm \sqrt{(16+16C_T)}}{8} = \frac{1+\sqrt{(1+C_T)}}{2}$$

and $$\eta_F = \frac{2}{1+\sqrt{(1+C_T)}}$$

The curve is shown in Fig. 7.9(b) from which it will be seen that a maximum efficiency of 100 per cent could only (theoretically) be achieved if the thrust force were zero.

Since $$C_T = \frac{T}{\frac{1}{2}\rho A U^2}$$

the expression

$$\eta_F = \frac{2}{1+\sqrt{(1+C_T)}}$$

shows that the efficiency of a propeller increases with increase in diameter $D$, increase in density $\rho$ and forward velocity $U$, and with decrease in thrust $T$. However, cavitation sets a limit to the diameter of propellers operating in water, for when the fluid pressure at any point is equal to the vapour pressure corresponding to the temperature of the water, bubbles of vapour alternately form and collapse causing pitting of the blades and also a decrease in efficiency. Similarly, shock waves set a limit on diameter of propellers operating in air. Such waves occur when the tip speeds of propeller blades approach the velocity of sound.

(b) At 15 knot in salt water of relative density 1·025, the dynamic pressure is:

$$p_0 = \tfrac{1}{2}\rho U^2 = \tfrac{1}{2} \times 1\ 025\ \frac{\text{kg}}{\text{m}^3} \times 225\ \text{knot}^2 \left[\frac{0\text{·}514\ \text{m}}{\text{knot s}}\right]^2$$

$$= 30\text{·}5\ \text{kN/m}^2$$

Hence, the thrust coefficient

$$C_T = \frac{T}{\tfrac{1}{2}\rho A U^2} = \frac{50\ \text{kN}}{30\text{·}5\ \text{kN/m}^2 \times \pi \text{m}^2} = 0\text{·}522$$

The corresponding ideal efficiency is thus:

$$\eta_F = \frac{2}{1+\sqrt{(1+C_T)}} = \frac{2}{1+\sqrt{(1\text{·}522)}} = \frac{2}{2\text{·}234} = 0\text{·}895\ \text{or } \mathbf{89\text{·}5}\ \textit{per cent}$$

and, therefore, the theoretical shaft (or brake) power required is:

$$P_B = \frac{TU}{\eta_F} = \frac{50\ \text{kN} \times 15\ \text{knot}}{0\text{·}895} \left[\frac{0\text{·}514\ \text{m}}{\text{knot s}}\right]\left[\frac{\text{kW s}}{\text{kN m}}\right]$$

$$= \mathbf{431}\ \text{kW}$$

In practice, however, the propeller would probably have a conversion efficiency of, say, 65 per cent so that a power of **663 kW** would need to be delivered along the propeller shaft.

## Exercises on Chapter 7

1. A set of turbine nozzle blades deflects steam through an angle of 70° and accelerates it from 100 to 2 000 m/s. The outlet area of nozzles is 5 000 mm² and at this section the specific volume of the steam is 1·25 m³/kg. Determine the components of the force on the nozzle ring in directions parallel and normal to the original direction of flow.

2. A 50 mm diameter jet of water having a velocity of 52·5 m/s is inclined at 20° to the direction of motion of a ring of impulse turbine blading. The blade inlet angle is 35° and the outlet angle is 40°. Friction reduces the water velocity relative to the blade at outlet to 0·85 of the relative velocity at inlet. Water enters the blades without shock. Density of water is 1 Mg/m³.

Draw the velocity diagram and obtain the correct blade velocity. Calculate the power developed at that velocity.

3. A jet of water 80 mm diameter with velocity 20 m/s flows tangentially on to a stationary vane which deflects it through 120°. What is the magnitude and direction, referred to the direction of the jet, of the resultant force on the vane?

If this jet flows on to a series of vanes similarly oriented with regard to it, but moving in the direction of the jet with a velocity of 10 m/s determine:

(a) the force on the system of vanes in the direction of motion,
(b) the work done per second,
(c) the efficiency.

4. A 90° bend is in a water pipe of 250 mm diameter in which the gauge pressure is 700 kN/m². The rate of flow is 250 litre/s. Calculate the magnitude and direction of the dynamic force and the total force necessary to anchor the bend.

5. A nozzle, having a coefficient of velocity of 0·96, operates under a head of 90 m of water and directs a 50 mm diameter jet of water onto a ring of axial-flow impulse blades, which have an inlet angle of 40° measured relative to the direction of blade motion and which turn the water through an angle of 105° during its passage over the blades (i.e. blade angle at outlet = 35°). Because of friction the velocity of the water relative to the blades at outlet is only 0·85 of that at inlet. The blade speed is to be 18 m/s and the water is to flow on to the blades without shock.

Draw the velocity diagram and determine:

(a) the angle which the line of the jet should make with the direction of motion of the blades,
(b) the power developed by the blade ring.

6. A 60° bend reduces a pipe from 150 mm to 75 mm diameter. The pressure in the 150 mm pipe is 350 kN/m² and the flow is 85 litre/s. Find the forces in the bend parallel and normal to the 150 mm pipe.

7. Estimate how many jets are needed for a Pelton wheel which is to develop a brake power of 9·1 MW at 500 r.p.m. The head of water entering the nozzles, which have a coefficient of velocity 0·98, is 264 m. Assume that the diameter of each jet is not to exceed $\frac{1}{9} \times$ wheel diameter and that the bucket velocity is 0·45 × jet velocity. Estimate, also, the diameter of the jets, the diameter of the wheel, and the quantity of water required. Assume that the ratio of the powers delivered to the wheel and the jet is 0·87.

8. The buckets of a Pelton wheel deflect a jet which has a velocity 60 m/s through an angle of 160°. Neglecting friction estimate the speed of the buckets so that the efficiency shall be a maximum and find this efficiency.

Assuming that 15 per cent of the velocity of water relative to buckets is lost in friction, find the efficiency when the bucket speed is 27 m/s. If the brake power is 205 kW find the diameter of the jet.

9. A lifeboat driven by reaction jets discharging astern is found to have a resistance to motion of 3·33 kN when moving at 25 knots. The cross-sectional area of the jets at discharge is 100 cm² and water enters the pumping system from the sides of the vessel. Pipe losses are assumed to be 5 per cent of the jet energy relative to the ship. Find the absolute jet velocity.

If the pumps are 73 per cent efficient, find the brake power of the driving engines. If the latter are 31 per cent efficient, determine the overall efficiency of the propelling system.

(Sea water weighs 10 kN/m³, and 1 knot equals 0·514 m/s.)

10. Assuming the Rankine–Froude momentum theory for the flow of fluid through a propeller, show that the outflow factor is twice the inflow factor ($a$) and that the ideal efficiency is $1/1 + a$.

11. Estimate the thrust force of an airscrew 3 m diameter travelling at a forward speed of 480 km/h when absorbing a shaft power of 900 kW at a height where the relative density of air is 0·6.

Assume the efficiency of the propeller is 85 per cent of the ideal efficiency and the density of the air at sea level is 1·22 kg/m³.

## Chapter 8

# *Behaviour of ideal and viscous fluids*

> *'But first I shall test by experiment before I proceed further, because my intention is to consult experience first and then with reasoning show why experience is bound to operate in such a way. And this is the true rule by which those who analyse the effects of nature must proceed.'* LEONARDO

## 8.1. Hydrodynamics and D'Alembert's paradox

The fundamental ideas on which the previous chapters are based were largely established by the beginning of the eighteenth century. Archimedes had long ago grasped the principle of buoyancy, and Stevinus the concept of fluid pressure and its variation with depth. Newton had given a formal statement of the laws by which motion could be studied and had introduced the subtle reasoning of the calculus to facilitate this. Bernoulli and Euler had applied these sweeping ideas to the motion of fluids, and so laid the foundations of fluid mechanics. Great technological advances, which foreshadowed the industrial revolution, were to be made during the eighteenth century. These advances owed little to fundamental science however; Watt's steam-engine was a triumph of the mechanical arts rather than a wonder of science. Whilst Brindley and Smeaton were building canals, fluid mechanics was becoming more and more abstract in the hands of such mathematicians as D'Alembert and Lagrange. By postulating a perfect fluid imagined capable of exerting pressure but not shear, and which could not be compressed, Lagrange developed what we now term 'classical hydrodynamics'. This is an analytical study of the behaviour of a 'perfect' fluid when flowing in or around such boundaries as can be described in analytical terms. The object is to deduce the equations for the stream-lines and, if this can be done, the kinematics of the flow is completely described. Fig. 8.1(a) shows the flow round a cylinder obtained in this way, together with a typical flow pattern obtained by the observation of smoke filaments introduced into a gas flow.

The spacing between the stream-lines is so chosen that the rate of flow, $Q$, through each stream-tube is the same; with this convention the width of the stream-tube is an inverse measure of the velocity since the area, $A = Q/U$.

Since a perfect fluid cannot, by assumption, exert a shear stress, there can be no friction between the elements. Hence, the constant, $H$, in Bernoulli's equation does not change along each stream-tube, and we may use this fact to deduce the pressure from the velocity. Fig. 8.1(b) shows the corresponding pressure distribution deduced from the stream-line pattern, together with a typical curve for a real fluid.

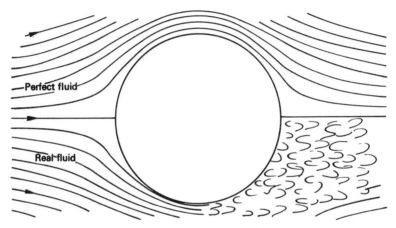

**Fig. 8.1(a).** *Flow round a cylinder*

The undisturbed pressure $p_0$ of the free stream has been chosen as a datum value; pressures in excess of this are plotted radially outwards, and those less than the datum value plotted inwards on a polar basis. The considerable widening of the stream-tube near the nose of the cylinder suggested in Fig. 8.1(a) shows that the 'damming-up' in this region causes a reduction in velocity and a consequent increase of pressure. The crowding together of the stream-lines where the cylinder approaches its maximum thickness causes the velocity of the fluid to increase, and the pressure to fall. The subsequent re-compression of the fluid downstream is, however, hardly realized at all in practice. In a real fluid, separation of the main body of the flow leaves a dead-water region or wake, see Fig. 8.1.(a). Across the wake the average pressure has an almost constant reduced value, see Fig. 8.1(b).

Thus if the lower half cylinder is imagined to be split up into a number of horizontal slices, each will experience a 'suction' on its tail, as indicated at B', and may well, in addition, suffer an over-pressure at its forward end, as at A'.

The sum, or integral of these increased and reduced pressures, when resolved along the direction of motion is referred to as the *form drag* of the body. For a perfect fluid this is obviously zero, since the flow pattern, and hence the pressure distribution, are symmetrical about a vertical line through the centre of the cylinder, see Fig. 8.1(b). Because of this, the drag caused by the pressure at A for example is balanced by an equal and opposite force at B. This conclusion of zero drag with an ideal fluid is not restricted to the particular case of a cylinder—it is typical of most *potential flows*, as they are called.

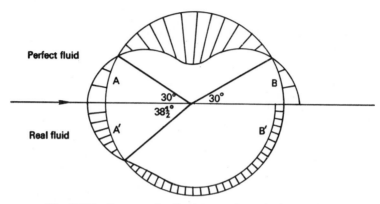

**Fig. 8.1(b).** *Pressure distribution round a cylinder*

The early eighteenth century has been called the 'Age of Reason' and the development of hydrodynamics is an example of the culminating spirit of that age. Like Euclid, the geometers of that time started from a few simple axioms (in this case a frictionless, incompressible fluid was postulated) and by rigorous mathematical argument they followed the consequences of these assumptions wherever they might lead. We may quote D'Alembert himself concerning the seemingly spectacular failure to predict resistance. 'I do not see then, I admit, how one can explain the resistance of fluids by the theory in a satisfactory manner. It seems to me that this theory, dealt with and studied with profound attention, gives, at least in most cases, resistance absolutely zero: a singular paradox which I leave to geometricians to explain.'

This failure can be traced, not of course to the logic of classical hydrodynamics, which is impeccable, but to the fundamental assumptions. The latter are both the source of its power and of its weakness. Recent studies with *real* fluids show that fluid friction due to viscous shear caused by the relative velocity or velocity gradient across adjoining layers exerts a controlling influence over most flow patterns, even though the friction forces themselves may be both localized and small. Consequently, no theory based on the assumption of a frictionless fluid can, in itself, interpret the behaviour of real fluids. When, however, the stream-line pattern is suitably modified to conform with the observed behaviour of real fluids, as in modern aerofoil theory (which finds many applications in fluid mechanics), an extremely powerful analytical tool results.

It may be noted here that owing to viscous friction two-dimensional un-symmetrical flow can be created round a cylinder by spinning it as indicated in Fig. 8.1(c). This causes a dynamic lift since the velocities of the stream-tubes

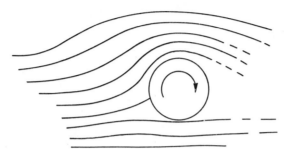

**Fig. 8.1(c).** *Unsymmetrical flow round a circular cylinder*

above are greater than the velocities below and hence the pressures below are greater than the pressures above the cylinder. This generation of lift normal to the undisturbed flow due to spinning is called the Magnus effect. If the fluid were perfect (inviscid) the spinning of the cylinder would not cause dynamic lift since the flow pattern would remain symmetrical as shown on the ideal side of Fig. 8.1(a). A more complicated case of dynamic force caused by spinning is that created on a spinning cricket or tennis ball to produce swerve or a curved flight through the air.

## 8.2. Shear stresses and viscous action

We may demonstrate the fact that a real fluid can experience and transmit friction forces by a very simple experiment. If a circular bowl containing water for example is rotated about its axis, the fluid is set in motion also and, if the bowl is stopped, the rotation of the fluid is eventually arrested. From this we can only conclude that there is some mechanism by which a tangential, i.e. a shear force, can be exerted by the bowl on the fluid and by adjacent layers of fluid on each other. The intensity of these friction forces in any given circumstances is a measure of what we term the *viscosity* of the fluid and this property is possessed to a marked degree by such substances as treacle and 'heavy' oils. The terms 'heavy' and 'light', or 'thick' and 'thin' respectively imply, in this connection, a large and a small viscosity, respectively. The fact that a swirl can be generated, even in gases, by a similar experiment to that described above, suggests that all fluids possess a capacity for transmitting and experiencing shear stresses, although in some cases these forces may be relatively small. No matter how small their magnitude, however, careful experiments, with e.g. miniature Pitot probes, show that real fluids always tend to 'stick'

to a solid boundary. That is, the layer of fluid in immediate contact with
a solid boundary always acquires the velocity of the latter, no matter
what conditions may exist in the main body of the stream. It is neglect of
this property of real fluids which accounts for the apparently anomalous
behaviour of a perfect (i.e. non-viscous) fluid assumed in hydrodynamics.
Viscosity is fundamentally due to cohesion and interaction between fluid
molecules. It causes fluids moving over solid boundaries to adhere to the

Perfect fluid                   Real fluid

**Fig. 8.2(a).** *Flow past a solid boundary*

surface and allows motion only by a sliding action between adjacent
layers. The sliding action induces shear stresses, and continuous power
is required to maintain the flow—the power dissipated by viscous action
ultimately reappears in the form of heat. Fig. 8.2(a) shows the difference
between the transverse velocity distributions of a perfect and a real
fluid as they move past a solid boundary.

   If we mount two concentric cylinders so that one can be rotated relative
to the other, and fill the space between them with, say, water the

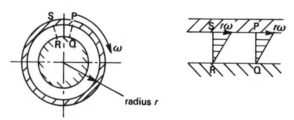

radius *r*

**Fig. 8.2(b).** *Velocity distribution across viscous film*

velocity distribution across the fluid will be somewhat as shown in
Fig. 8.2(b).

   The magnitude of the torque required to drive the outer cylinder at a
particular speed is a measure of how *viscous* the fluid is. In practice,
certain precautions are necessary to ensure that the flow is in fact as
assumed, but when these are taken, one fundamental method of measuring viscosity results. It remains to define the latter precisely, in such terms
that its value is a property of the fluid itself, and is not dependent on the
apparatus used. We may do this by a simple analogy with strength of
materials. If we can treat PQRS of Figs. 8.2(b) and 8.2(c) as a single fluid
element, it will, after a short time, have been deformed into the shape
P′ QRS′, i.e. it will have suffered a shear strain $\varphi$. As in strength of
materials, this is brought about by the shear stresses ($\tau$) acting on the

faces of the element as shown in Fig. 8.2(c). The only difference between the solid and fluid is that, whereas in the former $\tau$ tends to cause a fixed strain $\varphi$, in the latter it tends to cause a fixed *strain-rate d$\varphi$/dt* or $\dot{\varphi}$.

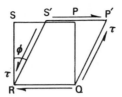

**Fig. 8.2(c).** *Shear stress and strain*

Just as certain solids, which are termed *elastic*, deform an *amount* which is proportional to the stress, i.e.:

$$\tau \propto \varphi$$

so certain fluids, which are termed *Newtonian*, deform at a *rate* which is proportional to the stress, i.e.:

$$\tau \propto \dot{\varphi}$$

The constant of proportionality in the former case is a measure of the rigidity of the solid, i.e. we write

$$\tau = C\varphi$$

and refer to $C$ as the *modulus of rigidity*. In the latter case the constant of proportionality is the coefficient of dynamic *viscosity* and, in fluid mechanics, is denoted by the Greek letter $\eta$ i.e.:

$$\tau = \eta\dot{\varphi}$$

The majority of fluids with which engineers are primarily concerned, such as water, air and oils, are Newtonian in character and it is to these that fluid mechanics is customarily restricted. Other substances, such as greases, pastes and waxes, which do not behave in the above fashion, are studied under the heading of 'rheology'.

Since coefficient of viscosity is defined as a stress (i.e. a force per unit area) divided by an angular rate (i.e. angle per unit time) the dimensions of $\eta$ are:

$$\eta = \frac{\tau}{\dot{\varphi}} = \left[\frac{F/L^2}{1/T}\right] = \left[\frac{FT}{L^2}\right] \quad \text{or} \quad \left[\frac{M}{LT}\right]$$

using square brackets [ ] to denote 'has the dimensions of'.

Hence, typical units are Ns/m² or kg/ms. The unit:

$$\frac{10^{-1}\,\text{kg}}{\text{sm}} \quad \text{or} \quad \frac{10^{-1}\,\text{Ns}}{\text{m}^2}$$

is referred to as a poise, in honour of the French physician Poiseuille who determined, by methodical experiment, the laws of discharge for flow through capillary tubes. This work had been anticipated by the engineer Hagen, but had apparently escaped attention at that time. The Hagen–Poiseuille law for viscous flow through pipes will be referred to in Section 8.5.

## 8.3. Laminar flow and viscosity

For the simple type of laminar flow illustrated by Figs. 8.2(b) and 8.3(a), we may imagine successive layers of fluid to slide over one another, as might a pack of cards when it is sheared. The shear stress between any two layers, $\delta y$ apart, moving with velocities $u$ and $(u + \delta u)$, as shown in Fig. 8.3(a), may be deduced by determining the deformation which takes

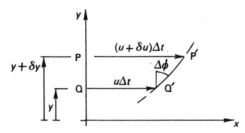

Fig. 8.3(a). *Velocity gradient across laminar flow*

place in a time $\Delta t$ say. The points P and Q in adjacent layers will, during this time, move to P′ and Q′ through distances $(u + \delta u)\, \Delta t$ and $u\Delta t$ respectively. Hence the deformation during this time is:

$$\Delta \varphi = \frac{(u + \delta u)\, \Delta t - u\Delta t}{\delta y}$$

and the *rate* of deformation for such *laminar flows* is:

$$\dot{\varphi} = \frac{\Delta \varphi}{\Delta t} = \frac{\delta u}{\delta y}$$

This expression becomes more precise as $\delta y$ tends to zero; hence, the shear stress between any two layers in a laminar flow may be written:

$$\tau = \eta\dot{\varphi} = \eta\frac{du}{dy}$$

This is the Newtonian definition of coefficient of dynamic viscosity, namely (shear stress/velocity gradient) and from which it will be seen that large viscous stresses may arise in a laminar flow wherever the transverse velocity gradient is large, and the magnitude of the stress is proportional both to this and to the viscosity of the fluid. These stresses are of fundamental importance in determining the behaviour of film lubricated bearings and dash-pots, besides having a major influence on fluid resistance generally. The velocity gradient and, therefore, the shear stress often has its largest value at the boundary to which the fluid sticks and is a deciding factor in the determination of resistance to flow.

EXAMPLE 8.3 (i)

*Calculate the viscous shear stress on a flat plate when air of viscosity* 17·83 *mg/sm flows parallel to the plate with a velocity distribution given by the numerical formula* $u = 40\,000(y - 164y^2)$ *in which u is in m/s and y is the distance from the plate in metres.*

The corresponding physical formula giving velocity distribution is:

$$\frac{U}{m/s} = 40\,000\,\frac{Y}{m}\left(1 - 164\,\frac{Y}{m}\right)$$

where                      $U = u$ m/s   and   $Y = y$ m

Thus                      $U = \frac{40\,000}{s}\,Y\left(1 - 164\,\frac{Y}{m}\right)$

$$\frac{dU}{dY} = \frac{40\,000}{s}\left(1 - 328\,\frac{Y}{m}\right)$$

Hence, on the plate where $Y = 0$,   $\dfrac{dU}{dY} = \dfrac{40\,000}{s}$

Therefore the shear stress,

$$\tau = \eta\,\frac{dU}{dY} = \left(17\cdot83 \times 10^{-6}\,\frac{kg}{sm}\right)\left(\frac{4 \times 10^4}{s}\right)$$

$$= 0\cdot7132\,\frac{kg}{m\,s^2}\left[\frac{N\,s^2}{kg\,m}\right] = \mathbf{0\cdot713\,2\ N/m^2}.$$

*Note.* One need not necessarily use the physical formula for velocity provided one realizes that when $y = 0$, $du/dy = 40\,000$/s and is not just 40 000. However, if one forgot the 's', the 'unity bracket' [Ns²/kg m] needed later on to interconnect force (N) with mass (kg) would reveal the omission and induce a student to reconsider his working.

EXAMPLE 8.3 (ii)

*A cylinder of* 75 mm *diameter is mounted concentrically in a drum of* 76·5 mm *internal diameter. Oil fills the space between them to a depth of* 200 mm. *The torque required to rotate the cylinder in the drum is* 4 Nm *when the speed of rotation is* 480 rev/min.

*Assuming that 'end effects' are negligible, calculate the coefficient of viscosity of the oil, expressing your answer in S.I. units.*

Viscous shear stress $\tau = \eta\,\dfrac{du}{dy}$

or                      $\dfrac{F}{A} = \eta\,\dfrac{u}{y}$     referring to Fig. 8.3(b).

Torque $T = FR = \eta A\,\dfrac{uR}{y} = \eta\,\dfrac{A}{y}\,\omega R^2$

Hence    $\eta = \dfrac{Ty}{AR^2\omega}$

$$= \frac{4\,000\text{ N mm} \times 0.75\text{ mm}}{(\pi \times 75\text{ mm} \times 200\text{ mm})(37.5\text{ mm})^2} \times \frac{\text{s}}{8\text{ rev}}\left[\frac{\text{rev}}{2\pi}\right]\left[\frac{10^6\text{ mm}^2}{\text{m}^2}\right]$$

$= 0.893\text{ Ns/m}^2$    or    **$0.893$ kg/sm**

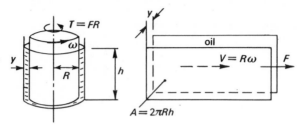

**Fig. 8.3(b).**  *Viscous shear*

EXAMPLE 8.3 (iii)

(a) *Show that the friction torque of a film lubricated journal bearing, when running on a light load, so that the shaft runs concentrically (Fig. 8.3(c)) is given by:*

$$M = \frac{\pi\eta\omega D^3 L}{2C}$$

*if the effect of end leakage and curvature of the surface may be neglected. In the above expression $\eta$ is the viscosity, $\omega$ the speed, $D$ the diameter, $L$ the length, and $C$ the diametral clearance.*

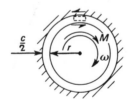

**Fig. 8.3(c).**  *Film lubricated bearing on no-load*

(b) *In order to determine the clearance in such a bearing it is set running under the above conditions, and when rotating at 1 200 rev/min the friction torque is found to be 2.94 Nm. Determine C if $D = 63.5$ mm and $L = 107$ mm, $\eta$ being 0.03 kg/ms.*

(a) By definition:

$$\tau = \eta\frac{\delta u}{\delta y}$$

in which

$$\delta u = r\omega$$

and

$$\delta y = \frac{C}{2}$$

Since the shear stress $\tau$ exerts a force $F = \tau \pi DL$ on the wetted area at a radius $r = D/2$:

$$M = Fr = \frac{\pi D^2 L}{2}\tau$$

i.e.

$$M = \frac{\pi D^2 L}{2}\frac{\eta D\omega}{2(C/2)}$$

$\therefore$

$$M = \frac{\pi\eta\omega D^3 L}{2C}$$

(b)

$$C = \frac{\pi\eta\omega D^3 L}{2M} = \frac{\pi\times 0\cdot 03\,\dfrac{\text{kg}}{\text{ms}}\times 40\,\dfrac{\pi}{\text{s}}\times\left(\dfrac{6\cdot 35}{10^2}\,\text{m}\right)^3\times 107\,\text{mm}}{2\times 2\,940\,\text{N mm}}$$

$$= \frac{55}{10^6}\,\text{m}\quad\text{or}\quad 0\cdot 055\,\text{mm}$$

EXAMPLE 8.3 (iv)

(a) *In connection with theoretical two-dimensional flow in a journal bearing for film lubrication, derive the equation for coefficient of friction known as the Petroff equation.*

(b) *A test of such a bearing at a speed of 100 rev/min using an oil of viscosity 30 cP and p = 12 kPa resulted in a value of $\mu$ = 0·01 being recorded. What is the relevant ratio of diameter to clearance?*

(c) *What is 'oiliness' as distinct from viscosity?*

(a) Shear stress ($\tau = \eta(du/dy)$) on moving cylindrical surface is:

$$\frac{F}{A} = \eta\frac{u}{C/2}$$

referring to Fig. 8.3(d).

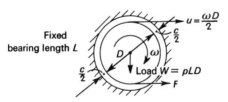

Fig. 8.3(d). *Journal bearing*

Thus the shear force on cylindrical surface is:

$$F = (\pi DL)\frac{\eta D\omega}{C} = \pi\eta\frac{D^2}{C}\omega L$$

$$\text{Coefficient of friction } \mu = \frac{\text{shear force}}{\text{normal load}} = \frac{F}{W}$$

i.e.
$$\mu = \frac{\pi\eta D^2\omega L}{pLDC} = \frac{\pi\eta}{p}\left(\frac{D}{C}\right)\omega$$

which is the Petroff equation.

(*b*) The viscosity

$$\eta = 30 \text{ cP} = 30\left(10^{-3}\frac{\text{kg}}{\text{ms}}\right) = 0.03\frac{\text{kg}}{\text{ms}}$$

Hence,

$$\frac{D}{C} = \frac{\mu p}{\pi\eta\omega} = \frac{0.01}{\pi}\times\frac{12 \text{ kN}}{\text{m}^2}\times\frac{\text{ms}}{0.03 \text{ kg}}\times\frac{60 \text{ s}}{100 \text{ rev}}\left[\frac{\text{rev}}{2\pi}\right]\left[\frac{\text{kg m}}{\text{Ns}^2}\right] = 121.5$$

(*c*) 'Oiliness' is that property which causes a difference in frictional resistance when different lubricants of the same coefficients of viscosity are used in identical tests. It is most relevant when the conditions for film lubrication are not, or cannot, be satisfied.

## 8.4. Kinematic viscosity

As both dynamic viscosity ($\eta$) and density ($\rho$) are fluid properties, the ratio ($\eta/\rho$) is a fluid property also. It is denoted by the Greek symbol $v$ (Nu) and, as its dimensions can be expressed completely in terms of those of length and time, it is referred to as the *kinematic* viscosity:

$$v = \frac{\eta}{\rho} = \left[\frac{M/LT}{M/L^3}\right] = \left[\frac{L^2}{T}\right]$$

Typical units for $v$ are thus m²/s or cm²/s, the latter being referred to as a stoke (St $= 10^{-4}$ m²/s) to perpetuate the name of the English mathematical physicist Sir George Stokes who among other things solved the difficult analytical problem of the resistance of a sphere, moving slowly through a viscous fluid. He found this to be:

$$F = 3\pi\eta ud$$

where $u$ is the velocity of the sphere, of diameter $d$, moving through fluid of viscosity $\eta$. This suggests that by timing the fall of a ball through a fluid we may determine the viscosity of the latter. The method is however of very limited application because, for reasons we shall examine later, the above expression is appropriate only to slow 'creeping' flows.

EXAMPLE 8.4 (i)

*Determine the rate at which air bubbles may be expected to rise through oil having a kinematic viscosity of 35 centi-stokes (cSt) if their diameter is (a) 0.1 mm (b) 1 mm.*

If the bubbles are sufficiently small, surface tension will maintain them in a virtually spherical form. They will experience an upthrust or buoyancy force equal to the weight of fluid displaced, i.e.:

$$B = w \times \tfrac{4}{3}\pi r^3$$

where $w = \rho g$ is the weight per unit volume of the oil. The motion will be steady when this is equal and opposite to their resistance, which we shall assume is given by Stokes' law:

$$F = 3\pi\eta u d$$

$$\therefore \qquad \rho g \times \tfrac{4}{3}\pi(\tfrac{d}{2})^3 = 3\pi\eta u d$$

i.e.

$$u = \frac{1}{18}\frac{\rho g d^2}{\eta} = \frac{1}{18}\frac{g d^2}{v}$$

(a)

$$u = \frac{9\;807\;\dfrac{mm \times 0.1^2\;mm^2}{s^2}}{18 \times 35\;cSt}\left[\frac{cSt \times 10^6}{m^2/s}\right]\left[\frac{m^2}{10^6\;mm^2}\right]$$

$$= \frac{9\;807 \times 0.01}{18 \times 35}\;mm/s = \mathbf{0.156}\;\mathbf{mm/s}$$

(b) Since the terminal velocity increases as the square of the diameter, the terminal velocity of a 1 mm air bubble is given by:

$$u_b = \left(\frac{d_b}{d_a}\right)^2 u_a = \left(\frac{1\;mm}{0.1\;mm}\right)^2 \times 0.156\;mm/s$$

$$= \mathbf{15.6}\;\mathbf{mm/s}$$

We may draw two conclusions from the above example. Firstly, the practical one, that whilst large air bubbles clear themselves rapidly from a vented hydraulic system, the smaller bubbles are very slow moving. Secondly the terminal velocity depends not so much on the dynamic viscosity $\eta$, but rather on the kinematic viscosity $v = (\eta/\rho)$. This is because the problem is essentially one of a balance between viscous forces (proportional to $\eta$) and buoyancy forces (proportional to $\rho$). Similarly in other problems it is the ratio of the viscous forces to the inertia forces that is important, and this again depends on $(\eta/\rho)$, i.e. on $v$. Thus although, in terms of $\eta$, water is much more viscous than air, typical values being:

$$\eta_w = 1.0\;cP = 10^{-3}\;kg/sm \text{ at } 20°C, \text{ and standard atmospheric}$$
pressure

$$\eta_a = 0.0181\;cP \text{ at } 20°\,C, \text{ and standard atmospheric pressure}$$

the very much lower density of the latter makes its kinematic viscosity about 15 times greater than the corresponding value for water, e.g.

$$v_w = 1.01\;cSt = 1.01 \times 10^{-6}\;m^2/s \text{ at } 20°\,C, \text{ and standard}$$
atmospheric pressure

$$v_a = 15.0\;cSt \text{ at } 20°\,C, \text{ and standard atmospheric pressure}$$

Many of the viscometers used for determining the viscosity of liquids consist essentially of a reservoir from which the fluid is drawn by gravity through some form of restriction or capillary tube. The time taken for a prescribed quantity of fluid to pass through the latter gives a measure of the viscosity. The Redwood viscometer is a robust instrument of this type. It is frequently used by engineers for checking the viscosity of oils,

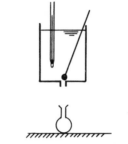

**Fig. 8.4(a).** *Redwood viscometer*

which may simply be expressed in a scale of 'Redwood Seconds', see Fig. 8.4(a). A laboratory instrument that is used for more accurate fundamental evaluations of the properties of a given fluid is shown in Fig. 8.4(b).

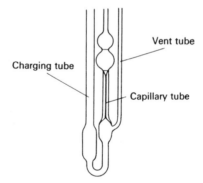

Charging tube

Vent tube

Capillary tube

**Fig. 8.4(b).** *Ubbelohde kinematic viscometer*

As, in each case, the flow is established by gravity forces working on the fluid against the viscous resistance of the latter, such instruments determine the kinematic viscosity $\nu$, rather than $\eta$. To fix ideas, if two oils of the same dynamic viscosity $\eta$, but different densities, were tested in a Redwood viscometer, the denser fluid would be drawn through faster, so that its viscosity in 'Redwood Seconds' would be less, and this would correspond to its lower value of $\eta/\rho$.

The viscosity of liquids, which may in part be traced to the cohesive forces between the molecules, consequently *decreases* as the magnitude of the latter drops with increasing temperature. Fig. 8.4(c) shows the variation of viscosity with temperature for some representative liquids.

The viscosity of gases can be explained in terms of momentum transfer on a molecular scale. Since the latter increases as the temperature rises,

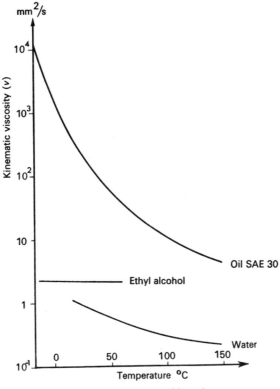

**Fig. 8.4(c).** *Viscosity of liquids*

the viscosity of gases *increases* with temperature, see Fig. 8.4(d). Pressure tends to increase the dynamic viscosity ($\eta$ *not* $\nu$) of both liquids and gases but its effect on liquids can often be ignored.

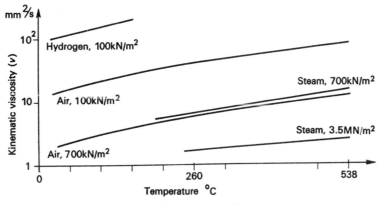

**Fig. 8.4(d).** *Viscosity of gases*

## 8.5. Hagen–Poiseuille law

An alternative fundamental method for determining the viscosity of a
fluid is to determine the rate of flow through a capillary tube that can be
maintained by a known pressure difference, acting against the viscous
resistance exerted by the pipe walls. Provided that the flow is sufficiently
slow, the flow is laminar, i.e. the fluid may be considered to consist of a
large number of hollow cylinders stacked one inside the other. As each

Fig. 8.5(a). *Laminar flow through a pipe*

advances at a different speed, sliding takes place between them as it does
between the sections of a telescope when it is extended. The difference of
speed $\delta u$ between two adjacent layers, $\delta r$ apart, causes a shear stress be-
teeen them which is proportional to the viscosity of the fluid and to the
transverse velocity gradient $\delta u/\delta r$.

As the velocity at the walls is zero, i.e. the fluid 'sticks' at the boundary,
the velocity distribution must be somewhat as shown in Fig. 8.5(a).

If we consider the equilibrium of a central plug of fluid, of radius $r$
and of length $l$, we shall conclude that it is the difference of pressure
$(p_1 - p_2)$ across its ends which supplies the driving force to overcome the
friction acting on its outer surface, see Fig. 8.5(b).

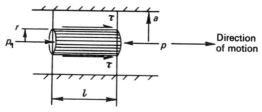

Fig. 8.5(b). *Driving force on a central plug of fluid*

As $(p_1 - p_2)$ acts on an area $\pi r^2$, and the shear stress (assumed positive
till proved otherwise) acts on $\pi r l$, the equilibrium equation for the plug is:

$$(p_1 - p_2)\pi r^2 + \tau 2\pi r l = 0$$

i.e.

$$\tau = -\frac{r}{2l}(p_1 - p_2)$$

As shown in Section 8.3 the shear stress $\tau$ equals $\eta\, du/dr$, so that this
equation, when integrated with the condition of no slip at the boundaries

is:

$$\eta \int_u^0 du = -\frac{(p_1 - p_2)}{2l} \int_r^a r dr$$

$$-\eta u = -\frac{(p_1 - p_2)(a^2 - r^2)}{4l}$$

i.e.

$$u = \frac{(p_1 - p_2)(a^2 - r^2)}{4\eta l}$$

This is the equation of a parabola and indicates that an imaginary stack of fluid annuli will, after a time $t$, have been drawn out into paraboloids at its ends as suggested by Fig. 8.5(c). To approach nearer to reality, the number of annuli must be imagined to increase without limit.

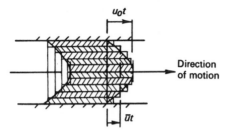

Fig. 8.5(c). *Drawing-out of a plug in laminar flow*

Since the volume of a paraboloid is one-half of the base area times its height, a volume of fluid equal to $\frac{1}{2}\pi a^2 u_{max} t$ will have passed through each cross-section in a time $t$. The maximum velocity occurs at the centre of the pipe, where $r = 0$ so that:

$$u_{max} = \frac{(p_1 - p_2)a^2}{4\eta l} = u_0$$

Hence

$$u = u_0\left(1 - \frac{r^2}{a^2}\right)$$

The corresponding equation for the mean velocity is thus:

$$\bar{u} = \frac{(p_1 - p_2)a^2}{8\eta l} = \frac{u_0}{2}$$

so that the volumetric flow rate is:

$$Q = \pi a^2 \bar{u} = \frac{\pi a^4 (p_1 - p_2)}{8\eta l} = \frac{\pi a^2 u_0}{2}$$

This indicates that the pressure drop per unit length of pipe is proportional to the discharge and experiments confirm this, so long as the flow remains laminar, as assumed. The agreement with experiment is in fact so good that most absolute viscosity measurements are made in this way. That is, by observing the rate of discharge $Q$ through capillary tubes of known bore $D = 2a$, for a known pressure drop per unit length $(p_1 - p_2)/l$, the value of $\eta$ may be deduced from the above so-called Hagen–Poiseuille law.

EXAMPLE 8.5 (i)

*The velocity along the centre line of a 150 mm diameter pipe conveying oil under laminar flow conditions is 3 m/s. The viscosity of the oil is 1·2 P and its relative density is 0·9.*

   *Assuming that the velocity distribution across the pipe is parabolic, obtain:*
   *(a) the rate of volumetric flow,*
   *(b) the shear stress in the oil at the pipe wall,*
   *(c) Also verify that the flow is laminar.*

Referring to Fig. 8.5(a), and using the Hagen–Poiseuille law, we have:

(a)

$$Q = \frac{\pi}{2} u_0 a^2 = \frac{\pi}{2} \times \frac{3 \text{ m}}{\text{s}} \left(\frac{7 \cdot 5 \text{ m}}{10^2}\right)^2 = \frac{265}{10^4} \text{ m}^3/\text{s}$$

$$= 26 \cdot 5 \text{ dm}^3/\text{s} \quad \text{or} \quad \textbf{26·5 litre/s}$$

(b) Also,

$$u = u_0 \left(1 - \frac{r^2}{a^2}\right)$$

and

$$\tau = \eta \frac{du}{dr} = -\eta u_0 \frac{2r}{a^2}$$

Hence at the wall where $r = a$, the shear stress

$$\tau = -\frac{2\eta u_0}{a} = -2 \left(\frac{10^2}{7 \cdot 5 \text{ m}}\right) \left(\frac{1 \cdot 2 \text{ kg}}{10 \text{ ms}}\right) \left(\frac{3 \text{ m}}{\text{s}}\right) \left[\frac{\text{N s}^2}{\text{kg m}}\right]$$

$$= 9 \cdot 6 \text{ N/m}^2 \quad \text{or} \quad \textbf{9·6 Pa}$$

(c) Since the mean velocity $\bar{u} = u_0/2$, the Reynolds number is:

$$Re = \frac{\bar{u}d}{\eta} \rho = \frac{1 \cdot 5 \text{ m}}{\text{s}} \times 0 \cdot 15 \text{ m} \left(\frac{900 \text{ kg}}{\text{m}^3}\right) \left(\frac{\text{ms}}{0 \cdot 12 \text{ kg}}\right)$$

$$= 1688$$

which is less than 2 300. Hence the flow is laminar (see Section 11.3).

EXAMPLE 8.5 (ii)

*Calculate the power required to pump oil of relative density 0·9 at 1·5 m/s through 100 tubes in parallel, each 3 m long and 12·5 mm internal diameter. Ignore inlet losses but allow for exit losses. Dynamic viscosity of oil is 0·1 Ns/m².*

$$Q = \frac{\pi a^4 (p_1 - p_2)}{8 \eta l} = \pi a^2 \bar{u}$$

Hence, relating to one tube:

$$(p_1 - p_2) = \frac{8\eta l \bar{u}}{a^2} = 8 \left(0 \cdot 1 \frac{Ns}{m^2}\right) (3m) \left(1 \cdot 5 \frac{m}{s}\right) \left(\frac{2 \times 10^3}{12 \cdot 5 \, m}\right)^2$$

$$= 92 \cdot 2 \text{ kN/m}^2$$

Friction head in each tube

$$h_f = \frac{p_1 - p_2}{w_{\text{oil}}} \quad (\text{where } w_{\text{oil}} = \rho_o g = s \rho_w g)$$

$$= 92 \cdot 2 \frac{kN}{m^2} \times \frac{1}{0 \cdot 9 \times 9 \cdot 807} \frac{m^3}{kN}$$

$$= 10 \cdot 45 \text{ m}$$

Kinetic head of oil leaving each tube is, very approximately

$$= \frac{\bar{u}^2}{2g} = \frac{2 \cdot 25 m}{19 \cdot 614} \simeq 0 \cdot 12 \text{ m} \quad (\text{see Exercise 13, however})$$

Total head loss, $h_L$, per tube $\simeq 10 \cdot 45 \text{ m} + 0 \cdot 12 \text{ m} \simeq 10 \cdot 57 \text{ m}$

Power required is $P = w_0 Q h_L$ where $Q$ is the rate of volumetric flow pumped through the tubes,

i.e. $\quad P = 0 \cdot 9 \times 9 \cdot 807 \frac{kN}{m^3} \left(100 \frac{\pi}{4} \times \frac{1 \cdot 563}{10^4} m^2 \times 1 \cdot 5 \frac{m}{s}\right) 10 \cdot 57 \text{ m} = \mathbf{1 \cdot 72 \text{ kW}}$

## 8.6. Laminar uni-directional flow between stationary parallel plates

The discharge between two fixed parallel plates of infinite width (i.e. no side flow) may be deduced by methods similar to those used in the previous section, provided that the flow is steady and laminar. We assume

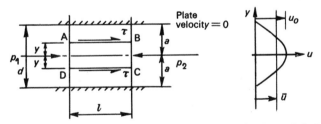

**Fig. 8.6(a).** *Laminar uni-directional flow between fixed parallel plates*

that the fluid is incompressible, of constant viscosity and that it does not slip on the plates. This analysis is of importance in connection with the performance of dash-pots and in leakage problems generally. Fig. 8.6(a) represents a section through the flow along the direction of motion and in this case the central plug ABCD is rectangular in shape, and of breadth $b$, say. For steady flow the resultant force on it is zero, i.e.

$$(p_1 - p_2)2yb + \tau 2lb = 0$$

Substituting $\eta \, du/dy$ for $\tau$ and integrating from a layer $y$, where the velocity is $u$, to the surface of the plate where the velocity is zero we obtain:

$$\tau = \eta \frac{du}{dy} = -\frac{(p_1 - p_2)y}{l}$$

and

$$\int_u^0 du = -\frac{(p_1 - p_2)}{\eta l} \int_y^a y \, dy$$

$\therefore$

$$u = \frac{(p_1 - p_2)(a^2 - y^2)}{2\eta l}$$

Hence, the velocity distribution is again parabolic and it has a maximum value on the centre plane of fluid, where $y = 0$, i.e.

$$u_0 = \frac{(p_1 - p_2)a^2}{2\eta l}$$

which is twice the corresponding value for a circular pipe of radius $a$. Since the area under a parabola is 2/3 that of the enclosing rectangle, the mean velocity is:

$$\bar{u} = \frac{(p_1 - p_2)a^2}{3\eta l}$$

The corresponding flow rate is thus, for a width $b$:

$$Q = \bar{u} b 2a$$
$$= \frac{2(p_1 - p_2)a^3 b}{3\eta l}$$

Noting that the gap $d = 2a$, i.e.

$$Q = \frac{(p_1 - p_2)d^3 b}{12\eta l} = -\frac{bd^3}{12\eta} \left(\frac{dp}{dx}\right)$$

where

$$\left(\frac{p_1 - p_2}{l}\right) \simeq -\left(\frac{dp}{dx}\right) = \text{downward slop of the hydraulic gradient.}$$

The intensity of the shear stress is:

$$\tau = \eta \frac{du}{dy} = \eta \left(-\frac{y(p_1 - p_2)}{\eta l}\right)$$
$$= -\frac{y}{l}(p_1 - p_2)$$

Hence, on the plates the intensity of the shear stress is:

$$\tau = -\frac{a}{l}(p_1 - p_2) = -\frac{3\bar{u}\eta}{a} = -\frac{6\eta\bar{u}}{d}$$

## 8.7. Laminar uni-directional flow between parallel plates having relative motion

*Making the assumptions that:*

(i) *there is no side flow or end leakage then* $p = f(x, y)$ *in Fig. 8.7(a) for plates of infinite width,*

hence $$\delta p(x, y) = \frac{\partial p}{\partial x} \delta x + \frac{\partial p}{\partial y} \delta y$$

(ii) *the flow is laminar and uni-directional then* $p(x, y) = p(x)$ *so that*

$$\frac{\partial p}{\partial y} = 0,$$

*i.e. there is no lateral pressure to induce transverse flow.*

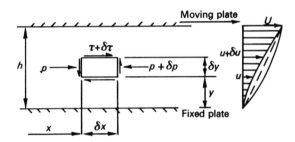

**Fig. 8.7(a).** *Laminar uni-directional flow with a moving surface of infinite width*

(iii) *the flow is fully developed in the sense that* $u(x, y) = u(y)$ *so that*

$$\frac{\partial u}{\partial x} = 0$$

(iv) *the fluid is Newtonian, then, if* $u = u(y)$

$$\tau = \eta \frac{du}{dy}$$

(v) *the fluid is incompressible and of constant viscosity, so that* $\rho$ *and* $\eta$ *remain constant.*

(vi) *there is no slip at the boundaries, then* $u = 0$ *when* $y = 0$ *and when* $y = h$.

In steady laminar flow there are no inertia forces—the only forces acting on an element being pressure and viscous shear as shown in Fig. 8.7(a).

Hence, equating horizontal forces:

$$\delta p . \delta y = \delta \tau . \delta x$$

and since $\tau = \eta(du/dy)$ the constant pressure gradient in the direction of flow is:

$$\frac{dp}{dx} = \frac{d\tau}{dy} = \eta \frac{d^2 u}{dy^2}$$

from which, since $(dp/dx)$ is independent of $y$, we deduce that

$$\frac{du}{dy} = \frac{1}{\eta}\left(\frac{dp}{dx}\right)y + C_1$$

and

$$u = \frac{1}{\eta}\left(\frac{dp}{dx}\right)\frac{y^2}{2} + C_1 y + C_2.$$

This general equation shows that the velocity profile across a thin film depends on the pressure gradient (see Fig. 8.7(b)).

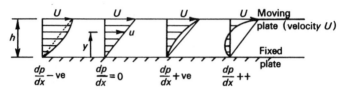

$$\frac{dp}{dx} - \text{ve} \qquad \frac{dp}{dx} = 0 \qquad \frac{dp}{dx} + \text{ve} \qquad \frac{dp}{dx} + +$$

**Fig. 8.7(b).** *Velocity distribution according to pressure gradient*

The boundary limits are (i) $y = 0$, $u = 0$, hence $C_2 = 0$, and (ii) $y = h$, $u = U$, hence

$$C_1 = \frac{U}{h} - \frac{1}{\eta}\left(\frac{dp}{dx}\right)\frac{h}{2}$$

Thus, the velocity at any height $y$ above the base is:

$$u = \frac{U}{h}y - \frac{1}{2}\frac{y}{\eta}\left(\frac{dp}{dx}\right)(h-y) = U\frac{y}{h} - \frac{h^2}{2\eta}\left(\frac{y}{h} - \frac{y^2}{h^2}\right)\left(\frac{dp}{dx}\right)$$

and the rate of volumetric flow per unit width is:

$$\int_0^h u\,dy = \frac{U}{h}\frac{h^2}{2} - \frac{1}{2\eta}\left(\frac{dp}{dx}\right)\left(\frac{h^3}{2} - \frac{h^3}{3}\right) = \frac{Uh}{2} - \frac{h^3}{12\eta}\left(\frac{dp}{dx}\right)$$

and the mean velocity is

$$\bar{u} = \frac{U}{2} - \frac{1}{12}\frac{h^2}{\eta}\left(\frac{dp}{dx}\right)$$

The diagrams shown in Fig. 8.7(b) show how the velocity distribution changes according to the pressure gradient in the direction of motion.

EXAMPLE 8.7 (i)

*A plunger, 200 mm diameter and 200 mm long, slides in a vertical cylinder with radial clearance 0·25 mm and carries a vertical load of 500 kN. If the cylinder is filled with oil of $\eta = 0\cdot115$ kg/ms, estimate the pump power necessary to maintain the load assuming that the plunger remains concentric in the cylinder. Also, estimate the approximate rate at which the plunger will sink when the supply valve is shut.*

Referring to Fig. 8.7(c), the mean velocity of fluid flowing through the clearance space is:

$$\bar{u} = \frac{U}{2} - \frac{1}{12}\frac{h^2}{\eta}\left(\frac{dp}{dx}\right)$$

If the plunger is stationary $U = 0$ and

$$-\frac{dp}{dx} = \frac{p}{l} = \frac{4W}{\pi d^2 l}$$

hence, the mean velocity through the clearance area is:

$$\bar{u} = \frac{1}{12}\frac{h^2}{\eta}\frac{4W}{\pi d^2 l}$$

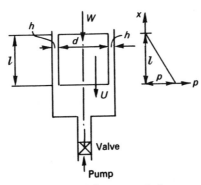

**Fig. 8.7(c).** *Oil-supported plunger*

and the rate of volumetric flow of fluid flowing through the clearance space in order to maintain the plunger stationary is:

$$Q = \pi dh\bar{u} = \frac{Wh^3}{3\eta dl}$$

The power required is:

$$P = Qp = \frac{4W^2}{3\pi\eta l}\left(\frac{h}{d}\right)^3 = \frac{4 \times 25 \times 10^{10} \text{ N}^2}{3\pi \times 0{\cdot}115 \dfrac{\text{kg}}{\text{sm}} \times 0{\cdot}2 \text{ m}}\left(\frac{0{\cdot}25}{200}\right)^3\left[\frac{\text{kg m}}{\text{N s}^2}\right]$$

$$= \frac{10^{12}}{\pi \times 0{\cdot}069}\left(\frac{1}{8}\right)^3 \frac{1}{10^6}\frac{\text{Nm}}{\text{s}} = \textbf{9 kW}$$

If the valve is closed, and assuming the same mean velocity of oil in the clearance space as when the plunger was maintained stationary, the velocity of the sinking plunger will be:

$$\frac{Q}{(\pi/4)d^2} = \frac{4}{3\pi}\frac{W}{\eta l}\left(\frac{h}{d}\right)^3 = \frac{4 \times (5 \times 10^5 \text{ N})}{3\pi \times 0{\cdot}115 \dfrac{\text{kg}}{\text{ms}} \times 0{\cdot}2 \text{ m}}\left(\frac{1}{8}\right)^3 \frac{1}{10^6}\left[\frac{\text{kg m}}{\text{N s}^2}\right]$$

$$= \frac{2}{111}\text{ m/s or } 18 \text{ mm/s}$$

## 8.8. Boundary layer and wake

The drag on a body is comprised of a difference in pressure on the upstream and downstream surfaces called *pressure drag* or form drag, and upon the viscous shear at the fluid boundaries—called *skin friction*.

As the viscous shear stresses in laminar flow are proportional to the transverse velocity gradient, their effect will be most marked where the latter ($du/dy$) is large. Experiments show that in most cases of flow past a slender body or flat plate, although the layers adjacent to the boundary are brought to rest and those close to it are slowed down considerably, this effect decays rapidly across the flow. Hence the flow pattern may be split up into two regions as shown in Fig. 8.8(a), namely, a thin *boundary*

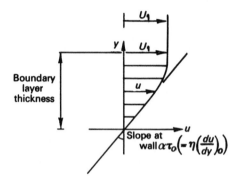

Fig. 8.8(a). *Boundary layer*

*layer* in which friction is important, and a region beyond this in which the total head in Bernoulli's equation remains constant. The concept of the boundary layer is due to Prandtl who has contributed so much to modern fluid mechanics. He introduced the idea in 1904 and it is now fundamental to aerodynamics in particular and is also invaluable in fluid mechanics generally. The concept of the boundary layer provides a link between ideal and real flow. The fluid layer which has its velocity affected by boundary shear is called the *boundary layer*.

Prandtl suggested that *outside* a thin layer and the wake into which it is shed, a real fluid will behave as would a perfect (i.e. frictionless) fluid. If the boundary layer thickens owing to the continual action of shear stress slowing down fluid particles, the velocity gradient $du/dy$ and hence the local shear stress $\tau$ (causing skin friction drag) decreases.

In practice the boundary layer does not generally remain laminar but becomes unstable as it thickens. The layers of fluid first acquire a sinuous motion and then finally form a confused mass of small eddies causing dissipation of energy into heat and cross-currents causing transfer of fluid and momentum. The boundary layer is then said to have become *turbulent* and the region in which the breakdown of the laminar flow occurs is referred to as the *transition*. Although transition to turbulence is accompanied by a considerable thickening of the boundary layer, as shown in Fig. 8.8(b), it may still remain, on an aircraft wing for example, only a fraction of an inch in thickness. Transition to turbulence is accompanied by a corresponding increase in skin friction. For a plate which is 'hydraulically' smooth, the latter continues to be exerted next to the

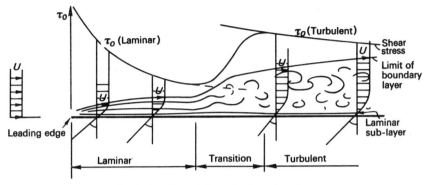

**Fig. 8.8(b).** *Instability of a boundary layer*

boundary by viscous action, in a very thin film, known as the laminar sub-layer.

If the pressure (which may be assumed virtually constant across the thickness of the boundary layer) is *decreasing in the direction of motion* (i.e. $dp/dx$ is negative as in Example 8.7(i) (p. 186), the pressure gradient is said to be *favourable* and if *increasing* (i.e. $dp/dx$ is positive as in Fig. 8.7(a)) (p. 185) the pressure gradient is said to be adverse.

Applying Bernoulli's equation to the main body of the flow, a *favourable* pressure gradient corresponds to an *accelerating* stream, and an *adverse* gradient to a *decelerating* stream.

When a gradient is favourable, the elements of the boundary layer are being drawn forward by the stream with a pressure behind them.

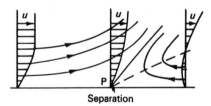

**Separation**

**Fig. 8.8(c).** *Separation of a laminar boundary layer*

Such conditions reduce the effectiveness of the viscous action so that the boundary layer remains thin and is stabilized. When the gradient is adverse, the pressure and viscosity both retard the boundary layer, causing it to thicken rapidly. If the pressure is sufficiently high downstream, a return flow near the boundary may even be established, see Fig. 8.8(c). The point P on the surface where the layer comes to rest and the return flow is first encountered is said to be the point of *laminar separation*. The effect of separation and wake is to impair the process of reconversion of kinetic energy into pressure energy, thus increasing the dissipation of energy into heat.

As an adverse pressure gradient encourages early transition to turbulence the latter often occurs before separation. As a turbulent boundary layer possesses more energy than the corresponding laminar one it is capable of moving against a greater pressure gradient without separating.

**Fig. 8.8(d).**  *Separation delayed by transitions to turbulence*

Referring back to Fig. 8.1(b) (p. 168), the region of almost constant reduced pressure at the rear of the cylinder called the *wake*, corresponds to that part which lies beyond the separation point. If this separation can be delayed, the wake is narrowed and the drag (which is largely pressure drag caused by the reduced pressure in the wake for bluff bodies) may be reduced. For example, under certain conditions, the drag of a sphere may be reduced by using a thin wire ring as a 'trip-wire' to make the boundary layer turbulent, see Fig. 8.8(d), and to delay separation.

The flow pattern, and hence the drag, of a body is determined by the viscous and inertia forces which cause skin-friction drag and pressure drag, respectively. The latter is the contribution to total drag due to pressure-difference over and round the body. Viscosity effects can change the flow pattern and cause turbulence and eddies which change the pressure distribution.

**Fig. 8.8(e).**  *Flow round streamlined body*

Drag reduction due to premature transition is exceptional, and can occur only with bluff bodies. The design of streamlined bodies is such as to have the point of separation as far downstream along the body as possible so that the boundary layer and wake remain thin and pressure is recovered downstream along the body. In this case drag is due only to skin friction caused by shear stress in the boundary layer. When a shape is streamlined, the expansion of the stream-tubes downstream of the point of maximum thickness is comparatively slow, see Fig. 8.8(e). Consequently, the corresponding deceleration of the free stream is gradual, and the associated adverse pressure gradient, very slight. Hence the boundary layer does not separate prematurely, and the form drag (i.e. that due to the pressure acting round it) is reduced to negligible proportions. The drag of such a body is almost entirely due to skin-friction, and its value critically dependent on the transition point. The reason for this is the much greater frictional resistance of a turbulent boundary layer than that associated with laminar flow. Appreciable areas of laminar flow can be maintained however only in very favourable circumstances and in practice may prove a Will-o'-the-wisp; free-stream turbulence, or slight imperfections on the surface, cause premature transition, and it is conservative to assume that the boundary layer is turbulent throughout.

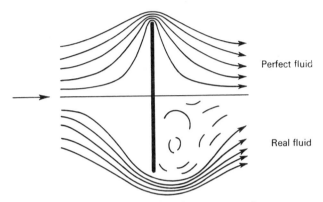

**Fig. 8.8(f).** *Flow round flat plate set normal to stream*

Even with a turbulent boundary layer however the reduction in drag due to streamlining is considerable. In a typical case it may be only one-twentieth of that of a flat plate of the same frontal area placed broadside on to the stream. The resistance of the latter is entirely due to the pressure differences acting across the plate. The dynamic pressures acting on the upstream side of the plate are not balanced by a subsequent pressure recovery downstream, as they would be in a perfect (i.e. non-viscous or *inviscid*) fluid, due to the separation of the boundary layer at the edges of the plate. Consequently a wide eddying wake is formed (see Fig. 8.8(f)) and the pressure on the downstream face of the plate is reduced below that of the free stream.

In terms of the ideas developed in this chapter, the resistance of a plate set normal to a flow is entirely *form* (i.e. pressure) drag, whilst that of a plate set along the flow is entirely *skin friction* (i.e. shear) *drag*. The drag in each case may be traced to the action of viscosity which causes a boundary layer and wake.

## 8.9. Real and potential flows

When dealing with real fluids it has not proved feasible to obtain results with a comprehensive theory that takes account of pressure, viscous, and dynamic effects—the analytical difficulties are too formidable. It is only by assuming that one of these effects may be neglected that a workable theory results. For example, 'potential' flows (i.e. ideal or theoretical flows based on an inviscid fluid) neglect viscous actions and merely provide a framework of reference against which the behaviour of a real fluid may be compared. When such comparisons are made marked discrepancies frequently occur. These are particularly noticeable in the case of bluff obstacles to an oncoming flow, for in real fluids the main body of the flow frequently separates from the boundaries leaving a dead-water region or wake filled with irregular eddies. As a further example of such a separation phenomenon we may compare the simple potential flow solution for the flow through a slit or orifice with the flow pattern that actually occurs with a real fluid. According to the concepts

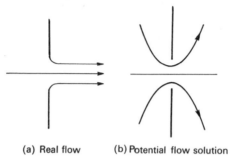

(a) Real flow          (b) Potential flow solution

**Fig. 8.9(a).**  *Flow through a slit or orifice*

of the previous section, boundary layer separation occurs at the lip with a real fluid, so that the fluid emerges as a contracting jet. Thus a real fluid shows no tendency to approach the potential flow pattern indicated in the second diagram of Fig. 8.9(a) which again suggests a symmetry between the flows on the upstream and downstream sides of the slit or orifice. The potential flow solution does however provide a clue that such a flow is physically impossible and this is to be found in the conditions at the edge of the slit which, if sharp, gives rise to an infinite velocity, and hence an infinitely negative pressure. This can be avoided by postulating that the fluid does not follow the boundary downstream of the slit, but forms a surface of discontinuity which gives rise to a profiled jet similar to that developed in a real fluid. The artifice of modifying potential flow solutions so that non-physical infinite velocities are avoided at sharp edges frequently overcomes the discrepancies that otherwise occur. It is an hypothesis of this type, concerning the flow at the trailing edge of an aerofoil that may be said to have laid the foundation of theoretical aerodynamics, see Fig. 8.9(b).

A further example of the uses of surfaces of discontinuity to avoid the infinite velocities encountered in potential flow round sharp edges, is Kirchhoff's solution to the flow round a plate set normal to an oncoming flow as shown in Fig. 8.9(c).

There is no doubt that sketch (b) of Fig. 8.9(c) represents a much closer approximation to the flows experienced with a real fluid, see Fig. 8.8(f), but the surfaces of discontinuity used to divide the main body of the flow from the dead-water region are found in practice to be non-steady and

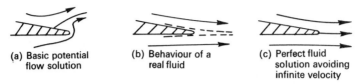

(a) Basic potential    (b) Behaviour of a    (c) Perfect fluid
flow solution          real fluid            solution avoiding
                                             infinite velocity

**Fig. 8.9(b).** *Conditions at the trailing edge of an aerofoil*

break up forming an eddying wake. Even without referring to the action of viscosity, which is a further complicating factor in the behaviour of real fluids, we may gain some insight into the reasons for this non-steadiness in real flows by subjecting Kirchhoff's solution to closer

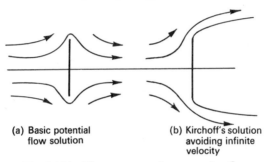

(a) Basic potential                    (b) Kirchoff's solution
flow solution                              avoiding infinite
                                           velocity

**Fig. 8.9(c).** *Plate set normal to oncoming flow*

scrutiny. We may liken it to the statement that a ruler balanced vertically on the tip of our finger is in equilibrium. For although conditions can, hypothetically remain as assumed, it may be shown that in each case the slightest disturbance will cause them to move further away from their

**Fig. 8.9(d).** *Instability of a surface of discontinuity*

equilibrium configuration. The unstable elements in Kirchhoff's solution are the surfaces of discontinuity which, consisting as they do of rows of vortices, tend to break up to form eddies, see Fig. 8.9(d). The conclusions reached concerning the flow in this particular problem are of general application and may be summarized as follows:

In most problems the solutions obtained by the straightforward application of potential flow theory seldom lead to results that are representative of the behaviour of real fluids. In many cases such flows may, by the application of some mathematical ingenuity based on physical insight, be made more realistic. One powerful method for achieving this is by the suitable introduction of vortices into the flow which may, for example, be employed to avoid an infinite velocity at a sharp edge. When this is done the vortices shed downstream will simulate the wake phenomena that are encountered in practice. Just as the wakes of real fluids are unsteady, so the systems of free vortices employed in theoretical work may be fundamentally unstable in the configuration assumed. This instability will cause such vortex sheets to roll up or coalesce and, particularly in the wake of an aerofoil, it is a matter of nice judgement as to which representation gives the most realistic approximation.

## Exercises on Chapter 8

1. If $u = ay + by^2$ represents the velocity of air in the boundary layer of a surface, $a$ and $b$ being constants and $y$ the perpendicular distance from the surface, calculate the viscous shear stress on the surface when the speed of the air relative to the surface is 75 m/s at a distance of 1·5 mm from the surface and 105 m/s when 3 mm from the surface. The viscosity of the air is $18 \times 10^{-6}$ kg/ms.

2. The distribution of velocity ($u$) in m/s with radius ($r$) in mm in a smooth tube of 25 mm bore follows the law $u = 2440 - kr^2$, the flow being laminar and the velocity at the pipe surface being zero. The fluid has a dynamic viscosity of $2·65 \times 10^{-4}$ kg/ms.

Determine:

(a) the rate of flow
(b) the shearing force between the fluid and the wall per metre length of tube.

3. Define clearly what is meant by the coefficient of viscosity and describe, briefly, any method of measuring absolute viscosity. A shaft 105 mm diameter rotates concentrically in a cylindrical bearing 150·15 mm diameter and 125 mm long. If the torque required to maintain a speed of 50 rev/min with the clearance space full of oil is 54 Nm, what is the dynamic viscosity of the oil?

4. A wind-tunnel test gave the following velocities parallel to the surface at a distance $y$ from an aerofoil.

| $y$ (mm) | 0·127 | 0·254 | 0·371 | 0·508 | 0·635 | 0·762 |
|---|---|---|---|---|---|---|
| $u$ (m/s) | 1·36 | 18·0 | 23·2 | 26·8 | 29·2 | 30·2 |

Plot a graph of $u$ against $y$ and estimate the velocity gradient at the surface. Hence obtain the intensity of skin friction on the aerofoil at this point, assuming $\eta$ to be $17·8 \times 10^{-6}$ kg/ms.

5. A piston 119·6 mm diameter and 140 mm long moves concentrically in a cylinder 120 mm diameter. If lubricating oil which fills the space between

them has a coefficient of viscosity of 0·65 poise, calculate the speed in mm/s with which the piston will move when the axial load on the piston is 8·5 N.

6. Use Stokes' Law $F = 3\pi\mu ud$ to find the dynamic viscosity of glycerine of weight 12·3 kN/m³ if a 3 mm diameter of steel ball of weight 77 kN/m³ falls through the fluid at a steady rate of 35 mm/s.

7. Deduce an expression for torque required to rotate a shaft on a bearing against the viscous resistance of an oil film. Calculate the torque if the bearing is 75 mm diameter, 125 mm long and the shaft rotates at 50 rev/min on a film of oil 0·05 mm thick. Dynamic viscosity of oil is 2 poise or 0·2 kg/sm.

8. Define the terms poise and stoke used for the measurement of viscosity. Give the units of measurement for each.

A 101·6 mm diameter journal rotates concentrically in a bearing having a diameter of 101·98 mm and length of 142·4 mm. The annular space between journal and bearing is filled with oil and it takes 1·12 kW to drive the journal at 2 400 rev/min.

Find the viscosity of the oil expressing its value in poise.

9. A 75 mm diameter shaft rotates concentrically in a journal 125 mm long with a radial clearance of 0·038 mm.

Find the shear stress on the shaft, the friction torque at 750 rev/min and the power absorbed if the viscosity of the oil is 1·5 poise.

10. Prove that in laminar flow of an incompressible fluid through a tube of circular section of radius $a$

(i) the axial velocity at radius $r$ is $U_0(1 - r^2/a^2)$, where $U_0$ is the velocity on the axis of the tube.

(ii) the rate of volumetric flow crossing any section is

$$\frac{\pi a^2}{8\eta} \frac{\Delta P}{L}$$

where $\eta$ is the coefficient of viscosity of the fluid and $\Delta P$ is the pressure drop between two sections distance $L$ apart.

(iii) the mean velocity is $U_0/2$.

(iv) Calculate the ratio of the total kinetic energy of the fluid to that of uniform flow at the same mean velocity.

11. An oil of kinematic viscosity 500 mm²/s and relative density 0·93 is discharged through a pipe 5 000 mm long and 30 mm diameter, the pressure drop being 69 kN/m². Calculate the rate of discharge.

12. Stating carefully the main assumptions involved develop, from first principles, an expression for the rate of steady flow, under laminar conditions, of a viscous, incompressible fluid through a rectangular passage of width $b_2$, very small depth $h$ and of length $L$ in the direction of the flow, under a differential pressure $p$. Show also that the intensity of shear stress on the wall of the passage is $6\eta\bar{u}/h$, where $\eta$ is the coefficient of viscosity and $\bar{u}$ is the mean velocity of flow. Neglect end and side effects.

13. Show that in laminar flow between extensive stationary flat plates the velocity distribution is parabolic. State all assumptions made.

Show also that the mean kinetic energy per unit mass of fluid is $1 \cdot 543 u_m^2/2$, where $u_m$ is the mean velocity.

14. Derive a formula for the rate at which pressure falls with length when a fluid flows in a straight direction between two parallel plates at a velocity below the critical.

Oil is supplied to a foot-step bearing under pressure and, when the shaft is at rest, flows out radially between two parallel surfaces, discharging at atmospheric pressure as shown in Fig. 8.7(a) (p. 185).

Calculate the flow in litre/h given the data:

Entry pipe diameter 10 mm; shaft diameter 110 mm; clearance space 0·05 mm; viscosity of oil 12 poise; supply pressure 294 kN/m².

15. A dash-pot has a 75 mm diameter cylinder and a piston 75 mm long. Calculate the required clearance between cylinder and piston so that when oil of viscosity 0·336 Ns/m² is used, a load of 100 kN may be lowered at a speed of 5 mm/s.

16. Viscous flow takes place between two stationary parallel surfaces whose length in the direction of flow is $L$ and distance apart is $h$. Show that, if the width of the surfaces in a direction transverse to the direction of flow is large compared with $h$ so that end effects can be neglected, the pressure drop in the direction of flow is:

$$P = 12 \frac{\eta L \bar{u}}{h^2}$$

where $\eta$ denotes dynamic viscosity of fluid and $\bar{u}$ is the mean velocity of flow between the surfaces.

A piston 150 mm long is fitted concentrically in a 250 mm diameter cylinder and carries a load of 200 kN. This is kept in a steady position by pumping oil having viscosity 0·1 Ns/m² into the cylinder below the piston. If the width of the annular space between the piston and cylinder wall is 0·3 mm, find the rate of leakage past the piston.

17. Show that the rate of change of pressure at any point along laminar flow between parallel surfaces is given by

$$\frac{dp}{dx} = \eta \left( \frac{d^2 u}{dy^2} \right)$$

where $\eta$ is the dynamic viscosity of the fluid and $u$ is the velocity at distance $y$ from one of the surfaces.

A piston having diameter $D$ and length $L$ moves concentrically in an oil dash-pot, the constant radial clearance being $h$. The piston moves with constant velocity $u$.

Neglecting forces arising from viscous stress in the oil adjacent to the piston wall, show that the resistance to the motion of the piston per unit area of the piston is given by

$$p = 12 \frac{\eta u L}{h^3} \left( \frac{D}{4} + \frac{h}{2} \right)$$

18. The piston working in a cylindrical dash-pot is 100 mm diameter and 75 mm deep and has a clearance of 0·5 mm all round. It is completely immersed in oil of viscosity 0·12 Ns/m².

Calculate the force required to move the piston down the cylinder at a speed of 250 mm/min.

State any assumptions made and deduce any formula used.

19. Two flat circular plates are at a small distance apart with their axes coincident, and are immersed in a large volume of liquid.

Show that if the plates are pulled apart along their common axis, the force of separation is proportional to the absolute viscosity, to the fourth power of the diameter, and to the speed of separation, and inversely proportional to the cube of the distance apart. Derive first an expression for the viscous flow between stationary flat parallel surfaces.

## Chapter 9

# *Viscous films of varying thickness*

'*The movement of friction is divided into parts
of which one is simple and all the others are
compound. Simple when the object is dragged
along a plane smooth surface without any
intervention. This is the only friction that
creates fire when it is powerful—that is to
say it generates fire—as can be seen at
waterwheels when the water is removed
between the whetted iron and the wheel.*'

LEONARDO

## 9.1. Lubricated bearings

When rubbing or sliding takes place between two solid surfaces, friction
and wear result. The subject of Tribology is concerned with problems of
this kind; the latter can largely be overcome, provided that a fluid film
can be maintained between the two surfaces under potentially damaging
conditions. Oil is the usual fluid but gases, including air may be used,
particularly in high-speed bearings.

Although metallic contact is prevented in pressure-fed bearings by
continuous pumping, a self-sustaining fluid film provides a simpler and
more reliable solution. It will be found however that to achieve this with
journal bearings some eccentricity of the shaft centre is needed in order
that a load-supporting fluid-wedge may be formed. It was Osborne
Reynolds who formulated a theory of lubrication (1886) following the
studies made by Beauchamp Towers. Sommerfeld in Germany (1904),
and Petroff in Russia further extended Reynolds' theory of the pressure-
sustaining property of a converging oil film. Michell applied the fluid
wedge concept to the design of bearings with tilting pads particularly
for shafts which carry end thrust.

## 9.2. Two-dimensional flow in viscous films

In Section 8.7 (p. 185) it was shown that for unidirectional viscous flow between two parallel surfaces, an expression for the transverse velocity distribution could be obtained from the laminar flow expression:

$$\tau = \eta \frac{du}{dz}$$

and the neglect of inertia forces, which required that:

$$\frac{dp}{dx} = \frac{d\tau}{dz} = \eta \frac{d^2u}{dz^2}$$

We thus obtain the result:

$$u = \frac{1}{\eta}\frac{dp}{dx}\frac{z^2}{2} + C_1 z + C_2$$

on the assumption that $p = p(x)$ varied only with distance along the film and that the velocity varied only in the transverse direction $z$, so that $u = u(z)$. Although these assumptions are not exact for a fluid film in a

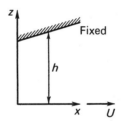

**Fig. 9.1(a).** *Viscous film*

journal bearing for example, the above expressions are acceptable for thin films (thickness $h$ small), so long as both the taper ($dh/dx$) and the curvature ($1/r$) of the film are also small. With the latter assumption ($h \ll r$) we can consider $x$ simply as the distance along the film, with the further restriction that we can apply the results obtained only if $dh/dx$ is also small.

In order to reduce such problems to ones of steady flow, we suppose (see Fig. 9.1(a)) that the plane surface $z = 0$ moves with a velocity $U$ in the direction of $x$ and that the upper surface, defined by $z = h(x)$, is at rest. For any cross section of the film, the boundary conditions are, in this case:

$$u = U \text{ when } z = 0, \quad u = 0 \text{ when } z = h.$$

This assumption of no slip at the boundaries leads to:

$$u = U\left(1 - \frac{z}{h}\right) - \frac{h^2}{2\eta}\left(\frac{z}{h} - \frac{z^2}{h^2}\right)\left(\frac{dp}{dx}\right)$$

The corresponding expression for the velocity distribution in terms of distance from the *fixed* boundary, $y$ say, is obtained by noting that:

$$y = h - z$$

and we thus rediscover the expression given in Section 8.7 (p. 186):

$$u = U\frac{y}{h} - \frac{h^2}{2\eta}\left(\frac{y}{h} - \frac{y^2}{h^2}\right)\left(\frac{dp}{dx}\right)$$

where it was also shown that the corresponding mean velocity:

$$\bar{u} = \frac{U}{2} - \frac{h^2}{12\eta}\left(\frac{dp}{dx}\right) = \frac{Q}{Bh}$$

The assumption of two-dimensional flow implies that the film is of sufficient breadth $B$ for the effect of side leakage to be neglected. The induced velocities then give rise to a volumetric flow rate $Q$ in the direction of $x$.

(a) PRESSURE GRADIENT

At any section $dp/dx$ is constant and its sign determines whether the velocity profile is convex or concave.

(i) **At the section where** $dp/dx = 0$ the oil is being dragged along rather than being pressure-impelled or resisted:

$$u = U\frac{y}{h} \quad \text{and} \quad \bar{u} = \frac{U}{2}$$

in which $y$ is measured from the fixed surface, see Fig. 9.2(a).

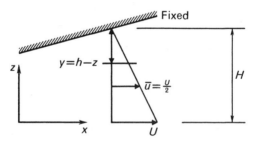

**Fig. 9.2(a).** *Viscous film where* $dp/dx = 0$

We shall use the corresponding value of $h$ as a reference thickness, $H$ say, so that:

$$u = \frac{y}{H}U$$

and redefine as a positive shear stress:

$$\tau = \eta\frac{du}{dy} = \eta\frac{U}{H}$$

is thus a constant for all layers across this section. The volumetric flow rate:

$$Q = \bar{u} \times \text{cross-sectional area} = \tfrac{1}{2}UHB$$

(ii) **At sections where** $dp/dx$ **is negative** $\bar{u} > \tfrac{1}{2}U$ and because $Q = \tfrac{1}{2}UHB = \bar{u}hB$ then $h < H$, as in Fig. 9.2(b).

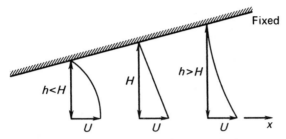

Fig. 9.2(b). *Viscous film of varying thickness*

(iii) **At sections where** $dp/dx$ **is positive** $\bar{u} < \tfrac{1}{2}U$ and $h > H$. see Fig. 9.2(b).

(iv) **Pressure gradient.** We can thus see how pressure builds up in a a fluid wedge, and subsequently falls again as $h$ becomes less than $H$, as shown in Fig. 9.2(b). We may relate $dp/dx$ to $h/H$ by noting that:

$$\frac{Q}{B} = \tfrac{1}{2}UH = \bar{u}h = h\left\{\tfrac{1}{2}U - \frac{1}{12}\frac{h^2}{\eta}\left(\frac{dp}{dx}\right)\right\}$$

A general expression for the pressure gradient is thus:

$$\frac{dp}{dx} = \tfrac{1}{2}U(h - H)\frac{12\eta}{h^3} = 6\eta U\left(\frac{1}{h^2} - \frac{H}{h^3}\right)$$

which may be expressed non-dimensionally as:

$$\frac{H^2}{6\eta U}\frac{dp}{dx} = \left(\frac{H}{h}\right)^2 - \left(\frac{H}{h}\right)^3$$

(v) **Maximum pressure gradient.** The pressure gradient is a maximum or a minimum when:

$$\frac{d^2p}{dx^2} = 0$$

i.e. when:

$$\frac{d}{dh}\left(\frac{dp}{dx}\right)\frac{dh}{dx} = 0$$

This corresponds either to the section where $dh/dx = 0$, i.e. where the film thickness is a minimum or to that at which:

$$\frac{d}{dh}\left(\frac{dp}{dx}\right) = 0$$

For non-zero values of $H$, $\eta$ and $U$ it is such that:

$$\frac{d}{d(h/H)}\left\{\frac{1}{(h/H)^2}-\frac{1}{(h/H)^3}\right\}=0$$

i.e.

$$h=\tfrac{3}{2}H$$

The corresponding value of $dp/dx$ is thus:

$$\left(\frac{dp}{dx}\right)_{max}=2\frac{\eta U}{h^2}=\frac{8}{9}\frac{\eta U}{H^2}$$

## (b) VELOCITY DISTRIBUTION

(i) **Points of zero velocity.** These occur where $y=y_0$ say, as shown in Fig. 9.2(c) and are such that:

$$0=U\frac{y_0}{h}-\frac{h^2}{2\eta}\left(\frac{y_0}{h}-\frac{y_0^2}{h^2}\right)\left(\frac{dp}{dx}\right)$$

i.e.

$$U=\frac{h^2}{2\eta}\left(1-\frac{y_0}{h}\right)\left(\frac{dp}{dx}\right)$$

or

$$y_0=h-\frac{2\eta U}{h(dp/dx)}$$

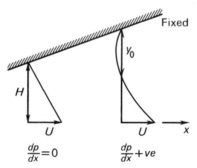

Fig. 9.2(c). *Diverging viscous film*

(ii) **Points of reverse flow on the fixed surface.** The flow direction reverses where $y_0=0$, such as the point R shown on Fig. 9.2(d), where $h=h_r$ say. The value of $h_r$ is given by:

$$h_r^2=\frac{2\eta U}{(dp/dx)}$$

Also, since:

$$\frac{Q}{B}=\tfrac{1}{2}UH=h_r\bar{u}_r=h_r\left\{\tfrac{1}{2}U-\frac{1}{12}\frac{h_r^2}{\eta}\left(\frac{dp}{dx}\right)\right\}=h_r\frac{U}{3}$$

it follows that $h_r=\tfrac{3}{2}H$ and this is already known to correspond to that at which $dp/dx$ is a maximum. Further, along the fixed surface (where $y=0$):

$$\frac{du}{dy}=\frac{U}{h}-\frac{h^2}{2\eta}\left(\frac{1}{h}-2\frac{y}{h^2}\right)\left(\frac{dp}{dx}\right)=\frac{U}{h}-\frac{1}{h}\left(\frac{h^2}{2\eta}\frac{dp}{dx}\right)$$

Noting that, at the point R:

$$\frac{h^2}{2\eta U}\frac{dp}{dx} = \frac{h_r^2}{2\eta U}\frac{dp}{dx} = 1$$

the value of $du/dy$ at R is seen to be zero. Thus R is a point at which the shear stress $\tau (= \eta\, du/dy)$ is zero and the pressure gradient:

$$\frac{dp}{dx} = \frac{2\eta U}{h_r^2} = \frac{8}{9}\frac{\eta U}{H^2}$$

is a maximum.

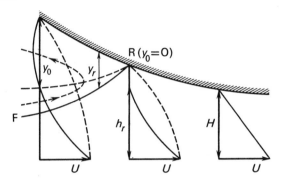

**Fig. 9.2(d).** *Reverse flow in a viscous film*

(iii) **Locus of points of return flow.** For sections which lie to the left of R on Fig. 9.2(d) there is a return flow in the region adjacent to the fixed surface. This must be fed by some of the fluid from below and we may define the total thickness of the layer involved in the return flow process by $y_r$ in the expression:

$$\frac{Q_r}{B} = 0 = \int_0^{y_r} u\,dy$$

Thus below the locus of the points of depth $y_r$ (FR on Fig. 9.2(d)), all the fluid which enters from the left emerges on the right. Above this line there is a continuous circulatory process. Substituting for $u$, we deduce:

$$\int_0^{y_r} u\,dy = \frac{U}{h}\frac{y_r^2}{2} - \frac{h^2}{2\eta}\left(\frac{y_r^2}{2h} - \frac{y_r^3}{3h^2}\right)\left(\frac{dp}{dx}\right) = 0$$

Hence:

$$U = \frac{h^2}{\eta}\left(\frac{1}{2} - \frac{y_r}{3h}\right)\left(\frac{dp}{dx}\right)$$

and:

$$y_r = \frac{3}{2}\left\{h - \frac{2\eta U}{h(dp/dx)}\right\} = \frac{3}{2}y_0$$

## (c) SHEAR STRESS

### (i) Boundary shears

$$\tau = \eta\frac{du}{dy} = \eta\frac{U}{h} - \tfrac{1}{2}(h-2y)\left(\frac{dp}{dx}\right)$$

On the moving surface, where $y = h$:

$$\tau_h = \eta\frac{U}{h} + \tfrac{1}{2}h\left(\frac{dp}{dx}\right)$$

On the stationary surface, where $y = O$, the shear stress is:

$$\tau_0 = \eta\frac{U}{h} - \tfrac{1}{2}h\left(\frac{dp}{dx}\right)$$

(ii) **Shear stress reversed on the moving surface.** If the shear stress on the moving surface is zero:

$$0 = \eta\frac{U}{h} + \tfrac{1}{2}h\left(\frac{dp}{dx}\right)$$

Denoting the film thickness where this occurs by $h_o$:

$$\frac{dp}{dx} = -\frac{2\eta U}{h_o^2}$$

Substituting this into the expression:

$$\frac{Q}{B} = \tfrac{1}{2}UH = h\bar{u} = h\left\{\tfrac{1}{2}U - \frac{1}{12}\frac{h^2}{\eta}\left(\frac{dp}{dx}\right)\right\}$$

$$\tfrac{1}{2}UH = h_o\left\{\tfrac{1}{2}U + \frac{1}{12}\frac{h_o^2}{\eta}\left(\frac{2\eta U}{h_o^2}\right)\right\} = \tfrac{2}{3}h_o U$$

i.e.                $h_o = \tfrac{3}{4}H$

(iii) **Shear stress reversal on the fixed surface.** We have already noted that the value of $\tau\ (= \eta\ du/dy)$ is zero at the point of reverse flow (R) and that the corresponding film thickness at this section is such that:

$$h_r^2 = \frac{2\eta U}{(dp/dx)}$$

so that:

$$h_r = \tfrac{3}{2}H$$

## EXAMPLE 9.2 (i)

*At a certain point in a lubricated journal running under steady conditions the film thickness (h) is 12·5 μm and the pressure gradient is 1·8 GN/m³. The maximum pressure occurs where the film thickness (H) is 5·0 μm. Estimate the pressure gradient at a point where the film thickness ($h_2 = h_r$) is 7·5 μm. Also estimate the shear stress on the journal at the point (R) of reverse flow.*

Referring to Fig. 9.2(d) we may deduce that, since:

$$\frac{Q}{B} = \bar{u}h = \bar{u}_r h_r = \bar{u}_H H = \tfrac{1}{2}UH = \left\{\tfrac{1}{2}U - \frac{1}{12}\frac{h^2}{\eta}\left(\frac{dp}{dx}\right)\right\}h$$

then:

$$U = \frac{h^3\,(dp/dx)}{6\eta\,(h-H)} = \frac{(12\cdot5\ \mu m)^3 \times 1\cdot8\ \text{GN/m}^3}{6\eta \times 7\cdot5\ \mu m} = \frac{0\cdot0781}{\eta}\frac{\text{N}}{\text{m}}$$

Reverse flow occurs when the thickness of the film is

$$\tfrac{3}{2}H = 7\cdot5\ \mu m = h_r$$

Hence:

$$\bar{u}_r = \frac{H}{h_r}\bar{u}_H = \frac{2}{3}\frac{U}{2} = \frac{U}{3} = \frac{U}{2} - \frac{1}{12}\frac{h_r^2}{\eta}\left(\frac{dp}{dx}\right)_r$$

and:

$$\left(\frac{dp}{dx}\right)_r = \frac{U}{6}\times\frac{12\eta}{h_r^2} = \frac{0\cdot0781\ \text{N}}{6\eta\ \ \text{m}}\times\frac{12\eta}{(7\cdot5\mu m)^2} = 2\cdot78\ \text{GN/m}^3$$

The shear stress on the stationary bearing at the point of reverse flow is zero, but on the shaft journal moving at a velocity $U$ it is, at $y = h = h_r$:

$$\tau_{h_r} = \eta\frac{U}{h_r} + \tfrac{1}{2}h_r\left(\frac{dp}{dx}\right)_r = \left(\frac{dp}{dx}\right)_r h_r$$

i.e.   $$\tau_{h_r} = 2\cdot78\ \text{GN/m}^3 \times 7\cdot5\ \mu m = 20\cdot84\ \text{kN/m}^2$$

## EXAMPLE 9.2 (ii)

*The relatively long cylindrical plunger of a slow-running pump delivers oil at a constant pressure. When the temperature of the oil is 15°C the leakage flow past the plunger is $28 \times 10^{-6}$ m³/min. Estimate the leakage flow when the temperature of the oil is 70°C assuming that the plunger remains concentric with the bore.*

*At 15°C the diametral clearance is 25 μm and at 70°C it is 50 μm.*

*At 15°C the dynamic viscosity of the oil is 1 kg/ms and at 70°C it is 0·05 kg/ms.*

Neglecting the velocity of the slow running plunger compared with that of the leaking oil:

$$\bar{u} = -\frac{1}{12}\frac{h^2}{\eta}\left(\frac{dp}{dx}\right)$$

in which $dp/dx$ is assumed constant, see Fig. 9.2 (e).

Hence the leakage:

$$Q = a\bar{u} = -\frac{\pi dh^3}{12\eta}\left(\frac{dp}{dx}\right)$$

in which the negative sign implies that $Q$ takes place in the opposite direction to $x$ and hence $U$

$$\frac{Q_2}{Q_1} = \left(\frac{h_2}{h_1}\right)^3\left(\frac{\eta_1}{\eta_2}\right)$$

Thus:

$$Q_2 = 28 \times 10^{-6} \text{ m}^3/\text{min} \left(\frac{50}{25}\right)^3 \left(\frac{1}{0\cdot05}\right) = 4\cdot48 \times 10^{-3} \text{ m}^3/\text{min}$$

$$= \textbf{4}\cdot\textbf{48} \text{ litre/min}$$

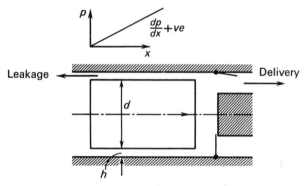

**Fig. 9.2(e).** *Leakage past a plunger*

EXAMPLE 9.2 (iii)

*A dashpot consists of a cylindrical plunger* 75 mm *long and* 25 mm *diameter sliding in a cylinder with a closed end. The gear is to have a time-lag of* 15 s *for a* 50 mm *movement of the plunger when a force of* 90 N *is applied. Estimate the necessary clearance when using oil of dynamic viscosity* 0·15 kg/sm *assuming that the plunger is (a) concentric (b) touches one side of the cylinder.*

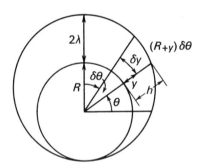

**Fig. 9.2(f).** *Dashpot with eccentric plunger*

Referring to Fig. 9.2(f) if $\lambda$ is relatively small

$$h = \lambda(1+\sin\theta)$$

$$u = U\frac{y}{h} - \frac{h^2}{2\eta}\left(\frac{y}{h} - \frac{y^2}{h^2}\right)\left(\frac{dp}{dx}\right)$$

The flow through the element is:

$$\delta Q = u(R+y)\delta\theta\delta y$$

Hence:

$$Q = \int_0^{2\pi} \int_0^h u(R+y)\,dy\,d\theta$$

and, neglecting $U$ relative to $u$:

$$Q = -\frac{1}{2\eta}\left(\frac{dp}{dx}\right)\int_0^{2\pi}\int_0^h (hy - y^2)(R+y)\,dy\,d\theta$$

$$= -\frac{1}{2\eta}\left(\frac{dp}{dx}\right)\int_0^{2\pi} (\tfrac{1}{6}Rh^3 + \tfrac{1}{12}h^4)\,d\theta$$

Neglecting the second term of this integral because $h$ is small relative to $R$, we have:

$$Q = -\frac{R}{12\eta}\left(\frac{dp}{dx}\right)\int_0^{2\pi} \lambda^3(1 + 3\sin\theta + 3\sin^2\theta + \sin^3\theta)\,d\theta$$

$$= -\frac{R\lambda^3}{12\eta}\left(\frac{dp}{dx}\right)[2\pi + 3\pi] = -\frac{5\pi R\lambda^3}{12\eta}\left(\frac{dp}{dx}\right)$$

and:

$$\bar{u} = \frac{Q}{2\pi R\lambda} = -\frac{5}{2}\left\{\frac{\lambda^2}{12\eta}\left(\frac{dp}{dx}\right)\right\}$$

$$= \tfrac{5}{2} \times \text{mean velocity for a concentric plunger.}$$

(a) For a concentric plunger of length $L$:

$$a\bar{u} = Q = -\frac{2\pi R\lambda^3}{12\eta}\left(\frac{P}{L}\right) = -\pi R^2\frac{dx}{dt} = -\frac{\pi}{4}\times 625 \text{ mm}^2 \times \frac{50 \text{ mm}}{15 \text{ s}}$$

$$= -1\,640 \text{ mm}^3/\text{s}$$

the negative sign indicating backward flow. Hence:

$$\lambda^3 = -\frac{6Q\eta}{\pi R}\left(\frac{L}{P}\right)$$

in which:

$$\frac{P}{L} = \frac{90 \text{ N}}{491 \text{ mm}^2} \times \frac{1}{75 \text{ mm}} = 2\cdot45\times10^{-3} \text{ N/mm}^3$$

i.e.

$$\lambda^3 = \frac{6\times1\,640 \text{ mm}^3/\text{s}}{\pi\times12\cdot5 \text{ mm}}\times0\cdot15\frac{\text{kg}}{\text{ms}}\times\frac{\text{mm}^3}{2\cdot45\times10^{-3} \text{ N}}\left[\frac{\text{N s}^2}{\text{kg m}}\right]$$

$$= 15\,350\frac{\text{mm}^5}{\text{m}^2} = 15\,350\times10^{-6} \text{ mm}^3$$

∴

$$\lambda = 24\cdot8\times10^{-2} \text{ mm} = \textbf{0·248 mm}$$

(b) If the plunger touches one side:

$$\lambda_e^3 = -\frac{12Q\eta}{5\pi R}\left(\frac{P}{L}\right) = \frac{2}{5}\lambda_c^3$$

i.e.

$$\lambda_e = 0{\cdot}737\,\lambda_c \quad \text{and} \quad 2\lambda_e = \mathbf{0{\cdot}356}\text{ mm}$$

EXAMPLE 9.2 (iv)

*A load of* 1 MN *is maintained in a testing machine by a plunger of* 300 mm *diameter and* 150 mm *long in a cylinder having a radial clearance of* 0·125 mm. *Estimate the pump power necessary, assuming that the plunger touches the cylinder. Neglect the resistance of the supply pipes and assume the dynamic viscosity of the oil to be* 0·15 kg/ms *or* 0·15 Ns/m².

Referring to Fig. 9.2(f), $2\lambda = 0{\cdot}25$ mm and the leakage past the eccentric plunger is given by:

$$Q_e = \frac{5}{2}Q_c = -\frac{5}{2}\left\{\frac{\pi}{6}\frac{R\lambda^3}{\eta}\left(\frac{dp}{dx}\right)\right\}$$

in which:

$$\frac{dp}{dx} = -\frac{W}{AL} = -\frac{W}{\pi R^2 L}$$

Hence:

$$Q_e = \frac{5}{12}\frac{W\lambda^3}{\eta RL}$$

The power:

$$\rho g Q_e \Delta H = Q_e \Delta p = Q_e \frac{W}{A}$$

assuming zero pressure at outlet from the plunger.

Hence the pump power necessary is:

$$\frac{5}{12}\frac{W^2\lambda^3}{\eta\pi R^3 L} = \frac{5}{12\pi}\times\frac{\text{m}^2}{0{\cdot}15\text{ Ns}}\times\frac{1\text{ MN}^2}{150\text{ mm}}\times\left(\frac{0{\cdot}125}{150}\right)^3$$

$$= \frac{5}{12\pi}\times\frac{10^2\text{ m}^2}{15\text{ Ns}}\times\frac{10^{12}\text{ N}^2}{15\times10^{-2}\text{ m}}\times\frac{125}{216}\times10^{-9}$$

$$= \frac{1}{36\pi}\times\frac{125}{15\times216}\times10^7\frac{\text{Nm}}{\text{s}}$$

$$= 3{\cdot}4\times10^3\frac{\text{Nm}}{\text{s}} = \mathbf{3{\cdot}4}\text{ kW}$$

## 9.3. Journal bearings

If the load on a journal bearing is sufficiently light for the shaft to run concentrically, the friction torque is simply given by:

$$M = \frac{2\pi\eta\omega R^3 B}{c}$$

in which $B$ is the width of the bearing and $c$ is the *radial* clearance. This expression was established in Section 8.3 (p. 175), where it was also shown that the effective 'coefficient of friction' is given by Petroff's equation:

$$\mu = \frac{\pi \eta \omega}{p}\left(\frac{R}{c}\right) = \frac{2\pi \eta \omega B R^2}{Wc}$$

This may be used to obtain information about the effective clearance in a bearing from running tests on no-load, so long as the speed is sufficiently high for a continuous fluid film to have formed. The corresponding power absorbed under these conditions is:

$$M\omega = \frac{2\pi \eta \omega^2 R^3 B}{c}$$

If the load on the shaft is not negligible then it is necessary, in bearings that are not pressure-fed, for the oil film to form a convergent fluid wedge. This creates the pressure in the film which is needed to support the load. Thus the shaft must run eccentrically, as indicated in Fig. 9.3(a),

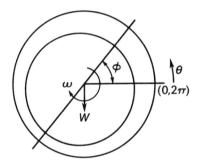

**Fig. 9.3(a).** *Attitude angle in a journal bearing*

which also shows the attitude angle $\varphi$ to be opposite that which would occur in the case of a non-lubricated bearing operating with dry friction. In the latter case the bearing would probably overheat and seize up under heavy load. By concentrating on the wedging action of a fluid film we are ignoring conditions where the fluid film diverges. Theory would predict high negative pressures in this region, but these are largely non-physical since a fluid cannot maintain a state of tension under steady conditions. Fig. 9.3(b) shows a journal bearing in which the half-clearance $c$ is small compared with $R$ so that the clearance ratio $c/R$ is small and the distances along the film in the direction of motion are defined by $\delta x = R\delta\theta$. The position of the shaft is defined by its eccentricity $e$, or its eccentricity ratio $\epsilon = e/c$.

Choosing a reference position as shown in the figure, the film thickness at any angle $\theta$ is given by:

$$h = c + e \cos \theta$$

and, in particular:

$$H = c + e \cos \Theta$$

where $\Theta$ is the angle where the pressure is a maximum, so that $dp/dx = 0$. The equations of the previous sections apply so long as side-leakage

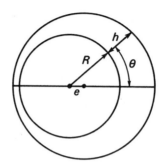

**Fig. 9.3(b).** *Journal bearing*

and viscosity variations may be neglected. Thus the point of reverse flow occurs when the film thickness is $h_r = \frac{3}{2}H$, i.e. when

$$c + e \cos \theta_r = \frac{3}{2}(c + e \cos \Theta)$$

i.e. when:

$$\cos \theta_r = \frac{c + 3e \cos \Theta}{2e} = \frac{1 + 3\epsilon \cos \Theta}{2\epsilon}$$

**Pressure gradient.** The general equation for a long journal is, for no side-leakage:

$$\frac{dp}{dx} = 6\eta U \left( \frac{1}{h^2} - \frac{H}{h^3} \right) = 6\eta U \left( \frac{h - H}{h^3} \right)$$

see Section 9.2 (p. 201).
Hence:

$$\frac{dp}{dx} = \frac{dp}{d\theta} \cdot \frac{d\theta}{dx} = \frac{1}{R} \frac{dp}{d\theta} = \frac{6\eta U e (\cos \theta - \cos \Theta)}{(c + e \cos \theta)^3}$$

i.e.

$$\frac{dp}{d\theta} = \frac{6\eta U R \epsilon}{c^2} \left\{ \frac{\cos \theta - \cos \Theta}{(1 + \epsilon \cos \theta)^3} \right\}$$

which applies to long bearings in which side-leakage may be neglected, so long as the viscosity of the oil remains constant.

From the general equation it follows that the maximum or minimum pressure gradient occurs when:

$$\frac{d}{dx} \left( \frac{dp}{dx} \right) = \frac{d}{dh} \left( \frac{dp}{dx} \right) \frac{dh}{dx} = 0$$

$dh/dx = 0$ relates to the positions of maximum and minimum clearance and

$d^2p/dhdx = 0$ refers to the condition where $h^3 - 3h^2 (h - H) = 0$, i.e. when $h = h_r = 3H/2$ as in Section 9.2. (p. 201).

Integrating the equation for $dp/d\theta$ we obtain the following non-dimensional expression in which $p_o$ is the pressure at $\theta = 0$ and $\theta = 2\pi$ for a bearing full of fluid:

$$\frac{p - p_o}{6\eta UR/c^2} = \frac{\epsilon \sin \theta \, (2 + \epsilon \cos \theta)}{(2 + \epsilon^2) \, (1 + \epsilon \cos \theta)^2}$$

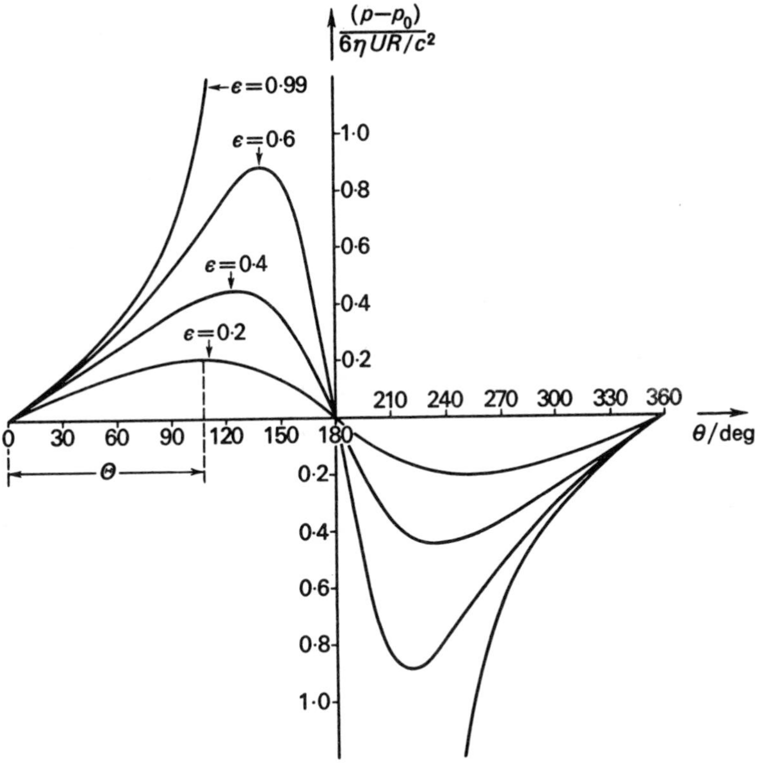

Fig. 9.3(c). *Non-dimensional pressure variation against angle for variable* $\epsilon = e/c$

From this it follows that the maximum pressure occurs when:

$$\cos \theta = -\frac{3\epsilon}{2 + \epsilon^2} = \cos \Theta$$

Thus for values of $\epsilon = 0\cdot6, 0\cdot4$ and $0\cdot2$, $\Theta \simeq 180° \mp 40°$, $180 \mp 50°$ and $180° \mp 73°$ respectively, which will be seen to be the angles where peak positive and negative pressures occur on the graphs of Fig. 9.3(c). Only the upper positive parts of these antisymmetrical curves are normally taken into account in practice since the reduced pressure normally allows the ingress of air from each side, which disrupts the oil film in this region. Even if this could be prevented, the film would eventually cavitate as zero pressure was approached.

Using the result that $\cos \theta = -3\epsilon/(2+\epsilon^2)$ for the maximum pressure it follows that:

$$\frac{(p-p_0)_{\max}}{6\eta UR/c^2} = \frac{\epsilon}{4(2+\epsilon^2)}\sqrt{\left(\frac{4-\epsilon^2}{1-\epsilon^2}\right)^3}$$

The graph of this non-dimensional equation Fig. 9.3(d) shows that

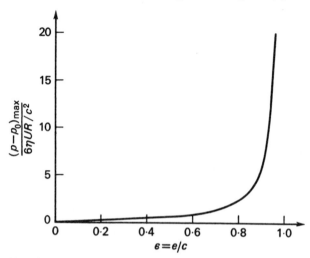

**Fig. 9.3(d).** *Non-dimensional plot of maximum pressure against eccentricity ratio*

heavy loads are supported with the surfaces nearly in contact (i.e. when $e/c = \epsilon$ approaches unity). Taking the particular case of $\epsilon \simeq 0.6$, $\Theta = 180° \mp 40°$ and $(p-p_0)_{\max}/(6\eta UR/c^2) = 0.86$. These are shown on Fig. 9.3(e) which is a polar diagram illustrating the pressure variation and velocity distribution of the fluid. For clockwise angular motion of the journal the upper half of the pressure curve is negative and therefore normally inoperative in practice. Since any air entrained in the oil separates out, the oil film ceases to be continuous. The negative pressure loop may then be ignored and the theory usually proceeds by considering only the positive loop of the half bearing.

Thus for $\epsilon \simeq 0.6$ the point of maximum pressure occurs when $\theta = 220° = \Theta$ and the clearance $h = H$ when $dp/d\theta = 0$. The point of reverse flow occurs when $h = h_r = \frac{3}{2}H$, i.e. when:

$$2+2\epsilon \cos \theta_r = 3(1+\epsilon \cos \Theta) = 3 - \frac{9\epsilon^2}{2+\epsilon^2}$$

For $\epsilon \simeq 0.6$, as shown in Fig. 9.3(e), $\theta_r = 180° \mp 72°$. For the lower half of a bearing the force on an elemental area $BR\delta\theta$ is $(p-p_0)BR\delta\theta$. Hence the total vertical component of the force on the half-bearing of breadth $B$ is, neglecting side leakage:

$$F_V = BR \int_{\pi}^{2\pi} (p-p_o) \sin\theta d\theta = BR\frac{6\eta UR}{c^2}\frac{\epsilon}{(2+\epsilon^2)}\int_{\pi}^{2\pi}\frac{(2+\epsilon\cos\theta)\sin^2\theta d\theta}{(1+\epsilon\cos\theta)^2}$$

i.e. $$F_V/B = \frac{6\pi\eta UR^2\epsilon}{c^2(2+\epsilon^2)(1-\epsilon^2)^{1/2}}$$

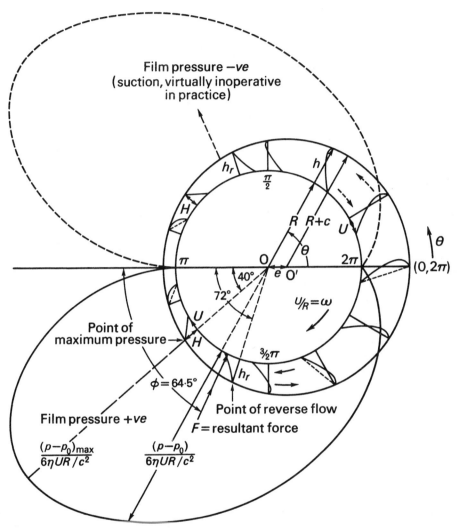

**Fig. 9.3(e).** *Polar diagram indicating theoretical variation of pressure and velocity distribution*

Similarly, the horizontal component is:

$$F_H = BR\int_{\pi}^{2\pi} (p-p_o)\cos\theta d\theta$$

i.e.
$$F_H/B = \frac{12\eta U R^2 \epsilon^2}{c^2(2+\epsilon^2)(1-\epsilon^2)}$$

Hence the resultant force per unit axial length is:

$$F/B = 6\eta U \left(\frac{R}{c}\right)^2 \frac{\epsilon}{(2+\epsilon^2)(1-\epsilon^2)} \sqrt{\{\pi^2(1-\epsilon^2)+4\epsilon^2\}}$$

Referring to Fig. 9.3(f):

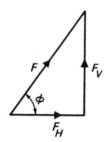

**Fig. 9.3(f).** *Attitude angle*

$$\tan \varphi = \frac{\pi}{2\epsilon}\sqrt{(1-\epsilon^2)}$$

Hence, for the particular case when $\epsilon = 0\cdot6$ (Figs. 9.3(e and f))

$$\tan \varphi = 2\pi/3 \quad \text{or} \quad \varphi = 64\cdot5° \quad \text{and:}$$

$$\left(\frac{F}{2RB}\right)\left(\frac{c}{R}\right)^2 \frac{1}{3\eta\omega} = \frac{\epsilon\{\pi^2(1-\epsilon^2)+4\epsilon^2\}^{1/2}}{(2+\epsilon^2)(1-\epsilon^2)} = 1\cdot11$$

Thus for a bearing with no side-leakage the force/bearing area is:

$$\frac{F}{2BR} = 3\omega\eta \left(\frac{R}{c}\right)^2 f(\epsilon)$$

where $\omega$ is the angular speed of the shaft.

The shaft sets itself at the attitude angle $\varphi$ as shown in Fig. 9.3(g) which, incidentally, is opposite to that of dry friction.

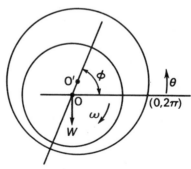

**Fig. 9.3(g).** *Attitude angle*

The load $W$ is carried by the vertical component which, per unit of projected cross-sectional area, is:

$$P = \frac{F_V}{2BR} = 3\omega\eta\left(\frac{R}{c}\right)^2\frac{\pi\epsilon}{(2+\epsilon^2)(1-\epsilon^2)^{1/2}}$$

Hence, if $\epsilon = 0.6$:

$$\frac{\text{Load}}{2BR}\left(\frac{c}{R}\right)^2\frac{1}{3\omega\eta} = 1.0$$

Alternatively:

$$\eta\frac{N}{P}\left(\frac{R}{c}\right)^2 = S = \frac{(2+\epsilon^2)(1-\epsilon^2)^{1/2}}{6\pi^2\epsilon}$$

in which $S$ is known as the Sommerfeld variable.

**In summarizing** we may say that although journal bearings should normally be made as long as possible, in practice some side or end leakage is inevitable. The actual load-carrying capacity and performance can generally be determined only by practical tests. In planning and interpreting the latter we take account of simple lubrication theory which shows that for no side leakage and constant viscosity:

(a) the load carrying capacity

$$W \propto \eta\omega\left(\frac{R}{c}\right)^2\psi(\epsilon) \quad\text{or}\quad P = \eta\frac{N}{S}\left(\frac{R}{c}\right)^2$$

where $S$ is the Sommerfeld variable.

(b) the friction force $f \propto \eta BR(U/c)$ for a particular eccentricity.

(c) the 'coefficient of friction' of the bearing $\mu = f/W \propto c/R$ for a particular $\epsilon$

(d) the power required to overcome the friction drag on the shaft journal is $fU = \mu W(2\pi RN) \propto cWN$ where the constant of proportionality depends on $\epsilon = \varphi(S)$

(e) friction always results in heat generation and in practice there are limits on the allowable operating temperature. Since the power required to overcome friction may alternatively be written $fU \propto \eta \times PR^2BN$, it follows that:

(i) the rate of dissipation of energy of a particular bearing, for which $BR$ is fixed, is proportional to $PN$ for a given $\epsilon$

(ii) the rate of heating of bearings running at the same speed and with the same loading ($N$ and $P$ constant) is proportional to the volume for a given $\epsilon$. Hence the larger the bearing the greater the heat transfer required to keep it within its allowable operating temperature.

EXAMPLE 9.3 (i)

*A journal* 100 mm *in diameter and* 300 mm *long runs in a bearing having a diametral clearance of* 0.5 mm *and carries a load of* 45 kN *when running at*

*600 rev/min and lubricated with oil of dynamic viscosity* 3·7 P. *Estimate the load permissible on a geometrically similar journal of* 150 mm *diameter, running at* 800 rev/min *when oil of dynamic viscosity* 7·4 P *is used, if it is to run at the same attitude. The effect of side-leakage may be neglected.*

For a geometrically similar bearing running at the same attitude, the eccentricity ratio $\epsilon = e/c$ will be the same. According to half-bearing theory:

$$W \propto \eta \omega B R \left(\frac{R}{c}\right)^2 \psi(\epsilon)$$

Hence for geometrically similar bearings, operating under similar conditions in the sense that $e_2/c_2 = e_1/c_1$:

$$\frac{W_2}{W_1} = \frac{\eta_2 \omega_2 R_2^2}{\eta_1 \omega_1 R_1^2} = 2 \times \tfrac{8}{6} \times (\tfrac{3}{2})^2 = 6$$

The permissible load on the 150 mm diameter bearing, when running at 800 rev/min = 270 kN and since $R_2/R_1 = 2 = B_2/B_1$, this bearing would need to be 600 mm long.

## 9.4. Tilting-pad bearings

It is an essential feature of Reynolds' theory that a fluid film can sustain a load between two surfaces which are in relative motion only if they are slightly inclined to one another. We have seen how such a fluid wedge can form naturally in a journal bearing by a slight eccentricity of the shaft. The fluid wedge concept has led to the evolution of slipper-pad bearings, particularly for shafts which carry a large end thrust. A typical example is a screw-propelled ship, in which the thrust of the propellers may be transmitted through a Michell thrust bearing of this type to the hull of the ship. In such bearings a series of pads, each subtending a circular arc, are arranged within an annulus normal to the direction of thrust, see Fig. 9.4(a). By suitably pivoting each pad, they are able to take up the necessary small angle of inclination to the moving surface. The fluid film can then be maintained under pressure—the angle of tilt being dependent on pressure, speed and viscosity.

Fig. 9.4(b) illustrates a tilting-pad of this type. Since only the relative motion is important, we reduce the problem to one of steady flow by considering the slipper to be at rest, and suppose that the adjacent surface, situated in the plane $z = 0$ moves with a velocity $U$ in the $x$ direction. We further assume that the fluid motion is essentially two-dimensional, thus neglecting the effect of side leakage. We suppose that the lower face of the slipper is a plane surface extending from $x = 0$ to $x = l$ and that the film thickness varies from $h = h_1$ where $x = 0$ to $h_2$ when $x = l$. Thus:

$$h = h_1 + mx, \quad h_2 = h_1 + ml$$

in which the slope $m$ is small—typically of the order of $10^{-3}$ in practice.

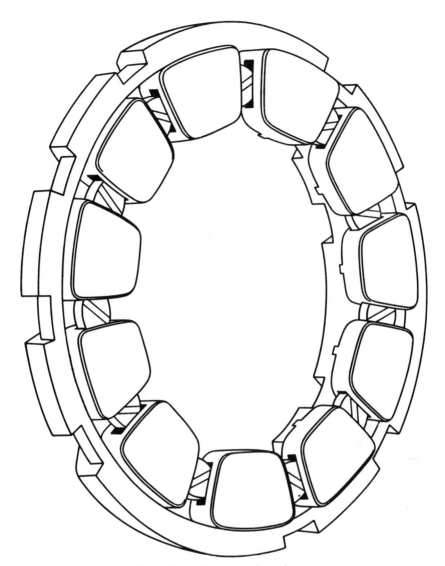

**Fig. 9.4(a).** *Tilting-pad thrust bearing*

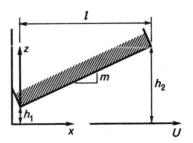

**Fig. 9.4(b).** *Viscous film of varying thickness*

Referring back to Section 9.2 (p. 201), the condition that the flow across each section of the film is the same leads to:

$$h^3 \left(\frac{dp}{dx}\right) = 6\eta U(h - H)$$

in which $H$ corresponds to the peak pressure. Noting that $dh/dx = m$:

$$\frac{dp}{dh} = \frac{dx}{dh}\frac{dp}{dx} = \frac{6\eta U}{m}\left(\frac{1}{h^2} - \frac{H}{h^3}\right)$$

which may be integrated to relate the pressure to the film thickness:

$$p = \frac{6\eta U}{m}\left(-\frac{1}{h} + \frac{H}{2h^2} + C\right)$$

If the datum pressure for the bearing is $p = p_o$ for $h = h_1$ and $h = h_2$:

$$\frac{p_o}{6\eta U/m} = -\frac{1}{h_1} + \frac{H}{2h_1^2} + C = -\frac{1}{h_2} + \frac{H}{2h_2^2} + C$$

$$\therefore \qquad \frac{1}{h_1} - \frac{1}{h_2} = \frac{H}{2}\left(\frac{1}{h_1^2} - \frac{1}{h_2^2}\right)$$

i.e. $$H = \frac{2h_1 h_2}{h_1 + h_2}$$

Alternatively, $$H = \frac{h_1 + h_2}{2} - \frac{(h_1 - h_2)^2}{2(h_1 + h_2)}$$

Thus the film thickness at the point of maximum pressure is always less than the mean film thickness, showing that the pressure tends to peak towards the trailing edge. Since absolute pressures are relatively of no importance we may take $p_o = 0$ to obtain the following expressions for the excess pressure within the film:

$$p = -\frac{6\eta U}{m}\left\{\frac{1}{h} - \frac{1}{h_1} - \frac{H}{2}\left(\frac{1}{h^2} - \frac{1}{h_1^2}\right)\right\}$$

$$= -\frac{6\eta U}{m}\left\{\frac{1}{h} - \frac{1}{h_2} - \frac{H}{2}\left(\frac{1}{h^2} - \frac{1}{h_2^2}\right)\right\}$$

We expect $U$ to be negative in these expressions for positive values of $m$ since only then does the incoming fluid form a converging fluid film. The extensive occurrence of reciprocals suggests the intermediate substitutions:

$$\rho = \frac{1}{h}, \ \rho_1 = \frac{1}{h_1}, \ \rho_2 = \frac{1}{h_2}, \ \rho_H = \frac{1}{H}$$

so that:

$$\frac{H}{2} = \frac{1}{2\rho_H} = \frac{1}{\rho_2 + \rho_1}$$

The former of the above expressions for $p$ then becomes:

$$p = -\frac{6\eta U}{m}\left\{(\rho - \rho_1) - \frac{\rho^2 - \rho_1^2}{\rho_2 + \rho_1}\right\} = -\frac{6\eta U}{m}\frac{(\rho - \rho_1)(\rho_2 - \rho)}{\rho_2 + \rho_1}$$

i.e. $$p = -\frac{6\eta U}{m}\frac{(h_1 - h)(h - h_2)}{(h_2 + h_1)h^2}$$

which confirms that an excess of pressure in the film requires that $h_1 > h_2$.

The total thrust per unit width, normal to the face of the pad is, neglecting side leakage:

$$F = \int_0^l p\,dx = \frac{1}{m}\int_{h_1}^{h_2} p\,dh = \frac{6\eta U l^2}{(h_1^2 - h_2^2)(h_2 - h_1)}\int_{h_1}^{h_2}\frac{(h_1 - h)(h - h_2)dh}{h^2}$$

$$= \frac{6\eta U l^2}{(h_1 - h_2)^2(h_1 + h_2)}\int_{h_1}^{h_2}\left\{-\frac{h_1 + h_2}{h} + \frac{h_1 h_2}{h^2} + 1\right\}dh$$

$$= \frac{6\eta U l^2}{(h_1 - h_2)^2}\left\{\log_e\frac{h_1}{h_2} + \frac{h_1 h_2}{h_1 + h_2}\left(\frac{1}{h_1} - \frac{1}{h_2}\right) + \frac{h_2 - h_1}{h_2 + h_1}\right\}$$

$$= \frac{6\eta U l^2}{(h_1 - h_2)^2}\left\{\log_e\frac{h_1}{h_2} - 2\left(\frac{h_1 - h_2}{h_1 + h_2}\right)\right\}$$

The line of the resultant force $F$ on the pad acts, at a distance $\bar{x}$ from the origin given by

$$F\bar{x} = \int_0^l px\,dx = \frac{1}{m^2}\int_{h_1}^{h_2} p(h - h_1)dh$$

This is most conveniently integrated by parts, noting that $p = 0$ at both $h = h_1$ and $h = h_2$:

$$F\bar{x} = -\frac{h_1}{m^2}\int_{h_1}^{h_2} p\,dh + \frac{1}{m^2}\left[\frac{ph^2}{2} - \int\frac{h^2}{2}\frac{dp}{dh}dh\right]_{h_1}^{h_2}$$

$$= -\frac{h_1}{m}F - \frac{1}{2m^2}\int_{h_1}^{h_2} h^2\frac{dp}{dh}dh = -\frac{h_1}{m}F - \frac{3\eta U}{m^3}\int_{h_1}^{h_2}\left(1 - \frac{H}{h}\right)dh$$

$$= \frac{h_1 F l}{h_1 - h_2} - \frac{3\eta U l^3}{(h_1 - h_2)^2}\left\{1 - \frac{2h_1 h_2}{(h_1^2 - h_2^2)}\log_e\frac{h_1}{h_2}\right\}$$

$$\therefore \quad \frac{\bar{x}}{l} = \frac{h_1}{h_1 - h_2} - \frac{\frac{1}{2}(h_1^2 - h_2^2) - h_1 h_2\log_e\dfrac{h_1}{h_2}}{(h_1^2 - h_2^2)\log_e\dfrac{h_1}{h_2} - 2(h_1 - h_2)^2}$$

The shear stress referred to the axes $z$ and $x$ is:

$$\eta\frac{du}{dz} = -\frac{\eta U}{h} - \frac{1}{2}(h - 2z)\frac{dp}{dx}$$

so that, on the moving surface, where $z = 0$:

$$\tau = -\frac{\eta U}{h} - 3\eta U\left(\frac{1}{h} - \frac{H}{h^2}\right)$$

Thus the friction force is:

$$f = -\int_{0}^{l} \eta\frac{du}{dz}\,dx = \frac{\eta U}{m}\int_{h1}^{h2}\left(\frac{4}{h} - \frac{3H}{h^2}\right)dh$$

$$= \frac{\eta Ul}{h_1 - h_2}\left\{4\log_e\frac{h_1}{h_2} - \frac{6(h_1 - h_2)}{h_1 + h_2}\right\}$$

The results of this section give the pressures and the forces acting on a tilting pad as a function of film thickness $h$. The latter is, for a plane pad, linearly related to distance along the film by:

$$h = h_1 + mx, \quad h_2 = h_1 + ml$$

Michell suggested measuring distances from the point of intersection of the lines representing the fixed and moving surfaces. We then have a

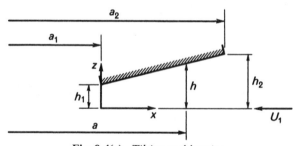

**Fig. 9.4(c).** *Tilting-pad bearing*

new coordinate ($a$) which is simply proportional to the corresponding film thickness ($h$):

$$ma = h, \quad ma_1 = h_1, \quad ma_2 = h_2, \quad ma_H = H$$

These substitutions for $h$ enable the forces and pressures to be related to $a$ noting that, for positive values of $m$, the direction of relative motion is in practice opposite to that assumed. We thus follow Michell and couple the use of $a$ with the introduction of $U_1 = -U$ to define the relative motion to keep both $m$ and the $a$'s positive.

EXAMPLE 9.4 (i)

*Deduce expressions for the pressure p, the mean pressure $\bar{p}$ and the position of the centre of pressure on a tilting-pad in terms of distances (a) measured from the intersection of the lines representing the fixed and moving surfaces. Neglect the effect of side leakage and assume that the viscosity remains constant.*
(*a*) The pressure is given by:

$$p = -\frac{6\eta U}{m}\left\{\frac{1}{h}-\frac{1}{h_1}-\frac{H}{2}\left(\frac{1}{h^2}-\frac{1}{h_1^2}\right)\right\}$$

For a convergent fluid wedge write $U_1 = -U$, see Fig. 9.4(c):

$$p = \frac{6\eta U_1}{m^2}\left\{\frac{1}{a}-\frac{1}{a_1}-\frac{a_H}{2}\left(\frac{1}{a^2}-\frac{1}{a_1^2}\right)\right\}$$

Noting that:

$$H = \frac{2h_1h_2}{h_1+h_2}\ ,\quad a_H = \frac{2a_1a_2}{a_1+a_2}$$

$$p = \frac{6\eta U_1}{m^2(a_1+a_2)}\left\{\frac{a_1+a_2}{a}-1-\frac{a_2}{a_1}-\frac{a_1a_2}{a^2}+\frac{a_2}{a_1}\right\}$$

$$= \frac{6\eta U_1}{m^2(a_1+a_2)}\left(\frac{a_1+a_2}{a}-\frac{a_1a_2}{a^2}-1\right)$$

It may alternatively be expressed as:

$$p = -\frac{6\eta U}{m}\frac{(h_1-h)(h-h_2)}{(h_2+h_1)h^2} = \frac{6\eta U_1}{m^2}\frac{(a_1-a)(a-a_2)}{(a_2+a_1)a^2}$$

(*b*) The mean pressure is:

$$\bar{p} = \frac{6\eta U l^2}{(h_1-h_2)^2}\left\{\log_e\frac{h_1}{h_2}-2\left(\frac{h_1-h_2}{h_1+h_2}\right)\right\}\frac{1}{a_2-a_1}$$

in which $(h_1-h_2)^2 = m^2(a_1-a_2)^2 = m^2 l^2$. Writing $U_1 = -U$:

$$\bar{p} = -\frac{6\eta U_1}{m^2(a_2-a_1)}\left\{\log_e\frac{a_1}{a_2}-2\frac{a_1-a_2}{a_1+a_2}\right\}$$

$$= \frac{6\eta U_1}{m^2}\left\{\frac{1}{a_2-a_1}\log_e\frac{a_2}{a_1}-\frac{2}{a_1+a_2}\right\}$$

(*c*) The centre of pressure is such that:

$$\frac{\bar{x}}{l} = \frac{h_1}{h_1-h_2}-\frac{\frac{1}{2}(h_1^2-h_2^2)-h_1h_2\log_e\dfrac{h_1}{h_2}}{(h_1^2-h_2^2)\log_e\dfrac{h_1}{h_2}-2(h_1-h_2)^2}$$

measured from end 1:

$$\therefore \quad \frac{\bar{x}}{l} = \frac{a_1}{a_1-a_2}-\frac{\frac{1}{2}(a_1^2-a_2^2)-a_1a_2\log_e\dfrac{a_1}{a_2}}{(a_1^2-a_2^2)\log_e\dfrac{a_1}{a_2}-2(a_1-a_2)^2}$$

Noting that $l = a_2-a_1$ and writing $\bar{a} = \bar{x}+a_1$

$$\bar{a} = \frac{1}{2}\frac{a_2^2-a_1^2-2a_1a_2\log_e\dfrac{a_2}{a_1}}{2(a_1-a_2)+(a_1+a_2)\log_e\dfrac{a_2}{a_1}}$$

EXAMPLE 9.4 (ii)

*Tabulate the film thicknesses at the leading and trailing edges of a tilting pad* 10 mm *long when carrying a load of* 1 N *per* mm *width, if the viscosity of the fluid is* 1 P *and the relative velocity of the moving surface is* 1 m/s. *Hence, or otherwise shows that the film thickness at the trailing edge is greatest when* $a_2/a_1$ *is about* 2·2.

The thrust per unit width may be expressed as:

$$F = \frac{6\eta U_1}{m^2}\left(\log_e \frac{h_2}{h_1} + 2\frac{1 - \frac{h_2}{h_1}}{1 + \frac{h_2}{h_1}}\right) = \frac{6\eta U_1}{m^2} f(h_2/h_1)$$

in which $h_2/h_1 = a_2/a_1$. Thus:

$$m = \sqrt{\left\{\left(\frac{6\eta U_1}{F}\right)f(a_2/a_1)\right\}} = \sqrt{\frac{6 \times \frac{1\,\text{g}}{10\,\text{mm s}} \times 1\frac{\text{m}}{\text{s}}}{\frac{1\ \text{N}}{\text{mm}}}\left[\frac{\text{N s}^2}{10^3\,\text{g m}}\right]}\sqrt{f(a_2/a_1)}$$

$$= 0{\cdot}024\ 6\ \sqrt{f(a_2/a_1)}$$

Hence:

| $a_2/a_1$ | 1·2 | 1·4 | 1·6 | 1·8 | 2·0 | 2·2 | 2·4 | 2·6 | 2·8 | 3·0 |
|---|---|---|---|---|---|---|---|---|---|---|
| $10^4 m$ | 5·6 | 14·0 | 23·2 | 32·2 | 41·6 | 49·0 | 57·0 | 64·4 | 71·7 | 78·6 |

Now:
$$h_2 - h_1 = m \times 10 \text{ mm}$$

$$h_2 = \left(\frac{a_2}{a_1}\right) \times h_1$$

∴
$$h_1 = \frac{10m}{\frac{a_2}{a_1} - 1} \text{ mm} \quad \therefore \quad h_2 = h_1 + 10m \text{ mm}$$

Hence:

| $a_2/a_1$ | 1·2 | 1·4 | 1·6 | 1·8 | 2·0 | 2·2 | 2·4 | 2·6 | 2·8 | 3·0 |
|---|---|---|---|---|---|---|---|---|---|---|
| $h_1/10^{-2}$ mm | 2·80 | 3·50 | 3·83 | 4·00 | 4·07 | 4·08 | 4·07 | 4·04 | 3·98 | 3·92 |
| $h_2/10^{-2}$ mm | 3·36 | 4·90 | 6·14 | 7·20 | 8·14 | 8·98 | 9·76 | 10·47 | 11·17 | 11·78 |

Thus for a given load and speed, the greatest clearance is maintained between the surfaces for a value of $a_2/a_1$ of about 2·2 if viscosity variations and side leakage are neglected. It will also be noted that in practice this figure may be rounded off to 2 and that the film thicknesses are so small that a high degree of surface finish is needed even for relatively modest bearing loads.

EXAMPLE 9.4 (iii)

Calculate the volumetric flow rate per unit width between each pair of moving surfaces in the previous example, assuming that $a_2/a_1 = 2\cdot2$.

$$\frac{Q}{B} = \tfrac{1}{2}\,UH$$

in which $H$ is the film thickness at the section where the pressure is a maximum ($dp/dx = 0$). The latter is given by:

$$H = \frac{2h_1h_2}{h_1+h_2} = \frac{2\times4\cdot08\times10^{-2}\ \text{mm}}{1+2\cdot2}\times2\cdot2$$

$$= 5\cdot61\times10^{-2}\ \text{mm}$$

$$\therefore\quad \frac{Q}{B} = \tfrac{1}{2}\times1\,\frac{\text{m}}{\text{s}}\times5\cdot61\times10^{-2}\ \text{mm}\left[\frac{10^3\ \text{mm}}{\text{m}}\right]$$

$$= 28\ \text{mm}^3/\text{s per mm.}$$

EXAMPLE 9.4 (iv)

*Find the pivot position for a tilting pad to operate with a film thickness at its leading edge which is twice that at the trailing edge. Neglect viscosity variations and side leakage.*

$$\bar{a} = \tfrac{1}{2}\,\frac{a_2^2-a_1^2-2a_1a_2\,\log_e\dfrac{a_2}{a_1}}{2(a_1-a_2)+(a_1+a_2)\,\log_e\dfrac{a_2}{a_1}} = \frac{a_1}{2}\,\frac{3-4\log_e2}{3\log_e2-2} = 1\cdot43a_1$$

For there to be no unbalanced moment about the pivot, the latter must be located at $(\bar{a}-a_1)$, i.e. at $0\cdot43a_1$ ahead of the trailing edge. For $a_2 = 2a_1$ it follows that $l = a_1$. Hence the pivot must lie $0\cdot43l$ ahead of the trailing edge or $0\cdot07l$ aft of the mid-point of the pad and experience shows that the equilibrium configuration is stable.

Each pivot position thus corresponds to a particular value of $a_2/a_1$ and hence $h_2/h_1$. This implies that if the load is varied, the film thickness so adjusts itself that the lines representing the fixed and moving surfaces always intersect at the same point. If the mean film thickness is $\bar{h}$ it follows that:

$$\frac{\bar{h}}{l} \propto \sqrt{\frac{\eta U}{F}}$$

or:

$$F \propto \frac{\eta U l^2}{\bar{h}^2}$$

in which $F$ is the thrust per unit width. If $f$ is the corresponding friction force per unit width:

$$f \propto \frac{\eta U l}{\bar{h}}$$

**224**    *Viscous films of varying thickness*

Hence the 'coefficient of friction' of the bearing:

$$\mu \propto \frac{f}{F} \propto \frac{\bar{h}}{l} \propto \sqrt{\frac{\eta U}{F}}$$

showing that there are advantages in using fluid of the lowest viscosity which is compatible with the maintenance of an adequate film thickness for the load to be carried.

If $R$ denotes the mean radius of the circumferential centre-line of the pads, the power required to overcome friction:

$$fU \propto \mu F(2\pi RN) \propto \frac{\bar{h}}{l} FRN$$

The power dissipated by friction causes heating and if the latter is expressed per unit area of each pad it is given by:

$$\frac{fU}{l} \propto \frac{\eta U^2}{h} \propto \frac{\eta R^2 N^2}{h}$$

where $N$ is the revolutions of the collar per unit of time.

EXAMPLE 9.4 (v)

*A model of a Michell bearing dissipates* 1·5 kW *in heating when running at* 1 000 rev/min *under load. A geometrically similar bearing three times as large as the model runs at* 750 rev/min. *The latter is cooled by oil of density* 1 Mg/m³, *specific heat capacity* 2 kJ/kgK *and the dynamic viscosity is* 0·15 kg/ms, *delivered through a pipe* 12·5 mm *diameter and* 30 m *long. Estimate the power of the pump necessary if the temperature rise is not to exceed* 50 K *when the full scale bearing is operating with the same tilt of the pads. Assume the dynamic viscosity remains constant.*

The heating per unit of projected area of each pad is proportional to $\eta R^2 N^2/\bar{h}$ and hence the rate of heating of each pad of the bearing $\propto \eta R^2 N^2 Bl/\bar{h}$. For geometrically similar bearings the ratio of corresponding linear dimensions is the same, e.g.

$$B_2/B_1 = l_2/l_1 = R_2/R_1$$

Hence the rate of heating:

$$Q \propto \eta L^3 N^2$$

i.e.

$$\frac{Q_2}{Q_1} = \frac{\eta_2}{\eta_1}\left(\frac{L_2}{L_1}\right)^3\left(\frac{N_2}{N_1}\right)^2$$

in which $L$ is a representative linear dimension,

i.e.

$$Q_2 = 1·5 \text{ kW} \times 1 \times 3^3 \times 0·75^2 = 22·8 \text{ kW}$$

and this is the rate at which energy is received by the oil to raise the temperature by a maximum of 50 K. Hence the volumetric flow rate:

$$V = \frac{Q_2}{\rho s \delta T} = 22·8 \frac{\text{kJ}}{\text{s}} \times \frac{\text{m}^3}{1 \text{ Mg}} \times \frac{\text{kgK}}{2 \text{ kJ}} \times \frac{1}{50 \text{ K}} = 228 \times 10^{-6} \text{ m}^3/\text{s}$$

and the mean velocity in the pipe is:

$$\frac{228 \times 10^{-6}}{\pi} \times \frac{4 \times 10^6 \; \text{m}}{12 \cdot 5^2 \; \text{s}} = 1 \cdot 86 \, \frac{\text{m}}{\text{s}}$$

Reynolds number:

$$Re = \frac{\rho d \bar{u}}{\eta} = \frac{1 \; \text{Mg}}{\text{m}^3} \times \frac{\text{m s}}{0 \cdot 15 \; \text{kg}} \times \frac{12 \cdot 5 \; \text{m}}{10^3} \times 1 \cdot 86 \, \frac{\text{m}}{\text{s}} = 155$$

so that the flow in the pipe is laminar. Hence:

$$\Delta p = \frac{32 \eta \bar{u} l}{d^2} = 32 \times 0 \cdot 15 \, \frac{\text{kg}}{\text{ms}} \times 1 \cdot 86 \, \frac{\text{m}}{\text{s}} \times \frac{30 \; \text{m} \times 10^6}{(12 \cdot 5 \; \text{m})^2} \left[ \frac{\text{N s}^2}{\text{kg m}} \right]$$

$$= 1 \cdot 59 \; \text{MN/m}^2$$

The pump power needed is thus:

$$P = V \Delta p = 228 \times 10^{-6} \, \frac{\text{m}^3}{\text{s}} \times 1 \cdot 59 \, \frac{\text{MN}}{\text{m}^2} = 362 \, \frac{\text{Nm}}{\text{s}} = 0 \cdot 362 \; \text{kW}$$

## 9.5. Bearings of finite width

For bearings of conventional proportions, the area open to side-leakage normally exceeds the inlet and outlet areas for the zone over which the fluid wedge is effective. Consequently a more complete theory requires that account be taken not only of the oil flow in the direction of $U_1$ ($x_1$ say) but also of lateral flow in the directions perpendicular to $U_1$, $x_2$ say.

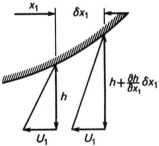

**Fig. 9.5(a).** *Converging viscous film*

A more comprehensive analysis would thus require that account be taken of pressure variations in both the $x_1$ and $x_2$ directions, as well as the fact that the fluid has velocity components $u_1$ and $u_2$ in each of these two directions.

We may, however, obtain some indications of the effect of side leakage by noting that it will be most marked from the regions where $p$ is largest, i.e. where $\partial p / \partial x_1$ is substantially zero. It can be shown that the distribution of $u_1$ across the film is then linear, as in Section 9.2. Referring to Figure 9.5(a) the flow rate induced across the section $x_1 = $ constant is $\frac{1}{2}U_1 h$ per unit width in the direction of $x_2$. If therefore we consider a fixed volume standing on the base $\delta x_1 \delta x_2$, relative movement of the two surfaces directly induces an inflow $\frac{1}{2}U_1(h + [\partial h/\partial x_1] \, \delta x_1) \, \delta x_2$. This exceeds the corresponding outflow by an amount:

$$\frac{1}{2}U_1 \frac{\partial h}{\partial x_1} \delta x_1 \delta x_2$$

This accumulation of fluid is prevented by sideways flow. It can be shown that the net flow across a plane $x_2 = $ constant is given, as in the two dimensional case, by:

$$\frac{Q}{B} = -\frac{1}{12}\frac{h^3}{\eta}\frac{\partial p}{\partial x_2}$$

in which $B = \delta x_1$. Thus fluid is withdrawn from the fixed control volume by the excess of the outflow across the plane $(x_2 + \delta x_2) = $ constant over the inflow across the plane $x_2 = $ constant:

$$\frac{\partial}{\partial x_2}(Q\delta x_1)\delta x_2 = -\frac{1}{12}\frac{\partial}{\partial x_2}\left(\frac{h^3}{\eta}\frac{\partial p}{\partial x_2}\right)\delta x_1 \delta x_2$$

Thus constancy of volume requires that:

$$\frac{1}{2}U_1 \frac{\partial h}{\partial x_1} = -\frac{1}{12}\frac{\partial}{\partial x_2}\left(\frac{h^3}{\eta}\frac{\partial p}{\partial x_2}\right)$$

i.e.

$$\frac{\partial}{\partial x_2}\left(h^3 \frac{\partial p}{\partial x_2}\right) = -6\eta U_1 \frac{\partial h}{\partial x_1}$$

Noting that $h$ is independent of $x_2$:

$$\frac{\partial^2 p}{\partial x_2^2} = -\frac{6\eta U_1}{h^3}\frac{\partial h}{\partial x_1}$$

∴

$$\frac{\partial p}{\partial x_2} = -\frac{6\eta U_1}{h^3}\frac{\partial h}{\partial x_1} x_2 + \varphi(x_1)$$

in which $\varphi(x_1) = 0$ since $\partial p/\partial x_2 = 0$ on the line of symmetry, which we make $x_2 = 0$. Integrating again and assuming that $p = 0$ when $x_2 = \pm (B/2)$ we obtain:

$$p = \frac{3\eta U_1}{h^3}\left(\frac{B^2}{4} - x_2^2\right)\frac{\partial h}{\partial x_1}$$

Apart from the usual assumptions for laminar flow in thin films, this result is exact only for lateral flow along lines for which the pressure gradient $\partial p/\partial x_1 = 0$. It will be approximately true for all sections of very narrow bearings—for which the side flow much exceeds that entrained by the moving surface, so that $\partial p/\partial x_2 \gg \partial p/\partial x_1$.

EXAMPLE 9.5 (i)

*Obtain an approximate expression for the pressure distribution in a narrow bearing, taking account of side leakage.*

The film thickness in a journal bearing is, referring to Fig. 9.3(b):

$$h = c + e \cos \theta = c(1 + \epsilon \cos \theta)$$

and since $R\theta = x$, then:

$$\frac{dh}{dx_1} = \frac{d\theta}{dx_1} \frac{dh}{d\theta} = -\frac{c}{R} \epsilon \sin \theta$$

When this is substituted into the above equation for pressure in the film we obtain for $U_1 = -R\omega$:

$$p = \frac{3\eta\omega}{c^2} \left( \frac{B^2}{4} - x_2^2 \right) \frac{\epsilon \sin \theta}{(1 + \epsilon \cos \theta)^3}$$

This shows that for a short bearing (i.e. when there is end leakage) the pressure distribution is parabolic and symmetrical in the axial ($x_2$) direction —as distinct from long journal bearings for which the effect of end leakage was neglected.

## Exercises on Chapter 9

1. (a) Deduce expressions for the mean velocity of flow and the velocity distribution across a viscous film of uniform thickness between two plates, one of which is moving with constant velocity. Clearly state the assumptions made.

(b) Assuming these formulae apply to a film whose thickness varies very gradually, show that the thickness of the film at a point of reverse flow is $h_r = (3/2)H$, where $H$ is the film thickness at a point where the pressure gradient is zero.

(c) Also show that the shear stresses on the two boundary plates are given by the expressions

$$\eta \frac{U}{h} \pm \frac{h}{2} \left( \frac{dp}{dx} \right)$$

(d) What is the shear stress on the stationary surface at a point of reverse flow?

2. (a) Show that the approximate power needed to maintain a plunger or ram stationary when concentric in a cylinder may be written

$$\frac{4}{3\pi} \frac{W^2 h^3}{\eta L d^3}$$

where $W$ is the load carried, $h$ is the uniform thickness of the film of oil of dynamic viscosity $\eta$, and $d$ and $L$ are the diameter and length of the ram, respectively.

(b) If the delivery valve of the pump is closed, deduce an expression giving the rate at which the ram will recede.

3. Show that for a dashpot in which the plunger is maintained concentric

and in which the flow or leakage of oil along the clearance space is laminar, the force exerted is proportional to the rate of movement of the plunger.

4. A plunger 200 mm diameter and 200 mm long slides in a vertical cylinder with a uniform radial clearance 0·25 mm and supports a vertical load of 500 kN. If the cylinder is filled with oil of dynamic viscosity 0·15 kg/ms estimate the rate at which the loaded plunger will sink when the supply valve is shut neglecting the speed of the plunger relative to the mean speed of the oil in the concentric clearance space.

5. The uniform radial clearance round a piston 75 mm long and 50 mm diameter which moves vertically in an oil dash-pot is 1·25 mm. The piston descends 37·5 mm in 40 s under its own weight and a piston of the same dimensions but 0·9 N heavier descends 37·5 mm in 25 s. Estimate the dynamic viscosity of the oil and state the assumptions made.

6. At a certain point in a journal bearing the thickness of the oil film is 2·5 μm and the linear velocity of the journal is 90 m/min. Calculate the values of shear stress on the journal and on the bearing at this point when the flow due to pressure is −0·5 that due to viscous drag of oil of dynamic viscosity 0·15 kg/ms.

7. At a certain section in the film of a lubricated journal bearing the thickness is 3·75 μm and the shear stresses on the journal and bearing are both equal to 7·5 N/mm². Estimate the shear stresses on the journal and bearing at a point where the film thickness is 2·5 μm stating the assumptions made.

8. Derive an expression for maximum pressure on a Michell pad, neglecting side leakage.

The collar speed in a Michell thrust bearing is 21 m/s and the entry and exit distances between the 60 mm long pad and the collar are 0·05 mm and 0·02 mm, respectively. Calculate the peak pressure if the dynamic viscosity of the lubricant is 0·01 kg/ms.

9. A tilting-pad in a Michell thrust bearing is 200 mm long, and its leading and trailing edges are 0·075 mm and 0·025 mm, respectively, from a flat collar moving at the rate of 1·2 m/s. Estimate, per cm width of pad, the load which the pad can sustain, the drag force on the pad and the power required to maintain the motion if the oil has a viscosity 0·7 P.

## Chapter 10

# Similarity and dimensional analysis

*'In order to write with the true science of the
flight of birds through the air it is necessary first
to give the science of winds, which we shall prove
by the motion of water inside itself; and the
understanding of this science will lead to the
required knowledge.'*  LEONARDO

## 10.1. Dynamic similarity

Discounting the effect of surface tension, which is generally small, the
motion of a fluid element is controlled by any one or combination of
three sets of forces. These are respectively caused by viscosity, gravity
and pressure difference. Their resultant on each particle of fluid causes
its mass accelerations, both along, and at right angles to, its direction of
motion.

We may, by similarity arguments, determine how any force would
vary from one system to another, when each executes kinematically
similar motions (see Section 5.2). Fig. 10.1(a) shows two similar aerofoils
at similar attitudes, with similar flows round them, i.e. the only difference
between the two systems is their size, as expressed by the length $l$. The two
corresponding elements at $P_1$ and $P_2$ will each have masses proportional
to $\rho l^3$ where $\rho$ is the fluid density. If the flows are kinematically similar
(i.e. the particles move on similar paths with velocities that are always in
the same ratio) the corresponding accelerations are proportional to
$U^2/l$, where $U$ is a reference velocity, that of the approaching stream say,
see Section 5.2.

From the equation of motion therefore each particle must experience
forces which are proportional to:

$$F = ma$$

i.e.
$$F \propto \rho l^3 \frac{U^2}{l} \propto \rho l^2 U^2$$

As this is true for all particles (including those surrounding the body) the forces which arise in kinematically similar flows are, if dynamic effects are appreciable, such that:

$$F \propto \rho l^2 U^2$$

This is a condition for *dynamic similarity*. For *all* the forces which arise to be in this ratio, then the viscous actions for example must also be in this ratio at corresponding points. In Section 10.3 it is shown that this requires the same value of the Reynolds number ($\rho l U/\mu$). *Thus one*

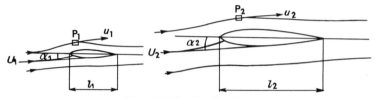

**Fig. 10.1(a).** *Similar flows*

*requirement for dynamic similarity is that the Reynolds numbers shall be the same.*

Dynamic similarity implies that the resultant force experienced by each of the aerofoils shown in Fig. 10.1(a) must be such that:

$$F_1 = k_{F}\rho_1 l_1^2 U_1^2$$
$$F_2 = k_{F}\rho_2 l_2^2 U_2^2$$

Such constants as $k_F$ are referred to as Newtonian force coefficients, and if dynamic similarity can be achieved, they will have the same value on both scales. It is the aim of many model tests in fluid mechanics to obtain the values of such force coefficients, which give, e.g., the lift or the drag, the thrust or the skin friction, of different shapes under various conditions. If we can be confident that the flow pattern round a model is similar to that on full scale (geometric similarity being a prerequisite for dynamic similarity), the values of $k_F$ may be used to predict forces in the latter case. In addition, it may enable results obtained, say, in a wind tunnel, to be applied to the design of hydraulic installations in which the working fluid may be oil or water. Thus similarity studies enable us to express experimental results in such a form that they may be given the widest possible application. The coefficients so obtained are equally relevant to any similar flow, independent of size, speed and scale.

EXAMPLE 10.1 (i)

*A one-tenth scale torpedo model was tested in a wind tunnel using compressed air having density of 25·5 kg/m³ and kinematic viscosity of 0·67 cSt or 0·67 × 10⁻⁶ m²/s.*

*The resistance was found to be 8 N when the air velocity was 30 m/s. Find the resistance and speed of the full-scale torpedo when moving, under dynamically similar conditions, through sea water having density of* 1 Mg/m³ *and kinematic viscosity of* 1·58 mm²/s.

For dynamically similar flows round geometrically similar models the Reynolds numbers must be the same, i.e.

$$\frac{ul}{v} = \frac{u_m l_m}{v_m}$$

Hence

$$u = u_m \frac{l_m}{l} \frac{v}{v_m} = 30 \frac{\text{m}}{\text{s}} \left(\frac{1}{10}\right) \left(\frac{0\cdot158 \times 10}{0\cdot67}\right) = 7\cdot08 \text{ m/s}$$

i.e. speed of torpedo in sea water is

$$u = 7\cdot08 \frac{\text{m}}{\text{s}} \left[\frac{\text{knot s}}{0\cdot514 \text{ m}}\right] = \textbf{13·76 knot}$$

Also, the force coefficients for the model and full-scale torpedo are the same when *Re* is the same, i.e.

$$\frac{F}{F_m} = \frac{\rho}{\rho_m} \left(\frac{l}{l_m}\right)^2 \left(\frac{u}{u_m}\right)^2 = \left(\frac{1\ 000}{25\cdot5}\right) \left(\frac{10}{1}\right)^2 \left(\frac{7\cdot08}{30}\right)^2 = 219\cdot5$$

and the force resisting the full-scale torpedo in sea water will be

$$F = 219\cdot5 \times 8 \text{ N} = \textbf{1 756 N}$$

## 10.2. Lift and drag coefficients

According to a definition by the B.S.I., an aerofoil is a body so shaped as to produce aerodynamic reaction normal to its motion through the air without excessive drag.

The *lift* on an aerofoil is defined as the component of the resultant force

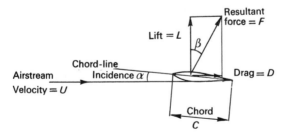

**Fig. 10.2(a).** *Lift and drag on an aerofoil*

which acts at right angles to its direction of motion. In fluid mechanics we reduce the problem to one of steady flow by imagining the aerofoil to be stationary, and the air to move towards it, as in a wind-tunnel test. From this point of view the lift, *L*, is the component force at right angles to the undisturbed stream, as shown in Fig. 10.2(a). *The symbol L, usually used to denote this force, should not be confused with the reference length* ($l$) *of the previous section.* In aerodynamics it is customary to use the symbol $c$ and not $l$ to denote the chord of the aerofoil, see Fig. 10.2(a), the *span* being denoted by $b$ and the *plan area bc* by $S$. The *drag*, i.e. the force component along the direction of motion (or along the direction of the free stream), is represented by $D$ and this has been considerably exaggerated in the figure. Drag forces are also important in rotary machines such as fans, compressors, turbines and propellers. In order to describe the attitude of an aerofoil, a datum line is chosen and this is usually defined to pass through the centres of curvature of the leading and trailing edges. The angle which this makes with the direction of the undisturbed stream is referred to as the *incidence*.

In terms of the notation of this section, the resultant force on all geometrically similar aerofoils at a particular incidence, if the flows round them are dynamically similar, is:

$$F \propto \rho c^2 U^2$$

For aerofoils of considerable width, such as a section completely spanning a wind tunnel, it is convenient to express the forces which act on it per unit plan area $S$ and in the case of isolated fuselages, submarines, torpedoes, etc., the maximum cross-sectional area normal to the stream is used. The resultant aerodynamic force $F$ varies with incidence both as regards magnitude, direction and line of action. Thus, the position of the centre of pressure—which is the point at which the line of action of the resultant force $F$ cuts the chord line (Fig. 10.2(a)), varies with incidence. Noting that, for geometrically similar aerofoils, $S$ is proportional to $c^2$, we may re-write the above expression as:

$$F \propto \rho S U^2$$

For a particular incidence, $L = F \cos \beta$ and $D = F \sin \beta$, hence the components of force $F$ may be written as:

$$L \propto \rho S U^2$$

and
$$D \propto \rho S U^2$$

For dynamically similar flows round similar aerofoils $L$ and $D$ may be expressed in terms of corresponding coefficients $k_L$ and $k_D$, which will have the same values on all scales:

$$L = k_L \rho S U^2$$

and
$$D = k_D \rho S U^2$$

It is, however, now usual to express results in terms of coefficients $C_L$ and $C_D$, which are twice $k_L$ and $k_D$, respectively, i.e. to write:

$$\text{Lift } L = C_L \tfrac{1}{2} \rho S U^2$$

and
$$\text{Drag } D = C_D \tfrac{1}{2} \rho S U^2$$

$C_L$ and $C_D$ being termed *lift* and *drag coefficients* respectively. These two coefficients are dimensionless ratios later shown (see Section 10.3) to be functions of Reynolds' number. The factor $\frac{1}{2}\rho U^2$ is the dynamic pressure of the air stream and may be thought of as a reference pressure—e.g. pressure distribution measurements are often expressed in terms of a pressure coefficient

$$C_P = \frac{p - p_0}{\frac{1}{2}\rho U^2}$$

where $p_0$ is the static pressure of the free stream and $p$ is the pressure at any point on a blade or aerofoil.

Thus, the lift per unit plan area $L/S$ (i.e. the mean lifting pressure) is $C_L$ times the dynamic pressure of the air stream.

EXAMPLE 10.2 (i)

*A rectangular aerofoil of 300 mm chord and 1·5 m span was tested in a wind tunnel in which the air speed was 36 m/s. The resultant force was measured to be 310 N inclined at an angle of 4° to the vertical or axis of lift. Calculate the coefficients of lift and drag taking the density of air as $\rho = 1\cdot25$ kg/m³.*

$$L = F\cos\beta = 310 \text{ N} \times \cos 4° = 310 \text{ N} \times 0\cdot9926 = 308 \text{ N}$$
$$D = F\sin\beta = 310 \text{ N} \times \sin 4° = 310 \text{ N} \times 0\cdot06976 = 21\cdot6 \text{ N}$$
$$\text{Wing area } S = 1\cdot5 \text{ m} \times 0\cdot3 \text{ m} = 0\cdot45 \text{ m}^2$$

Reference pressure

$$p = \tfrac{1}{2}\rho u^2 = \tfrac{1}{2} \times 1\cdot25 \, \frac{\text{kg}}{\text{m}^3} \times 1\,296 \, \frac{\text{m}^2}{\text{s}^2} = 810 \text{ N/m}^2$$

Hence, coefficient of lift is

$$C_L = \frac{L}{pS} = \frac{308}{810 \times 0\cdot45} = 0\cdot845$$

and the coefficient of drag is

$$C_D = \frac{D}{pS} = \frac{21\cdot6}{810 \times 0\cdot45} = 0\cdot059$$

Fig. 10.2(b) shows typical curves for the variation of the lift and drag coefficients with incidence.

For a symmetric section the lift curve passes through the origin, but a cambered section (i.e. one that is flexed upwards towards its mid-chord) has a *no-lift angle* $\alpha_0$ which is negative.

The pressure distribution round an aerofoil is shown in Fig. 10.2(c). On the upper surface it will be seen that the pressure is mostly less than atmospheric while underneath it tends to alternate about the pressure of the surrounding atmosphere. Fig. 10.2(b) shows that the lift coefficient $C_L$ rises linearly with incidence for small departures from the angle of no-lift $\alpha_0$, but that the slope of the curve falls as the *stall* is approached—*the angle of stall being that angle of incidence corresponding*

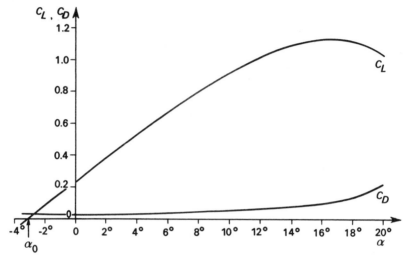

**Fig. 10.2(b).** *Lift and drag coefficients*

*to maximum lift coefficient.* This occurs when the boundary layer separates, causing the lift to fall and the drag to rise. The reason for the separation is that, on the upper surface, the boundary layer is moving against an adverse pressure gradient downstream of the point of maximum suction, as indicated in (b) of Fig. 10.2(c).

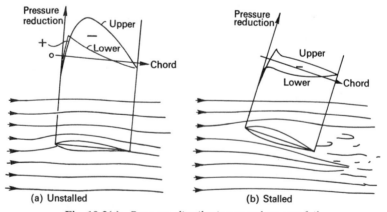

**Fig. 10.2(c).** *Pressure distribution round an aerofoil*

As the lift is increased so the boundary layer thickens due to the increased pressure gradient against which it moves. Finally, break-away of the flow occurs, leaving a broad eddying wake. The maximum lift coefficients that can normally be obtained without the use of high lift devices is of the order of 1·5, and this corresponds to an angle of incidence which is about 15° above the no-lift angle.

As stalling sets an upper limit to $C_L$ the maximum lift that can be developed is:

$$L_{max} = (C_L)_{max}\tfrac{1}{2}\rho SU^2$$

Flight at low speeds (e.g. landing and take-off) and low air densities (i.e. high altitudes) thus tend to call for large $C_L$'s, as do manoeuvres which require large values of $L$. Stalling limits performance under all these conditions.

It should be noted that the under surface contributes little to the lift and, in some cases, may even experience a downward force.

Theoretical studies show that, for a boundary layer which is vanishingly thin, the theoretical slope of the $C_L/\alpha$ curve is, for thin aerofoils:

$$\frac{dC_L}{d\alpha} = 2\pi$$

In practice, the slope is more nearly 6, i.e. a change of incidence of 0·1 radian (i.e. 5·73°) causes a change in $C_L$ of about 0·6.

The drag coefficient $C_D$ has a minimum value near the angle of zero lift and is small and fairly constant over a range of about 5°. After that it rises gradually until the stall point, beyond which it increases rapidly. In terms of the two components of the drag force—that due to skin friction and that due to pressure differences—the form drag, separation and the consequent eddies cause a failure to maintain pressure recovery towards the trailing edge. Consequently the downstream side of the aerofoil experiences a reduced pressure from the wake and the form drag increases. Such separation or stalling may occur on the blading of machinery as well as on aircraft. This is particularly likely on pumps, fans and compressors where the pressure is higher downstream. Conditions are analogous to those in a diverging hydraulic channel in which, if the angle of divergence is too great, separation and turbulence, with consequent dissipation of energy into heat, are induced.

EXAMPLE 10.2 (ii)

*Estimate the brake-power of the engines required to propel an aeroplane at 500 km/h if the total wing area is 60 m² and the wing loading is 2 kN/m². The efficiency of the propellers may be taken as 75 per cent and the lift/drag ratio is 10.*

Weight of aeroplane is

$$W = 60 \text{ m}^2 \times 2 \text{ kN/m}^2 = 120 \text{ kN} = L$$

$$\text{Drag } D = \frac{L}{10} = 12 \text{ kN}$$

Thrust power required from propellers $= \text{Drag} \times \text{Forward velocity}$

$$= 12 \text{ kN} \times \frac{500 \text{ km}}{3 \cdot 6 \text{ ks}} \left[ \frac{\text{W s}}{\text{Nm}} \right]$$

$$= 1\ 667 \text{ kW}$$

Hence, brake-power required from engines:

$$= \frac{\text{Thrust power of propellers}}{\text{Efficiency of propellers}}$$

$$= 1\ 667 \text{ kW} \times \tfrac{4}{3} = \mathbf{2 \cdot 2 \text{ MW}}$$

### Example 10.2 (iii)

*The overall drag coefficient of an aircraft of weight $W$ and wing area $S$ is given by $C_D = a + bC_L^2$ where $a$ and $b$ are constants. Derive formulae for:*
  (i) *minimum drag ($D_1$) in horizontal flight,*
  (ii) *the speed ($U_1$) at which minimum drag occurs,*
  (iii) *the power ($P_1$) expended in overcoming minimum drag in horizontal flight.*

(i) $$C_L = \frac{W}{\tfrac{1}{2}\rho U^2 S} = \frac{W}{pS}$$

Hence,

$$C_D = a + \frac{bW^2}{p^2 S^2}$$

and the drag force

$$D = C_D . \tfrac{1}{2}\rho U^2 S = pSC_D$$

i.e.    $$D = aSp + \frac{bW^2}{S}\frac{1}{p} = \text{parasitic drag } D_s + \text{induced drag } D_i$$

Therefore,

$$\frac{dD}{dp} = aS - \frac{bW^2}{S}\frac{1}{p^2}$$

and

$$\frac{d^2 D}{dp^2} = \frac{2bW^2}{Sp^3}$$

The latter is positive for all values of $p$; hence the minimum drag, which occurs when $dD/dp = 0$, i.e. when

$$p^2 = \frac{b}{a}\frac{W^2}{S^2} = p_1^2$$

is    $$D_1 = aS \left( \frac{W}{S}\sqrt{\frac{b}{a}} \right) + \frac{bW^2}{S} \left( \frac{S}{W}\sqrt{\frac{a}{b}} \right) = D_{s_1} + D_{i_1} = aSp_1 + \frac{bW^2}{p_1 S}$$

$$= W\sqrt{(ab)} + W\sqrt{(ab)} = \mathbf{2W\sqrt{(ab)}}$$

This shows that, for minimum drag, the parasitic drag and induced drag both equal $W \sqrt{(ab)}$.

(ii) The dynamic pressure of the free stream (i.e. the reference pressure) for minimum drag is

$$\tfrac{1}{2}\rho U^2 = p_1 = \frac{W}{S}\sqrt{\frac{b}{a}}$$

Hence the speed for minimum drag is

$$U_1 = \left(\frac{2W}{\rho S}\right)^{1/2}\left(\frac{b}{a}\right)^{1/4}$$

Also, when the drag is a minimum ($D_1$) the ratio $W/D_1$ is a maximum, i.e. since the weight $W$ remains constant the maximum lift/drag ratio occurs at minimum drag speed $U_1$.

(iii) The power required to overcome drag $D_i$ at speed $U_1$ is:

$$P_1 = D_1 U_1 = 2W(ab)^{1/2}\left(\frac{2W}{\rho S}\right)^{1/2}\left(\frac{b}{a}\right)^{1/4}$$

i.e.       $$P_1 = \frac{(2W)^{3/2}a^{1/4}b^{3/4}}{(\rho S)^{1/2}}$$

## EXAMPLE 10.2 (iv)

*An aeroplane weighing 100 kN has a wing area of 60 m² and a drag coefficient (based on wing area) $C_D = 0.029\ 3 + 0.039\ 8\ C_L^2$. Taking the density of air to be 1·25 kg/m³ and assuming horizontal flight, calculate:*

*(a) the minimum drag and the corresponding speed and power expended,*
*(b) the total drag at 175 knot and power expended,*
*(c) the speed, drag and power when minimum power is expended.*

From Example 10.2 (iii): (a) minimum drag

$$D_1 = 2W \sqrt{(ab)} = 2 \times 100 \text{ kN } \sqrt{(0.029\ 3 \times 0.039\ 8)} = \textbf{6·826 kN}$$

Also, for minimum drag:

$$\tfrac{1}{2}\rho U_1^2 = p_1 = \frac{W}{S}\sqrt{\frac{b}{a}}$$

the speed

$$U_1 = \left(\frac{2W}{\rho S}\right)^{1/2}\left(\frac{b}{a}\right)^{1/4} = \left(\frac{2 \times 10^5 \text{ N}}{1\cdot25 \text{ kg/m}^3 \times 60 \text{ m}^2}\left[\frac{\text{kg m}}{\text{N s}^2}\right]\right)^{1/2}\left(\frac{398}{293}\right)^{1/4}$$

$$= \left(\frac{10^4 \text{ m}^2}{3\cdot75 \text{ s}^2}\right)^{1/2}(1\cdot36)^{1/4}$$

$$= 55\cdot7 \text{ m/s} \quad \text{or} \quad \textbf{108 knot}$$

Therefore, the power expended for minimum drag is:

$$P_1 = D_1 U_1 = 6\cdot826 \text{ kN} \times 55\cdot7 \text{ m/s} = \textbf{380 kW}$$

(b) The drag at any speed $U$ is

$$D = aSp + \frac{bW^2}{S}\frac{1}{p} = \left(\frac{D_{s_1}}{p_1}\right)p + (D_{i_1} \cdot p_1)\frac{1}{p} = D_{s_1}\left(\frac{U}{U_1}\right)^2 + D_{i_1}\left(\frac{U_1}{U}\right)^2$$

But since

$$D_{s_1} = D_{i_1} = D_1/2 = W\sqrt{(ab)}$$

the drag

$$D = \frac{D_1}{2}\left\{\left(\frac{U}{U_1}\right)^2 + \left(\frac{U_1}{U}\right)^2\right\} = \frac{6 \cdot 826}{2}\text{ kN}\left\{\left(\frac{175}{108}\right)^2 + \left(\frac{108}{175}\right)^2\right\} = \textbf{10·26 kN}$$

and the power expended when flying at 175 knot or 91·7 m/s is

$$P = DU = 10 \cdot 26 \text{ kN} \times 91 \cdot 7 \text{ m/s} = \textbf{941 kW}$$

(c)

$$P = DU = \frac{D_1}{2}\left\{\frac{U^3}{U_1^2} + \frac{U_1^2}{U}\right\}$$

Hence

$$\frac{dP}{dU} = \frac{D_1}{2}\left\{\frac{3U^2}{U_1^2} - \frac{U_1^2}{U^2}\right\} = 0$$

whence $3U^4 = U_1^4$.

Power is a minimum when

$$U = \frac{U_1}{\sqrt[4]{3}} = 0 \cdot 76 U_1$$

Hence the speed for minimum power is $U = 0 \cdot 76 \times 108$ knot = **82** knot.

The drag for minimum power is:

$$D = \frac{D_1}{2}\left\{\left(\frac{U}{U_1}\right)^2 + \left(\frac{U_1}{U}\right)^2\right\} = \frac{6 \cdot 826}{2}\text{ kN}\left\{\left(\frac{82}{108}\right)^2 + \left(\frac{108}{82}\right)^2\right\} = \textbf{7·88 kN}$$

and the minimum power is

$$7 \cdot 88 \text{ kN} \times 82 \text{ knot} \left[\frac{0 \cdot 514 \text{ m}}{\text{knot s}}\right] = \textbf{332 kW}$$

## 10.3. Conditions for dynamic similarity

Most of our present ideas on similarity stem from Osborne Reynolds' studies of the nature of flow through pipes (see *Transactions of Royal Society*, 1883), a subject which will be considered in more detail later— see Chapter 11. This reason alone makes the laminar pipe flow problem (see Section 8.5, p. 180) an appropriate starting-point for discussing the precise conditions for similar flow patterns to be established.

Fig. 8.5(b) (p. 180) showed a fluid element moving under the combined influence of a pressure difference across its ends and a shear stress round its circumference. Fig. 10.3(a) shows two similar elements in such a flow, that is each has a radius and length proportional to the corresponding pipe diameter, $D$, which may be taken as the reference length. Hence:

$$\frac{\delta r_1}{\delta r_2} = \frac{r_1}{r_2} = \frac{x_1}{x_2} = \frac{D_1}{D_2}$$

and, in general, for all such elements:

$$r \propto x \propto D$$

If $\Delta p_1$ and $\Delta p_2$ are the pressure differences, the corresponding forces $F_p$ due to such pressure changes ($\Delta p$) are such that:

$$\frac{F_{p_1}}{F_{p_2}} = \frac{\Delta p_1 2\pi r_1 \delta r_1}{\Delta p_2 2\pi r_2 \delta r_2} = \frac{\Delta p_1 D_1^2}{\Delta p_2 D_2^2}$$

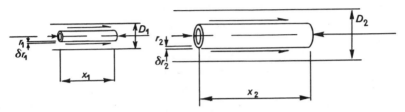

**Fig. 10.3(a).** *Geometrically similar elements in flow through pipes*

In general we may conclude that whenever we consider corresponding elements in similar flows, each will experience pressure forces:

$$F_p \propto \Delta p D^2$$

It may be noted that so long as the fluid is incompressible, the absolute pressure has no intrinsic effect as it is only *differences* of pressure that produce a resultant.

Turning now to the effect of the shear stresses round the *outer* circumference the viscous or shear force $F_\eta = 2\pi r x \tau$, is similarly related to the shear stress ($\tau$) at that radius, by:

$$F_\eta \propto \tau D^2$$

in which expression, the laminar shear:

$$\tau = \eta \frac{\delta u}{\delta r}$$

For kinematically similar flows $\delta u$ will be proportional to the reference velocity ($U$, say) and $\delta r$ will similarly be proportional to the reference length $D$. Hence:

$$\tau \propto \frac{\eta U}{D}$$

so that the viscous force:

$$F_\eta \propto \tau D^2 \propto \eta U D$$

Although the two results $F_p \propto \Delta p D^2$ and $F_n \propto \eta U D$ have been established with reference to pipe-flow, it will be found that the same argument applies to any similar elements undergoing similar motions. The motion need not be one of uniform velocity, i.e. it may involve accelerations, in which case, the resultant force $F$ must, for similar motions, be $\propto \rho D^2 U^2$, or $\rho l^2 U^2$ where $l$ is any convenient reference length (see Section 10.1 (p. 230) and 10.2 (p. 233)).

If the viscous effects $(F_n)$ were to make a relatively greater contribution to this resultant force $F$ in one system than in another, the two flows would not be truly dynamically similar. Thus, for dynamic similarity the viscous forces must bear the same ratio to the resultant in each system: $F_n \propto F$, i.e. $\eta U l \propto F$, where $l$ is the reference length such as the diameter in the case of a pipe or the length in the case of a ship.

Similar considerations apply to the gravity forces $(F_g = mg = \rho g \times \text{vol})$ if these are important as in wave-making during ship propulsion: $F_g \propto F$, i.e. $\rho g l^3 \propto F$.

So long as these two conditions are satisfied, the component remaining, namely $F_p$ due to pressure difference, will of necessity make a proportional contribution:

$$F_p \propto F, \text{ i.e. } \Delta p l^2 \propto F$$

Thus, if viscosity, gravity and pressure difference are the only factors influencing the problem (i.e. surface tension may be neglected) **the two incompressible flows will be dynamically similar—i.e. all forces will be proportional to $\rho l^2 U^2$—so long as $F/\eta U l$ and $F/\rho g l^3$ are, respectively, the same in each.**

This implies that the values of:

$$\frac{F}{\eta U l} \propto \frac{\rho l^2 U^2}{\eta U l} \text{ or } \frac{\rho l U}{\eta}$$

and

$$\frac{F}{\rho g l^3} \propto \frac{\rho l^2 U^2}{\rho g l^3} \text{ or } \frac{U^2}{lg}$$

must, respectively, be the same in each flow.

*The non-dimensional parameter $(\rho l U/\eta) = (U l/\nu)$, which is a Reynolds number (denoted by the symbol Re) is a measure of the relative importance of dynamic (or inertia) and viscous (or shear) forces.* It is so named in honour of Osborne Reynolds, who first established its importance in determining the nature of fluid flow through pipes (see Section 11.3, p. 287).

*The non-dimensional quantity $U^2/lg$ is similarly a measure of the ratio which the dynamic actions bear to the gravitational forces.* This has been referred to earlier (see Example 15.2 (ii) p. 101), and $U/\sqrt{(lg)}$ is termed the *Froude number (Fr)* in honour of William Froude who first established its importance in determining the resistance of ships.

Since the force coefficients $k_F$ in equations such as $F = k_F \rho l^2 U^2$ on two different scales may be expected to be the same only if *Re* and *Fr* have the same respective values in the two cases it follows that $k_F$ depends on the Reynolds number *Re* and on the Froude number *Fr*. These two (supplemented for compressible flows by Mach number *Ma*) are the most important non-dimensional parameters in fluid mechanics.

In terms of symbols this may be written:

$$k_F = \frac{F}{\rho l^2 U^2} = \varphi\,(Re,\,Fr)$$

the right-hand side denoting 'some unspecified function of *Re* and *Fr*'. This is often referred to as **Rayleigh's Law** since Lord Rayleigh was mainly responsible (1914) for the systematic application of similarity arguments to fluid mechanics. He obtained the formula by dimensional arguments, and his method will be outlined later (see Section 10.7, p. 256).

*If there are no free surface effects (i.e. no wave-making), Rayleigh's formula simplifies to*:

$$k_F = \frac{F}{\rho l^2 U^2} = \varphi\,(Re)$$

It would appear from this, for example, that if a wind tunnel is to be used for force measurements on a model, it should be so operated that the Reynolds number is the same as that on the full scale. If this condition for dynamic similarity is not satisfied, there is no assurance that the force coefficients so obtained will correspond to those on full scale. Also, if the results are to be strictly comparable, the bodies must be geometrically similar even down to the roughness of the surfaces.

*Thus, the flows of fluids around geometrically similar bodies are dynamically similar if the forces acting at points in similar positions in the two patterns of flow have the same ratio—a condition which results in different laws of similarity depending on what kind of forces are induced. As we have seen, if only the viscous and inertia forces are important then only Reynolds' number need be the same for dynamical similarity. If gravitational forces (as in wave-making) are also important, then the law for dynamical similarity of flow round geometrically-similar bodies involves both Reynolds and Froude numbers.*

EXAMPLE 10.3 (i)

*An aeroplane is to be built to fly at high altitude where the density and coefficient of viscosity of air is 0·2 and 1/1·25 times the corresponding value at sea-level, respectively.*

*Before building the prototype, tests are carried out on a model of linear scale ratio 1/50 in a compressed-air tunnel operated at 20 atmospheres. Estimate the speed of air in the tunnel for dynamic similarity between the model tested in the compressed-air tunnel at sea-level and the proposed aeroplane travelling at 640 km/h at high altitude.*

Using suffixes $a$ for altitude and $s$ for sea-level, $m$ for model and $p$ for prototype we have:

$$\frac{\rho_a}{\rho_s} = 0\cdot2 \quad \text{and} \quad \frac{\eta_a}{\eta_s} = \frac{1}{1\cdot25}$$

For the prototype flying at altitude $\rho_p = \rho_a = 0\cdot2\rho_s$, and for the model tested at 20 atmospheres

$$\rho_m = 20\rho_s$$

Hence, $\rho_p/\rho_m = 1/100$ and, assuming $\eta$ is independent of pressure,

$$\frac{\eta_m}{\eta_p} = 1\cdot25$$

For dynamical similarity

$$\left(\frac{Ul\rho}{\eta}\right)_m = \left(\frac{Ul\rho}{\eta}\right)_p$$

i.e.

$$U_m = \frac{\rho_p\,\eta_m}{\rho_m\,\eta_p}\frac{l_p}{l_m} U_p = \frac{1}{100} \times 1\cdot25 \times 50 \times 640 \text{ km/h}$$

$$= \textbf{400 km/h} \quad or \quad \textbf{216·3 knot}$$

## EXAMPLE 10.3 (ii)

*The power absorbed by the circulating pump of a cooler is found to be 7·5 kW. Estimate the pump power necessary for a geometrically-similar cooler with all linear dimensions increased 75 per cent and designed to have three times the rate of heat-transfer with the same temperature conditions as before. Assume the resistance to flow of fluid through the coolers is proportional to the square of the mean velocity in accordance with Darcy formula.*

$$\text{Power } P = \rho g \dot{V} \Delta H = \rho g \left(\frac{\pi}{4}d^2\right) \bar{u} \times \left(4\frac{fl}{d}\frac{\bar{u}^2}{2g}\right)$$

i.e.
$$P \propto ld u^3$$

Heat-transfer rate $\dot{Q} = \dot{m}C\Delta T = \rho a u C \Delta T$

i.e.
$$\dot{Q} \propto d^2 u \Delta T$$

Hence, if $\Delta T$ is the same in the two cases,

$$\frac{\dot{Q}_1}{\dot{Q}_2} = \frac{u_1}{u_2}\left(\frac{d_1}{d_2}\right)^2$$

i.e.
$$\frac{1}{3} = \frac{u_1}{u_2}\left(\frac{4}{7}\right)^2 \quad or \quad u_2 = 3\left(\frac{4}{7}\right)^2 u_1$$

Similarly,
$$\frac{P_2}{P_1} = \left(\frac{l_2}{l_1}\right)\left(\frac{d_2}{d_1}\right)\left(\frac{u_2}{u_1}\right)^3$$

$$= (\tfrac{7}{4})^2\,(\tfrac{4}{7})^6 \times 27 = (\tfrac{4}{7})^4 \times 27 = 2\cdot88$$

Hence, $P_2 = 2\cdot88 \times 7\cdot5 \text{ kW} = \textbf{21·6 kW}$.

EXAMPLE 10.3 (iii)

*The force coefficient for resistance of a surface ship is a function of Reynolds and Froude numbers. At what speed should a geometrically-similar model of an ocean-going ship be tested in order that the conditions for dynamical similarity are satisfied? Why cannot this be achieved in practice?*

The force coefficient

$$k_F = \frac{F}{\rho l^2 U^2} = \varphi(Re, Fr)$$

must be the same for both full-scale and the model if dynamical similarity is to be achieved. Hence, using suffixes 1 for the model, we must have

$$\frac{U_1 l_1}{\nu_1} = \frac{Ul}{\nu} = Re$$

i.e.

$$\frac{U_1}{U} = \frac{l}{l_1}\frac{\nu_1}{\nu}$$

and

$$\frac{U_1}{\sqrt{(gl)_1}} = \frac{U}{\sqrt{(gl)}} = Fr$$

i.e.

$$\frac{U_1}{U} = \sqrt{\frac{l_1}{l}}$$

Thus, if both are to be satisfied as is necessary for dynamical similarity,

$$\frac{U_1}{U} = \frac{l}{l_1}\frac{\nu_1}{\nu} = \left(\frac{l_1}{l}\right)^{1/2} \quad \text{or} \quad \frac{\nu_1}{\nu} = \left(\frac{l_1}{l}\right)^{3/2}$$

i.e. the model must be tested in a fluid of kinematic viscosity $\nu_1$ which is $(l_1/l)^{3/2}$ times the kinematic viscosity of sea water, and the speed $U_1$ must be $\sqrt{(l_1/l)}$ times that of the proposed full-scale ship.

In practice no two liquids (liquids being essential for gravity wave-making) are available which have sufficiently different kinematic viscosities to satisfy the combined conditions of equal $Re$ and equal $Fr$ for the geometrically similar ships. Hence, it is customary to base ship-model tests on equal Froude numbers and then to take viscous forces (or skin friction) into account by computation.

## 10.4. Prediction of ship resistance

From Example 10.3 (iii) it can be seen that it is impossible to achieve dynamic similarity. Froude overcame the difficulty of predicting the total resistance ($R$) of a ship by splitting the resistance into three parts between which, he argued, there would be little interaction. He considered it to consist of a wave-making resistance $R_W$, and eddy-making resistance $R_E$ and a skin-friction resistance $R_F$, each of which was subject to different scaling laws. It has become customary to adopt a simplified version of his argument which, although lacking the subtlety of the original, leads to the same result, In this, the resistance is considered to consist of two parts due, respectively, to the shear stresses and the pressure differences which act round the hull. The former is referred to as the skin friction resistance $R_F$, and the latter as the residuary resistance $R_R$, see Fig. 10.4(a). (In aerodynamics the pressure drag is referred to as the form drag. It rises sharply at speeds which are sufficiently high for compressibility to cause a pressure-wave drag, which is qualitatively analogous to the gravity-wave drag of a ship.) Assuming for the moment that the skin

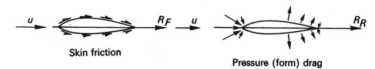

**Fig. 10.4(a).** *Skin friction and residuary resistance*

friction, $R_F$, can be estimated by some other method (as it can), the residuary resistance, $R_R$, of a ship model may be obtained by deducting $R_F$ from the total model resistance $R$. The latter must, of course, be measured at the same Froude number, i.e. at the corresponding speeds given by:

$$u \propto \sqrt{l}$$

The residuary resistance of the ship $R_R$ is by Froude's method obtained through scaling according to the law for dynamic similarity. Thus, if suffix $m$ is used to denote the model values:

$$\frac{R_R}{(R_R)_m} = \frac{\rho l^2 u^2}{(\rho l^2 u^2)_m}$$

The skin friction resistance $R_F$ of the ship when added to $R_R$, so obtained, gives the full-scale resistance:

$$R = R_R + R_F$$

It is usual to estimate the skin friction resistance of both ship and model from towing tests on planks, such as those carried out by Wm. Froude. The latter tests were made in the first tank built for model experiments at Chalston Cross, Torquay, in 1871.

A working rule which embodies these results is:

$$R_F = fSU^{1 \cdot 825}$$

in which $S$ is the wetted area and $U$ is the forward speed. For surfaces in salt water, the coefficient $f$ is, as a function of length, given by:

| $l$ | 15 | 30 | 150 | 270 | m |
|---|---|---|---|---|---|
| $f$ | 0·458 | 0·439 | 0·418 | 0·411 | N/m² knot$^{1 \cdot 825}$ |

and these values may be transformed to the corresponding values for fresh water by multiplying by the relative density:

$$\sigma = \frac{\rho_S}{\rho_F} = 1\cdot026$$

The above values incorporate the effect of roughness but other more elaborate methods have been proposed, by which this is allowed for separately. The departure from the velocity squared law of the skin friction resistance, suggested by the above expression, may be attributed to the combined scale effects due to changes in relative roughness and Reynolds number. Scale effects are minimized by testing models of the largest practicable size, usually not less than 3 m long.

EXAMPLE 10.4 (i)

*A ship having a length of hull 150 m is to travel at 10 m/s. Calculate the Froude number and the speed at which a 1:25 model should be towed through water.*

$$\text{(Froude number)}^2 = Fr^2 = \frac{U^2}{lg} = \frac{10^2}{150 \times 9\cdot81} = 0\cdot0679$$

When two geometrically similar ships generate similar flow patterns it is usually the inertia and gravity forces which dominate—i.e. the Froude number which is the ratio of the two is more important than Reynolds number, the ratio of inertia to viscous forces.

Thus $Fr$ for the prototype is the same as for the model, i.e.

$$\frac{U_m}{U_p} = \sqrt{\frac{l_m}{l_p}} = \sqrt{\frac{1}{25}} = \frac{1}{5} \quad \text{or} \quad U_m = \frac{10}{5} \text{ m/s} = 2 \text{ m/s}$$

EXAMPLE 10.4 (ii)

*The resistance of a model ship 5 m long and 2 m² wetted area is 42 N when tested at a speed of 5·3 knot in sea water. Estimate the propulsive power required to overcome the resistance of a geometrically similar ocean-going ship 180 m long. The skin friction constants for the model and ship may be taken as 0·478 and 0·415 N/m²(knot)$^{1\cdot825}$, respectively.*

The total resistance $R = $ skin friction resistance $R_F +$ residual resistance $R_R$. Hence, using suffix 1 for the model, its residual resistance is:

$$R_{R1} = R_1 - R_{F1} = R_1 - f_1 S_1 U_1^n$$
$$= 42 \text{ N} - 0\cdot478 \times 2 \times (5\cdot3)^{1\cdot825} \text{ N} = 22 \text{ N}$$

For the same Froude number, we have:

$$Fr^2 = \frac{U_1^2}{gl_1} = \frac{U^2}{gl}$$

and the corresponding speed for the full-scale ship is:

$$U = U_1 \sqrt{\frac{l}{l_1}} = 5\cdot3 \text{ knot} \sqrt{\left(\frac{180}{5}\right)} = 31\cdot8 \text{ knot}$$

The ratio of the residual resistances of the ship is:

$$\frac{R_R}{R_{R1}} = \frac{\rho U^2 S}{\rho U_1^2 S_1} = \frac{l}{l_1}\left(\frac{l}{l_1}\right)^2 = \left(\frac{l}{l_1}\right)^3 = (36)^3$$

Hence, the residual resistance of the full-scale ship is:

$$R_R = (36)^3 \times R_{R1} = (36)^3 \times 22 \text{ N} = 1\cdot026 \text{ MN}$$

The skin friction of the full-scale ship is:

$$R_F = fSU^{1\cdot825} = 0\cdot415 \times (36)^2 \times 2 \times (31\cdot8)^{1\cdot825} \text{ N} = 0\cdot581 \text{ MN}$$

Thus, the total resistance of the full-scale ship is:

$$R = 1\cdot607 \text{ MN}$$

and the propulsive power is:

$$P = RU = 1\cdot607 \text{ MN} \times 31\cdot8 \text{ knot} \left[\frac{0\cdot514 \text{ m}}{\text{knot s}}\right]\left[\frac{\text{MWs}}{\text{MNm}}\right]$$

$$= \mathbf{26\cdot25} \text{ MW}.$$

## 10.5. Dimensional analysis and non-dimensional products

The method of dimensional analysis (see Chapter 5) will be reintroduced here as it provides a powerful and general method whereby several factors and variables can be grouped together and correlated as non-dimensional parameters. The latter often point the way towards experimental work in which variation in several variables can all be accommodated on single (unique) graphs by using the non-dimensional parameters or groups of variable as ordinates and abscissae. For theoretically insoluble problems such experiments are necessary to find the numerical values of the constants or non-dimensional coefficients involved. It is to Lord Rayleigh that we are largely indebted for the systematic application to fluid mechanics of what is now termed *dimensional analysis*.

We may conveniently develop the principles of dimensional analysis from the following proposition or principle of dimensional homogeneity, which is considered to be self-evident:

**It is characteristic of physical equations that only like quantities, i.e. those having the same dimensions, are added or equated.**

In order to get a grasp of the method of dimensions let us take a simple kinematic problem to which we already know the answer.

Suppose that *s* is the distance travelled by a body moving with constant acceleration, it can be expressed as:

$$s = ut + \tfrac{1}{2}at^2$$

As the left-hand side has the dimensions of a length, each of the terms on the right-hand side must also have these dimensions. That this is so may easily be checked, for using square brackets [ ] in conjunction with an equal sign to mean 'has the dimensions of', we have:

$$ut = \left[\frac{L}{T}\right] \times [T] = [L]$$

and

$$\tfrac{1}{2}at^2 = \left[\frac{L}{T^2}\right] \times [T^2] = [L]$$

This principle of *dimensional homogeneity*, as it is called, implies that any physical equation can be expressed non-dimensionally. For, if all the terms in it have the same dimensions, we have merely to divide through by any one of them, such as $ut$ in the simple equation:

$$s = ut + \tfrac{1}{2}at^2$$

to obtain:

$$\left(\frac{s}{ut}\right) = 1 + \tfrac{1}{2}\left(\frac{at}{u}\right)$$

The quantities enclosed within the brackets are termed non-dimensional groups or products and may each be considered as a single variable or coefficient. Anticipating the '$\Pi$ Theorem' we may mention here that the symbol $\Pi$ (capital 'pi') is nowadays (see Section 10.8, p. 265) frequently used to denote such products, so that the above equation becomes:

$$\Pi_1 = 1 + \tfrac{1}{2}\Pi_2$$

in which

$$\Pi_1 = \left(\frac{s}{ut}\right) \quad \text{and} \quad \Pi_2 = \left(\frac{at}{u}\right)$$

This enables us to draw a single graph (which is in this case a straight line) to display *all* the information contained in the original equation, see Fig. 10.5(a).

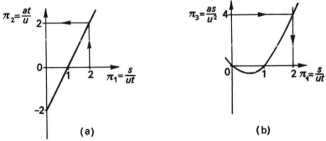

(a)                 (b)

**Fig. 10.5(a).** *Motion with constant acceleration*

EXAMPLE 10.5 (i)

*Use Fig. 10.5(a) (p. 247) to determine the constant acceleration which must be given to a particle projected at 3 m/s in order that it may travel 6 m in the first second.*

The value of $\Pi_1$ is to be:

$$\Pi_1 = \left(\frac{s}{ut}\right) = \frac{6 \text{ m}}{3 \text{ m/s} \times 1 \text{ s}} = 2$$

From the graph, the corresponding value of $\Pi_2$ is also 2. Hence:

$$\Pi_2 = \left(\frac{at}{u}\right) = 2$$

i.e.

$$a = 2 \times \frac{3 \text{ m/s}}{1 \text{ s}} = 6 \text{ m/s}^2$$

EXAMPLE 10.5 (ii)

*Let us suppose that in the previous problem we had been given the acceleration (6 m/s²) and had been required to find the time taken.*

In this case we could use Fig. 10.5(a) (a) only on a trial and error basis, as both $\Pi_1$ and $\Pi_2$ involve $t$. The difficulty may be overcome by forming an alternative non-dimensional product which eliminates $t$ such as:

$$\Pi_3 = (\Pi_1\Pi_2) = \frac{at}{u} \times \frac{s}{ut} = \left(\frac{as}{u^2}\right)$$

Just as $\Pi_2$ is a definite function of $\Pi_1$, which is written as:

$$\Pi_2 = f(\Pi_1)$$
$$= 2(\Pi_1 - 1) \text{ in this case} \qquad (i)$$

so, to each value of $\Pi_1$ there is a definite value of $(\Pi_1\Pi_2)$, i.e. of $\Pi_3$:

$$\Pi_3 = F(\Pi_1)$$

i.e.

$$\left(\frac{as}{u^2}\right) = F\left(\frac{s}{ut}\right)$$

The form of the function denoted by F( ) is, in fact:

$$\Pi_3 = \Pi_1\Pi_2 = \Pi_1 \times 2(\Pi_1 - 1) \qquad (ii)$$

and this is shown in Fig. 10.5(a) (b). Thus, for the problem considered:

$$\Pi_3 = \frac{as}{u^2} = \frac{6 \text{ m/s}^2 \times 6 \text{ m}}{3^2 \text{ m}^2/\text{s}^2} = 4$$

and the corresponding value of $\Pi_1$ is, from the graph, 2:

$$\Pi_1 = \frac{s}{ut} = 2$$

∴

$$t = \tfrac{1}{2} \times \frac{6 \text{ m}}{3 \text{ m/s}} = 1 \text{ s}$$

We may draw the following conclusions from the above example:

(i) *By working in terms of non-dimensional parameters or coefficients (namely, $\Pi$'s), the number of variables can be reduced to a minimum.*

(ii) *This may enable a simple graph, or its equivalent, to be produced which relates the $\Pi$'s.*

(iii) *Sometimes it is convenient to formulate an alternative parameter which does not contain a certain variable. This may be done by multiplying or dividing powers of the original non-dimensional groups. The new ($\Pi$) variable so obtained is uniquely related to the original ones, and may be used instead of one of the latter.*

Had we been able to forecast that in the above examples $\Pi_2$ (or $\Pi_3$) were uniquely related to $\Pi_1$ (Equations (i) and (ii)), then the form of the curve could have been deduced by experiment, for any measurements on bodies moving with constant acceleration would, when plotted on Fig. 10.5(a) (a) or (b) lie on the curves shown. The latter, so obtained, could be used equally well to solve problems by the above method. **Dimensional analysis does in fact enable the non-dimensional products (i.e. the $\Pi$'s) to be formulated without a complete analysis being necessary. Problems which defy theoretical solution can thus be solved with the aid of curves obtained by a limited number of experiments.**

EXAMPLE 10.5 (iii)

*Use Rayleigh's method of dimensional analysis or the principle of dimensional homogeneity to relate the distance ($s$) travelled by a body as a function of time ($t$) to its initial velocity ($u$) and its constant acceleration ($a$).*

The fact that we already know the answer by a different method for constant acceleration need not be brought to mind—the method is just the same whether or not a complete analytical solution exists.

If we consider that $s$ depends only on $t$, $u$ and $a$, i.e.

$$s = f(t, u, a)$$

in which the right-hand side is, for example, some number times powers of $t$, $u$ and $a$. More generally it may consist of any number of terms of this type; that is, it may be expressed as a power series, a typical term of which could be:

$$Ct^p u^q a^r$$

The coefficient $C$ has no dimensions, i.e. it is an unspecified number (or numeric) as are the indices $p$, $q$ and $r$.

The principle of dimensional homogeneity places certain restrictions on the values of $p$, $q$, and $r$, however, since each term must have the same dimensions, namely those of the left-hand side (i.e. L). The dimensions of a typical term

$$Ct^p u^q a^r \quad \text{are} \quad \left[ T^p \left( \frac{L}{T} \right)^q \left( \frac{L}{T^2} \right)^r \right]$$

i.e. L is raised to the power of $(q+r)$ and T to the power $(p-q-2r)$. If we equate these to the corresponding dimensions of the right-hand side we have:

L:                                        $1 = q+r$

T:                                        $0 = p-q-2r$

These may be partially solved by expressing both $p$ and $r$ in terms of $q$ for example:

$$1 = q+\frac{p-q}{2}$$

i.e.                                      $p = 2-q$

and                                       $r = 1-q$

Thus the expression for $s$ must consist of terms of the type:

$$C\frac{t^2}{t^q}u^q\frac{a}{a^q} = Cat^2\left(\frac{u}{at}\right)^q$$

As $q$ may take *any* value without violating the principle of dimensional homogeneity, the latter can tell us no more than:

$$s = C_0at^2\left(\frac{u}{at}\right)^0 + C_1at^2\left(\frac{u}{at}\right)^1 + C_2at^2\left(\frac{u}{at}\right)^2 + \ldots$$

i.e.        $$\left(\frac{s}{at^2}\right) = C_0 + C_1\left(\frac{u}{at}\right) + C_2\left(\frac{u}{at}\right)^2 + \ldots$$

The right-hand side simply represents some unspecified function of $(u/at)$, so that the equation is better written as:

$$\left(\frac{s}{at^2}\right) = F\left(\frac{u}{at}\right)$$

That this is so may be checked by dividing the expression:

$$s = ut+\tfrac{1}{2}at^2$$

by $at^2$, which results in:

$$\left(\frac{s}{at^2}\right) = \left(\frac{u}{at}\right)+\frac{1}{2}$$

*Thus, although dimensional analysis is able to predict that $(s/at^2)$ is uniquely related to $(u/at)$, it is unable to state the form of the relationship between them.* A small number of experimental measurements would, however, suffice to determine that the law relating them is, in fact, linear.

As was mentioned at the end of Example 10.5 (ii), we may multiply or divide powers of the original non-dimensional products to formulate alternative ones. For example, as there is a definite value of $(s/at^2)$ for each value of $(u/at)$ there is, equally well, a definite corresponding value of their quotient:

i.e. $$\frac{(s/at^2)}{(u/at)} = \left(\frac{s}{ut}\right) = \varphi\left(\frac{u}{at}\right)$$

where $\varphi$ again denotes 'some function of' the quantity within the bracket.

The parameter $(s/ut)$ was previously denoted by $\Pi_1$ and was said (Equation (i) of Example 10.5 (ii)) to be uniquely related to $\Pi_2 = (at/u)$. This result has thus been rediscovered by dimensional analysis, since $\Pi_2$ is simply the reciprocal of $(u/at)$.

Dimensional analysis sometimes yields or suggests the form of the relationship between the variables in problems which are too difficult of analysis by any other method. We shall therefore make more extensive use of the method to suggest the conditions under which similarity may be expected.

## 10.6. Conditions for dynamic similarity over notches and weirs

In Chapter 5 a number of simple relationships were developed to relate the head and the flow over notches and weirs, on the assumption that the flows were similar. As vee-notches always present the same shape to the oncoming flows the latter are likely to depend on the density $(\rho)$ and viscosity $(\eta)$ of the fluid, on the head $(H)$ over the notch and on the acceleration $(g)$ due to the pull of gravity, i.e.

$$Q = f(\rho, \eta, H, g)$$

A typical term in a power series for the right-hand side will be $C\rho^a\eta^b H^c g^d$ in which $C$, $a$, $b$, $c$, $d$, are unspecified numbers. Dimensional homogeneity, however, requires that:

$$Q = C\rho^a\eta^b H^c g^d$$

i.e. $$\left[\frac{L^3}{T}\right] = \left[\frac{M}{L^3}\right]^a \left[\frac{M}{LT}\right]^b [L]^c \left[\frac{L}{T^2}\right]^d$$

the dimensions of $\eta$ being $[M/LT]$ as shown in Section 8.2.
Equating indices of M, L and T in turn results in the three equations:

M: $\qquad\qquad 0 = a+b$

L: $\qquad\qquad 3 = -3a-b+c+d$

T: $\qquad\qquad 1 = b+2d$

As there are only three equations for the four unknowns a complete solution is not possible and three of the unknowns must be expressed in terms of the fourth, say $d$. Thus:

$$b = 1 - 2d$$
$$a = -b = 2d - 1$$
$$c = 3 + 3a + b - d$$
$$= 3 + (6d - 3) + 1 - 2d - d$$
$$= 1 + 3d$$

Hence, the expression for $Q$ involves terms of the type:

$$C \frac{\rho^{2d}}{\rho} \frac{\eta^1}{\eta^{2d}} H^{1+3d} g^d$$

i.e.

$$C \frac{\eta}{\rho} H \left( \frac{\rho^2 H^3 g}{\eta^2} \right)^d \quad \text{or} \quad C \frac{\eta}{\rho} H \left( \frac{\rho}{\eta} H^{3/2} g^{1/2} \right)^{2d}$$

in which $C$ and $d$ may take *any* numerical values without violating the principle of homogeneity. Thus, the right-hand side is simply $C(\eta/\rho)H$ times some unspecified function of $(\rho/\eta)H^{3/2}g^{1/2}$.

i.e.

$$Q = \frac{\eta}{\rho} H \left\{ C_0 + C_1 \left( \frac{\rho H^{3/2} g^{1/2}}{\eta} \right)^2 + C_2 \left( \frac{\rho H^{3/2} g^{1/2}}{\eta} \right)^4 + \dots \right.$$

$$\therefore \qquad \frac{Q\rho}{\eta H} = \varphi \left( \frac{\rho H^{3/2} g^{1/2}}{\eta} \right)$$

in which it is easily verified that the two groups of variables form non-dimensional products.

If it is required to eliminate $\rho/\eta$ from the left-hand side then the non-dimensional group $\{(\rho/\eta)H^{3/2}g^{1/2}\}$ can be used as a divisor to give:

$$\frac{Q}{H^{5/2} g^{1/2}} = \psi \left( \frac{\rho g^{1/2} H^{3/2}}{\eta} \right)$$

which result would follow directly if, from the three indices equations, $a$, $c$, and $d$ were found in terms of $b$.

## Example 10.6 (i)

*It is required to find the flow of fluid of relative density 0·8 and dynamic viscosity eight times that of water, over a vee-notch with a depth of 300 mm by running water over a similar test notch. The flow over the test notch is known to be $0·83H^{2·5}$ m³ of water per s where H is the depth in metres.*

*Estimate the corresponding depth of water in the test notch and hence the probable rate of flow of fluid over the working notch.*

For geometrically similar vee-notches

$$\frac{Q}{H^{5/2} g^{1/2}} = \psi \left( \frac{\rho g^{1/2} H^{3/2}}{\eta} \right)$$

for all liquids.

Hence if for the particular fluid (of relative density) 0·8 we make $\rho g^{1/2} H^{3/2}/\eta$

equal to the same group for water then:

$$\left(\frac{H_w}{H}\right)^{3/2} = \left(\frac{\rho}{\rho_w}\right)\left(\frac{\eta_w}{\eta}\right) = 0\cdot8\times\frac{1}{8} = 0\cdot1$$

and
$$H_w = (0\cdot1)^{2/3}H = 0\cdot215H$$

Therefore, if the depth of fluid is $0\cdot3$ m, the depth of water must be

$$H_w = 0\cdot215\times0\cdot3 \text{ m} = 0\cdot0645 \text{ m} = \mathbf{64\cdot5} \text{ mm}$$

and hence, the quantity of water

$$Q_w = 0\cdot83H^{2\cdot5} = 0\cdot83(0\cdot0645)^{2\cdot5} = 0\cdot877 \text{ dm}^3/\text{s}$$

Similarly,

$$\frac{Q}{H^{5/2}g^{1/2}} = \frac{Q_w}{H_w^{5/2}g^{1/2}}$$

Hence,

$$\frac{Q}{Q_w} = \left(\frac{H}{H_w}\right)^{5/2} = \left(\frac{1}{0\cdot215}\right)^{5/2} = \frac{1}{0\cdot0214}$$

and
$$Q = \frac{0\cdot877}{0\cdot0214} \text{ dm}^3/\text{s} = \mathbf{41} \text{ litre/s}$$

EXAMPLE 10.6 (ii). *Rectangular or Broad-crested Weirs*

*Assuming the weirs to be wide enough (breadth B) for end effects to be neglected the discharge per unit width will be the same at all vertical sections. Hence, we must seek a corresponding expression for $Q/B$ which has the dimensions $L^3/TL$, i.e.*

$$\frac{Q}{B} = \left[\frac{L^2}{T}\right]$$

On the assumption that for a fluid of density $\rho$ and coefficient of viscosity $\eta$, $Q/B$ depends on $\rho$, $\eta$, $g$ and head $H$ we may write $Q/B = f(\rho, \eta, g, H)$ where $f(\ )$ implies 'some unspecified function of'—a typical term in the series being $C\rho^a\eta^bg^cH^d$.

Hence:

$$\frac{L^2}{T} = \left(\frac{M}{L^3}\right)^a \left(\frac{M}{LT}\right)^b \left(\frac{L}{T^2}\right)^c L^d$$

| | | |
|---|---|---|
| $0 = a+b$ | $\therefore a = -b$ | and $a = 2c-1$ |
| $2 = -3a-b+c+d$ | $b = 1-2c$ | $b = 1-2c$ |
| $1 = b+2c$ | $d = 2+3a+b-c$ | $d = 2+(6c-3)+(1-2c)-c$ |
| | | $= 3c$ |

Therefore, the expression for $Q/B$ involves terms of the type:

$$C\rho^{2c-1}\eta^{1-2c}g^cH^{3c} \quad \text{or} \quad C\frac{\eta}{\rho}\left(\frac{\rho}{\eta}g^{1/2}H^{3/2}\right)^{2c}$$

and hence:

$$\frac{Q\rho}{B\eta} = \varphi\left(\frac{\rho}{\eta}g^{1/2}H^{3/2}\right)$$

or alternatively:

$$\frac{Q}{Bg^{1/2}H^{3/2}} = \psi\left(\frac{\rho g^{1/2}H^{3/2}}{\eta}\right)$$

If the flow pattern is the same on a model as on the full-scale equipment, conditions in the two installations are said to be *dynamically similar*. The first requirement for this is that all lengths, i.e. quantities of dimension [L], on the model must be in the same ratio as the corresponding lengths full scale. This ensures that the two systems are *geometrically similar*.

Further, the motions must be kinematically similar, i.e. all quantities having the same dimensions in terms of [L] and [T] must be in the same ratio. Finally, for the systems to be *dynamically similar* all corresponding

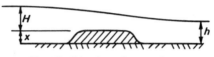

**Fig. 10.6(a).** *Broad-crested weir*

forces (of dimension F) are in the same ratio. All other quantities of the same kind (i.e. of the same dimensions) must also be in the same ratio as the corresponding quantities full scale. If these conditions can be satisfied, then all the non-dimensional groups (i.e. all the ratio of quantities whose resultant dimensions are of the same kind, e.g. $Q/(Bg^{1/2}H^{3/2})$) will have the same values for both model and full-scale installation.

In the case of rectangular weirs the 'constant' $K$ used in $Q = KBH^{3/2}$ depends on such factors as the shape of the approach channel, the depth of water upstream, the sharpness of the edge on the plate, and even the roughness of the plate. Hence, as with all metering devices, direct calibration, or the careful reproduction of a calibrated installation is always desirable.

In the case of submerged broad-crested weirs (Fig. 10.6(a)), which incidentally are less vulnerable to floating debris, etc., than sharp-edged rectangular weirs, the law $Q \propto Bg^{1/2}H^{3/2}$ is valid only under certain conditions since $x$ and $h$ can also influence the flow pattern (see Chapter 14, p. 413).

EXAMPLE 10.6 (iii)

*A model weir built to one-sixteenth full scale, is found to discharge 0·5 m³/min when the head over it is 75 mm. Assuming dynamically similar conditions, find:*
*(a) the corresponding head over the full-scale weir,*
*(b) the corresponding discharge over the latter.*

(a) Since the breadth, $B_1$, of the weir and the head, $H_1$, over it are both quantities having the same dimension, namely, length, their ratio must have the same value on full scale for dynamic similarity. They also satisfy the requirement that the model must be geometrically similar to the full scale before dynamical similarity can operate. Thus,

$$\frac{B_1}{H_1} = \frac{B_2}{H_2}$$

or
$$H_2 = \frac{B_2}{B_1} H_1 = 16 \times 0.075 \text{ m} = \mathbf{1.2} \text{ m}$$

(b) Since dynamical similarity requires all corresponding non-dimensional groups to have the values for both model and full-scale weir the important one for this problem taken from Section 5.6 is:

$$\frac{Q}{Bg^{1/2}H^{3/2}} = \text{constant}$$

i.e.
$$\frac{Q_2}{Q_1} = \frac{B_2}{B_1}\left(\frac{H_2}{H_1}\right)^{3/2}$$

or since:
$$\frac{B_2}{B_1} = \frac{H_2}{H_1}$$

then:
$$\frac{Q_2}{Q_1} = \left(\frac{H_2}{H_1}\right)^{5/2}$$

Hence:
$$Q_2 = (16)^{5/2} \times 0.5 \text{ m}^3/\text{min}$$
$$= \mathbf{512} \text{ m}^3/\text{min}$$

The group $Q/(Bg^{1/2}H^{3/2})$ is referred to as a 'non-dimensional co-efficient' denoted, say, by the symbol $k$. Another is $(\rho/\eta)g^{1/2}H^{3/2}$ since the dimensions cancel and leave a pure number or numeric. The latter is related to Reynolds' number $(ud/\nu)$. Another non-dimensional quantity is Mach number (which is the ratio of local velocity to the velocity of sound under the same conditions of pressure and temperature) to be dealt with later.

**Dynamic similarity requires that all such non-dimensional coefficients shall have the same value on both model and full scales.** Similarly, each flow pattern corresponds to particular values of $k$. The idea of a number being associated with a particular flow pattern is not perhaps as unfamiliar as it might at first appear; for example, in this age of jet propulsion, rockets, etc., few people can have missed hearing the term 'Mach number'. Its importance lies in the fact that it is a non-dimensional coefficient which, in part, *defines* the nature of the *flow pattern*. When its value exceeds unity, shock waves can form, just as, downstream of a weir, standing waves may develop (see Section 14.4, p. 418). The latter cannot occur until a similar non-dimensional coefficient, based on the ratio of the speed of the water to the speed of a gravity wave, reaches unity.

Hydraulics has, in the past, been somewhat scathingly referred to as a 'science of coefficients'. This remark is equally appropriate to modern fluid mechanics but with the difference that no greater compliment can be paid to it, *for by expressing our results in non-dimensional terms, i.e. as coefficients, we are able to give the greatest possible generality to them.*

For example, the contraction and discharge coefficients

$$c = \frac{Q}{A \sqrt{(2gH)}}$$

obtained by passing water through an orifice are equally relevant to one of a different size, operating at a different pressure, and transmitting a different fluid, so long as the flow patterns are similar.

## 10.7. Rayleigh's formula for resistance

It was dimensional arguments which led Lord Rayleigh to a very general expression for the force or drag caused by the relative motion between a body and a viscous fluid. He outlined a method whereby the many inter-related variables in problems of fluid flow could be grouped. Thus observations may be made in experiments which do not necessitate the keeping of all variables constant except two, plotted as abscissa and ordinate on graphs as was necessary in the pre-Rayleigh era. In many cases it is impossible to study the effect of each variable individually, but the Method of Dimensions enables non-dimensional groups of variables to be formulated and used as co-ordinates so that the effect of each variable can be seen.

We may suppose that a drag force $F$ (which could be the resistance of a submerged body as, for example, a submarine or torpedo in motion) will depend on the density ($\rho$) and the coefficient of viscosity ($\eta$) of the fluid. It will, in addition, depend on the size ($l$) and speed ($u$) of the relative motion between the fluid and the body—i.e. the variable force $F$ depends on or is a function of the four independent variables each of which could be changed.

Thus we start by *assuming* that $F = \varphi(\rho, \eta, l, u)$. A typical term in a power series for the right-hand side will be: $C\rho^p \eta^q l^r u^s$ in which $C$, $p$, $q$, $r$ and $s$ are unspecified numbers.

Dimensional homogeneity, however, requires that:

$$F = C\rho^p \eta^q l^r u^s$$

i.e.
$$\left[\frac{ML}{T^2}\right] = \left[\frac{M}{L^3}\right]^p \left[\frac{M}{LT}\right]^q [L]^r \left[\frac{L}{T}\right]^s$$

Equating indices of M, L and T in turn results in three equations: $1 = p+q$ and $1 = -3p-q+r+s$ and $-2 = -q-s$. As there are only three equations for the four unknowns, the equations can be only partially solved. The first and third enable $p$ and $s$ to be expressed in terms of $q$ directly, so that it is probably simplest to work in terms of $q$: i.e.

$$p = 1-q; \quad s = 2-q \quad \text{and} \quad r = 1+3p-s+q = 2-q$$

Hence, the expression for $F$ involves terms of the type:

$$C\left(\frac{\rho}{\rho^q}\right)(\eta^q)\left(\frac{l^2}{l^q}\right)\left(\frac{u^2}{u^q}\right) \quad \text{or} \quad C\rho l^2 u^2 \left(\frac{\rho l u}{\eta}\right)^{-q}$$

in which $C$ and $(-q)$ may take *any* numerical values, without violating the principle of dimensional homogeneity. Thus, the right-hand side is simply $C\rho l^2 u^2$ times some unspecified function of $\rho l u/\eta$,

i.e.
$$F = \rho l^2 u^2 \left\{C_0 + C_1\left(\frac{\rho l u}{\eta}\right) + C_2\left(\frac{\rho l u}{\eta}\right)^2 + \ldots\right\}$$

or
$$\frac{F}{\rho l^2 u^2} = \varphi\left(\frac{\rho l u}{\eta}\right)$$

or the force coefficient:

$$k_F = \varphi(Re)$$

As has previously been noted, this very general expression for force caused by relative motion between a body and a fluid is generally referred to as Rayleigh's formula. Although any of the variables may take on different values in a series of tests, a plot of the force coefficient $k_F$ against Reynolds' number $Re$ will yield points falling on one graph or a unique curve and thus the effect of a number of variables can be judged from one curve.

Alternative force coefficients may be obtained by different choices of the index left unknown (e.g. $p$ or $s$ instead of $q$) or by multiplying each side of the equation already found by $Re$ or its square. This will yield:

$$\frac{F}{\eta l u} = \varphi_1(Re) \quad \text{and} \quad \frac{F\rho}{\eta^2} = \varphi_2(Re)$$

*The omission of a relevant variable in the original assumption will cause scatter of points and necessitate another analysis after the missing variable has been included.*

On the other hand, the inclusion of a variable which is really unimportant to the problem (although not realised as negligible in the original assumption—for instance surface tension in any but very small rates of flow over notches) will, as a result of dimensional analysis, produce a superfluous non-dimensional group which includes the unimportant variable. All the groups will by the look of the resulting equation, appear to have an influence on the results of experiment or series of tests. However, the results of practical tests will show the variations of the unimportant quantity to have no effect. That is, it has no influence (however much it is varied) on the single or unique curves plotted to relate the other parameters or non-dimensional groups of variables.

In the particular cases of lift and drag on an aerofoil assumed moving in an incompressible liquid (see Section 10.2, p. 232) Rayleigh's formula is usually expressed as:

drag coefficient
$$C_D = \frac{D}{\frac{1}{2}\rho S u^2} = \varphi_1(Re)$$

and lift coefficient

$$C_L = \frac{L}{\frac{1}{2}\rho S u^2} = \varphi_2(Re)$$

i.e. the coefficients $C_D$ and $C_L$ are functions of Reynolds' number $Re = \rho u c / \eta$ where the representative length $c$ is that of the chord (Fig. 10.2(a), p. 231)

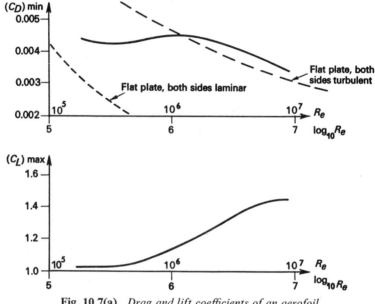

**Fig. 10.7(a).** *Drag and lift coefficients of an aerofoil*

Thus, variation in density, velocity and viscosity of air, and size of aerofoil can all be accommodated on single graphs by plotting $C_D$ and $C_L$ against $Re$ or log $Re$ as in Fig. 10.7(a) which fully specifies the two functions $\varphi_1$ and $\varphi_2$, and eliminates the need for analytical or mathematical equations.

We thus see that dimensional analysis and dynamic similarity are two tools which, because of the resulting dimensionless ratios, enable the effect of many variables to be studied together. They prove to be of great value in organizing and correlating the quantities involved in complex, and mathematically insoluble, fluid flow problems. The way is thus pointed to solutions via experimental data to which reference must always ultimately be made. Such experiments are to find the values of the non-dimensional coefficients such as $k_F$, $C_D$ and $C_L$. In essence, therefore, dimensional analysis suggests the groupings of variables—i.e. it reveals the non-dimensional parameters for experimental work, and the results obtained tell us whether the variables included are unimportant or not.

## EXAMPLE 10.7 (i)

*If power $P_0$ is required to overcome the drag on a capsule flying at a low speed $U_0$ at sea-level deduce a formula for power and estimate the power and speed ratios when the capsule moves under dynamically similar conditions, at altitudes:*

*(i) of  7 500 m where $\rho/\rho_0 = 0\cdot448$ and $\eta/\eta_0 = 0\cdot868$.*
*and (ii) of 15 000 m where $\rho/\rho_0 = 0\cdot152$ and $\eta/\eta_0 = 0\cdot807$.*

Power = Force × Velocity or $P \propto \rho l^2 \, U^2 \, \varphi(Re) \times U$. For the same capsule moving under dynamically similar conditions at different altitudes, we have $l$ and $Re$ the same, i.e.

$$\frac{U\rho}{\eta} = \frac{U_0 \rho_0}{\eta_0}$$

Hence,

$$\frac{U}{U_0} = \frac{\eta}{\eta_0}\frac{\rho_0}{\rho} \quad \text{and} \quad \frac{P}{P_0} = \frac{\rho}{\rho_0}\left(\frac{U}{U_0}\right)^3 = \left(\frac{\rho_0}{\rho}\right)^2 \left(\frac{\eta}{\eta_0}\right)^3$$

Therefore, (*i*) at 7 500 m

$$\frac{P}{P_0} = \frac{(0\cdot868)^3}{(0\cdot448)^2} = 3\cdot26 \quad \text{and} \quad \frac{U}{U_0} = \frac{0\cdot868}{0\cdot448} = \mathbf{1\cdot94}$$

and (*ii*) at 15 000 m

$$\frac{P}{P_0} = \frac{(0\cdot807)^3}{(0\cdot152)^2} = 22\cdot75 \quad \text{and} \quad \frac{U}{U_0} = \frac{0\cdot807}{0\cdot152} = \mathbf{5\cdot3}$$

## EXAMPLE 10.7 (ii)

*Show, by applying the method of dimensions, that the resistance, F, to the motion of a sphere having diameter D, and moving with uniform velocity U through a fluid having density $\rho$ and viscosity $\eta$ may be expressed by*

$$F = \frac{\eta^2}{\rho} \, \varphi \left(\frac{\rho U D}{\eta}\right)$$

*When the velocity is very small it is found that conditions are such that*

$$\varphi \left(\frac{\rho U D}{\eta}\right) = 3\pi \left(\frac{\rho U D}{\eta}\right)$$

*Hence find the viscosity in poise of a liquid through which a steel ball of diameter 1·0 mm falls, with uniform velocity, a distance of 200 mm in 10 s. The relative density of the liquid is 0·90 and of the steel, 7·8.*

If $F$ is a function of $\eta$, $\rho$, $D$ and $U$ then we have seen (p. 257) that the method dimensions yields the formula:

$$F = \frac{\eta^2}{\rho} \, \varphi \left( \frac{\rho U D}{\eta} \right) = \frac{\eta^2}{\rho} \, \varphi (Re)$$

Thus, if it is found that at very small velocities:

$$\varphi \left( \frac{\rho U D}{\eta} \right) = 3\pi \left( \frac{\rho U D}{\eta} \right)$$

then under these conditions:

$$F = \frac{\eta^2}{\rho} \times 3\pi \, \frac{\rho U D}{\eta} = 3\pi \eta U D$$

which is Stokes' law.

The resultant force on a sphere of relative density $d_s$ falling in a liquid of relative density $d_l$ is:

$$F = \rho_w g (d_s - d_l) \times \text{vol.} = \rho_w g (d_s - d_l) \frac{4}{3} \pi \frac{D^3}{8}$$

Thus in the case given,

$$F = \frac{9 \cdot 807 \text{ kN}}{\text{m}^3} (7 \cdot 8 - 0 \cdot 9) \frac{\pi}{6} (1 \text{ mm})^3 = 35 \cdot 5 \text{ μN}$$

Hence, the coefficient of viscosity of the liquid is:

$$\eta = \frac{F}{3\pi U D} = \frac{35 \cdot 5 \text{ μN}}{3\pi \times 20 \text{ mm/s} \times 1 \text{ mm}}$$

$$= \frac{35 \cdot 5}{6\pi} \times 10^{-1} \frac{\text{Ns}}{\text{m}^2} \left[ \frac{\text{P m}^2}{10^{-1} \text{ Ns}} \right] = 1 \cdot 886 \text{ P}$$

EXAMPLE 10.7 (iii)

*Use the method of dimensional analysis to deduce Darcy's formula for loss of head in a horizontal pipe using k to denote the depth of roughness indentations on the surface in contact with the fluid.*

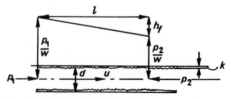

**Fig. 10.7(b).** *Loss of head in a rough pipe*

Referring to Fig. 10.7(b) the pressure drop $p_1 - p_2$ is a measure of the resistance to flow assumed to depend on length $l$, diameter $d$, density $\rho$ and viscosity $\eta$ of the fluid, velocity, $u$ and roughness $k$, i.e. $p_1 - p_2 = \psi \, (l, d, \rho, \eta, u, k)$ or $p_1 - p_2 = C l^a d^b \rho^c \eta^r u^s k^t$ plus a series of similar terms. Sufficiently far from the inlet, it is obvious, and proved by experiment, that the pressure drop is directly proportional to length; hence the exponent of the length is unity, i.e. $a = 1$.

Hence:

$$\left[\frac{ML}{L^2T^2}\right] = \left[\frac{M}{LT^2}\right] = \left[L^1 L^b \left(\frac{M}{L^3}\right)^c \left(\frac{M}{LT}\right)^r \left(\frac{L}{T}\right)^s L^t\right]$$

from which:

$$
\left.\begin{aligned}
1 &= c + r \\
1 &= 3c + r - 1 - b - s - t \\
2 &= r + s
\end{aligned}\right\} \quad \text{and}
$$

$$
\begin{aligned}
c &= 1 - r \\
s &= 2 - r \\
b &= 3c + r - 2 - s - t \\
&= 3 - 3r + r - 2 - 2 + r - t \\
&= -1 - r - t
\end{aligned}
$$

i.e. 
$$p_1 - p_2 = Cl^1 d^{-1-r-t} \rho^{1-r} \eta^r u^{2-r} k^t + \text{similar terms}$$

$$= C\frac{l}{d}\rho u^2 \left(\frac{\eta}{d\rho u}\right)^r \left(\frac{k}{d}\right)^t + \text{similar terms}$$

$$\frac{p_1 - p_2}{w} = \frac{l}{d}\frac{u^2}{2g}\varphi\left\{\left(\frac{u d \rho}{\eta}\right), \left(\frac{k}{d}\right)\right\}$$

Using $h_f$ to denote $(p_1 - p_2)/w$:

$$h_f = \lambda \frac{l}{d}\frac{u^2}{2g} = 4f\frac{l}{d}\frac{u^2}{2g}$$

This is the well-known Darcy formula in which

$$\lambda = 4f = \varphi\left(Re, \frac{k}{d}\right)$$

The effect of roughness has been explored experimentally and resulted in Nikuradse's curves shown in Fig. 11.5(a).

EXAMPLE 10.7 (iv)

*Two tanks with vertical sides are connected by a pipe 50 mm diameter and 60 m long. If the surface areas of water in the two tanks are 200 m² and 50 m², respectively, calculate the time taken for the difference in levels to drop from 9 m to 4 m. Assume turbulent flow from the larger tank (1) in the connecting pipe and f = 0·008.*

The difference in surface levels $z_1 - z_2 = h_f = H$,

i.e. 
$$H = 4f\frac{l u^2}{d 2g} = \frac{0\cdot016}{9\cdot807 \text{ m/s}^2} \times \frac{60 \text{ m}}{50 \text{ mm}} \times u^2 = 1\cdot958 \frac{\text{s}^2}{\text{m}} u^2$$

Hence,

$$u = 0\cdot715 \sqrt{\left(H\frac{\text{m}}{\text{s}^2}\right)}$$

Rate of volumetric flow

$$\dot{Q} = au = \frac{\pi}{4} \times \frac{25}{10^4} \text{ m}^2 \times 0.715 \sqrt{\left(H\frac{\text{m}}{\text{s}^2}\right)}$$

$$= \frac{14.05}{10^4} \text{ m}^2 \sqrt{\left(H\frac{\text{m}}{\text{s}^2}\right)}$$

Also,

$$\dot{Q} = -A_1\frac{dz_1}{dt} = +A_2\frac{dz_2}{dt} \quad \text{and} \quad H = z_1 - z_2$$

Hence,

$$dH = dz_1 - dz_2 = -\dot{Q}\delta t\left\{\frac{1}{A_1} + \frac{1}{A_2}\right\} = -\frac{\dot{Q}\delta t}{A_1}\left(1 + \frac{A_1}{A_2}\right)$$

Thus

$$dt = -\frac{A_1\delta H}{\dot{Q}\left(1 + \dfrac{A_1}{A_2}\right)} = -\frac{200 \text{ m}^2 \ 10^4}{(1+4) \times 14.05 \text{ m}^2} \frac{\delta H}{\sqrt{(H \text{ m/s}^2)}}$$

i.e.

$$\int dt = -\frac{40}{14.05} \times 10^4 \frac{\text{s}}{\sqrt{\text{m}}} \int_{9m}^{4m} \frac{dH}{\sqrt{H}}$$

time

$$= -\frac{40}{14.05} \times 10^4 \frac{\text{s}}{\sqrt{\text{m}}} 2[\sqrt{(4 \text{ m})} - \sqrt{(9 \text{ m})}]$$

$$= +\frac{80 \times 10^4}{14.05}(3-2)\text{s}\left[\frac{\text{h}}{3.6 \times 10^3\text{s}}\right]$$

$$= \mathbf{15.82 \text{ h}}$$

## EXAMPLE 10.7 (v)

*An aeroplane of 60 kN estimated gross weight is being designed to fly normally at 10 000 m altitude where the relative air density is 0·374 and the temperature −44°C. A scale model of the aeroplane is to be tested in a compressed-air tunnel working at 24 atmospheres and 15°C. If the condition for dynamical similarity is to be satisfied in the test, what will be the lift force to be measured on the model?*

*Take coefficient of viscosity $\eta$ proportional to $(T)^{3/4}$ where T is in kelvins; the effect of Mach number may be neglected.*

Aerodynamic lift force $L = \rho u^2 l^2 \varphi(Re)$, where $Re = ul\rho/\eta$ which must be the same for the model and full-scale aeroplane for dynamical similarity.

Hence, the ratio of the lift on the model to that on the full-scale aeroplane is:

$$\frac{L_m}{L} = \frac{\rho_m}{\rho}\left(\frac{\eta_m}{\eta}\frac{\rho}{\rho_m}\right)^2, \quad \text{if} \quad Re = \frac{ul\rho}{\eta} = \frac{u_m l_m \rho_m}{\eta_m}$$

$$= \left(\frac{\eta_m}{\eta}\right)^2 \frac{\rho}{\rho_m}$$

Also,

$$\frac{\eta_m}{\eta} = \left(\frac{T_m}{T}\right)^{3/4} = \left(\frac{288}{229}\right)^{3/4} = 1.18$$

and if $\rho_0$ is the density of air on the ground at 15°C,

$$\frac{\rho_m}{\rho} = \frac{\rho_m \rho_0}{\rho_0 \rho} = \frac{24}{0.374} = 64.17$$

Hence,

$$\frac{L_m}{L} = \frac{(1.18)^2}{64.17} = 0.021\ 98$$

i.e. the lift force on the model

$$= 0.02198 \times 60 \text{ kN}$$
$$= \mathbf{1.319} \text{ kN}$$

It should be noted that the method of dimensions determines the arrangement of variables into non-dimensional groups. The method merely *correlates* non-dimensionally the physical quantities that we have *assumed* are involved, e.g. in establishing Rayleigh's formula

$$F = \rho l^2 U^2 \varphi(Re) \quad \text{or} \quad k_F = \varphi(Re)$$

the method states, in effect, that if $F$ depends on $\rho$, $\eta$, $l$ and $U$ only, *then* $k_F$ depends on $Re$ only, and in a way which can normally only be revealed by experiment. In this connection we may quote Bouasse concerning analytical methods generally. He pointed out that mathematics is the art of drawing inevitable conclusions from passively accepted data. It can do no more than render explicit what is implicit in its postulates, and is incapable of turning out anything transcending them in validity. Thus if we lack sufficient understanding to forecast correctly the fundamental quantities involved in a given problem, we shall be led to false, or incomplete, conclusions. Successful application of the method of dimensional analysis depends entirely on an ability to sift the *essential physical quantities* or *variables* from those which are of trivial importance in any given case.

For example, in establishing Rayleigh's formula the omission of $g$ from our list of variables influencing $F$ implied that gravity did not affect the problem, so that the expression for $k_F$ does not include a non-dimensional product containing it (i.e. a Froude number). Whilst this may be true for, say, the resistance of an aircraft, it is not valid for a ship which, of course, makes waves against gravitational pull. Similarly, the final expression says nothing concerning the shape of the body, simply because we did not imply that this had an effect. To do so it would have been necessary to have included, not only a reference length $l$, but also all the other lengths which describe its shape, $l_1, l_2 \ldots$, say. This would render unmanageable the method of dimensional analysis we have outlined above. As the aim is simply to correlate all the variables non-

2

64

*Similarity and dimensional analysis*

dimensionally this task can equally well proceed by inspection. Each new variable introduced into the original equation must have its non-dimensional counterpart. Thus, since $g$ can be made nondimensional by dividing it by $U^2/l$, and $l_1, l_2 \ldots$, etc., by dividing each by $l$, the expression:

$$\left(\frac{F}{\rho l^2 U^2}\right) = \psi\left(\frac{\rho l U}{\eta}\right)$$

can be elaborated to:

$$\left(\frac{F}{\rho l^2 U^2}\right) = \psi\left(\frac{\rho l U}{\eta}, \frac{U^2}{lg}, \frac{l_1}{l}, \frac{l_2}{l} \ldots\right)$$

i.e. 
$$k_F = \psi(Re, Fr, k_1, k_2 \ldots)$$

in which $k_1$, $k_2$ etc., are shape coefficients. In geometrically similar systems these latter will all have the same value, so that the value of $k_F$ depends only on $Re$ and $Fr$ in such cases.

EXAMPLE 10.7 (vi)

*The flow of a gas in a uniform circular duct is to be simulated by means of water flow in a quarter-scale transparent model. The full-scale velocity is to be 25 m/s.*

*Find: (a) the corresponding water velocity in the model, (b) the pressure drop to be expected per unit length of the full-scale duct if the pressure drop per m length of the model is 40 kN/m².*

*One kilogramme of gas under conditions in the full-scale duct occupies 0·62 m³, and the dynamic viscosity of water is 62 times that of the gas.*

Using Rayleigh's formula:

$$\frac{F}{\rho l^2 U^2} = \psi\left(\frac{\rho l U}{\eta}\right)$$

or

$$\frac{D}{\rho U^2} = \psi\left(\frac{\rho l U}{\eta}\right)$$

where $D$ is the drag force per unit of wetted area of ducting; we may write $Re = \rho l U/\eta$ as the Reynolds number which, if made the same for water flow in the model and gas flow in the full-scale duct yields $D_w/D = (\rho_w/\rho)(U_w/U)^2$ where the suffix $w$ refers to the water duct.

(a) For the same Reynolds number we deduce that the water velocity in the model must be

$$U_w = U\frac{\eta_w}{\eta}\frac{l}{l_w}\frac{\rho}{\rho_w}$$

$$= 25 \text{ m/s} \times 62 \times 4 \times \frac{1}{0·62 \times 10^3} = \mathbf{10 \text{ m/s}}$$

(b) Hence:

$$\frac{D_w}{D} = \frac{\rho_w}{\rho}\left(\frac{U_w}{U}\right)^2 = (620)\left(\frac{10}{25}\right)^2 = 99·2$$

Also,

$$D \times \pi dl = \Delta P \times \frac{\pi}{4} d^2$$

Hence,

$$\frac{\Delta P}{l} = \frac{4D}{d}$$

∴

$$\frac{\Delta P/l}{(\Delta P/l)_w} = \frac{D}{D_w} \frac{d_w}{d} = \frac{1}{99 \cdot 2} \times \frac{1}{4}$$

Hence, for the gas duct,

$$\Delta P/l = \frac{40}{396 \cdot 8} \text{ kN/m}^2 \text{ per m}$$

$$= 100 \cdot 8 \text{ Pa/m}$$

## 10.8. Non-dimensional products and the Π theorem

Mechanical problems generally involve three dimensions, such as M, L, and T, and this implies that three measures or units are needed, in terms of which the quantities involved may be expressed. Any three dimensions may be chosen, so long as they are mutually independent; those of length, density and velocity would for example serve equally well and so would force, length and time.

EXAMPLE 10.8 (i)

*Express the dimensions of force* $(F)$, *and viscosity* $(\eta)$ *in terms of those of length* $(L)$ *density* $(D)$ *and velocity* $(V)$.

(*a*) The dimensions of $F$ in terms of M, L and T are:

$$F = \left[ \frac{ML}{T^2} \right]$$

The following identities of dimensions which follow from the definitions of density and velocity:

$$D \equiv \left[ \frac{M}{L^3} \right]$$

$$V \equiv \left[ \frac{L}{T} \right]$$

enable us to formulate 'unity brackets' for converting dimensions:

$$\left\{ \frac{DL^3}{M} \right\} \equiv 1$$

$$\left\{ \frac{VT}{L} \right\} \equiv 1$$

Hence:

$$F = \left[\frac{ML}{T^2}\right]\left\{\frac{DL^3}{M}\right\}\left\{\frac{V^2T^2}{L^2}\right\} = [DL^2V^2]$$

(*b*) The dimensions of viscosity are similarly:

$$\eta = \left[\frac{M}{LT}\right]$$

and may be converted into those of L, D and V in the same way:

$$\eta = \left[\frac{M}{LT}\right]\left\{\frac{DL^3}{M}\right\}\left\{\frac{VT}{L}\right\} = [DLV]$$

In order to express a quantity non-dimensionally it is expressed, not in terms of a standard unit, but in terms of a unit chosen from within the problem. For example, in geometric similarity, all lengths, such as $l_1$, are expressed as some number times a reference length, $l$, say, i.e. as $(l_1/l)$. If we generalize this idea we may, in mechanical problems, choose up to three of the quantities as reference (or primary) units and so express all the others in units derived from these. This provides an alternative method for re-casting any equation into a non-dimensional form. It has the advantage that we may take each variable in turn, thus avoiding simultaneous equations.

For example, in the problem considered in Section 10.7 p. 263 (which would be unwieldy to tackle by Rayleigh's method) it was supposed that:

$$F = f(\rho, \eta, l, U, g, l_1, l_2 \ldots )$$

To re-write this non-dimensionally we choose certain of the variables to act as 'primary units' from which we derive 'units' for all the others. With one eye on the previous examples we choose $l$, $\rho$ and $U$ in the present case, and draw up a table as follows:

| Quantity | Symbol | Dimensions | Reference group |
|---|---|---|---|
| force | $F$ | $DL^2V^2$ | $\rho l^2 U^2$ |
| density | $\rho$ | D | $\rho$ |
| viscosity | $\eta$ | DLV | $\rho l U$ |
| length | $l, l_1, l_2\ldots$ | L | $l$ |
| velocity | $U$ | V | $U$ |
| gravitational acceleration | $g$ | $V^2/L$ | $U^2/l$ |

When each quantity is expressed in terms of its appropriate unit, we obtain the following non-dimensional groups:

$$\left(\frac{F}{\rho l^2 U^2}\right), \left(\frac{\eta}{\rho l U}\right), \left(\frac{g}{U^2/l}\right), \left(\frac{l_1}{l}\right), \left(\frac{l_2}{l}\right) \dots$$

Knowing that the original equation can be re-expressed as a relationship between the non-dimensional products we have as before:

$$k_F = \varphi(Re, Fr, k_1, k_2 \dots)$$

We may note that both the original and the final expressions need not be written in an *explicit* form, i.e. with one variable separated from the others. In general we suspect that a relationship exists between certain physical quantities which are thought to be important in a physical problem, namely: $q_1, q_2, q_3 \dots$ , that is, the complete equation is:

$$f(q_1, q_2, q_3, \dots) = 0$$

Dimensional arguments show that this may be re-expressed as a relationship between non-dimensional products $\Pi_1, \Pi_2, \Pi_3, \dots$ , formed by those $q$'s, i.e.

$$\varphi(\Pi_1, \Pi_2, \Pi_3, \dots) = 0$$

Information as to the nature of the function $\varphi$ can only be obtained via experiments or a theory of the phenomena in question.

Because three fundamental or primary units (such as M, L, T) are involved in mechanical problems, we may choose up to three of the $q$'s as primary quantities for forming the 'units' with which the remainder are combined to form the $\Pi$'s. If $n$ is the number of $q$'s there will generally be $(n-3)$ in number of $\Pi$'s.

This is a particular (the mechanical engineering) application of what is usually called 'the $\Pi$ theorem' which states that if $p$ be the number of fundamental or primary units in some physical phenomenon which involves $n$ physical ($q$) quantities then:

$$f(q_1, q_2, \dots q_n) = 0$$

and the $\Pi$ theorem states that the relation can alternatively be written:

$$\varphi(\Pi_1, \Pi_2, \Pi_3 \dots \Pi_{n-p}) = 0$$

where each $\Pi$ is an independent dimensionless group of some of the $q$'s.

The $(n-p)$ dimensionless group or $\Pi$'s can be found by (i) selecting a number of variables which include all the fundamental units and equal in number to the latter, and (ii) setting up dimensionless equations combining the variables previously selected (in (i)) with each of the other variables in turn.

In mechanical engineering (where $p = 3$) it may be possible to form only two, or even one, primary unit; in these cases there will be either

$(n-2)$ or $(n-1)$ $\Pi$'s. These facts are often useful when non-dimensional groups are formed by inspection, as in such cases we cannot easily be certain that the appropriate number of groups has been formed.

### EXAMPLE 10.8 (ii)

*A shaft rotates in a lubricated bearing. Assuming that the tangential force F opposing the rotation of the shaft depends on angular speed N of the shaft, load W on the shaft, diameter D of shaft and on the dynamic viscosity $\eta$ of the lubricant, find the non-dimensional groups or parameters and an equation connecting them (i) using the $\Pi$ method, (ii) using Rayleigh's method.*

(i) Using the $\Pi$ method we see that the number of variables involved is $n = 5$, namely $F$, $N$, $W$, $D$, $\eta$, and the number of fundamental primary units is $p = 3$, namely M, L, T. Consequently, since $n-p = 2$ there are two $\Pi$'s or non-dimensional groups which obey the equation

$$\varphi(\Pi_1, \Pi_2) = 0 \quad \text{or} \quad \Pi_1 = \psi(\Pi_2)$$

Thus, selecting $N$, $W$, and $D$ as the three variables which between them contain the fundamental units M, L and T, we may find $\Pi_1$, and $\Pi_2$ by combining the three chosen variables with the two others in turn, namely:

$$\Pi_1 = N^a W^b D^c F$$

which dimensionally is

$$\left(\frac{1}{T}\right)^a \left(\frac{ML}{T^2}\right)^b L^c \left(\frac{ML}{T^2}\right)$$

hence:

$$b+1 = 0 \quad \therefore b = -1$$

$$b+c+1 = 0 \quad \therefore c = 0 \quad \left.\right\} \text{therefore } \Pi_1 = \frac{F}{W}$$

$$a+2b+2 = 0 \quad \therefore a = -2+2 = 0$$

and $\Pi_2 = N^d W^e D^h \eta$ which dimensionally is

$$\left(\frac{1}{T}\right)^d \left(\frac{ML}{T^2}\right)^e L^h \left(\frac{M}{LT}\right)$$

hence:

$$e+1 = 0 \quad \therefore e = -1$$

$$e+h-1 = 0 \quad \therefore h = 2 \quad \left.\right\} \text{therefore } \Pi_2 = \frac{N}{W} D^2 \eta$$

$$d+2e+1 = 0 \quad \therefore d = -1+2 = 1$$

Thus, dimensional analysis by the $\Pi$ method gives

$$\frac{F}{W} = \psi \left(\frac{\eta N D^2}{W}\right)$$

which is a useful equation for studying lubricated bearings—experimental data being necessary to determine the function. If three other variables had been chosen containing the fundamental units M, L and T, a different equation would have resulted, e.g. if $D$, $\eta$, and $N$ had been chosen the resulting equation would have been

$$\frac{F}{\eta N D^2} = \varphi\left(\frac{\eta N D^2}{W}\right)$$

(ii) Using Rayleigh's method, we assume that force $F$ depends on $N$, $W$, $D$ and $\eta$, i.e. $F = \psi(N, W, D, \eta)$.

A typical term of the function would be $C N^a W^b D^c \eta^d$ in which $C$ is a number or numeric having no dimensions. Thus, the dimensions of this typical term

$$\left(\frac{1}{T}\right)^a \left(\frac{ML}{T^2}\right)^b (L)^c \left(\frac{M}{LT}\right)^d$$

must be those of a force $(F)$ namely $ML/T^2$.

Equating corresponding indices we get:

$$b+d = 1 \quad \therefore b = 1-d$$
$$b+c-d = 1 \quad \therefore c = 1+d-b = 2d$$
$$a+2b+d = 2 \quad \therefore a = 2-2b-d = d$$

Hence, the expression for $F$ involves terms of the type

$$C N^d W^{1-d} D^{2d} \eta^d \quad \text{or} \quad CW\left(\frac{ND^2\eta}{W}\right)^d$$

in which $C$ and $d$ may take any numerical values, i.e.

$$F = W\left\{C_0 + C_1\left(\frac{ND^2\eta}{W}\right) + C_2\left(\frac{ND^2\eta}{W}\right)^2 + \ldots\right\}$$

or

$$\frac{F}{W} = \psi\left(\frac{\eta ND^2}{W}\right)$$

as before.

EXAMPLE 10.8 (iii)

*Apply dimensional analysis to slipper-pad or Michell bearings*

Since inertia effects are negligible in the case of slipper-pads inclined at very small angles fluid density $(\rho)$ plays no part in relating the attitude of the slippers to the thrust or load carried. Thus, we may name: length $(L)$ and width or breadth $(B)$ and slope $(\theta)$ of pad; thickness $(h)$ and dynamic viscosity $(\eta)$ of film; relative speed $(U)$ of bearing surfaces as independent variables on which the load $(W)$ depends,

i.e.                $W = \psi_1(L, B, \theta, h, \eta, U)$

The $\Pi$ theorem $(n-p = 7-3 = 4)$ shows that this is equivalent to an equation containing four dimensionless groups involving three primary or basic dimensions M, L, T.

*Load or Thrust.* These four groups (or $\Pi$'s) can be chosen in any way provided that they are independent of each other, e.g.

$$\text{let}\quad \Pi_1 = \frac{W}{\eta UL}, \quad \Pi_2 = \frac{B}{L}, \quad \Pi_3 = \theta, \quad \Pi_4 = \frac{h}{L},$$

$$\text{then}\qquad\qquad \frac{W}{\eta UL} = \psi_2\left(\frac{B}{L}, \theta, \frac{h}{L}\right)$$

Hence, qualitatively, we may deduce that

(i) for given values of $\dfrac{B}{L}$ and $\theta$, $\dfrac{W}{\eta UL}$ is determined by $\dfrac{h}{L}$.

(ii) for a given position and size of slipper pad (i.e. for a specified $\theta$, $B$, $L$, $h$) $W \propto \eta U$; hence if $\eta$ or $U$ are, say, doubled then $W$ would be doubled In practice, however, each pad is pivoted towards the trailing edge to allow it to stabilize itself by taking its own inclination $\theta$ and film thickness $h$.

*Coefficient of Friction* $(\mu)$ of the bearing is the ratio of $(f)$ the friction force between the fluid and slipper pad faces and the thrust or load $(W)$. As compared with dry friction between solids in which $\mu$ is often assumed to be independent of both load and speed, in fluid friction $\mu$ is dependent on both, i.e.

$$\mu = \frac{f}{W} = \psi_3(L, B, \theta, h, \eta, U).$$

Because of the basic unit [T] of time no $\Pi$ can be formed with either $\eta$ or $U$ in combination with any of the other variables. Hence, the $\Pi$ theorem leads to

$$\frac{f}{W} = \mu = \psi_4\left(\frac{B}{L}, \theta, \frac{h}{L}\right).$$

Also, if the flow is assumed two-dimensional (i.e. side leakage is neglected) the breadth $B$ has no influence on $f/W$,

$$\text{i.e.}\qquad\qquad \mu = \varphi_5\left(\theta, \frac{h}{L}\right)$$

EXAMPLE 10.8 (iv)

(a) *Neglecting compressibility, show that the thrust of propellers or airscrews can be expressed in the form:*

$$\frac{F}{\rho U^2 D^2} = \varphi\left(\frac{UD}{\nu}, \frac{U}{ND}\right) \quad \text{or} \quad k_F = \varphi(Re, J)$$

*in which it is assumed that thrust $F$ depends on the forward velocity $U$,* diameter *D* of propeller rotating at speed *N*, and on the viscosity *η* and density *ρ* of the fluid.

(b) *In wind-tunnel tests on a* $\frac{1}{10}$ *scale model of an aircraft the propellers of the model are to be rotated at speed in order to determine the effect of their slip-stream on the model aeroplane. If the speed of the full-scale areoplane is* 500 km/h *or* 435 knot *and its propellers revolve at* 1 800 rev/min, *estimate the speed of the model propellers if the air speed in the tunnel is* 100 km/h *based on* $U/ND = J$ *being the same for the model and full-scale propellers.*

(a) The general term for thrust *F* may be written in the form $\rho^p U^q D^r \nu^s N^t$ the dimensions of which are

$$\left[\frac{M}{L^3}\right]^p \left[\frac{L}{T}\right]^q [L]^r \left[\frac{L^2}{T}\right]^s \left[\frac{1}{T}\right]^t$$

and must reduce to those of a force, namely, ML/T². Equating indices gives the results: $p = 1$, $q = 2-s-t$, $r = 2-s+t$. Hence, the general term becomes

$$\rho U^2 D^2 \left(\frac{UD}{\nu}\right)^{-s} \left(\frac{U}{ND}\right)^{-t}$$

and, therefore,

$$\frac{F}{\rho U^2 D^2} = \varphi\left(\frac{UD}{\nu}, \frac{U}{ND}\right)$$

or thrust coefficient

$$k_F = \varphi(Re, J)$$

where $U/ND = J$ is called the advance/diameter ratio of screw since $U/N$ is the distance a blade element advances during one revolution and alternatively called the effective pitch.

(b) *Re* and *J* should be constant for dynamical similarity between two geometrically similar propellers, but as this is difficult to arrange, it is usual to arrange for *J* to be constant, i.e.

$$\frac{U_m}{N_m D_m} = \frac{U}{ND}$$

and the rotary speed of the model propeller must be:

$$N_m = N\frac{U_m}{U}\frac{D}{D_m} = 1\ 800\ \text{rev/min} \times \frac{100}{500} \times 10 = \textbf{3600 rev/min}$$

EXAMPLE 10.8 (v)

*Find non-dimensional parameters or co-ordinates for studying the thrust of a screw propeller completely immersed in a liquid. Assume that the variables involved are: axial thrust F, propeller diameter D, velocity of advance U, angular speed N, density of fluid ρ, kinematic viscosity of fluid ν, and gravity g. There are seven variables n = 7 and three fundamental or primary units*

$p = 3$. Hence, there are four $(n-p = 4)$ dimensionless ratios which may be found by combining $D$, $U$, $\rho$ (which between them contain the three fundamental units M, L, and T) with each of the others in turn.

Thus $\Pi_1 = D^a U^b \rho^c F$, which dimensionally is

$$\left[ L^a \left(\frac{L}{T}\right)^b \left(\frac{M}{L^3}\right)^c \frac{ML}{T^2} \right]$$

consequently

$$a+b-3c+1 = 0 \left.\right\} a = 3c-b-1 = -2$$

$$c+1 = 0 \left.\right\} c = -1$$

$$b+2 = 0 \left.\right\} b = -2$$

i.e.

$$\Pi_1 = D^{-2} U^{-2} \rho^{-1} F = \frac{F}{\rho D^2 U^2}$$

Similarly

$$\Pi_2 = \frac{DN}{U}; \; \Pi_3 = \frac{Dg}{U^2} \quad \text{and} \quad \Pi_4 = \frac{\nu}{DU}$$

Thus, one solution is $\Pi_1 = \varphi(\Pi_2, \Pi_3, \Pi_4)$ or the thrust coefficient

$$\frac{F}{\rho D^2 U^2} = \varphi \left(\frac{DN}{U}, \frac{Dg}{U^2}, \frac{\nu}{DU}\right)$$

i.e.

$$k_F = \varphi(J, Fr, Re)$$

Experiment is now necessary to find the function. It may be that relation is

$$F = k_{Fp} D^2 U^2 \left(\frac{DN}{U}\right)^{x_1} \left(\frac{Dg}{U^2}\right)^{x_2} \left(\frac{\nu}{DU}\right)^{x_3}$$

in which case experimental data is needed to find $k_F$, $x_1$, $x_2$, $x_3$.

The above is *one* solution, but there are others dependent on which three variables are selected with which to combine the others each in turn.

### EXAMPLE 10.8 (vi)

(a) *Find an expression for the efficiency of a propeller in terms of non-dimensional coefficients and* (b) *show, with the aid of a sketch, why a propeller should have twisted blades*

(a) The thrust force $F$ and a non-dimensional thrust coefficient $k_F$ may be related as follows:

$$\frac{F}{\rho N^2 D^4} = f(Re, Ma, J) = k_F$$

and the torque $Q$ and non-dimensional torque coefficient $k_Q$ by:

$$\frac{Q}{\rho N^2 D^5} = \varphi(Re, Ma, J) = k_Q$$

Hence, the efficiency of a propeller moving with forward velocity $U$ and angular velocity $N$ is:

$$\eta = \frac{FU}{2\pi QN} = \frac{1}{2\pi}\left(\frac{k_F\rho N^2 D^4}{k_Q\rho N^2 D^5}\right)\left(\frac{U}{N}\right) = \frac{1}{2\pi}\frac{k_F}{k_Q}\left(\frac{U}{ND}\right) = \frac{k_F}{k_Q}\frac{J}{2\pi}$$

where $J$ is the advance/diameter ratio of the propeller or screw.

(b) For a section of the propeller at a particular radius, the velocity and force diagrams are shown in Fig. 10.8(a). Under steady conditions all blade

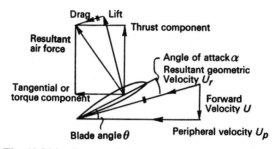

**Fig. 10.8(a).** *Force and velocity diagrams for propeller*

elements will follow helices and the resultant force exerted by the air on the particular cross-section of blade shown may be thought of as composed either of the two components lift added vectorially to drag, or as thrust force added vectorially to tangential (torque) force. The forces exerted by the blade on the air are, of course, equal and opposite to those shown, in Fig. 10.8(a). The blade angle at the section shown is

$$\theta = \alpha + \tan^{-1}\left(\frac{U}{U_p}\right)$$

Hence, for constant angle of attack ($\alpha$), the blade angle, $\theta$, must increase as the radius of section decreases since peripheral velocity $U_p$ is proportional to the radius whereas the forward velocity remains constant. Thus, the blade angle is large at the root of a propeller and should gradually reduce and the radius increases by twisting the blade from root to tip.

The so-called nominal blade angle for the entire propeller is taken at the section at 0·75 of the tip radius. The overall performance of propeller is, however, determined by wind-tunnel or water-tunnel tests, the data from which is plotted non-dimensionally usually using the advance-diameter ratio $J = U/ND$ as abscissa and thrust coefficient $k_F$ (or $C_T$) as ordinate.

## Exercises on Chapter 10

1. Give sketches indicating how the boundary layer changes in thickness and type (laminar, turbulent, etc.) when a flat plate is towed edgewise through a fluid.

A vane of aerofoil section moves through water at 2 m/s. The chord is inclined at 5° to the direction of motion and the corresponding values of $C_L$ and $C_D$ are 0·92 and 0·015 respectively. If the length of the chord is 400 mm, determine the resultant force on the vane per metre of length and the direction of the resultant relative to the chord.

2. The lift and drag coefficients of a rectangular aerofoil of 14 m span and 1·8 m chord when at a certain angle of incidence are 0·6 and 0·025 respectively. Calculate the power required to propel this aerofoil at 200 knot, through undisturbed air of density 1·1 kg/m³, and the lift obtained at this speed.

3. A wind-tunnel model is supported by long wires 1·6 mm diameter. Estimate its Reynolds number when the air speed is running at 60 m/s at atmospheric pressure. If, at this speed, the coefficient of drag is 1·1, estimate the drag force per metre of wire. Density and viscosity of air 1·23 kg/m³ and $17·8 \times 10^{-6}$ kg/ms.

4. Show that for dynamic similarity in the motion of geometrically similar bodies through a viscous, incompressible fluid:

(a) the linear size is proportional to $\eta/\rho u$
(b) the velocity is proportional to $\eta/\rho L$
(c) the density is proportional to $\eta/uL$

Hence show that the force resisting motion may be expressed as:

$$F = \frac{\eta^2}{\rho} \varphi(Re) = \eta u L \psi(Re)$$

5. The drag of a small aircraft component is to be found from a test on it in a water tunnel. The speed of the aircraft is to be 350 knot at an altitude where the density is 0·9 kg/m³ and coefficient of viscosity is $16·7 \times 10^{-6}$ kg/s m.

Estimate the water speed in the tunnel for dynamic similarity and the ratio of the drag in water to that in air.

For water $\rho = 1$ Mg/m³; $\nu = 1·14 \times 10^{-6}$ m²/s.

6. For laminar flow round a sphere Stokes showed that the drag in a fluid of coefficient of viscosity $\eta$ and velocity $u$ is $3\pi\eta u d$.

Find the coefficient of drag of the sphere based on the projected area $(\pi/4)d^2$.

7. Prove that the viscous resistance $F$ of a sphere of diameter $d$, moving at constant speed $u$ through a fluid of density $\rho$ and viscosity $\eta$, may be expressed as:

$$F = \frac{\eta^2}{\rho} \varphi \left( \frac{\rho u d}{\eta} \right)$$

Show that Stokes' result for low velocities, $F = 3\pi\eta u d$, is in agreement with this general formula.

A sample of emery powder was shaken in water contained in a glass beaker and then allowed to settle. It was found that the water cleared in 100 s when the depth was 180 mm. Calculate the minimum diameter of the particles, assuming them all spherical and taking the relative density of emery as 4·0 and the coefficient of viscosity of water as 0·012 poise.

8. Show that one form of the law governing the frictional loss in a pipe can be expressed by the formula $D/\rho u^2$ = a function of Reynolds number and show also that $D/\rho u^2$ is dimensionless, $D$ being the frictional drag force per unit area of wetted surface.

The frictional loss in a pipe carrying air is to be estimated from a suitable test on a similar pipe carrying water. Using the particulars given in the table, find the value of $u$ and the ratio of the pressure drops per unit length of pipe, water to air.

| Fluid | Pipe diam. (mm) | Velocity (m/s) | Density (kg/m³) | Dynamic viscosity (kg/s m) |
|---|---|---|---|---|
| Water | 50 | $u$ | 1 000 | 0·000 803 |
| Air | 100 | 12 | 1·25 | 0·000 018 5 |

9. (i) For tests made under dynamically similar conditions where size of model is a constant ratio of size of wind tunnel, show that the power required to drive the air through the tunnel decreases with increase of tunnel dimensions. Neglect compressibility effects.

(ii) For a given size of tunnel and dynamically similar conditions, find the ratio of the power required with air as the working fluid to that required when water is the working fluid.

For air $\rho = 1·225$ kg/m³; $\nu = 14·8 \times 10^{-6}$ m²/s.

For water $\rho = 1$ Mg/m³; $\nu = 1·15 \times 10^{-6}$ m²/s.

10. A sharp-edged orifice of diameter $d$ is used to measure the rate of flow of gas having density $\rho$ and viscosity $\mu$ from a large tank. Show, by applying the method of dimensions, that the rate of flow is given by:

$$Q = Cd^2 \sqrt{\frac{\Delta p}{\rho}}$$

in which $\Delta p$ is the pressure drop across the orifice, and the coefficient $C$ depends upon the value of the criterion $(Q\rho/\mu d)$.

Explain briefly why, provided flow is in the turbulent region, the coefficient $C$ tends to decrease as $(Q\rho/\mu d)$ increases.

11. A sharp-edged orifice having a diameter of 40 mm and discharge coefficient of 0·6 is used to measure the flow of air from a large tank. The pressure drop across the orifice is 12·5 mm of water, the barometric pressure is 755 mm mercury and the air temperature is 15°C. Assuming incompressible flow, find the discharge in kg/h. Take the gas constant $R$ in $PV = RmT$ for air as 288 J/kg K.

12. Explain why the Reynolds number is a criterion for dynamical similarity of flow of fluids.

The velocity of flow of water through a 75 mm diameter pipe is 5 m/s and the loss of head at this speed is 15 m of water for a length of 30 m. What is the corresponding velocity for flow of air through a pipe of similar material but 375 mm diameter, and what is the loss of head in inches of water per 30 m length at this speed?

Density of air $= 1 \cdot 23$ kg/m$^3$; viscosity of air $= 180 \times 10^{-6}$ poise; viscosity of water $= 15 \times 10^{-3}$ poise.

13. Show, by applying the method of dimensions, that a rational formula for rate of flow over a vee-notch is:

$$Q = H^2 \sqrt{(gH)} \varphi \left\{ \frac{H \sqrt{(gH)}}{\nu}, \theta \right\}$$

where $H$ is the head above the apex of the notch, $\nu$ is the kinematic viscosity of the liquid, $\theta$ is the notch angle and $g$ is the acceleration due to gravity.

Experiments on flow of water over a vee-notch show that the rate of discharge is given by the practical formula $Q = 1 \cdot 33 \, H^{2 \cdot 48}$ m$^3$/s and head $H$ being in metres.

Assuming the functional term in the above rational formula can be written in the form:

$$C \left( \frac{H \sqrt{(gH)}}{\nu} \right)^n$$

estimate the percentage error involved in using the practical formula for measuring the flow of oil whose kinematic viscosity is ten times that of water.

14. Show that when compressibility of a viscous fluid is taken into account, Rayleigh's formula becomes $F = \rho u^2 l^2 f(Re, Ma)$, where $Ma$ is Mach number and $Re$ is Reynolds number.

Assume the velocity of sound $a$ to be the criterion of compressibility.

15. Show that in the case of a cylindrical bearing without radial load, i.e. with uniform clearance such as a vertical locating bearing, the friction torque is given by:

$$T = k \frac{\eta D^3 NL}{C}$$

Show by the method of dimensions that if the expression is to apply to similar journal bearings carrying radial load $W$, $k$ must be replaced by

$$\varphi \left( \frac{\eta D^2 N}{W} \right)$$

where $\varphi$ is an unknown function which can be obtained by experiment. Describe how such an experiment could be carried out.

16. Show that for dynamic similarity the wave-making resistance of a ship in a given liquid is proportional to the cube of the representative length or the displacement of the ship.

17. A full-scale sea-going ship is to be 150 m long and to cruise at a speed of 20 knot. A geometrically similar model is made 3 m long and tested in

fresh water with a view to predicting the wave-making or residual resistance of the full-scale ship. Estimate the speed at which the model should be tested in fresh water and hence deduce the ratio of the wave-making resistances. Take the ratio of the densities of sea water to fresh water as 1·026.

18. Assuming that the variables involved in the flow of fluid over a notch or weir are $Q$, $H$, $\rho$, $g$, $\eta$ and surface tension $\sigma$ of the interface between the fluid and air, use the $\Pi$ method to deduce relationship between the three non-dimensional groups consequent upon choosing $Q$, $H$ and $\rho$ with which to combine the other three each in turn.

19. Using the Method of Dimensions, find an expression for $Q/\omega D^3$ assuming $Q$ to depend on pressure difference $\Delta p$, speed $\omega$, viscosity $\eta$, density $\rho$ and diameter of impeller $D$.

A centrifugal pump is to be designed to deliver 16 m³/min at 60 m head when handling a liquid hydrocarbon product of relative density 0·95 and viscosity 2·5 poise. The pump is to run at 1 450 rev/min. In order to check the performance of the full-scale pump a half-scale model is to be tested pumping a light oil of relative density 0·90 and viscosity 0·25 poise. At what speed should the model be driven? What should be its delivery head and volume flow at the design point if the specified performance is to be achieved with the full-scale pump?

## Chapter 11

# Steady flow in pipes and channels

*' But before you found a law on this case test it two or three times and see whether the experiments produce the same effects.'*

LEONARDO

## 11.1. Chezy's formula

The problem of finding the friction force which acts on a fluid when it is transmitted through a pipe or a channel has long been recognized as being of fundamental importance in hydraulic engineering. In many cases, for example with natural streams and rivers, artificial canals, sewers and pipes not running full, the free surface is subjected to atmospheric pressure only—i.e. the hydraulic gradient is the surface of the liquid being conveyed in the channel. A representative problem is that of steady flow down an open channel of uniform small slope $i$ ( $= \delta H/\delta L$). The fluid is impelled forward by the component of its weight, and a steady state is reached if this is balanced by the friction. Steady uniform flow occurs in a channel of constant cross-section where the terminal velocity has been reached, and steady non-uniform flow occurs in channels of irregular cross-section in which discharge does not change with time. If we denote the cross-sectional area of the stream by $A$, gravity will exert a force which, for geometrically similar systems, has a magnitude:

$$iW \propto i\rho g A L$$

see Fig. 11.1(a).

Once a steady state has been reached, the depth of flow in the channel and the mean velocity across the section will both have become constant and this corresponds to what is termed *normal flow*. Under these conditions the force per unit length of channel will have become constant. *If* the flows down two such channels are dynamically similar, then corresponding friction forces will (using $Y$, the wetted perimeter, as the reference length) have magnitudes proportional to $\rho Y^2 U^2$. The corresponding force per unit length of channel will thus have a magnitude $\rho YU^2$ so that, over a length $L$ the force will be:

$$F \propto \rho YLU^2$$

Equating this to the component of the weight in the direction of motion, we have:

$$\rho YLU^2 \propto i\rho gAL$$

i.e. $$U = K\sqrt{(gmi)}$$

in which $K$ is an undetermined coefficient and $A/Y$ has been denoted by $m$. The latter is referred to as the hydraulic mean depth since, if the same flow were imagined to be spread over a flat bed of width $Y$, it would have a depth $m$. Thus, **hydraulic mean de** ˙ **is the ratio of cross-sectional area to wetted perimeter.**

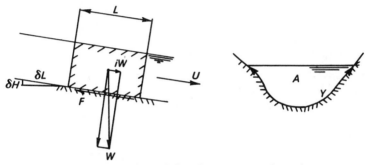

**Fig. 11.1(a).** *Normal flow down an open channel*

A formula of this type was established empirically as long ago as 1775 by Chezy, the factor $g$ being assimilated into the constant of proportionality:

$$U = C\sqrt{(mi)}$$

According to our derivation, $C$ may be expected to have the same value for all channels of a given shape, provided that the flows down them are dynamically similar. However, appropriate friction factors are more difficult to decide on for open channels than for flow in pipes, and the treatment of open channel flow is usually more empirical than for pipe-flow. Numerous hydraulic engineers have subsequently elaborated Chezy's formula by deducing working rules for $C$ under various conditions. Its value may be expected to depend on all those factors which cause dissimilarities between flows. These will be geometric parameters, such as those which describe the shape of the section and its roughness, together with such flow parameters as that which establishes the relative importance of dynamic and viscous effects. Such variables should appear non-dimensionally; otherwise the resulting formulae are unlikely to be valid over a much wider range than that of the particular experiments on which they were based. A good representative value of $C$ (which, it should be noted, is not non-dimensional) is $55\cdot2$ m$^{1/2}$/s,

i.e. $$U = (55\cdot2 \text{ m}^{1/2}/\text{s}) \sqrt{(mi)}$$

In cases of flow of water where the range of temperatures and velocities is small, Manning's formula, $U = Mm^{2/3}i^{1/2}$, where $M$ varies with roughness of surface is often preferred to Chezy's formula, see Example 11.5 (ii).

The usual problem of channels is one of design. The slope $i$ depends on the slope of the ground or rock in which the channel is made and is usually about 1 in 5000 to 10 000 in open earthen channels and may be as low as 1 in 100 000 in large rivers.

If slope $i$ and rate of volumetric flow $Q$ are known then, since

$$Q = AU = AC\sqrt{\left(\frac{A}{Y}i\right)}$$

the term $A\sqrt{(A/Y)}$ will be known assuming $C$ has been chosen to suit the roughness of channel facing. Several shapes of channel are possible each having the same area $A$, but one must be more suitable than another having regard to excavation costs and facing of the channel with, say, concrete. The mean velocity in large aqueducts may be about 1 m/s, for if the speed is too small suspended matter would be deposited and slimy growths would adhere to the sides. On the other hand, if the velocity is too large, erosion takes place. In sewers there are no deposits when the velocity is high, and the low velocities resulting from low discharge rates can to some extent be counteracted by making the sewers oval or egg-shaped in cross-section.

EXAMPLE 11.1 (i)

(a) *Find the shape of the symmetrical trapezoidal section of constant area $A$ and slope of sides $k$, which will require a minimum of facing.*

(b) *Estimate the dimensions of such a channel to convey 12 m$^3$ of water per second if the side slopes are 2 vertical to 1 horizontal and the bed of the channel slopes 1 in 2 000. The value of $C$ in Chezy's formula may be taken as $55\cdot5$ m$^{1/2}$/s.*

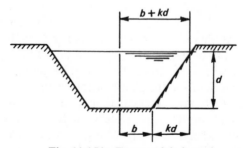

Fig. 11.1(b). *Trapezoidal channel*

(*a*) Referring to Fig. 11.1(b):

$$\text{Cross-sectional area } A = (2b+kd)d = \text{constant}$$
$$\text{Wetted perimeter } Y = 2b+2d \sqrt{(1+k^2)}$$

From which we deduce that

$$2b = \frac{A}{d} - kd$$

and

$$Y = \left(\frac{A}{d} - kd\right) + 2d \sqrt{(1+k^2)}$$

Hence,

$$\frac{dY}{dd} = -\frac{A}{d^2} - k + 2 \sqrt{(1+k^2)}$$

which is zero when

$$\frac{A}{d^2} + k = 2 \sqrt{(1+k^2)} = \frac{2b+kd}{d} + k$$

i.e. the perimeter $Y$ is a minimum when

$$b+kd = d \sqrt{(1+k^2)}$$

**Thus, minimum facing is required when half the top width is equal to the slant side,** in which case the hydraulic mean depth

$$m = \frac{A}{Y} = \frac{(2b+kd)d}{2b+2(b+kd)} = \frac{d}{2}$$

A trapezoidal channel will not silt up as quickly as a rectangular channel and is also easier to face.

(*b*)

$$\text{Velocity } U = C \sqrt{(mi)} = C \sqrt{\left(\frac{d}{2} i\right)}$$

and

$$\text{Area } A = [2d \sqrt{(1+k^2)} - kd]d$$

for minimum perimeter.

Hence,

$$Q = AU = [2 \sqrt{(1+k^2)} - k]d^2 C \sqrt{\left(\frac{d}{2} i\right)}$$

or

$$12 \text{ m}^3/\text{s} = [2 \sqrt{(1+\tfrac{1}{4})} - \tfrac{1}{2}]d^2 \times 55\!\cdot\!5 \text{ m}^{1/2}/\text{s} \sqrt{\left(\frac{d}{2} \times \frac{1}{2\,000}\right)}$$

$$12 \text{ m}^{5/2} = \frac{3\!\cdot\!98 \times 55\!\cdot\!5}{63\!\cdot\!2} d^{5/2}$$

and

$$d = (3\!\cdot\!43)^{2/5} \text{ m} = \mathbf{1\!\cdot\!637\ 2 \text{ m}}$$

$$b = \frac{d}{2} = \mathbf{0\!\cdot\!819 \text{ m}}$$

and half the top width = slant side

$$= d \sqrt{(1+k^2)} = d \sqrt{(1+\tfrac{1}{4})} = (\sqrt{5})\frac{d}{2}$$

$$= 1 \cdot 83 \text{ m}$$

## EXAMPLE 11.1 (ii)

*Using Manning's formula deduce the conditions (a) for maximum rate of volumetric flow and (b) for maximum velocity when liquid flows through a circular channel partly full.*

Let the water line subtend an angle $\theta$ at the centre of the circle as in Fig. 11.1(c), then $A = \tfrac{1}{2}r^2(\theta - \sin \theta)$ and the wetted perimeter $Y = r\theta$.

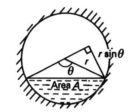

**Fig. 11.1(c).** *Circular channel*

(i) Using Manning's formula, $U = Mm^{2/3}i^{1/2}$, we may write

$$Q = AU = AM \left(\frac{A}{Y}\right)^{2/3} i^{1/2} = Mi^{1/2} \left(\frac{A^5}{Y^2}\right)^{1/3}$$

Then, if the slope $i$ of the pipe and water surface is constant and we require to know the particular value of $\theta$ which makes $Q$ a maximum we must differentiate $A^5/Y^2$ with respect to $\theta$ and equate to zero,

$$\frac{d(A^5/Y^2)}{d\theta} = \frac{5Y^2A^4(dA/d\theta) - 2A^5Y(dY/d\theta)}{Y^4}$$

which is zero when

$$5Y\frac{dA}{d\theta} = 2A\frac{dY}{d\theta}$$

i.e. when

$$5r\theta \times \frac{r^2}{2}(1 - \cos \theta) = 2 \times \tfrac{1}{2}r^2(\theta - \sin \theta) \times r$$

or

$$5\theta(1 - \cos \theta) = 2(\theta - \sin \theta)$$

or

$$3\theta = 5\theta \cdot \cos \theta - 2 \sin \theta$$

which, by trial and error, gives $\theta = 302°$ for maximum $Q$ rate of flow through a circular duct not under pressure. (Chezy's formula gives 308°.) The discharge rate is greater when $\theta = 302°$ than when full because the small increase in cross-sectional area of water when the channel is full does not compensate for the increase in drag or resistance caused by the increased wetted perimeter. Such conduits are not usually built to withstand much pressure and are usually designed to flow about three-quarters full as measured up the centre line of symmetry to allow a margin for excess flow.

(*b*) Maximum velocity occurs when $m^{2/3}$ is a maximum, or when $A/Y$ is a maximum, i.e. when

$$Y \frac{dA}{d\theta} = A \frac{dY}{d\theta}$$

or

$$r\theta \times \frac{r^2}{2}(1 - \cos\theta) = \frac{r^2}{2}(\theta - \sin\theta)r$$

or

$$\tan\theta = \theta$$

which, by trial, gives $\theta = 257°$. This is about 45° less than the angle for the maximum volumetric rate.


## 11.2. Darcy's formula

If we imagine normal flow down a uniform channel having a cross-section in the shape of an inverted key-hole, we may in the limit, consider it to be a pipe with a narrow continuous gauge glass which shows the hydraulic gradient, see Fig. 11.2(a). The slop of the latter is a measure of the loss

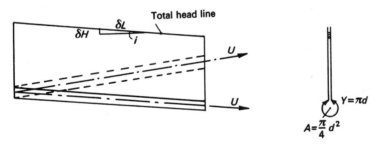

**Fig. 11.2(a).** *Flow down a uniform channel*

of total head per unit length due to friction, and we may transform the previous equation:

$$U = K \sqrt{(gmi)}$$

to give *i* directly:

$$i = \frac{U^2}{K^2 gm}$$

It should be noted that the free surface is no longer an essential feature of the flow, so that gravity no longer has any effect in controlling the flow pattern. That is, the loss of total head would be the same down the pipe shown dotted, if fluid were transmitted through it at the same rate, by some means other than gravity. As $(U^2/2g)$ is the kinetic head of the fluid moving through the pipe, it is natural to re-arrange the above equation so that the loss of total head is expressed in terms of this:

$$i = \frac{1}{k^2} \times \frac{1}{m} \times \frac{U^2}{2g}$$

The symbol '$f$' is variously used in fluids literature to denote either $(1/k^2)$ or $(4/k^2)$. We shall denote the former by $f$ as distinct from the latter, which will be denoted by the symbol $\lambda$. Thus $\lambda = 4f = 4/k^2$ and

$$i = \frac{f}{m}\frac{u^2}{2g} = \frac{\lambda}{4m}\frac{u^2}{2g}$$

The number 4 is introduced to compensate for that which occurs in the expression for $m$ in circular pipes *flowing full* for which

$$m = \frac{A}{Y} = \frac{\pi d^2/4}{\pi d} = \frac{d}{4}$$

As it is usual to consider pipes of non-circular section in terms of an 'equivalent diameter' we shall generally use the friction factor

$$\lambda = \frac{4}{k^2} = 4f$$

so that:

$$\frac{1}{k^2 m} = \frac{\lambda}{4}\frac{4}{d} = \frac{\lambda}{d}$$

for a pipe running full, and:

$$i = \frac{\lambda}{d}\frac{u^2}{2g} = \frac{4f}{d}\frac{u^2}{2g}$$

so that loss of head:

$$h = \frac{4fl}{d}\frac{u^2}{2g}$$

This is known as Darcy's formula, since, as a result of extensive measurements, Darcy suggested (1855) a velocity-squared law for the friction loss in conduits; consequently formulae of this type are associated with his name.

Since $u = K\sqrt{(gmi)} = k\sqrt{(2gmi)}$, and Chezy's formula is $u = C\sqrt{(mi)}$, the value of the friction factor $f$ corresponding to $C = 55\cdot2$ m$^{1/2}$/s in the Chezy formula is given by:

$$C^2 = 2k^2g = \frac{2g}{f} = \frac{8}{\lambda}g$$

Hence,

$$\lambda = \frac{8g}{C^2} = \frac{8 \times 9\cdot807 \text{ m/s}^2}{3\ 050 \text{ m/s}^2} = 0\cdot025\ 72$$

and
$$f = \frac{\lambda}{4} = 0.006\ 43$$

The Chezy formula is, however, frequently used in civil engineering —i.e. it is applied to large-scale installations. The above value of $f$ correlates only with pipes of a similar scale. In mechanical engineering values of $\lambda$ up to $0.04$ or $f = 0.01$ are more representative.

EXAMPLE 11.2 (i)

*When a horizontal pipe of cross-sectional area A runs full, relate the resistance coefficient f to the drag force F on the wall of the pipe of circumference Y and length l.*

Taking $1/k^2 = f = (mi2g/u^2)$, where $u$ is the mean velocity, $m$ is the hydraulic mean depth $(A/Y)$ and $i$ is the slope of the hydraulic gradient $[(p_1 - p_2)/wl]$, then

$$f = \frac{A}{Y} \frac{(p_1 - p_2)}{\rho g l} \frac{2g}{u^2} = \frac{A(p_1 - p_2)}{\frac{1}{2}\rho u^2\ Yl}$$

i.e.
$$f = \frac{F}{\frac{1}{2}\rho u^2 S} = \varphi(Re)$$

which is, of course, Rayleigh's formula (see Section 10.7, p. 258) in which $S = Yl = $ surface area in contact with the fluid, and $F$ is the drag force.

EXAMPLE 11.2 (ii)

*A cylindrical sewer is 1·5 m diameter and slopes 1 in 900. Estimate the rate of volumetric flow if the greatest depth of water is 0·95 m and the constant in Chezy's formula is 54 m$^{1/2}$/s.*

*Find the corresponding resistance coefficient in the Darcy formula.*

$$i = \frac{f}{m}\frac{u^2}{2g}$$

Referring to Fig. 11.2(b):

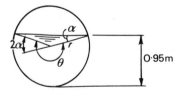

**Fig. 11.2(b)** *Circular sewer*

$$\sin \alpha = \frac{0.2}{0.75} = 0.266\ 7$$

∴
$$\alpha = 15° \ 30'$$

and
$$\theta = 180° + 2\alpha = 211°$$

Hence the wetted perimeter is

$$Y = r\theta = 0.75 \text{ m} \times 211 \text{ deg} \left[\frac{\pi}{180 \text{ deg}}\right] = 2.76 \text{ m}$$

Also, the cross-sectional area of the water is:

$$A = \frac{r^2}{2}(\theta + \sin 2\alpha)$$

$$= \frac{9}{32} \text{ m}^2 \left(211 \text{ deg} \left[\frac{\pi}{180 \text{ deg}}\right] + 0.515\right) = 1.18 \text{ m}^2$$

Hence, the hydraulic mean depth $m = A/Y = 0.4275$ m and the mean velocity

$$u = C \sqrt{(mi)} = 54 \text{ m}^{1/2}/\text{s} \sqrt{\left(\frac{0.427\ 5 \text{ m}}{900}\right)} = 1.177 \text{ m/s}$$

and the rate of volumetric flow is:

$$Q = Au = 1.18 \text{ m}^2 \times 1.177 \text{ m/s} = 1.39 \text{ m}^3/\text{s}$$

The resistance factor in the particular form of the Darcy formula given is $f = mi2g/u^2$, and Chezy's formula gives $C^2 = u^2/mi$.

Hence,     $f \times C^2 = 2g$

and     $f = \dfrac{2g}{C^2} = \dfrac{18.613 \text{ m/s}^2}{54 \times 54 \text{ m/s}^2} = 0.006\ 38$

## 11.3. Reynolds' experiments

It may be recalled that the Hagen–Poiseuille law established earlier (see Section 8.5, p. 181) predicted a pressure drop proportional to the velocity, whereas the Darcy formula (at least with a constant value of $f$), suggests it depends on the velocity-squared. Each has been proved by experiments, the former primarily at low velocities through capillary tubes, and the latter on large installations.

It was Osborne Reynolds who was largely responsible for reconciling these results, by means of two striking experiments at Manchester (see *Phil. Trans. of Royal Society*, 1886).

In his first experiment water was drawn through a glass tube at rates which could be controlled by a hand-cock as shown in Fig. 11.3(a). The water in the tank was allowed to stand for a few hours to ensure that there were no settling eddies still in existence before the valve was slowly opened. At the same time a stream of dye was drawn into the bell-mouth of the tube by syphon action from a ventilated auxiliary vessel. As the speed of the water increased, the dye began to show the formation of eddies near the outlet end of the tube. This break-up of the laminar flow spread towards the entry as the velocity was further increased. The velocity at which transition to turbulence began was called the *higher critical velocity* by Reynolds, the *lower critical velocity* corresponding to that at which the flow again became laminar as the cock was gradually closed.

In his second experiment, Reynolds measured the drop of pressure over a 1·5 m length of lead pipe of 6 mm, and later 13 mm, diameter when

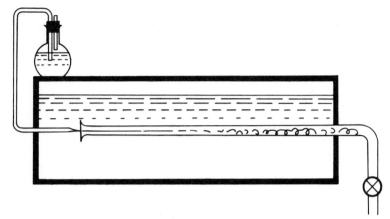

**Fig. 11.3(a).** *Reynolds' apparatus*

transmitting water. The test length was preceded by a 3 m settling length, as shown in Fig. 11.3(b), the fluid entering this in a disturbed state through the introduction of a right-angled bend. By using carbon disulphide as the manometric fluid for small pressure differences and mercury at the higher rates of flow it was found possible to make reliable measurements of the pressure drop both above and below the critical velocities. His

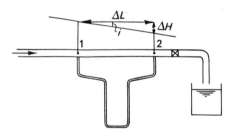

**Fig. 11.3(b).** *Reynolds' second experiment*

results showed that the slope of the hydraulic gradient was, as the Hagen–Pouiseuille law predicts, proportional to the mean velocity so long as the flow remained laminar. This was followed by an unsteady period, during which transition to turbulence was taking place until finally, as the discharge was increased still further, the resistance law became:

$$i \propto U^n$$

in which $n$ increased progressively from 1·8 to 2, see Fig. 11.3(c).

Fig. 11.3(d) shows the effect of the temperature of the fluid on the pressure drop, there being less resistance to flow at the higher temperature, this corresponding to a lower viscosity.

The lower critical velocity ($U_c$), i.e. that below which a disturbed flow will revert to laminar motion, was thus shown to depend on both the

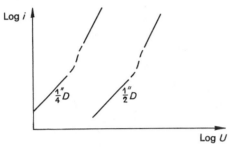

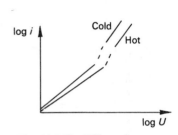

**Fig. 11.3(c).** *Effect of velocity on hydraulic gradient*

**Fig. 11.3(d).** *Effect of temperature on pipe resistance*

viscosity of the fluid and the size of the pipe. Reynolds showed that this velocity was such that:

$$Re = \frac{DU_c}{\nu} = 2\,300$$

in which $D$ is the diameter of the pipe and $\nu$ is the kinematic viscosity of the fluid. The quantity $(DU/\nu)$ is what we now term a Reynolds number (see page 240), a parameter which is of fundamental importance in most fluid-resistance problems. As it is a relative measure of the importance of inertia and viscous effects ($Re \propto$ inertia force/viscous force, see page 240), the critical Reynolds number may be interpreted as follows. So long as the dynamic actions are not too dominant, any disturbances will be damped out by viscous action. When however the former, which are proportional to $\rho D^2 U^2$, exceed a certain value relative to the latter, which are proportional to $\eta U D$, disturbances may be amplified resulting in a breakdown to turbulence. In pipe-flow it is found that this can occur when:

$$\rho D^2 U^2 > 2\,300\,\eta U D$$

i.e.

$$\frac{\rho D U}{\eta} > 2\,300$$

i.e.

$$\frac{D U}{\nu} > 2\,300$$

Example 11.3 (i)

*Oil of relative density 0·9 and dynamic viscosity 0·175 kg/ms is pumped through a 75 mm diameter pipe 750 m long at the rate of 3 kg/s. Taking the critical Reynolds number as 2 300 show that the critical velocity is not exceeded and calculate the pressure required at the pump if the pipe is horizontal. Also, estimate the power required to pump the oil.*

The critical velocity is

$$u_c = Re_c \frac{\eta}{d\rho} = \frac{2\,300 \times 0\cdot175}{0\cdot075\ \text{m}} \frac{\text{kg}}{\text{m s}} \times \frac{\text{m}^3}{0\cdot9 \times 10^3\ \text{kg}} = 6\ \text{m/s}$$

The actual velocity is

$$u = \frac{\dot{m}}{\rho a} = 3\,\frac{\text{kg}}{\text{s}} \times \frac{\text{m}^3}{900\ \text{kg}} \times \left(\frac{4}{\pi} \times \frac{16}{0\cdot09\ \text{m}^2}\right) = 0\cdot75\ \text{m/s}$$

Hence the flow is laminar and the rate of volumetric flow through a pipe of length *l* and radius *r* is

$$Q = \frac{\pi r^4 (p_1 - p_2)}{8\eta l}$$

as proved in Section 8.5.

From this we deduce that the pressure difference between the ends of the pipe is:

$$p_1 - p_2 = \frac{8\eta l}{\pi r^4} Q = \frac{8\eta l \dot{m}}{\pi r^4 \rho} = \frac{8}{\pi} \times 0\cdot175\,\frac{\text{kg}}{\text{m s}} \times \frac{750\ \text{m}}{(3\cdot75 \times 10^{-2}\ \text{m})^4} \times 3\,\frac{\text{kg}}{\text{s}} \times \frac{\text{m}^3}{900\ \text{kg}}$$

$$= 563 \times 10^3\,\frac{\text{kg}}{\text{ms}^2} \left[\frac{\text{N s}^2}{\text{kg m}}\right] = 563\ \text{kN/m}^2$$

Power required is $P = Q(p_1 - p_2) = \dot{m}(p_1 - p_2)/\rho$

$$= 3\,\frac{\text{kg}}{\text{s}} \times 563\,\frac{\text{kN}}{\text{m}^2} \times \frac{\text{m}^3}{900\ \text{kg}} \left[\frac{\text{W s}}{\text{N m}}\right] = 1\cdot88\ \text{kW}$$

## 11.4. Resistance of smooth pipes

If fluid enters a pipe through a bell-mouthed entry as shown in Fig. 11.4(a) a boundary layer of retarded fluid forms, and continues to grow, until

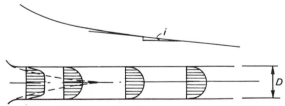

**Fig. 11.4(a).** *Boundary layer growth at pipe inlet*

eventually it fills the pipe entirely. The steady conditions that are established in this way will be either laminar or turbulent, according to the Reynolds number. The pressure drop per unit length, and hence the hydraulic gradient, will then have become constant. The corresponding resistance per unit length will have reached a steady value and, if the pipe is circular, there will be a constant shear stress acting over the walls of the pipe, $\tau_0$, say. The corresponding resistance of a length $L$ of pipe will then be given by:

$$F = \tau_0 S$$

in which $S$ is its wetted area, i.e. $\pi DL$.

According to Rayleigh's law, the resistance $F$ may be expressed in terms of a resistance coefficient such as:

$$f = \frac{F}{\frac{1}{2}\rho S U^2} = \varphi(Re)$$

(see Example 11.2 (i) (p. 285) and Section 10.7 (p. 258)) and this will, for similar pipes, depend only on the Reynolds number, i.e.

$$f = \varphi(Re) = \varphi\left(\frac{DU}{\nu}\right)$$

The coefficient $f$ is related to the shear stress at the wall by:

$$f = \frac{\tau_0}{\frac{1}{2}\rho U^2}$$

the reference velocity $U$ in the above equation being most conveniently chosen as the mean, i.e. $(Q/A)$, $A$ being the cross-sectional area of the bore. The shear stress at the wall determines the pressure gradient, and the relationship between them may be determined from Fig. 11.4(b). If the plug of fluid shown shaded is not accelerating:

$$\frac{\pi D^2}{4} \times \delta p + \pi D \delta L \tau_0 = 0$$

i.e.

$$\tau_0 = -\frac{D}{4}\frac{\delta p}{\delta L}$$

the negative sign indicating of course that $\delta p$ is, in fact, negative. Using $i$ to denote the *fall* in the hydraulic gradient per unit length (i.e. $i = -\delta H/\delta L$), and noting that $\delta p = w\delta H = \rho g \delta H$

$$\tau_0 = -\frac{wD}{4}\frac{\delta H}{\delta L} = \rho g \frac{Di}{4}$$

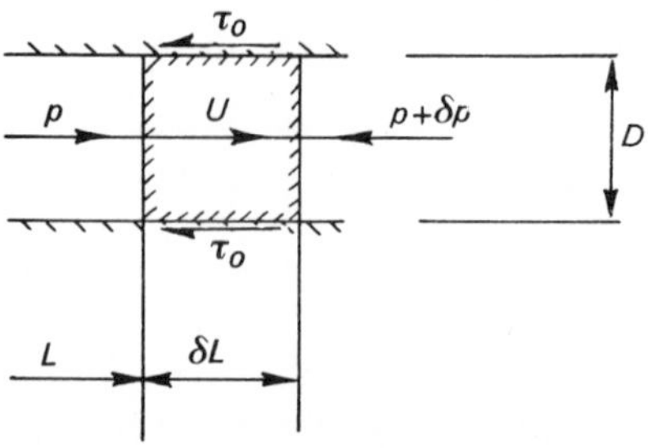

Fig. 11.4(b). *Wall shear and pressure gradient down pipe*

If we express this in terms of $f$, we rediscover Darcy's formula,

$$\tau_0 = f \tfrac{1}{2}\rho U^2 = \rho g \frac{Di}{4}$$

i.e.

$$i = \frac{4f}{D}\frac{U^2}{2g} = \frac{\lambda}{D}\frac{U^2}{2g}$$

The present argument suggests however that $f$, and hence $\lambda$, depend on $Re$, and experiments are needed to determine the form of the relationship. In this country, Stanton and Pannell carried out extensive tests at the National Physical Laboratory (*Proc. Roy. Soc.* (A), Vol. 91, p. 46, 1915)

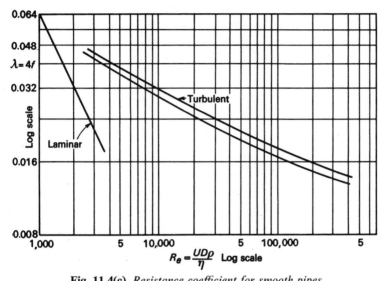

**Fig. 11.4(c)** *Resistance coefficient for smooth pipes*

measuring the resistance to flow of water, oil and air in pipes varying from 3 to 130 mm diameter. Their results confirmed the assumptions on which the analysis was based, all the measurements lying on a single curve, as shown in Fig. 11.4(c). The two branches of the curve correspond to laminar and turbulent flow respectively, the marked scatter in the results in the region of $Re = 2300$ being due to transition.

In order to find the hydraulic gradient corresponding to any particular flow rate, the Reynolds number must first be determined. The corresponding value of $\lambda = 4f$ can then be substituted in:

$$i = \frac{\lambda}{D}\frac{U^2}{2g}$$

and this method may be used, no matter whether the flow be laminar or turbulent. The laminar branch of the $\lambda$-$Re$ curve is in fact hyperbolic so that, since $Re$ is proportional to $U$,

$$\lambda Re \propto \lambda U = \text{constant}$$

and, alternatively, $f Re \propto f U = \text{constant}$.

Hence $fU^2$ in the Darcy formula becomes, for laminar flows, $(fU) \times U$, i.e. proportional to $U$. This is to be expected from the Hagen–Poiseuille law, which predicts a pressure drop proportional to the velocity. We can in fact deduce the theoretical equation for the laminar branch of the curve from this law. In terms of the notation of the present section, the latter is:

$$U = - \frac{D^2 \delta p}{32 \eta \delta L}$$

Recalling that:

$$\frac{\delta p}{\delta L} = - \frac{4\tau_0}{D}$$

$\therefore$
$$U = \frac{D\tau_0}{8\eta}$$

$\therefore$
$$\tfrac{1}{2}\rho U^2 = \frac{1}{16} \left( \frac{\rho D U}{\eta} \right) \tau_0$$

which becomes, when we denote $(\tau_0 / \tfrac{1}{2}\rho U^2)$ by $f$ and $(\rho D U / \eta)$ by $Re$:

$$f = \frac{16}{Re}, \quad \text{or} \quad \lambda \times Re = 64$$

This confirms the hyperbolic relationship referred to above and all the experimental points in the laminar range lie on this curve.

It has not proved possible to obtain a corresponding theoretical solution for *turbulent flows*, but Blasius' law, based on early experiments:

$$\lambda = \frac{0\cdot3164}{Re^{1/4}} = 4f$$

is found to fit the experimental results well, up to $Re = 10^5$. This implies that $\lambda \propto U^{-0\cdot25}$, so that $\lambda U^2$, and hence $i$, varies as $U^{1\cdot75}$ or $U^{7/4}$ in this range of turbulent flow. As the Reynolds number is further increased so the slope of the curve on the log-log plot becomes less marked, indicating that the index $\tfrac{1}{4}$ should decrease progressively if agreement with experiment is to be maintained.

The equation:

$$\frac{1}{\sqrt\lambda} = 2 \log_{10} (Re \sqrt(\lambda)) - 0\cdot8$$

although inconvenient if a graph is not drawn, is found to fit all the data at present available on smooth pipes (up to $Re = 3 \times 10^6$). This logarithmic law was established by Prandtl and von Kármán. Its form is suggested by theoretical studies and the observation that the velocity across a pipe tends to increase as the log of the distance from the wall. This law is found to apply from a point very near the wall to one very near the centre of the pipe, so that, in this range:

$$\frac{u}{u_0} = \log\frac{y}{y_0}$$

The 'constants' $y_0$ and $u_0$ depend in fact on the Reynolds number and, from dimensional considerations have the units of a distance and a velocity respectively, see Section 11.8 (p. 306).

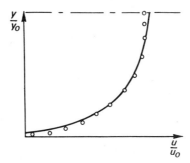

**Fig. 11.4(d).** *Logarithmic approximation to the velocity distribution*

EXAMPLE 11.4 (i)

*Calculate the resistance coefficient and the skin-friction drag force per metre length of straight pipe 6 mm diameter when air of density 1·80 kg/m³ and dynamic viscosity $18 \times 10^{-6}$ kg/ms flows along at mean speeds of (a) 3 m/s, and (b) 30 m/s.*

(a) Reynolds number

$$Re = \frac{u d \rho}{\eta} = 3\frac{\text{m}}{\text{s}} \times \frac{6}{10^3}\,\text{m} \times 1\cdot80\,\frac{\text{kg}}{\text{m}^3} \times \frac{\text{ms}}{18 \times 10^{-6}\,\text{kg}} = 1\,800$$

Hence, the flow is laminar obeying the law for mean velocity

$$\bar{u} = \frac{a^2}{8\eta}\left(\frac{p_1 - p_2}{l}\right)$$

where $a$ is the radius of pipe.

The velocity of a streamline at any radius $r$ is:

$$u = \frac{(a^2 - r^2)}{4\eta}\left(\frac{p_1 - p_2}{l}\right) = 2\bar{u}\left(1 - \frac{r^2}{a^2}\right)$$

Hence,

$$\frac{du}{dr} = -\frac{4\bar{u}r}{a^2}$$

and the viscous shear stress at the wall of the pipe is:

$$\tau_0 = \eta\left(\frac{du}{dr}\right)_{r=a} = 4\eta\frac{\bar{u}}{a} = \frac{\text{drag force } F}{\text{surface area } Yl}$$

where $Y$ is the circumference of the wall of the pipe.

The resistance coefficient

$$f = \frac{F}{\frac{1}{2}\rho\bar{u}^2 \, Yl} = \varphi(Re)$$

Hence, since

$$\frac{F}{Yl} = 4\eta\frac{\bar{u}}{a} = \tau_0$$

then

$$f = \frac{\tau_0}{\frac{1}{2}\rho\bar{u}^2} = \frac{16\eta\bar{u}}{(2a)\rho\bar{u}^2} = \frac{16}{Re}$$

$$= \frac{16}{1800} = \textbf{0·008 9}$$

and the drag force per unit length is:

$$\frac{F}{l} = f\tfrac{1}{2}\rho\bar{u}^2 \, Y = \frac{0·008\ 9}{2} \times 1·8\,\frac{\text{kg}}{\text{m}^3} \times 9\,\frac{\text{m}^2}{\text{s}^2} \times \pi\,\frac{6\ \text{m}}{10^3}\left[\frac{\text{N s}^2}{\text{kg m}}\right]$$

$$= \textbf{1·36} \times \textbf{10}^{-3}\ \text{N/m} \quad or \quad \textbf{1·36 mN/m}$$

(*b*) At a velocity of 30 m/s Reynolds number is:

$$Re = \frac{ud\rho}{\eta} = 18\ 000$$

Hence, the flow is turbulent. Blasius' formula

$$\lambda = \frac{0·316\ 4}{Re^{1/4}} = 4f$$

applies up to $Re = 10^5$, i.e.

$$f = \frac{0·079\ 1}{(18\ 000)^{1/4}} = \textbf{0·006 84}$$

and

$$\frac{F}{l} = f\tfrac{1}{2}\rho\bar{u}^2 \, Y = \frac{0·0068\ 4}{2} \times 1·8\,\frac{\text{kg}}{\text{m}^3} \times 900\,\frac{\text{m}^2}{\text{s}^2} \times \pi\,\frac{6\ \text{m}}{10^3}\left[\frac{\text{N s}^2}{\text{kg m}}\right]$$

$$= \textbf{104·5} \times \textbf{10}^{-3}\ \text{N/m} \quad or \quad \textbf{104·5 mN/m}$$

Thus the ratio of skin-friction drag forces per metre of pipe when the air velocities are 30 and 3 m/s is 104·5/1·36 = 77.

EXAMPLE 11.4 (ii)

*Oil having a relative density of 0·85 is pumped through a pipe (which may be assumed horizontal) 1 000 m long and 150 mm diameter at the rate of 1·25 m³/min. The shaft power of the pump is 4 kW and its efficiency is 65 per cent.*

*Assuming the resistance coefficient, f in the Darcy formula*

$$h_f = \frac{4fl}{d}\frac{u^2}{2g}$$

*is equal to* 16 Re$^{-1}$ *where Re is the Reynolds number, calculate the viscosity and kinematic viscosity of the oil expressing the former in poise and the latter in stokes. Hence find the Reynolds number and justify the assumption.*

For a horizontal pipe,

$$h_f = 4\left(\frac{16\eta}{ud\rho}\right)\frac{l}{d}\frac{u^2}{2g} = \frac{p_1 - p_2}{w}$$

Hence, the pressure difference over the whole length of the pipe is

$$p_1 - p_2 = 32\frac{\eta l u}{d^2}$$

and the power required to pump the oil through the horizontal pipe is:

$$P = Q(p_1 - p_2) = \frac{\pi}{4}d^2 u(p_1 - p_2) = 8\pi\eta l u^2$$

and the shaft power is:

$$P_s = \frac{P}{\epsilon} = \frac{8\pi\eta l u^2}{\epsilon}$$

Hence, the coefficient of viscosity is

$$\eta = \frac{\epsilon P_s}{8\pi l u^2}$$

i.e., since

$$u = \frac{Q}{a} = \frac{1\cdot25\ \text{m}^3}{60\ \text{s}} \times \frac{4}{\pi}\frac{10^4}{225\ \text{m}^2} = 1\cdot185\ \text{m/s}$$

then

$$\eta = \frac{0\cdot65 \times 4\,000\ \text{W}}{8\pi \times 1\,000\ \text{m}} \times \frac{1}{(1\cdot185)^2}\frac{\text{s}^2}{\text{m}^2}\left[\frac{\text{Nm}}{\text{Ws}}\right]\left[\frac{10\ \text{Pm}^2}{\text{Ns}}\right] = 0\cdot47\ \text{P}$$

$$\therefore \quad v = \frac{\eta}{\rho} = \frac{0\cdot74\ \text{Pm}^3}{0\cdot85 \times 10^3\ \text{kg}}\left[\frac{10^{-1}\ \text{kg}}{\text{P ms}}\right] = 0\cdot871 \times 10^{-4}\ \text{m}^2/\text{s} = 0\cdot871\ \text{St}$$

$$Re = \frac{ud}{v} = 1\cdot185\frac{\text{m}}{\text{s}} \times 0\cdot15\ \text{m} \times \frac{10^4\ \text{s}}{0\cdot871\ \text{m}^2} = 2\,040$$

Thus, since *Re* is less than 2 300 the flow is laminar and hence the assumption regarding the friction coefficient is justified.

EXAMPLE 11.4 (iii)

*The flow of a gas in a uniform duct is to be simulated by means of water flow in a ¼-scale transparent model.*

*The full-scale gas velocity is expected to be 25 m/s. Find: (a) the corresponding water velocity in the model, and (b) the pressure drop to be expected per metre length of the full scale duct if the pressure drop per m length of the model is 45 kN/m². The dynamic viscosity of water is 62 times that for the gas, and 1 kg of the gas under conditions in the full-scale duct occupies 0·72 m³.*

(a) If suffix 1 is used to denote the full-scale duct conditions we have

$$\frac{u_1 l_1 \rho_1}{\eta_1} = \frac{u l \rho}{\eta}$$

for dynamical similarity. Hence

$$u = u_1 \frac{l_1}{l} \frac{\rho_1}{\rho} \frac{\eta}{\eta_1} = 25 \frac{\text{m}}{\text{s}} (4) \left(\frac{1}{720}\right) \left(\frac{62}{1}\right)$$

i.e. velocity of water in the $\frac{1}{4}$-scale model is

$$u = \textbf{8·61} \text{ m/s}$$

(b) Using Rayleigh's formula

$$\frac{F}{\frac{1}{2}\rho u^2 S} = \varphi(Re)$$

for flow of fluids we may re-write it as:

$$\frac{F}{\frac{1}{2}\rho u^2 S} = \frac{A \times \Delta \text{P}}{\frac{1}{2}\rho u^2 Yl} = \varphi(Re) = \frac{m}{\frac{1}{2}\rho u^2} \frac{\Delta P}{l}$$

where $m$ is the hydraulic mean depth $A/Y$.

Thus, if $Re$ is the same for the model and full-scale duct:

$$\frac{\frac{\Delta P_1}{l_1}}{\frac{\Delta P}{l}} = \frac{\frac{\rho_1 u_1^2}{m_1}}{\frac{\rho u^2}{m}} = \frac{\rho_1}{\rho} \frac{m}{m_1} \left(\frac{u_1}{u}\right)^2$$

$$= \left(\frac{1}{720}\right) \left(\frac{1}{4}\right) \left(\frac{25{\cdot}0}{8{\cdot}61}\right)^2 = \frac{1}{342}$$

$$\therefore \quad \frac{\Delta P_1}{l_1} = \frac{1}{342} \times 45 \frac{\text{kN/m}^2}{\text{m}} = \frac{1}{7{\cdot}6} \text{ kN/m}^3 = \textbf{132} \text{ Pa per metre length}$$

## 11.5. Effect of roughness

Nikuradse extended the scope of the available data on pipe resistance by carrying out methodical series of tests on pipes which had been artificially roughened with sand grains. The phrase 'similar pipes' in this connection implies a particular size of the grains $(K)$, *relative* to the size of the pipe $(D)$, i.e. a fixed value of $K/D$. Similarity arguments thus suggest a unique curve for $f$ as a function of $Re$ for each value of $K/D$. Nikuradse's curves confirm this and are shown on Fig. 11.5(a), which is plotted on a semi-log basis. The Hagen–Poiseuille and Blasius equations provide a framework of reference for this plot on which the log scale telescopes the curves at high values of $Re$.

Two important conclusions may be drawn from these tests. Firstly there is a Reynolds number, depending on the value of $K/D$, below which a roughened pipe has the same resistance as a smooth one. i.e. it is 'hydraulically smooth'. Secondly there is a Reynolds number, again depending on $K/D$, above which a roughened pipe has a resistance which is independent of $Re$. In this range $\lambda$ becomes constant so that the resistance then increases as the velocity squared.

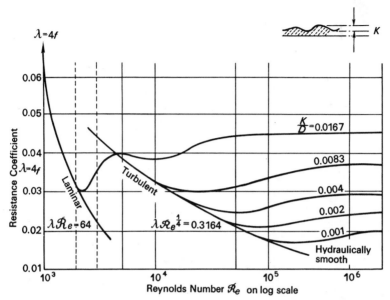

**Fig. 11.5(a).** *Nikuradse's tests on artificially roughened pipes*

The relationship between $\lambda$ and $K/D$ in this range, corresponding to the logarithmic form of the $\lambda - Re$ curve for smooth pipes, is found to be:

$$\frac{1}{\sqrt{\lambda}} = 2 \log_{10} \left(\frac{D}{K}\right) + 1 \cdot 14$$

Even when the flow is turbulent it is found that there is no slip at the boundary, i.e. the velocity tends to zero as the wall is approached. There is thus a thin film of fluid which moves so slowly that viscous effects predominate in this region, and this is known as the *laminar sub-layer* (see Section 11.8, p. 305). As the Reynolds number is increased, dynamic effects become progressively more dominant, and the thickness of this film is reduced. So long as the surface projections are embedded in this layer the wall is said to be hydraulically smooth, i.e. the roughness has no effect. The other extreme is reached when the sub-layer is so broken up by the projections that they act as bluff obstacles, the drag of which becomes almost independent of viscosity.

Nikuradse's tests are not representative of the behaviour of commercial pipes because the surface of the latter generally consists of protuberances of various sizes. Colebrook and White (1937 and 1939) therefore carried out tests with pipes that had been roughened with sand grains which were mixed in size. They found, as might be expected, that the large humps caused an early departure from the smooth-pipe value of $\lambda$ or $4f$. As the Reynolds number was increased, so the average roughness fell, causing a gradual transition to a constant value of $f$ which determines the 'equiv-

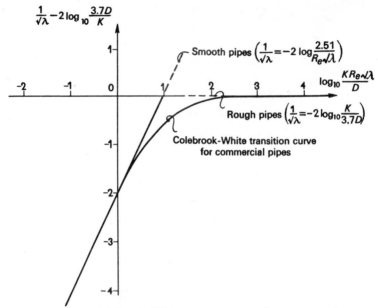

**Fig. 11.5(b).** *Colebrook–White transition curve for commercial pipes*

alent roughness' of the surface. They found that tests on commercial pipes showed a similar gradual transition and proposed the function:

$$\frac{1}{\sqrt{\lambda}} = -2\log_{10}\left[\frac{2 \cdot 51}{Re\sqrt{\lambda}} + \frac{K}{3 \cdot 7D}\right]$$

where $\lambda = 4f$ in the Darcy formula

$$h_f = 4f\frac{L}{D}\frac{U^2}{2g}$$

This curve is asymptotic to the 'smooth' and 'rough' laws, for if the pipe is smooth it reduces to:

$$\frac{1}{\sqrt{\lambda}} = -2\log\frac{2 \cdot 51}{Re\sqrt{\lambda}}$$

i.e.

$$\frac{1}{\sqrt{\lambda}} = 2\log Re\sqrt{(\lambda)} - 0 \cdot 8$$

and if roughness predominates it becomes:

$$\frac{1}{\sqrt{\lambda}} = -2\log\frac{K}{3 \cdot 7D}$$

i.e.

$$\frac{1}{\sqrt{\lambda}} = 2\log\frac{D}{K} + 1 \cdot 14$$

The Colebrook–White equation with its asymptotes, are shown on Fig. 11.5(b). Just as Nikuradse's results reduce to a single curve on this plot, we may carry out the inverse process and expand the Colebrook–

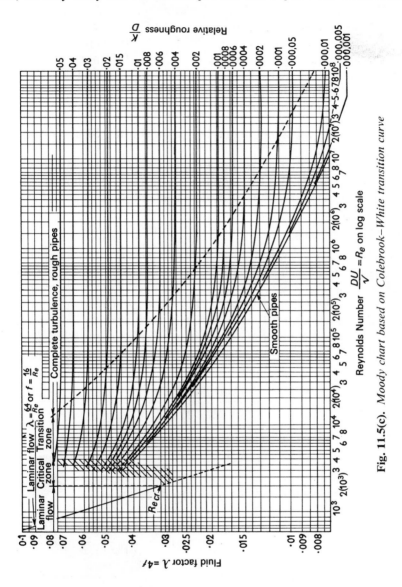

Fig. 11.5(c). *Moody chart based on Colebrook–White transition curve*

White curve into a plot which corresponds to that on which Nikuradse's results were originally shown, see Fig. 11.5(a).

This is often referred to as the Moody diagram (1944) and can be used to estimate the friction factor for any pipe once its equivalent roughness has been found.

The equivalent roughness cannot at present be inferred directly from roughness measurements, but only by finding the limiting value to which *f* tends from tests on a similar pipe. Representative values of *K* are as follows:

|  | *K* |
|---|---|
| Coated cast iron | 0·13 mm |
| Uncoated cast iron | 0·20 mm |
| Galvanized steel | 0·15 mm |
| Uncoated steel | 0·05 mm |
| Concrete (cast on steel forms) | 0·20 mm |
| Spun concrete | 0·08 mm |

Copper, brass and aluminium tube may be considered smooth, as may glass and plastic.

It is preferable that more direct data be used in practice, based on experience, especially as the formation of growths or slime may cause considerable increases in the equivalent roughness over a period of time.

EXAMPLE 11.5 (i)

*A circulating pump delivers* 90 m³/min *of water through a condenser containing* 4 800 *tubes of* 16·5 mm *external diameter and* 1·25 mm *thick and* 4 m *long. The supply and exit pipes can be taken as equivalent to* 15 m *of straight pipe* 600 mm *diameter. Estimate the loss of head due to friction and the pump power required. Assume f* = 0·003 *for the supply and exit pipes and* 0·006 5 *for the tubes. Also,* $\eta$ = 1·5 g/ms *and* $\rho$ = 1 Mg/m³. *Neglect losses other than pipe friction.*

Velocity in supply and exit pipes

$$= U = \dot{Q}/A = \frac{90 \text{ m}^3}{60 \text{ s}} \times \frac{4}{\pi} \left(\frac{1}{0 \cdot 60 \text{ m}}\right)^2 = 5 \cdot 31 \text{ m/s}$$

Hence,

$$Re = UD\frac{\rho}{\eta} = 5 \cdot 31 \frac{\text{m}}{\text{s}} \times 0 \cdot 6 \text{ m} \times 1 \frac{\text{Mg}}{\text{m}^3} \times \frac{\text{ms}}{1 \cdot 5 \text{ g}} = 2 \cdot 124 \times 10^6$$

i.e. the flow is turbulent and Darcy's formula applies.

Thus,

$$h_f = 4\frac{fl}{D}\frac{U^2}{2g} = 4 \times 0 \cdot 003 \times \frac{15 \text{ m}}{0 \cdot 6 \text{ m}} \times (5 \cdot 31)^2 \frac{\text{m}^2}{\text{s}^2} \times \frac{\text{s}^2}{19 \cdot 614 \text{ m}} = 0 \cdot 431 \text{ m}$$

Mean velocity in the condenser tubes:

$$u = \frac{\dot{Q}}{na} = \frac{90 \text{ m}^3}{60 \text{ s}} \times \frac{1}{4800} \times \frac{4}{\pi} \left[\frac{10^2}{1 \cdot 4 \text{ m}}\right]^2 = 2 \cdot 03 \text{ m/s}$$

Hence,

$$Re = ud\frac{\rho}{\eta} = 2 \cdot 03 \frac{\text{m}}{\text{s}} \times \frac{1 \cdot 4 \text{ m}}{10^2} \times \frac{10^6 \text{ s}}{1 \cdot 5 \text{ m}^2} = 1 \cdot 896 \times 10^4$$

i.e. the flow is turbulent and for each tube (all of which run into the same header)

$$h_f = 4\frac{fl\,u^2}{d\,2g} = 4\times0{\cdot}006\,5\times\frac{4\ \text{m}}{1{\cdot}4\times10^{-2}\ \text{m}}\times(2{\cdot}03)^2\,\frac{\text{m}^2}{\text{s}^2}\times\frac{\text{s}^2}{19{\cdot}614\ \text{m}} = 1{\cdot}56\ \text{m}$$

Hence, total friction head against which the pump operates is $H_f = (0{\cdot}431 +1{\cdot}56)\ \text{m} = 1{\cdot}991\ \text{m}$, and the power required is

$$P = w\dot{Q}H_f = \rho g\dot{Q}H_f = 10^3\,\frac{\text{kg}}{\text{m}^3}\times9{\cdot}807\,\frac{\text{m}}{\text{s}^2}\times\frac{90\ \text{m}^3}{60\ \text{s}}\times1{\cdot}991\ \text{m}\left[\frac{\text{W s}^3}{\text{kg m}^2}\right]$$

$$= 29{\cdot}3\ \text{kW}$$

EXAMPLE 11.5 (ii)

*Compare the Chezy–Manning expressions with those for the resistance of a rough pipe flowing full.*

If roughness governs:

$$\frac{1}{\sqrt{\lambda}} = 2\log\frac{D}{K}+1{\cdot}14$$

in which the $C$ in $U = C\sqrt{(mi)}$ is given by:

$$\lambda = \frac{8g}{C^2}$$

and

$$m = \frac{D}{4}$$

for a pipe flowing full.

Hence:

$$\frac{C}{\sqrt{(8g)}} = 2\log\frac{4\,m}{K}+1{\cdot}14$$

$$C = \left(2\log\frac{4m}{K}+1{\cdot}14\right)\sqrt{(8g)}$$

which suggests that $C$ is not a constant, but increases with hydraulic mean depth $(m)$, and decreases with increasing roughness $K$. These conclusions are in line with Manning's expression for $C$:

$$C = Mm^{1/6}$$

which shows $C$ to increase with $m$ and justifies his recommendation that $M$ should vary inversely with a factor which is a measure of the channel roughness.

## 11.6. Transport by turbulence

In turbulent flow the velocity at any point does not remain constant, either in magnitude or direction. We may, however, imagine it to consist of an underlying steady flow, governed by the time average velocity, together with a fluctuating component, caused by eddying. As the latter causes transverse displacements, or cross-currents, there is a constant mixing taking place across the streamlines of the underlying flow pattern. This turbulent mixing is important in many *transport* problems, as it exerts a controlling influence over the transfer of such widely different quantities as heat and sediment. It implies that the faster moving stream-tubes are constantly receiving fluid from its slower neighbours and vice versa. This is analogous to transferring a parcel between two moving vehicles. During the exchange, the slower vehicle is dragged forward and the faster is retarded. The analogy would be more complete if two parcels were to be exchanged simultaneously, since each stream-tube must, on average, lose as much material as it receives. In terms of mechanics, the mixing process has caused a *transport* of momentum, the faster moving stream constantly losing more momentum than it receives. This rate of loss of momentum causes an *apparent* or *turbulent* shear stress in addition to that caused by *viscous* action. Kinetic theory of gases suggests that the latter may similarly be explained by *molecular* rather than *molar* exchanges. The much larger mass transfer caused by the latter process implies that heat or momentum transfer tends to be much more vigorous in turbulent than in laminar flow. Consequently if turbulence occurs, the shear stresses associated with this are generally much greater than those due to viscous action. It is only in a very narrow region, immediately adjacent to a solid boundary, that viscous action is appreciable in turbulent flow. In this laminar sub-layer, as it is called, the turbulence ultimately dies out, leaving a very thin laminar film in contact with the boundary.

## 11.7. Velocity defect law

The previous section suggests that for turbulent flow through pipes, there is a central plug of fluid which moves forward independent of the action of viscosity except in so far as the latter controls the shear stress which is applied to its surface, see Fig. 11.7(a). Let us suppose that the magnitude of the latter is $\tau_0$ and that the density of the fluid is $\rho$. Denoting the maximum velocity, at the centre of the pipe by $u_c$ and the velocity at any other radius $r$ by $u$ Stanton and others have confirmed by experiment that the amount by which the fluid is retarded $(u_c - u)$ at any fraction of the radius $(r/r_0)$, is proportional to $\sqrt{(\tau_0/\rho)}$ and depends on nothing else. Thus for any given value of $(r/r_0)$:

$$(u_c - u) \propto \sqrt{(\tau_0/\rho)}$$

the constant of proportionality depending on, i.e. being 'a function of', $(r/r_0)$, i.e.

$$\frac{u_c - u}{\sqrt{(\tau_0/\rho)}} = \varphi\left(\frac{r}{r_0}\right)$$

Dimensional analysis indicates why $\sqrt{(\tau_0/\rho)}$ should appear in the denominator of the left-hand side, as this has the dimensions of a velocity and is, in fact, often referred to as the friction velocity $u_*$:

$$u_* = \sqrt{\frac{\tau_0}{\rho}} = \left[\frac{M}{LT^2}\right]^{1/2}\left[\frac{L^3}{M}\right]^{1/2} = \left[\frac{L}{T}\right]$$

Thus the above equation is a relationship between non-dimensional parameters. It is known as the velocity-defect law and the form of the

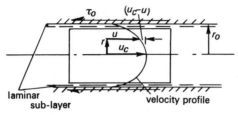

**Fig. 11.7(a).** *Effect of shear stress*

relationship, deduced from Stanton's experiments, is shown in Fig. 11.7(b). That the velocity distribution over the central core should be independent of the Reynolds number, when expressed in this way, adds force to the contention that the viscous shear is negligible, compared with that due to momentum transport in turbulent flow. It should also be

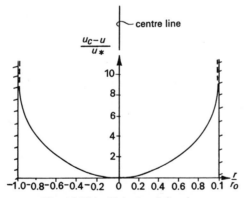

**Fig. 11.7(b).** *Velocity-defect law*

noted that the velocity defect $(u_c - u)$ depends only on $u_*$, so that the boundary, which may be smooth or rough, is important only in so far as it affects the wall shear $\tau_0$, and hence the value of:

$$u_* = \sqrt{\frac{\tau_0}{\rho}}$$

It was shown in Section 11.4 (p. 290) that:

$$f = \frac{\tau_0}{\frac{1}{2}\rho U^2}$$

∴ $$\lambda = 4f = \frac{8\tau_0}{\rho U^2}$$

Thus the friction velocity $u_*$ is related to the friction factor $\lambda$ and the mean velocity $U$ in Darcy's equation by:

$$\frac{u_*}{U} = \sqrt{\frac{\lambda}{8}} = \sqrt{\frac{f}{2}}$$

In fact $u_*$ was determined from Stanton's experiments in this way. Hence the velocity defect law may be re-expressed in terms of $\lambda$ as:

$$\frac{u_c - u}{U} = \left(\sqrt{\frac{f}{2}}\right)\varphi\left(\frac{r}{r_0}\right)$$

and the right-hand side of this equation is simply $\sqrt{(f/2)}$ times each of the ordinates of Fig. 11.7(b).

The way in which the velocity defect builds up in any particular case may be deduced from this equation. For example, for smooth pipes, as the

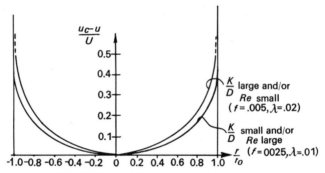

Fig. 11.7(c). *Effect of Re on defect law*

Reynolds number increases, so $f$ (and hence the factor by which each of the ordinates must be multiplied) falls. Thus the velocity distribution becomes more uniform at large Reynolds numbers, for any given mean velocity (i.e. $U$ fixed) see Fig. 11.7(c).

Similarly, if $f$ is increased by increasing the roughness, the velocity will be less uniformly distributed, other things being equal.

All the above results apply only to fully established turbulent flow, i.e. it presupposes a settling length of, say $5 \times 10^5 \, D/Re$, see Section 11.4.

## 11.8. Wall and universal velocity laws

In the previous Section attention was focused on the velocity distribution in the central region of the pipe. It remains to consider the velocity distribution in the immediate neighbourhood of the wall, i.e. the region including and immediately beyond the laminar sub-layer. In this region the effect of viscosity cannot, in general, be ignored so that we must seek some other simplification if any pretensions to a general result are to be obtained. The key to this lies in the fact that if we restrict our attention to the immediate vicinity of the wall we may assume it to be a region of sensibly uniform shear $\tau_0$ as shown in Fig. 11.8(a). Stated in other terms, we are ignoring the forces due to the pressure gradient which acts over the small areas at the ends of the region ABCD, compared with the nearly equal and opposite shear stresses that act along its length. We also assume that the fluid in this region is very little influenced by $r$; the fact that it is in contact with a boundary which is, in comparison with its thickness, of relatively large curvature is assumed to be of little sig-nificance—it might as well be flat. If the above rationalization of the

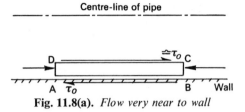

**Fig. 11.8(a).** *Flow very near to wall*

problem (due to von Kármán) is considered acceptable, the velocity $u$ at any small distance $y$ from a smooth wall will depend primarily on $\tau_0$, $\rho$ and $\eta$ (or $v$), i.e.:

$$u = \varphi(y, \tau_0, \rho, \eta)$$

Dimensional analysis enables us to reduce the number of variables, see Section 10.5, p. 249.

Recalling from the previous section that $\sqrt{(\tau_0/\rho)}$ has the dimensions of a velocity, we have merely to multiply this by $y$ to obtain a quantity having the dimensions $L^2/T$, which are those of $v$.

We thus obtain two non-dimensional groups:

$$\left(\frac{u}{u_*}\right), \quad \left(\frac{v}{u_* y}\right)$$

so that, for a smooth wall:

$$\left(\frac{u}{u_*}\right) = \varphi\left(\frac{y}{v/u_*}\right)$$

Hence, a unique curve should result for the variation of the velocity $u$ with the distance from the wall $y$, provided that the former is expressed in units of $u_*$ and the latter in units of $v/u_*$. Comparison with experiment shows that our highest hopes are well fulfilled as all the results lie on a single curve, independent of the pipe Reynolds number as shown in Fig. 11.8(b). Rather surprisingly, the law appears to apply well into the flow, beyond the immediate boundary region.

A good approximation to the experimental curve is the so-called 'one-seventh power law':

$$\left(\frac{u}{u_*}\right) = 8{\cdot}7 \left(\frac{y}{\nu/u_*}\right)^{1/7}, \quad \text{i.e. } u^7 \propto y$$

and, if it is assumed to apply well out towards the centre of the pipe can be correlated with Blasius' expression for $\lambda$:

$$\lambda = \frac{0{\cdot}316\,4}{Re^{1/4}} = 4f$$

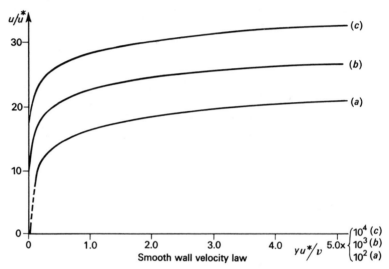

Fig. 11.8(b). *Experimentally determined smooth wall velocity law*

(see Section 11.4, p. 292). Similarly the logarithmic law for $\lambda$:

$$\frac{1}{\sqrt{\lambda}} = 2 \log_{10} Re \sqrt{(\lambda)} - 0{\cdot}8$$

referred to in the same section, is closely related to a logarithmic form for the velocity distribution:

$$\frac{u}{u_*} = 5{\cdot}75 \log_{10} \left(\frac{y}{\nu/u_*}\right) + 5{\cdot}5$$

This expression implies a straight line for the velocity profile when drawn on a semi-logarithmic basis, see Fig. 11.8(c). The experimental points are shown to follow this closely, except for very small values of $y$, i.e. in the laminar sub-layer. In this region both the power and the log laws break down as in the limit, the velocity distribution is simply given by:

$$\tau_0 = \eta \frac{du}{dy} = \frac{\eta u}{y}$$

for very small values of $y$. In terms of the parameters used in this section this laminar law may be re-written as:

$$\frac{\tau_0}{\rho} = \frac{\eta u}{\rho y} = \frac{\nu u}{y}$$

i.e.

$$\frac{u_*}{u} = \frac{\nu/u_*}{y}$$

$$\therefore \quad \left(\frac{u}{u_*}\right) = \left(\frac{y}{\nu/u_*}\right)$$

i.e. it is simply a straight-line relationship, which becomes an exponential curve on the semi-logarithmic plot. The intersection of the two curves on

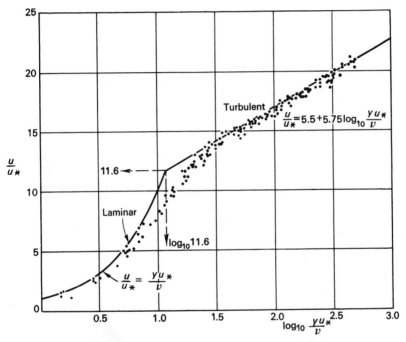

**Fig. 11.8(c).** *Universal velocity law for smooth pipes*

Fig. 11.8(c) conveniently defines the thickness of the laminar sub-layer and writing $y = h$ at this intersection:

$$\frac{h}{\nu/u_*} = 11 \cdot 6$$

It may be noted that it is a peculiarity of the logarithmic scale that the boundary lies at an infinite distance to the left on Fig. 11.8(c).

It was mentioned earlier that so long as the surface projections on the surface of the pipe are embedded in the laminar film the surface acts as if it were smooth. The allowable size of the projections for this to apply can now be seen to depend on $k/(\nu/u_*)$. From Nikuradse's results so long as this is less than 3 the friction factor $f$ follows the smooth pipe law. At the other extreme, if $k/(\nu/u_*) > 60$, $f$ becomes independent of $Re$ suggesting that the projections must then dominate the viscous effects. This would suggest that the wall velocity law must also become virtually independent of $\eta$ (or $\nu$) but dependent on $k$ instead, i.e. :

$$u = \varphi(y, \tau_0, \rho, k)$$

Expressing this non-dimensionally, as for the smooth wall law, we obtain, for a surface that is completely rough,

$$\left(\frac{u}{u_*}\right) = f\left(\frac{y}{k}\right)$$

Nikuradse's experiments confirm this result, and again show good agreement with a logarithmic law. Well out towards the centre of the pipe:

$$\frac{u}{u_*} = 5\cdot75 \log_{10}\left(\frac{y}{k}\right) + 8\cdot48$$

no matter what the relative roughness $(k/r_0)$, see Fig. 11.8(d).

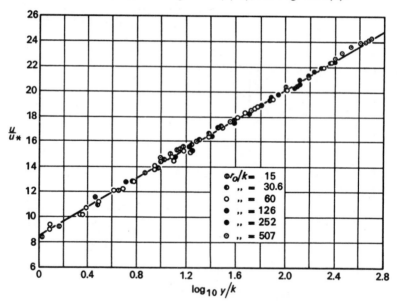

Fig 11.8(d). *Universal velocity law for rough pipes*

Accepting gratefully the fortuitous agreement into the central regions of the flow, the expressions:

$$\frac{u}{u_*} = 5\cdot75 \log_{10}\left(\frac{y}{\nu/u_*}\right) + 5\cdot5$$

$$\frac{u}{u_*} = 5\cdot75 \log_{10}\left(\frac{y}{k}\right) + 8\cdot48$$

are frequently referred to as the universal velocity laws for pipes which are respectively hydraulically smooth, or completely rough. If each were to apply right to the centre of the pipe where $y = r_0$ and $u = u_c$ they give:

$$\frac{u_c}{u_*} = 5{\cdot}75 \log_{10}\left(\frac{r_0}{\nu/u_*}\right) + 5{\cdot}5$$

$$\frac{u_c}{u_*} = 5{\cdot}75 \log_{10}\left(\frac{r_0}{k}\right) + 8{\cdot}48$$

Subtracting each from its progeny, each pair yields the same equation:

$$\frac{u_c - u}{u_*} = 5{\cdot}75 \log_{10}\left(\frac{r_0}{y}\right)$$

Noting that $(r + y) = r_0$, this may be re-written as:

$$\frac{u_c - u}{u_*} = -5{\cdot}75 \log_{10}\left(1 - \frac{r}{r_0}\right).$$

which is not only consistent with the velocity defect law (see Section 11.7, p. 302), but even suggests a form for the curve. This does in fact fit the experimental points well, as might be expected since the constants in the 'universal laws' were so chosen as to give good agreement right up to the pipe axis.

## Example 11.8 (i)

*Incompressible fluid is flowing through a pipe of circular cross-section. Calculate from first principles, the radius at which a single reading of the velocity of flow across the section would give the mean velocity when the flow is (i) wholly laminar, (ii) wholly turbulent. In case (i) assume that the velocity distribution is parabolic, and in (ii) that it is given by $u = u_0(y/a)^{1/7}$, where y is distance from the wall of the pipe, $u_0$ is the velocity on the axis, and a is the radius of the pipe.*

Referring to Fig. 11.8(e) we may write:

(i)
$$u = u_0\left(1 - \frac{r^2}{a^2}\right)$$

$$\delta Q = 2\pi r u \delta r$$

$$Q = \frac{2\pi u_0}{a^2}\int_0^a (a^2 - r^2)r\,dr = \frac{2\pi u_0}{a^2}\left(\frac{a^4}{2} - \frac{a^4}{4}\right) = \frac{\pi}{2}u_0 a^2$$

$$\text{mean velocity} = \frac{Q}{\pi a^2} = \frac{u_0}{2}$$

which occurs when

$$\left(1 - \frac{r^2}{a^2}\right) = \frac{1}{2}$$

i.e. when

$$r^2 = \frac{a^2}{2} \quad \text{or} \quad r = \frac{a}{\sqrt{2}} = 0\text{·}707a$$

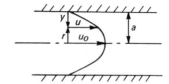

**Fig. 11.8(e).** *Flow through circular pipe*

(ii)
$$u = u_0 \left(\frac{y}{a}\right)^{1/7}$$

The rate of flow through an elemental ring is:

$$\delta Q = 2\pi(a-y)u\delta y$$

Hence:
$$Q = \frac{2\pi u_0}{a^{1/7}} \int_0^a (a-y)y^{1/7}dy$$

$$= \frac{2\pi u_0}{a^{1/7}} \left[\frac{7}{8}a^{15/7} - \frac{7}{15}a^{15/7}\right]$$

$$= 2\pi u_0 \frac{7\times 7}{8\times 15}a^2 = \frac{49}{60}\pi u_0 a^2$$

mean velocity $= \dfrac{Q}{\pi a^2} = \dfrac{49}{60}u_0 = 0\text{·}817u_0$

which occurs when

$$\left(\frac{y}{a}\right)^{1/7} = \frac{49}{60}$$

i.e. when

$$y = a\left(\frac{49}{60}\right)^7$$

or when
$$\frac{a-y}{a} = 1 - \left(\frac{49}{60}\right)^7 = 1 - 0\text{·}242\,6$$

i.e.
$$(a-y) \quad \text{or} \quad r = 0\text{·}7574a$$

## 11.9. Velocity distribution in open channels; secondary flow

When the flow has a free surface, the retarding effect of the air is very much less than that due to the walls, so that the velocity profile ceases to be symmetrical as in a circular pipe, flowing full. In such cases, and in many other installations where the ideal conditions visualized in the previous sections do not apply, a complete Pitot traverse, or its equivalent is necessary to determine the velocity profile. Fig. 11.9(a) shows typical profiles across an open channel and indicates that it is necessary to use graphical integration or numerical methods, such as Simpson's rules, in order to estimate the discharge.

**Fig. 11.9(a).** *Velocity profiles across an open channel*

The velocity contours for non-circular sections, with or without a free surface, generally tend to be 'attracted' towards corners, implying that a high velocity is maintained well towards such regions. Prandtl suggests that this is closely related to a transverse secondary flow as suggested in Fig. 11.9(b). Evidence of this secondary flow in open channels is provided

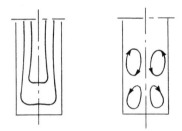

**Fig. 11.9(b).** *Secondary flow in non-circular section*

by the tendency of small floating objects to move into mid-stream. This movement away from regions of high wall shear towards the centre of the channel must be accompanied by a corresponding movement into regions where the shearing stress is less. Thus the shear stress at the boundary is largely evened out and this suggests that we are not much in error in using values of the friction factor based on circular pipes for non-circular sections.

## Exercises on Chapter 11

1. A trapezoidal channel is 0·9 m deep, 1·8 m wide at the top and 0·6 m wide at the bottom. The slope is 370 mm/km.

Find the hydraulic mean depth if the water in the channel is 0·6 m deep.

Calculate the velocity of flow if $C$ in Chezy's formula is 60·5 $m^{1/2}/s$, and the rate of flow in $m^3/h$.

2. A channel 5 m wide at the top and 2 m deep has sides sloping 2 vertically in 1 horizontal. The fall of the water along the channel is 1 in 1 000. Find

the rate of flow along the channel when the water is 1 m deep. $C$ in the Chezy formula may be taken as 53 m$^{1/2}$/s.

What will be the depth of water when the flow is doubled?

3. (*a*) A channel is to be constructed for conveying a maximum of 1140 m$^3$/h. The side slopes are to be 1 vertical to $1\frac{1}{2}$ horizontal and $C$ in Chezy's formula may be taken as 66 m$^{1/2}$/s. Determine the dimensions of the channel so that the length of the wetted perimeter shall be a minimum if the fall is 1 in 800.

(*b*) What is the least diameter of a cylindrical channel that will convey this quantity of water, assuming the same value of $C$ in Chezy's formula.

4. (*a*) Find the relation between the constant $f$ and $C$ in the alternative expressions for pipe flow

$$h_f = \frac{flu^2}{2gm} \quad \text{and} \quad u = C\sqrt{(mi)}$$

(*b*) A circular brick sewer is 1·22 m diameter and has a fall of 1 in 500. Calculate the rate of discharge of water when the depth of flow at the centre is 0·91 m assuming $C$ in Chezy's formula to be 52 m$^{1/2}$/s.

5. The water supply for a turbine passes through a conduit which for convenience has its cross-section in the form of a square with one diagonal vertical.

If the conduit is required to convey, under conditions of maximum discharge, 8·5 m$^3$/s when the slope is 1 in 4 900, determine its size, assuming the velocity of flow is given by Manning's formula $u = Mi^{1/2}m^{2/3}$ in which $M$ may be taken as 80 m$^{1/3}$/s.

6. The cross-section of an open channel consists of a semi-circular invert with vertical sides, the overall depth being equal to the maximum width. The channel is required to convey 85 m$^3$/min of water and, in order to guard against overflow, the depth of the stream is to be only 95 per cent of the maximum depth of the channel. The slope of the bed is 1 in 1 500 and $C$ in Chezy's formula may be taken as 60 m$^{1/2}$/s. Calculate the radius of the invert.

7. (*a*) Deduce the formula

$$f = \frac{2gmi}{u^2} = \frac{D}{\frac{1}{2}\rho u^2} = \varphi(Re) = NRe^n$$

where $D$ is the drag force per unit wall area. $N = 16$ and $n = -1$ for laminar flow, and take $N = 0·079$ and $n = -\frac{1}{4}$ for turbulent flow.

(*b*) Estimate the drop in pressure head per unit length of pipe if air of kinematic viscosity 14·8 mm$^2$/s flows at 40 m/s through a pipe 50 mm diameter.

8. A surface condenser consists of 2 000 tubes, each 3 m long and of smooth bore of 12·5 mm diameter.

Determine whether the flow is laminar or turbulent when 60 Mg of salt water per hour are being circulated. Estimate the loss of head across the tubes under these conditions if the kinematic viscosity of salt water is $10^{-6}$ m$^2$/s and the relative density is 1·026.

9. Air flows at the rate of 5 m$^3$/h through a long, straight, circular pipe of

12·5 mm diameter. The pressure drop between two points 3 m apart is equivalent to a head of 64 mm on a water manometer.

What would be the pressure drop in mm of water between two points 1·5 m apart in a long circular pipe of 25 mm diameter through which water flows at a rate such that the two flows are dynamically similar? What would be the rate of flow of water?

For water $\rho = 1$ Mg/m³ and $\nu = 1·145 \times 10^{-6}$ m²/s  or  1·145 mm²/s.

For air $\rho = 1·225$ kg/m³ and $\nu = 14·8 \times 10^{-6}$ m²/s.

10. If $p$ is the loss of pressure in a pipe of diameter $d$ and length $l$ due to the flow of fluid of dynamic viscosity $\eta$ and density $\rho$ at a velocity $u$, show that:

$$p = \rho u^2 \varphi \left( \frac{l}{d}, Re \right)$$

where $Re$ is the Reynolds number corresponding to the flow.

Hence show that for similar pipes the coefficient $f$ in the expression

$$\frac{4fl}{d} \frac{u^2}{2g}$$

is a function of the Reynolds number.

11. A weir spans the entire width of an open rectangular channel which conveys water at a depth of 2·25 m. The width of the channel is 10 m and the sill of the weir is 1 m above the bed of the channel. If the discharge over the weir is $Q = 3·2\ BH^{1·5}$ m³/s, where $B$ and $H$ are in metres, and if $u = C \sqrt{(mi)}$ for the channel, find the flow rate and the gradient of the bed in order that a steady flow of water may be maintained.

Take $C = 66$ m$^{1/2}$/s.

12. A tunnel to be constructed as part of a hydro-electric scheme has a circular cross-section and is required to convey water at the rate of 4 800 m³/min with a permitted friction loss of 10 m/km.

Determine the minimum diameter for the tunnel if the friction factor $f = 0·08 Re^{-0·23}$ in $h_L = (4fl/d)(u^2/2g)$. Assume the viscosity for water is $7·5 \times 10^{-4}$ kg/ms and density 10³kg/m³.

13. Oil of kinematic viscosity $1·8 \times 10^{-4}$ m²/s flows through a smooth pipe 150 mm diameter and 60 m long. What would be the frictional head loss when the mean velocity is (a) 1 m/s, (b) 4·5 m/s? Use the formula $h = (4fl/d)$ $(u^2/2g)$ in which the value of the coefficient $f$ can be taken as $f = 16/Re$ for laminar flow and $f = 0·08/Re^{1/4}$ for turbulent flow, where $Re$ represents the Reynolds number.

14. In an irrigation system a flow of 10 m³/s is to be carried on a longitudinal slope $i$ of 1 in 5 000. The channel is to be of trapezoidal cross-section, with a depth equal to the bed-width, and side slopes of 3 horizontal to 2 vertical. For the concrete lining of the channel, the value of $M$ in Manning's formula

$$u = Mm^{2/3}i^{1/2}$$

may be taken as 77 when SI units are used.

What should be the minimum dimensions of the channel?

## Chapter 12

# Flow and power transmission through pipe-lines

> '*Those who fall in love with practice without science are like a sailor who steers a ship without a helm or compass, and who never can be certain whither he is going.*'   LEONARDO

## 12.1. Expansion losses

In Section 8.8 it was explained that when fluid flows past an obstacle with a sharp edge, a surface of discontinuity springs from the latter. This surface (which separates the main-stream from the dead-water region), being unstable, rolls up, so forming an eddying wake. Fig. 12.1(a) suggests the conditions downstream of a sudden expansion, the flow pattern in practice changing continuously with respect to time. The eddying which occurs in these and other similar circumstances often represents a major percentage of the total loss in a hydraulic system. An estimate of the expansion loss may be made by means of a few simplifying assumptions. If the Reynolds number is reasonably high the velocity distribution upstream will, at least for steady flow, be fairly uniform. The vigorous turbulent mixing process which occurs downstream will also tend to produce a uniform velocity distribution. If we assume that there is no net transverse acceleration of the fluid between points P and Q, we may conclude that the pressure $p_1$, will be transmitted unchanged from P to Q. Hence the fluid contained between sections 1 and 2 experiences a force:

$$F = (p_1 - p_2)A_2$$

in the direction of motion, $A_2$ being the downstream area of the pipe. The corresponding equation of motion is (see Chapter 7, p. 145):

$$(p_1 - p_2)A_2 = \dot{m}(u_2 - u_1) = \rho A_2 u_2 (u_2 - u_1)$$

i.e.

$$\frac{p_1 - p_2}{\rho} = u_2(u_2 - u_1)$$

Defining the head loss $h_L$ between these sections from Bernoulli's equation, we write:

$$z_1 + \frac{p_1}{w} + \frac{u_1^2}{2g} = z_2 + \frac{p_2}{w} + \frac{u_2^2}{2g} + h_L$$

i.e.

$$h_L = \frac{p_1 - p_2}{w} + \frac{u_1^2 - u_2^2}{2g}$$

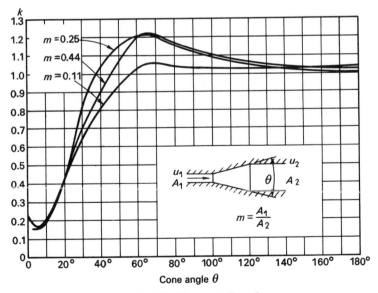

**Fig. 12.1(a).** *Sudden expansion*

Substituting for $(p_1 - p_2)/w$ from the equation of motion shows that a combination of momentum and energy results in an expression for the head loss:

$$h_L = \frac{2u_2(u_2 - u_1)}{2g} - \frac{u_2^2 - u_1^2}{2g}$$

i.e.

$$\boldsymbol{h_L = \frac{(u_1 - u_2)^2}{2g}}$$

Thus, when a stream of velocity $u_1$ impinges on slower moving fluid in a channel of uniform section, there is a so-called 'shock loss' in which the kinetic energy of their relative motion $(u_1 - u_2)$ is destroyed. When a pipe-line discharges into a large reservoir, so that $u_2 = 0$, the corresponding head loss is of course equal to its kinetic head $u^2/2g$.

**Fig. 12.1(b).** *Conical diffuser loss*

Expansion losses may be minimized by the use of a suitable diffuser, as in a Venturi meter. If this is of sufficiently slow taper, separation may be delayed (see Section 8.9, p. 193), and the extent of the dead-water region considerably reduced. Fig. 12.1(b) shows the results of Gibson's tests, in which the loss is expressed as some fraction $k$ of the sudden expansion loss, i.e.

$$h_L = k \frac{(u_1 - u_2)^2}{2g}$$

They suggest that the loss is not reduced appreciably until the included angle of the cone is less than about 30°. Any further reduction below about 6° causes an increase in the loss coefficient $k$, due to the additional wetted area on which the friction acts.

## 12.2. Losses at restrictions

There is an expansion loss or dissipation of energy, similar to that described in the previous section, whenever a high velocity stream impinges on slower moving fluid. For example, if a nozzle is installed in

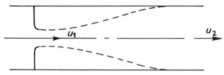

**Fig. 12.2(a).** *Expansion loss*

a pipe-line for metering purposes, as shown in Fig. 12.2(a) the expansion loss downstream is, as before:

$$h_L = \frac{(u_1 - u_2)^2}{2g}$$

If the nozzle is replaced by an orifice plate, the same relationship applies, so long as $u_1$ is taken to be that of the contracted jet, see Fig. 12.2(b).

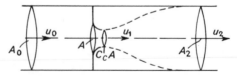

**Fig. 12.2(b).** *Loss at orifice plate*

Hence the expression for the loss may in general be written as:

$$h_L = \frac{(u_1 - u_2)^2}{2g}$$

in which, from continuity, i.e. equality of incoming and outgoing volumetric flow rates:

$$Q = u_1 C_c A = u_2 A_2$$

$$\therefore \qquad h_L = \left\{ \frac{1}{C_c} \frac{A_2}{A} - 1 \right\}^2 \frac{u_2^2}{2g} = k \frac{u_2^2}{2g}$$

The value of the contraction coefficient is very little dependent on conditions downstream and, if the orifice diameter is much smaller than the upstream diameter of the pipe, $C_c$ is very nearly equal to 0·6 (see Section 4.5, p. 83). The refinement of distinguishing between $C_c$ and $C_d$ is hardly justifiable in estimating pipe losses, particularly as we are working in terms of mean velocities, whereas in practice the velocity distribution (which depends on such factors as the Reynolds number) also has some influence. The corresponding values for the contraction coefficient given in Section 4.6, are in the notation of this Section:

| $A/A_0$ | 0·1 | 0·2 | 0·3 | 0·4 | 0·5 | 0·6 | 0·7 | 0·8 | 0·9 | 1·0 |
|---|---|---|---|---|---|---|---|---|---|---|
| $C_c$ | 0·624 | 0·632 | 0·643 | 0·659 | 0·681 | 0·712 | 0·755 | 0·813 | 0·892 | 1·00 |

Hence, if $A_0 = A_2$ (which is the case if the constriction is fitted in a pipe of uniform bore) the corresponding expression for $K$ is:

$$k = \left\{ \frac{1}{C_c} \frac{A_0}{A} - 1 \right\}^2$$

and hence its values are:

| $A/A_0$ | 0·1 | 0·2 | 0·3 | 0·4 | 0·5 | 0·6 | 0·7 | 0·8 | 0·9 | 1·0 |
|---|---|---|---|---|---|---|---|---|---|---|
| $k$ | | 47·8 | | | 7·8 | | 1·8 | | 0·29 | 0 |

EXAMPLE 12.2 (i)

*Two co-axial horizontal pipes, one 75 mm diameter and the other 100 mm diameter, are connected by flanges between which is bolted a plate with a 50 mm diameter sharp-edged orifice whose centre is on the centre line of the pipes. A mercury U-gauge is connected by water-filled tubes to tapping points upstream and downstream of the orifice where it may be assumed the velocity across the pipes is uniform. The gauge reads 320 mm of mercury when water flows through the orifice from the smaller to the larger pipe at the rate of $14 \times 10^{-3}$ m³/s or 14 litre/s. Assuming the only loss of head is that due to expansion of the streamlines on the downstream side of the orifice, calculate the contraction efficient of the orifice.*

Referring to Fig. 12.2(b),

$$h_L = \frac{(u_1 - u_2)^2}{2g} = \left\{ \frac{1}{C_c} \frac{A_2}{A} - 1 \right\}^2 \frac{u_2^2}{2g} = k \frac{u_2^2}{2g} = \frac{k}{A_2^2} \frac{Q^2}{2g}$$

Also $\qquad Q = A_0 u_0 = C_c A u_1 = A_2 u_2 = 14 \times 10^{-3}$ m³/s

From the energy equation

$$\frac{p_0}{w}+\frac{u_0^2}{2g}=\frac{p_2}{w}+\frac{u_2^2}{2g}+h_L$$

we deduce that

$$h_L = \left(\frac{u_0^2-u_2^2}{2g}\right) + \left(\frac{p_0-p_2}{w}\right) = \frac{Q^2}{2g}\left(\frac{1}{A_0^2}-\frac{1}{A_2^2}\right)+(s-1)h_m$$

Hence,

$$\frac{k}{A_2^2}\frac{Q^2}{2g} = \frac{Q^2}{2g}\left(\frac{A_2^2-A_0^2}{A_0^2A_2^2}\right)+(s-1)h_m$$

or

$$k = \left\{\left(\frac{A_2}{A_0}\right)^2-1\right\}+(s-1)h_m2g\left(\frac{A_2}{Q}\right)^2$$

$$= \left\{\left(\frac{4}{3}\right)^4-1\right\}+12{\cdot}6\times0{\cdot}32 \text{ m}\times19{\cdot}61 \text{ m/s}^2 \times \left(\frac{\pi}{4}\times\frac{10^{-2} \text{ m}^2\text{s}}{14\times10^{-3} \text{ m}^3}\right)^2$$

$$= 2{\cdot}16+24{\cdot}94 = 27{\cdot}1$$

Thus, since

$$k = \left\{\frac{1}{C_c}\frac{A_2}{A}-1\right\}^2 = 27{\cdot}1$$

then

$$\frac{1}{C_c} = 6{\cdot}20\,\frac{A}{A_2}$$

and

$$C_c = \frac{1}{6{\cdot}20}\times\left(\frac{10}{5}\right)^2 = \frac{4}{6{\cdot}20} = 0{\cdot}645$$

## 12.3. Contraction losses

The expression:

$$h_L = \left[\frac{1}{C_c}\frac{A_2}{A}-1\right]^2\frac{u_2^2}{2g}$$

developed in the previous section applies equally well if the upstream and downstream pipe areas $A_0$ and $A_2$ are different and, in addition, does not require that the orifice be sharp edged, see Fig. 12.3(a). In order to

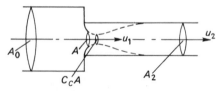

**Fig. 12.3(a).** *Contraction loss*

apply, it is necessary that the value of $C_c$ be known, and this depends not only on the shape of the orifice but also on the area ratio $A/A_0$.

If $A = A_2$, the expression corresponds simply to a contraction in the area of the pipe, see Fig. 12.3(b). If the shape of the contraction is faired, or bell-mouthed, so that overshooting of the stream is avoided (i.e.

$C_c = 1$) there will be no loss. In other cases, the loss will be given by:

$$h_L = \left[\frac{1}{C_c} - 1\right]^2 \frac{u_2^2}{2g}$$

since $A_2$ now equals $A$.

A widely used value for the entry loss into a pipe leading from a tank is $0.5\ u_2^2/2g$. This corresponds to a value of $C_c = 0.585$ in the above expression and is thus seen to be very close to the expected value of about

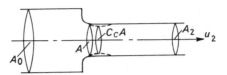

**Fig. 12.3(b).** *Loss due to contraction*

0·6 for an unfaired opening leading from a large reservoir. This suggests that we may use the table of values given in the previous section for $C_c$ as a function of $A/A_0$ to estimate the loss which occurs at any *sudden* contraction from $A_0$ to $A$:

| $A/A_0$ | 0·2 | 0·4 | 0·6 | 0·8 | 1·0 |
|---|---|---|---|---|---|
| $C_c$ | 0·632 | 0·659 | 0·712 | 0·813 | 1·0 |
| $h_L/(u_2^2/2g)$ | 0·34 | 0·27 | 0·16 | 0·05 | 0 |

EXAMPLE 12.3 (i)

*A horizontal pipe* 100 mm *diameter suddenly enlarges to* 150 mm *diameter. If water flows through it at the rate of* 1·8 m³/min *find:*
(a) *the loss of head due to the sudden enlargement,*
(b) *the difference in pressure in the two pipes,*
(c) *the same items if the same quantity of water were made to flow through in the opposite direction assuming $C_c = 0.66$,*
(d) *the change in pressure-head if the change in diameter were effected by a gradual cone without loss of head.*

(a)
$$h_L = \frac{(u_1 - u_2)^2}{2g}$$

where
$$u_1 = \frac{Q}{A_1} = \frac{1.8\ \text{m}^3}{60\ \text{s}} \times \frac{10^4}{\pi\ 25\ \text{m}^2} = 3.82\ \text{m/s}$$

and
$$u_2 = \left(\frac{d_1}{d_2}\right)^2 u_1 = \frac{4}{9} \times 3.82\ \text{m/s} = 1.69\ \text{m/s}$$

hence,
$$h_L = \frac{(2.13)^2}{19.61}\text{m} = \mathbf{0.232\ m}$$

(b)

$$\frac{p_1}{w}+\frac{u_1^2}{2g}+z_1 = \frac{p_2}{w}+\frac{u_2^2}{2g}+z_2+h_L$$

hence

$$\frac{p_2-p_1}{w} = \left(\frac{u_1^2-u_2^2}{2g}\right)-h_L$$

since $z_1 = z_2$, i.e.

$$\frac{p_2-p_1}{w} = \frac{3\cdot 82^2-1\cdot 69^2}{19\cdot 61}\ \text{m}-0\cdot 232\ \text{m} = 0\cdot 368\ \text{m head}$$

or     $p_2-p_1 = 9\cdot 807\ \text{kN/m}^3 \times 0\cdot 368\ \text{m} = \textbf{3\cdot 61 kPa}$

gain in pressure.

(c) If the direction of flow were reversed the velocity at the vena contracta would be

$$u_3 = \frac{a_1 u_1}{a_3} = \frac{u_1}{C_c}$$

hence

$$h_L = \frac{(u_3-u_1)^2}{2g} = \frac{u_1^2}{2g}\left\{\left(\frac{a_1}{a_3}\right)-1\right\}^2 = \left(\frac{1}{C_c}-1\right)^2\frac{u_1^2}{2g}$$

$$= \left(\frac{1}{0\cdot 66}-1\right)^2\times\frac{3\cdot 82^2}{19\cdot 61}\ \text{m} = \textbf{0\cdot 1975 m}$$

and     $\frac{p_2-p_1}{w} = \frac{u_1^2-u_2^2}{2g}+h_L = (0\cdot 598+0\cdot 198)\ \text{m} = 0\cdot 796\ \text{m}$

or the loss in pressure,

$$p_2-p_1 = 9\cdot 807\ \text{kN/m}^3\times 0\cdot 796\ \text{m} = \textbf{7\cdot 8 kPa}$$

(d) If the change of section were gradual without loss, then

$$\frac{p_2-p_1}{w} = \frac{u_1^2-u_2^2}{2g} = 0\cdot 598\ \text{m}$$

or the gain in pressure = **5·87 kPa**

## 12.4. Borda mouthpiece

As fluid approaches the mouth of an orifice it acquires velocity head at the expense of its pressure head. Consequently the pressure drops along the upstream face of the orifice plate shown in Fig. 12.4(a), giving rise to a 'force defect' or suction effect which adds to the unbalanced pressure driving fluid through the orifice. If the opening can be moved a sufficient distance from the reservoir wall, the velocity along the latter (and hence the suction effect) may be reduced to negligible proportions, see Fig. 12.4(b).

Such an opening is known as a Borda mouthpiece, or re-entrant orifice, and we may quite simply establish an equation of motion for the discharge through it, by equating the resultant force to the corresponding mass flow times its change of velocity (see Section 7.1, p. 138). If the

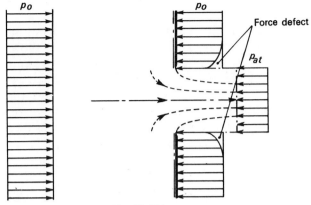

Fig. 12.4(a)

pressure is distributed hydrostatically over the vertical walls shown, and we assume for simplicity that the fluid is being discharged into the atmosphere, the resultant force, for a gauge pressure $p_0$ in the reservoir is:

$$F = p_0 A$$

If the contracted area of the emergent jet is $C_c A$, the corresponding mass flow is:

$$\dot{m} = \rho C_c A u$$

its corresponding velocity being $u$. Hence:

$$p_0 A = \rho C_c A u \times u$$

i.e.
$$p_0 = C_c \rho u^2$$

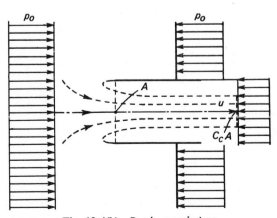

Fig. 12.4(b). *Borda mouthpiece*

From Bernoulli's equation however:

$$\frac{p_0}{w} = \frac{u^2}{2g}$$

i.e.

$$p_0 = \tfrac{1}{2}\rho u^2$$

Comparison of the two expressions obtained from momentum and energy considerations respectively indicates that $C_c$ theoretically equals 0·5. Experiments confirm this result so long as the re-entrant portion is long enough for the fluid velocity along the reservoir walls to be negligible.

If this condition is not satisfied, the momentum equation contains an additional term on the left-hand side, which depends on the magnitude of the force defect, or suction effect, so that:

$$p_0 < C_c\rho u^2$$

As the energy equation still applies, we deduce that in general:

$$\tfrac{1}{2}\rho u^2 < C_c\rho u^2$$

i.e.

$$C_c > \tfrac{1}{2}$$

Thus there is a lower limit to the contraction coefficient of an orifice, in addition to the upper limit of unity, which corresponds to a nozzle-like opening. Hence, neglecting friction:

$$\tfrac{1}{2} < C_c < 1$$

for all orifices or openings, whatever their shape.

In deducing the expression for a Borda mouthpiece it was assumed that the fluid discharged freely into the atmosphere. If there is a back-pressure downstream, the pipe leading from the reservoir eventually flows full, as shown in Fig. 12.4(c). The above analysis may still be applied

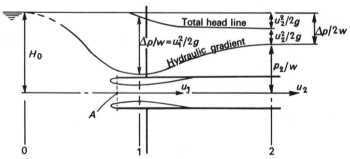

Fig. 12.4(c). *Borda mouthpiece with exit pipe*

between the reservoir and the vena contracta provided that the pressure may be assumed constant across the pipe at section 1. This can be justified in the dead-water region by an argument similar to that used concerning a similar region in Section 12.1. In developing the equations for the mouth-piece flowing full, the only difference is that the pressure drop is now denoted by $\Delta p$ instead of $p_0$, and $u_1$ replaces $u$. The contraction coefficient is thus still theoretically 0·5 and the subsequent loss which occurs downstream is, from the previous section:

$$h_L = \left[\frac{1}{C_c} - 1\right]^2 \frac{u_2^2}{2g} = \frac{u_2^2}{2g}$$

As there is negligible loss between sections 0 and 1 the hydraulic gradient at the latter section lies below the reservoir still-water surface a distance:

$$\frac{\Delta p}{w} = \frac{u_1^2}{2g}$$

The velocity at section 2 is, by continuity, one-half that at the vena contracta:

$$Q = u_1\tfrac{1}{2}A = u_2 A$$

so that the total head loss between sections 1 and 2:

$$h_L = \frac{u_2^2}{2g} = \frac{1}{4}\frac{u_1^2}{2g} = \frac{1}{4}\frac{\Delta p}{w}$$

When we add to this the kinetic head at this section, viz.:

$$\frac{u_2^2}{2g} = \frac{1}{4}\frac{\Delta p}{w}$$

we find that the hydraulic gradient at this section lies a distance below the still water surface given by:

$$\frac{1}{4}\frac{\Delta p}{w} + \frac{1}{4}\frac{\Delta p}{w} = \frac{1}{2}\frac{\Delta p}{w}$$

If the fluid at section 2 were to discharge straight into the atmosphere, the pressure head $p_2/w$ at this section would be zero, and that at section 1 would then lie a distance $H_0$ below atmospheric. Thus if the mouthpiece is made to flow full the kinetic head at the vena contracta is double that when flowing free. The discharge is consequently increased by the factor $\sqrt{2}$ through this reduction of pressure at the vena contracta. There is a physical limit to the latter, set by the liability of the fluid to cavitation. This tendency is aggravated if the exit section (which may represent for example the inlet to a pump) is already at a pressure below atmospheric.

## 12.5. The effect of bends

When fluid which is moving along a straight pipe encounters a bend, its momentum may be imagined to carry it to the outer wall of the latter where it splays out circumferentially from the region of high pressure developed there. These conclusions are based on paint streak patterns obtained experimentally and illustrated in Fig. 12.5(a). The inward

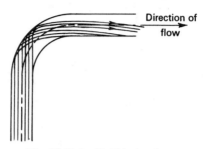

**Fig. 12.5(a).** *Fluid in bends*

spiralling of the fluid at the walls must be compensated by a corresponding outward flow as indicated in Fig. 12.5(b), so forming a double spiral. This persists for a considerable distance—say 50 to 75 diameters—downstream. The increase in pressure at the outside of the bend may cause a region of separation as indicated at A in Fig. 12.5(b), in addition to the major separation at B.

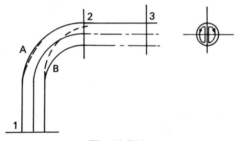

**Fig. 12.5(b)**

The total loss at a bend may be considered to consist of three parts. Firstly, that due to the dissipation of the energy of the spiralling motion downstream, secondly, that due to eddying in the regions of separation and, thirdly, that due to friction.

Apart from these direct effects, bends may also disrupt or modify the flow through fittings downstream. Thus the total loss through two bends in series is, in fact, generally less than the sum of that caused by each separately. It is somewhat difficult to correlate data on bend losses as there is no uniform agreement as to how they shall be measured and defined. There is very little loss between sections 1 and 2 in Fig. 12.5(b) where the stream is still contracting, and the total head line will vary somewhat as shown in Fig. 12.5(c).

We shall define the bend loss by means of this figure as the *additional* loss due to the presence of the bend, over that which would otherwise occur between sections 1 and 3 due to friction. The corresponding hydraulic gradient also suggested in the figure indicates that static pressure tappings either at, or downstream of, a pipe-bend may be grossly misleading, except when introduced for test purposes in order to estimate $h_L$ as shown.

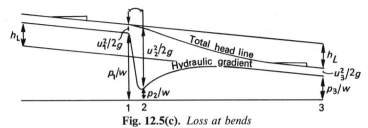

**Fig. 12.5(c).** *Loss at bends*

Fig. 12.5(d) shows some results obtained by Hofmann, which suggest that for circular pipes the effect of the Reynolds number on the loss is not large, especially for rough pipes. The curves indicate the effect of bend radius/pipe diameter (i.e. $R/D$) on the head loss $h_L$, expressed as a fraction of the upstream kinetic head $u^2/2g$. Experiments on rectangular sections, of width $W$, and depth $D$ in the plane of the bend, indicate that

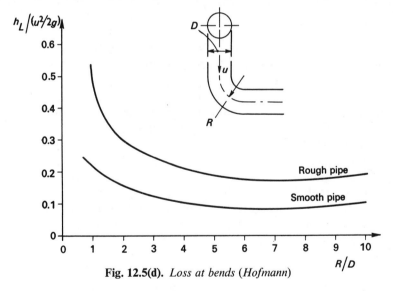

**Fig. 12.5(d).** *Loss at bends (Hofmann)*

the loss is reduced to a minimum by making both $R/D$ and $W/D$ large. By using splitters, or turning vanes, a bend of small aspect ratio (as $W/D$ is called) may effectively be replaced by a number of bends of large aspect ratio in parallel. Experiments confirm that the total loss due to these may well be considerably less than that due to the bend without guide vanes, careful design with bent sheet enabling the loss to be reduced to about $0.15 \; u^2/2g$ for right-angled bends.

Typical test results showing the bend loss for angles other than 90° are shown in Fig. 12.5(e).

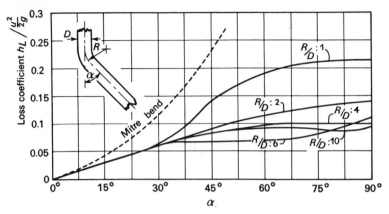

Fig. 12.5(e). *Loss coefficient for smooth bends*

## 12.6. Effect of pipe-fittings generally

The losses which occur at valves and pipe fittings depend critically on their geometry and no general data can usefully be given. The principles outlined in the previous sections still apply however. Any restriction will cause a jet to form and the area of the latter may be less than that of the opening due to overshooting. The contraction coefficient will normally lie between that for a Borda mouthpiece (0·5) and that for a nozzle (1·0). The kinetic head of the issuing jet will usually be destroyed in the subsequent re-expansion downstream. If it impinges directly on to slower moving fluid it is possible that only the kinetic energy of the relative motion may be destroyed. It is certainly optimistic to assume that there will be any re-compression, as in the diffuser of a Venturi, unless very careful attention is paid to its hydrodynamic design. Although a stream may be contracted quite rapidly and efficiently, efficient re-compression is difficult. These remarks apply primarily to hydraulic systems (i.e. at high Reynolds numbers with negligible compressibility effects). In such systems the head loss may be expressed as:

$$h_L = k\frac{u^2}{2g}$$

in which $u$ is some convenient reference velocity such as the mean value $Q/A$.

The value of $k$ depends primarily on the geometry, such as the valve setting for example, and is generally little influenced by scale effects. If the separation points for the overshooting actions are located by sharp edges, as in the valves shown in Figs. 12.6(a) and (b), the loss will almost certainly vary as the velocity squared. If, as in pipe bends for example, viscosity plays a more direct part in controlling the flow pattern, the loss may be expected to vary as $u^n$, the coefficient $n$ generally lying between the values 1·7 and 2. Such refinement is however seldom justifiable as such factors as interaction between fittings usually introduces considerable uncertainty into any estimation of circuit losses.

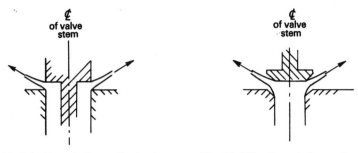

**Fig. 12.6(a).** *Ported sharp edged valve*   **Fig. 12.6(b).** *Seated sharp edged valve*

EXAMPLE 12.6 (i)

*A hydraulic accumulator, maintained at a constant pressure of 7 MN/m², supplies water to a ram 240 mm diameter through a pipe 40 mm diameter and 36 m long which contains eight 90° elbows and a valve whose port area is equivalent to a 25 mm circle. Estimate (i) the maximum speed at which the ram can move when not exerting thrust, and (ii) the thrust exerted when the speed is half this maximum. f = 0·007 5.*

Friction head in smooth pipe

$$= h_f = 4f\left(\frac{l}{d}\right)\frac{u^2}{2g} = 0{\cdot}03 \times \left(\frac{3\ 600}{24}\right)\frac{u^2}{2g} = 4{\cdot}5\,\frac{u^2}{2g}$$

Friction head in 8 elbow 90° bends

$$= 8\left(0{\cdot}2\,\frac{u^2}{2g}\right) = 1{\cdot}6\,\frac{u^2}{2g}$$

Friction head in valve

$$= \left(\frac{1}{0{\cdot}64}-1\right)^2\frac{u^2}{2g} = 0{\cdot}3\,\frac{u^2}{2g}$$

Hence, total friction head

$$= 6{\cdot}4\,\frac{u^2}{2g} = 8\ 290\,\frac{U^2}{2g},$$

where

$$u = \left(\frac{24}{4}\right)^2 U = 36\ U.$$

Thus, (i) for maximum speed ($U$) of ram,

$$\frac{Po}{w} = \frac{8290\ U^2}{19{\cdot}61\ \text{m/s}^2} = 7\ \frac{\text{MN}}{\text{m}^2} \times \frac{\text{m}^3}{9{\cdot}807\ \text{kN}}$$

$$\text{or}\quad U = \mathbf{1{\cdot}3}\ \text{m/s}.$$

(*ii*) Maximum force at zero speed is

$$\pi \times \frac{144 \text{ m}^2}{10^4} \times 7 \ \frac{\text{MN}}{\text{m}^2} = 317 \text{ kN}$$

Friction head at half maximum speed is

$$\frac{8\ 290}{19 \cdot 61 \text{ m/s}^2} \times (0 \cdot 65 \text{ m/s})^2 = 179 \text{ m}$$

Hence, pressure at ram is

$$P = 7 \text{ MN/m}^2 - 179 \times 9 \cdot 807 \text{ kN/m}^2 = 5 \cdot 244 \text{ MN/m}^2$$

and thrust at half maximum speed is

$$\pi \times \frac{144}{10^4} \text{ m}^2 \times 5 \cdot 244 \text{ MN/m}^2 = \textbf{237 kN}$$

## 12.7. The H–Q characteristic of a hydraulic line

The total head loss ($H_L$ say) through any particular line neglecting interaction, is:

$$H_L = \Sigma h_L = \Sigma \frac{4fl}{D} \frac{u^2}{2g} + \Sigma k \frac{u^2}{2g}$$

Each loss coefficient $k$ is sometimes expressed in terms of an equivalent length of pipe ($l_e$, say) such that, for each fitting:

$$l_e = \frac{kD}{4f}$$

If the pipe-line is of the same diameter ($D$) throughout, for example, the total loss may then be expressed as:

$$H_L = \left\{ \sum \frac{4fl}{D} + \sum \frac{4fl_e}{D} \right\} \frac{u^2}{2g}$$

i.e.
$$H_L = \frac{4fL}{D} \frac{u^2}{2g}$$

in which the total equivalent length of pipe $L$ is given by:

$$L = (\Sigma l + \Sigma l_e)$$

Such a representation, whilst being obviously convenient, is, however, difficult to justify unless the Reynolds number and the relative roughness are so large that $f$ has become constant. A constant value of $L$ (and hence $l_e$ for each fitting) implies that any reduction in $f$ (due e.g. to an increase in $Re$) is accompanied by a corresponding reduction in $k$ for each fitting. Whilst this may be partly true for bends, it is certainly not true for most expansions, and these generally make a large contribution towards the total loss.

In the original expression for the head loss, each $u$ is a velocity which is related to the discharge $Q$ by means of some corresponding reference area $A$ which is such that:

$$u = \frac{Q}{A}$$

*Since the discharge through a number of fittings in series is the same for each*:

$$H_L = \sum \frac{4fl(Q/A)^2}{D} \frac{}{2g} + \sum k \frac{(Q/A)^2}{2g}.$$

$$= \left\{ \sum \frac{4fl}{DA^2 2g} + \sum \frac{k}{A^2 2g} \right\} Q^2$$

i.e. $$H_L = KQ^2$$

in which $K$ is a measure of the resistance of the line, and is given by:

$$K = \frac{1}{2g} \left\{ \sum \frac{4fl}{DA^2} + \sum \frac{k}{A^2} \right\}$$

The above $H_L$–$Q$ relationship is as fundamental to the analysis of hydraulic circuits as is Ohm's law to electric circuits:

$$E = RI$$

In fact the head difference $H_L$ is analogous to the potential difference $E$, being dependent on the hydraulic resistance $(K)$ (instead of the electrical resistance $R$) and the hydraulic discharge $Q$ (instead of the electric current $I$).

The analogy must not however be pushed too far, as the hydraulic losses tend to vary quadratically with $Q$, whereas electrical losses vary linearly with $I$.

We may alternatively express the hydraulic loss as a pressure drop, $(\Delta p = wH_L)$ by defining a quantity $K' = wK$, so that:

$$\Delta p = K' Q^2.$$

Whilst it may be objected that the measures of hydraulic resistance $K$ (and $K'$) may depend on such factors as the Reynolds number, it should be borne in mind that the electrical resistance $R$ of a conductor also depends on the flow to some extent, due for example to heating effects changing its resistivity. The major difference between the hydraulic and electric laws, namely that the loss of potential ($H_L$, $\Delta p$ or $E$) tends to vary more nearly as the square of the discharge ($Q$) in hydraulic systems, whereas it tends to vary proportionately with the current ($I$) in electric circuits (see Figs. 12.7(a) and 12.7(b)), makes analysis of the former considerably more difficult.

Slow creeping flows are an important exception to this, for if the Reynolds number $Re$ is sufficiently low the flow will be laminar (see

Section 8.5, p. 181). In this range $f$ (and hence $K$) varies inversely as $Re$ (and hence $Q$)—see Section 11.4 (p. 292). Hence laminar flow is directly analogous to flow of electric charge, and the methods developed for

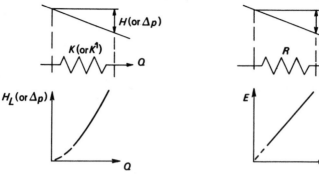

**Fig. 12.7(a).** *Loss of head with Q*

**Fig. 12.7(b).** *Loss of head analogy with electric current*

analysing circuits in the one case may be applied, without modification, to the other.

EXAMPLE 12.7 (i)

*Liquid is tapped from points along a supply main at the volumetric rate q per unit length of pipe. Estimate the loss of head due to friction in the main pipe if the supply of liquid is exhausted after a length l.*

After a length of pipe $x$ the volumetric rate of flow through the main pipe in Fig. 12.7(c) is:

$$Q = Q_i - qx = Au$$

and the loss of head in a length $\delta x$ is:

$$\delta h = \frac{4f\delta x}{d}\frac{u^2}{2g}$$

or

$$\delta h = \frac{4f\delta x}{d2gA^2}(Q_i - qx)^2 = \frac{4f\delta x}{2gA^2d}(Q_i^2 - 2qxQ_i + q^2x^2)$$

**Fig. 12.7(c).** *Supply main*

Hence, assuming $f$ to remain constant (otherwise a graphical or step-by-step method is necessary) we deduce that friction causes a loss of head up to any section $x$ of

$$h_x = \frac{4f}{2gA^2d}\left(Q_i^2 x - qx^2 Q_i + q^2\,\frac{x^3}{3}\right)$$

$$= \frac{4fx}{2gA^2d}\left(Q_i^2 - qQ_i x + q^2\,\frac{x^2}{3}\right)$$

Thus, in the case of a supply pipe of length $l$ in which the supply at the end is exhausted we have $Q_i = ql$ and

$$h_l = \frac{4fl}{2gA^2d}(q^2 l^2)\left\{1 - 1 + \frac{1}{3}\right\}$$

$$h_l = \frac{1}{3}\times\frac{4fl}{2gd}\left(\frac{ql}{A}\right)^2$$

$$= \tfrac{1}{3}\times\textbf{loss of head}$$

for a constant discharge $ql$.

## 12.8. Hydraulic lines in series and parallel

In the majority of hydraulic circuits the quadratic law of resistance $H_L = KQ^2$ may be applied to each line, so that, for steady flow through a series of lines with negligible interaction, the total head loss $H$ is obtained simply by summing the $K$'s:

$$H = (K_1 + K_2 + \ldots)\,Q^2$$

(see Fig. 12.8(a)). This applies since the discharge is the same through

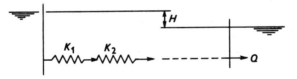

**Fig. 12.8(a).** *Hydraulic lines in series*

each, if there is no accumulation of fluid. If a number of lines are connected *in parallel* the total head loss through each must be the same, see Fig. 12.8(b).

$$\therefore \qquad H = K_1\,Q_1^2 = K_2\,Q_2^2 = \ldots$$

If we denote the total flow by $Q$:

$$Q = Q_1 + Q_2 + \ldots$$

i.e.

$$Q = \sqrt{\frac{H}{K_1}} + \sqrt{\frac{H}{K_2}} + \ldots$$

$$= \left\{\frac{1}{\sqrt{K_1}} + \frac{1}{\sqrt{K_2}} + \ldots\right\}\sqrt{H}$$

Thus the equivalent resistance of the lines ($K$, say) defined by:

$$H = KQ^2 \quad \text{or} \quad Q = \frac{1}{\sqrt{K}}\sqrt{H}$$

is given by comparing the above two $H$–$Q$ expressions and indicates that:

$$\frac{1}{\sqrt{K}} = \left\{\frac{1}{\sqrt{K_1}} + \frac{1}{\sqrt{K_2}} + \ldots\right\}$$

This may be compared with the equivalent expression for electric resistances in parallel:

$$\frac{1}{R} = \left\{\frac{1}{R_1} + \frac{1}{R_2} + \ldots\right\}$$

i.e. for lines in parallel the reciprocals of the resistances (i.e. the conductances) are additive. Electrical conductance is closely related to the

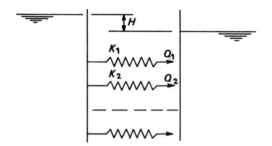

**Fig. 12.8(b).** *Hydraulic lines in parallel*

discharge coefficient used in hydraulics. For example, if the resistances shown in Fig. 12.8(b) are in fact a number of orifices of area $A_1$, $A_2$, etc., the total flow between the two tanks is, simply:

$$Q = C_{d_1}A_1 \sqrt{(2gH)} + C_{d_2}A_2 \sqrt{(2gH)} + \ldots$$

(see Section 4.5),

i.e. $$Q = \{C_{d_1}A_1 + C_{d_2}A_2 + \ldots\} \sqrt{(2gH)}$$

Denoting the total area $A_1 + A_2 \ldots$ by $A$, it is seen that the equivalent discharge coefficient $C_d$, defined by:

$$Q = C_d A \sqrt{(2gH)}$$

is given by:

$$C_d A = \{C_{d_1}A_1 + C_{d_2}A_2 + \ldots\}$$

i.e. the values of $C_d A$ for orifices in parallel are additive. These results suggest that we may conveniently define a discharge constant $C$ for a pipe-line by:

$$Q = C \sqrt{H}$$

For an orifice $C = C_dA \sqrt{(2g)}$, but in general it may be determined from the line resistance $K$ in:

$$H = KQ^2$$

Comparison of these two expressions shows that:

$$C = \frac{1}{\sqrt{K}}$$

This quantity, which has been shown to be additive for lines in parallel, has a resultant for lines *in series* given by:

$$K = K_1 + K_2 + \ldots$$

i.e.

$$\frac{1}{C^2} = \frac{1}{C_1^2} + \frac{1}{C_2^2} + \ldots$$

**We may summarize the above results by stating that hydraulic resistances ($K$) add for elements in series, and discharge constants ($C$) add for elements in parallel. $K$ and $C$ are defined respectively by:**

$$H = KQ^2 \quad \text{and} \quad Q = C\sqrt{H}$$

Just as it is possible to define an alternative resistance constant $K'$ to relate pressure to discharge, so we may define an alternative discharge constant $C'$:

$$p = K'Q^2$$
$$Q = C'\sqrt{p}$$

The resistance constants are related by:

$$K' = wK$$

and the discharge constants by:

$$C' = \frac{C}{\sqrt{w}}$$

EXAMPLE 12.8 (i)

*Two reservoirs* 10 km *apart have a difference of water level of* 30 m *and are connected together by a pipe-line.*

*For the first* 4 km *there is a single pipe sloping downwards* 8 m/km *and then* 6 km *of double pipe sloping* 1 m/km. *If the same size of pipe be used throughout what should be the diameter of the pipe so that the velocity may not exceed* 1·5 m/s. *Assuming constant* $f = 0.007\,5$ *in the formula*

$$h_L = \frac{4fl}{d}\frac{u^2}{2g}$$

Referring to Fig. 12.8(c) we may state that since the pipes are of equal diameter

$$Q_1 = 2Q_2, \quad \text{and} \quad u_1 = 2u_2$$

Also, considering a stream-line ABC we may state the energy equation:

Head at A = Head at C+Loss between A and C

$$z_A = D + u_c^2/2g + z_c + h_{f1} + h_{f2}$$

or $$z_A - (D + z_c) = H = \frac{u_c^2}{2g} + h_{f1} + h_{f2}$$

Hence, neglecting $u_c^2/2g$ relative to the friction losses we have:

$$H = \frac{4f}{2gd}(l_1 u_1^2 + l_2 u_2^2)$$

or $$H = \frac{4f}{d}\frac{u_1^2}{2g}\left(l_1 + \frac{l_2}{4}\right)$$

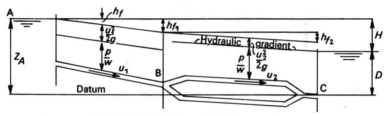

**Fig. 12.8(c).** *Single and double pipe lines*

Therefore,

$$d = \frac{4f}{H}\frac{u_1^2}{2g}\left(l_1 + \frac{l_2}{4}\right) = \frac{4 \times 0{\cdot}007\,5}{30 \text{ m}} \times \frac{2{\cdot}25}{19{\cdot}61} \text{ m} \left(4 + \frac{6}{4}\right) \text{ km}$$

$$= 0{\cdot}63 \text{ m} \quad or \quad 630 \text{ mm}$$

## EXAMPLE 12.8 (ii)

*A tank A with vertical sides and plan area 270 m² is connected with another tank B with vertical sides and plan area 360 m² by means of 3 horizontal pipes laid alongside in parallel, each 60 m long and of diameter d. The entry and exit of each pipe is sharp. Initially there is a difference of level of 3 m between the water surfaces in the two tanks. If water is now admitted to tank A at the rate of 18 m³/min, what must be the diameter d of the connecting pipes if the difference of level is to remain constant at 3 m?*

*Take the frictional coefficient for each pipe $\lambda = 0{\cdot}03$ in the formula*

$$h_L = \frac{\lambda l}{d}\frac{u^2}{2g}$$

The energy equation for one pipe (Fig. 12.8(d)) is:

$$z_1 = \left(z_2 + \frac{u^2}{2g}\right) + \text{Losses (entry and pipe friction)}$$

$$\therefore \qquad z_1 - z_2 = \frac{u^2}{2g} + \frac{1}{2}\frac{u^2}{2g} + \frac{\lambda l}{d}\frac{u^2}{2g} = \frac{u^2}{2g}\left(1 + \frac{1}{2} + \frac{\lambda l}{d}\right)$$

or
$$3\text{ m} = \frac{u^2}{2g}\left(\frac{3}{2}+\frac{0\cdot03\times60\text{ m}}{d}\right)\tag{1}$$

If the difference of levels is to remain constant at 3 m, then the rate of rise of level in the two tanks must be equal, i.e.

$$\frac{Q_i-Q_0}{A_1} = \frac{Q_0}{A_2}$$

Hence
$$Q_i = Q_0\left(\frac{A_1}{A_2}+1\right)$$

and
$$\frac{18\text{ m}^3}{60\text{ s}} = 3\times\frac{\pi}{4}d^2u\left(\frac{3}{4}+1\right) = \frac{21\pi}{16}d^2u\tag{2}$$

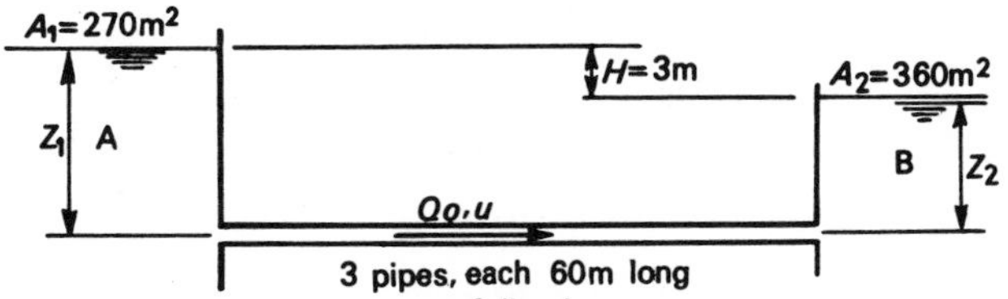

**Fig. 12.8(d).** *Pipes in parallel*

From Equations (1) and (2) we get:

$$\frac{3}{2}+\frac{1\cdot8\text{ m}}{d} = 3\text{ m}\times19\cdot61\frac{\text{m}}{\text{s}^2}\left(\frac{21}{16}\times\frac{60}{18}\pi d^2\right)^2\frac{\text{s}^2}{\text{m}^6}$$

i.e.
$$\frac{3}{2}+1\cdot8\left(\frac{\text{m}}{d}\right) = 110\times10^2\left(\frac{d}{\text{m}}\right)^4$$

Hence,
$$d = \mathbf{0\cdot18}\text{ m} \quad or \quad \mathbf{180}\text{ mm}$$

EXAMPLE 12·8 (iii)

*Two reservoirs, having a constant difference in water level of 60 m, are connected by a 200 mm diameter pipe, 4 km long. The pipe is tapped at a point distant 1·5 km from the upper reservoir, and water is drawn off at the rate of 2·5 m³/min. If the friction coefficient λ is 0·036 in formula*

$$h_L = \frac{\lambda l}{d}\frac{u^2}{2g}$$

*determine the rate at which water enters the lower reservoir, neglecting all losses except pipe friction.*
   Referring to Fig. 12.8(e), we see that:

$$Q_1 = Q_2+Q \quad or \quad u_1-u_2 = \frac{4Q}{\pi d^2}$$

i.e.
$$u_1-u_2 = \frac{4}{\pi}\times\frac{2\cdot5\text{ m}^3}{60\text{ s}}\times\frac{10^2}{4\text{ m}^2} = 1\cdot326\text{ m/s}\tag{1}$$

Also

$$H_1 = H_2 + \text{losses} \quad \text{or} \quad H_1 - H_2 = \frac{\lambda}{d2g}(l_1 u_1^2 + l_2 u_2^2)$$

Hence,    $$60 \text{ m} = \frac{0\cdot036}{0\cdot2} \frac{\text{s}^2}{\text{m} \times 19\cdot61} \frac{}{\text{m}} (1\cdot5 u_1^2 + 2\cdot5 u_2^2) \text{ km}$$

or    $$3 u_1^2 + 5 u_2^2 = 13\cdot07 \text{ m}^2/\text{s}^2 \tag{2}$$

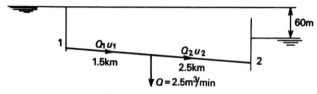

**Fig. 12.8(e).** *Tapped pipe*

From (1) and (2) $3(1\cdot326 + u_2)^2 + 5u_2^2 = 13\cdot07$ where $u_2$ is in m/s

i.e.    $$u_2^2 + 0\cdot996 u_2 - 0\cdot974 = 0 \quad \text{and} \quad u_2 = 0\cdot608 \text{ m/s}$$

Hence,    $$Q_2 = \frac{\pi}{4} \times 0\cdot04 \text{ m}^2 \times 36\cdot48 \text{ m/min} = \mathbf{1\cdot145} \text{ m}^3/\text{min}$$

## 12.9. Hydraulic circuit analysis

There are two basic ideas which may be used to analyse the flows through branched circuits. The first is that of continuity, which requires that the total inflow at any junction must equal the total outflow. The second is that the total head at each junction has a unique value and, if there are no pumps or motors in the circuit, the drop in head between any two points is *numerically* equal to $KQ^2$ and always occurs in the actual direction of $Q$. *The rules which have been developed for electric circuit analysis cannot be applied blindly to hydraulic circuits,* as the equation:

$$H = KQ^2$$

*applies only if the flow is actually in the direction assumed. Reversal of the flow in fact reverses the sign of H,* whereas, according to the above equation, $H$ is independent of the sign of $Q$, see Fig. 12.9(a).

In symbols we ought in fact to write:

$$H = KQ|Q|$$

in which $|Q|$ implies the *numerical* value of $Q$, *no matter what its algebraic sign may be.* In calculations it implies that if we write $KQ^2$ the algebra will be correct only if $Q$ is positive, i.e. in the direction assumed. If it is not, we shall have to re-work the problem with $Q$ in the right direction. In practice, however, the differences of levels ($H$) are usually known, which determine the directions of flow as in the following Example 12.9 (i).

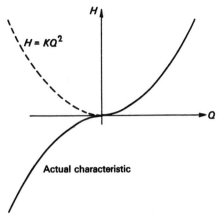

**Fig. 12.9(a).** *H/Q graph*

EXAMPLE 12.9 (i)

*Find the rate of volumetric flow in the three pipes of Fig. 12.9(b) if the levels in the reservoirs A, B and C are 60, 45 and 55 m, respectively, above a horizontal datum line. The pipes leading from the reservoirs have diameters 400, 240 and 320 mm, and their lengths 600, 912 and 1 120 m, respectively. Assume λ = 0·02 in the formula $h_f = (\lambda l/d)(u^2/2g)$ for each of the three pipes. Also find the head at the junction J. Neglect secondary losses.*

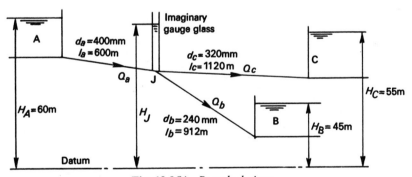

**Fig. 12.9(b).** *Branched pipes*

Since

$$Q_a = Q_b + Q_c$$

then

$$u_a d_a^2 = u_b d_b^2 + u_c d_c^2$$

i.e.

$$25u_a = 9u_b + 16u_c \qquad (1)$$

Also

$$H_A - H_J = \frac{\lambda l_a}{d_a}\frac{u_a^2}{2g} = \frac{0·02 \times 600 \text{ m}}{0·4 \text{ m}}\frac{u_a^2}{2g} = 30\frac{u_a^2}{2g} \qquad (2)$$

and
$$H_J - H_B = \frac{\lambda l_b}{d_b}\frac{u_b^2}{2g} = \frac{0 \cdot 02 \times 912 \text{ m}}{0 \cdot 24 \text{ m}}\frac{u_b^2}{2g} = 76\frac{u_b^2}{2g} \tag{3}$$

and
$$H_J - H_C = \frac{\lambda l_c}{d_c}\frac{u_c^2}{2g} = \frac{0 \cdot 02 \times 1\,120 \text{ m}}{0 \cdot 32 \text{ m}}\frac{u_c^2}{2g} = 70\frac{u_c^2}{2g} \tag{4}$$

Thus, adding (2) and (3) we get

$$H_A - H_B = 30\frac{u_a^2}{2g} + 76\frac{u_b^2}{2g}$$

or
$$g \times 15 \text{ m} = 15u_a^2 + 38u_b^2 \tag{5}$$

and adding (2) and (4) gives

$$H_A - H_C = 30\frac{u_a^2}{2g} + 70\frac{u_c^2}{2g}$$

or
$$g \times 5 \text{ m} = 15u_a^2 + 35u_c^2 \tag{6}$$

Substituting $u_b$ and $u_c$ from (5) and (6) in Equation (1) we get

$$25u_a = 9\sqrt{\left(\frac{g15 \text{ m} - 15u_a^2}{38}\right)} + 16\sqrt{\left(\frac{g5 \text{ m} - 15u_a^2}{35}\right)}$$

from which $u_a = 1 \cdot 2$ m/s. Hence from (5) $u_b = 1 \cdot 73$ m/s and from (6) $u_c = 0 \cdot 88$ m/s.

Finally

$$Q_a = \frac{\pi}{4}d_a^2 u_a = 0 \cdot 151 \text{ m}^3/\text{s}, \qquad Q_b = 0 \cdot 077 \text{ m}^3/\text{s} \qquad \text{and} \qquad Q_c = 0 \cdot 075 \text{ m}^3/\text{s}$$

and
$$H_J = H_A - 30\frac{u_a^2}{2g} = 60 \text{ m} - 2 \cdot 2 \text{ m} = 57 \cdot 8 \text{ m}$$

*Note:* Had $H_J$ proved to be less than $H_C$, it would have been necessary to re-work the problem with $Q_c$ reversed. The changeover in the direction of flow in $Q_c$ occurs when $u_a = 1 \cdot 81$ m/s, for this value corresponds to a head at $J$ given by:

$$H_A - H_J = 30\frac{u_a^2}{2g} = 5 \text{ m}$$

i.e. $H_J = 55$ m if $u_a = 1 \cdot 81$ m/s.

Thus if the flow through the line A is increased by reducing the resistance of line B, e.g. the head at J eventually drops to that in reservoir C. Flow to the latter then ceases. Any further reduction in the resistance of line B will cause flow out from reservoir C instead of into it.

## 12.10. Maximum power and efficiency of transmission

All hydraulic power transmission systems consist essentially of a source (such as a pump or reservoir) which supplies fluid to a machine (such as a ram or turbine) via some form of line in which the loss may be expressed as $KQ^2$. As a typical example we may consider a supply pipe conveying water from a reservoir to a Pelton wheel nozzle, see Fig. 12.10(a).

If the total head at the nozzle is $H_2$, and the delivery is $Q$, the power available ($P_2$, say) is $wQH_2$. On no flow, when $H_2 = H_1$, the power

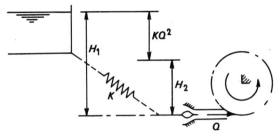

**Fig. 12.10(a).** *Transmission of hydraulic power*

available is of course zero. Referring to Fig. 12.10(b) (i), as the spear valve is opened, so the point $A$ moves down the $H$–$Q$ characteristic. The shaded area, which is proportional to the power available, is seen first to grow and then ultimately to decrease as $H_1$ falls. The available power will be a maximum at that point on the characteristic where, on going

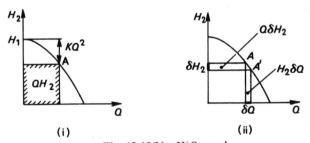

**Fig. 12.10(b).** *H/Q graphs*

from $A$ to $A'$ the added area $H_2\delta Q$ in Fig. 12.10(b) (ii) just equals the lost area $Q\delta H_2$, i.e.:

$$-Q\delta H_2 = H_2\delta Q$$

$$\therefore \quad \frac{dH_2}{dQ} = -\frac{H_2}{Q}$$

This may alternatively be proved by differentiating the logarithmic equation of $P = wQH$ which gives

$$\frac{\delta P}{P} = \frac{\delta Q}{Q} + \frac{\delta H}{H} = 0$$

at maximum power, i.e. when

$$\frac{dH}{dQ} = -\frac{H}{Q}$$

Noting that:

$$H_2 = H_1 - KQ^2$$

$$\frac{H_2}{Q} = \frac{H_1 - KQ^2}{Q}$$

and:

$$\frac{dH_2}{dQ} = -2KQ$$

Hence the condition for maximum power is:

$$2KQ = \frac{H_1 - KQ^2}{Q}$$

i.e.

$$H_1 = 3KQ^2$$

Since $KQ^2$ is the head loss, **the power available is a maximum when one-third of the static head $H_1$ is sacrificed in friction.** The corresponding values of $H_2$ and $Q$ are given by:

$$H_2 = \tfrac{2}{3}H_1 \quad \text{and} \quad Q^2 = \frac{H_1}{3K}$$

so that the maximum power available is:

$$P_2 = wQH_2 = w\sqrt{\left(\frac{H_1}{3K}\right)}\frac{2}{3}H_1$$

i.e.

$$P_{\max} = \frac{2}{3\sqrt{3}}\frac{w}{\sqrt{K}}H_1^{3/2}$$

**and this is obtained with an efficiency of 2/3, i.e. 66·7 per cent.**

It should be noted that in a Pelton wheel installation the kinetic energy of the issuing jet performs useful work. In many other applications, particularly in lines to positive-displacement machines, the valve simply acts as a throttle, causing additional loss. In such cases $K$ must include the throttling loss at the valve, together with the exit loss from the line, if the valve is well upstream of this section.

EXAMPLE 12.10 (i)

*Neglecting all losses except pipe friction find an expression for K referring to Fig. 12.10(a). Hence, deduce an alternative formula to the one given in Section 12.10 for the maximum power available in the jet. Also deduce an expression for the ratio of area of jet to area of supply line when the power delivered is a maximum. The frictional coefficient for the pipe-line is λ = 4f in Darcy's formula.*

If

$$h_f = \frac{4fl}{d_1}\frac{u_1^2}{2g} = \frac{\lambda l}{d_1 2g}\left(\frac{Q}{a_1}\right)^2 = KQ^2$$

we deduce that

$$K = \frac{\lambda l}{d_1 2g a_1^2}$$

Hence, the maximum power available in the jet is:

$$P_2 = wQH_2 = \frac{2}{3\sqrt{(3K)}} wH_1^{3/2}$$

$$= \frac{2w}{3\sqrt{3}} H_1^{3/2} \sqrt{\left(\frac{2gd_1 a_1^2}{\lambda l}\right)}$$

i.e.
$$P_2 = \frac{2\sqrt{2}}{3\sqrt{3}} wa_1 H_1^{3/2} \sqrt{\left(\frac{gd_1}{\lambda l}\right)}$$

Maximum power in jet occurs when $H_2 = \frac{2}{3}H_1$ or when $Q^2 = H_1/(3K)$, i.e.

$$(a_2 u_2)^2 = \frac{H_1}{3K}$$

Hence, since

$$H_1 = \frac{u_2^2}{2g} + \frac{\lambda l}{d_1} \frac{u_1^2}{2g}$$

$$= \frac{u_2^2}{2g}\left(1 + \frac{\lambda l}{d_1}\left(\frac{a_2}{a_1}\right)^2\right)$$

then for maximum power

$$a_2^2 \left(\frac{2gH_1}{1 + \frac{\lambda l}{d_1}\left(\frac{a_2}{a_1}\right)^2}\right) = \frac{H_1}{3\left(\frac{\lambda l}{2gd_1 a_1^2}\right)}$$

i.e.
$$1 + \frac{\lambda l}{d_1}\left(\frac{a_2}{a_1}\right)^2 = \frac{3\lambda l}{d_1}\left(\frac{a_2}{a_1}\right)^2$$

or
$$\left(\frac{a_2}{a_1}\right) = \sqrt{\left(\frac{d_1}{2\lambda l}\right)} \quad \text{or} \quad \sqrt{\left(\frac{d_1}{8fl}\right)}$$

## EXAMPLE 12.10 (ii)

*Referring to Fig. 12.10(a), the available head $H_1$ is 30 m and entry to the pipe-line (of length 300 m, diameter $d_1$ 150 mm and frictional coefficient $\lambda = 4f = 0\cdot03$) is bell-mouthed. The entry loss is $0\cdot05\ u_1^2/2g$, and there are four bends in the pipe each responsible for a loss of head of $0\cdot1\ u_1^2/2g$. Calculate the rate of discharge (a) without a nozzle, (b) with a nozzle 75 mm diameter and loss of head in the nozzle of $0\cdot05\ u_2^2/2g$, and (c) find the diameter of nozzle which will give maximum power in the jet, neglecting the nozzle loss.*

(a) Without nozzle, the loss of head is due to entry, friction and bends. Hence the energy equation is:

$$H = \frac{u_1^2}{2g} + \Sigma h_L = \frac{u_1^2}{2g}(1 + \Sigma k)$$

$$\therefore u_1 = \sqrt{\left(\frac{2gH}{1+\Sigma k}\right)} = \sqrt{\left(\frac{19\cdot61 \times 30 \text{ m}^2/\text{s}^2}{1 + 0\cdot05 + 0\cdot4 + \frac{0\cdot03}{0\cdot15} \times 300}\right)} = \sqrt{\left(\frac{588\cdot3 \text{ m}^2/\text{s}^2}{61\cdot45}\right)}$$

$$= 3\cdot08 \text{ m/s}$$

and
$$Q = a_1 u_1 = \frac{\pi}{4} \times 0\cdot15^2 \text{ m}^2 \times 3\cdot08 \text{ m/s} = \mathbf{3\cdot27} \text{ m}^3/\text{min}$$

(b) With a nozzle from which the jet issues with a velocity $u_2$,

$$H = \frac{u_2^2}{2g} + \text{losses} = \frac{u_2^2}{2g} + \frac{\lambda l}{d_1}\frac{u_1^2}{2g} + 0 \cdot 05\frac{u_1^2}{2g} + 4 \times 0 \cdot 1\frac{u_1^2}{2g} + 0 \cdot 05\frac{u_2^2}{2g}$$

$$= \frac{u_2^2}{2g}\left\{(1 + 0 \cdot 05) + \left(\frac{a_2}{a_1}\right)^2\left(\frac{\lambda l}{d_1} + 0 \cdot 05 + 0 \cdot 4\right)\right\}$$

Hence,

$$u_2 = \sqrt{\left[\frac{588 \cdot 3 \text{ m}^2/\text{s}^2}{1 \cdot 05 + \frac{1}{16}\left(\frac{0 \cdot 03 \times 300}{0 \cdot 15} + 0 \cdot 05 + 0 \cdot 4\right)}\right]} = 11 \text{ m/s}$$

and

$$Q = a_2 u_2 = \frac{\pi}{4} \times \frac{7 \cdot 5^2}{10^4} \text{ m}^2 \times 11 \text{ m/s} = \mathbf{2 \cdot 916 \text{ m}^3/\text{min}}$$

(c) For maximum power, noting that $l_e$ (Section 12.7) $= 302$ m.

$$\frac{a_2}{a_1} = \sqrt{\left(\frac{d_1}{2\lambda l_e}\right)} = \sqrt{\left(\frac{0 \cdot 15 \text{ m}}{2 \times 0 \cdot 03 \times 302 \text{ m}}\right)} \simeq \frac{1}{11}$$

Example 12.10 (iii)

*Calculate the power which can be delivered to a factory* 6 km *distant from a hydraulic power-house through* 3 *horizontal pipes each* 150 mm *diameter laid in parallel, if the inlet pressure is maintained constant at* 5 MPa *and the efficiency of transmission is* 94 per cent.

*If one of the pipes becomes unavailable what increase in pressure at the power station would be required to transmit the same power at the same delivery pressure as before, and what would be the efficiency of transmission under these conditions?*

*Take* $f = 0 \cdot 007\ 5$ *in the formula* $h_f = (4fl/d)(u^2/2g)$ *in both cases.*

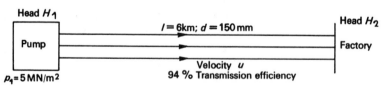

**Fig. 12.10(c).** *Pipes in parallel*

Power input to pipe-lines is $P_i = wQH_1$ and power output is $P_o = wQH_2$ where $Q$ is the rate of volumetric flow of water along the pipes and $H_1$ and $H_2$ are the total heads at entry and exit respectively. For one pipe $H_1 = H_2$ + loss, i.e.

$$\frac{p_1}{w} + \frac{u^2}{2g} = \frac{p_2}{w} + \frac{u^2}{2g} + \frac{4fl}{d}\frac{u^2}{2g}$$

Also, efficiency

$$\epsilon = \frac{H_2}{H_1} = 0 \cdot 94$$

Hence
$$H_1 = 0.94H_1 + \frac{4fl}{d}\frac{u^2}{2g}$$

or
$$0.06\left(\frac{p_1}{\rho} + \frac{u^2}{2}\right) = \frac{4fl}{d}\frac{u^2}{2}$$

i.e.
$$\frac{u^2}{2}\left\{\frac{4fl}{d \times 0.06} - 1\right\} = \frac{p_1}{\rho} \quad \frac{u^2}{2}\left\{\frac{4 \times 0.007\ 5}{0.15 \times 0.06} \times 6 \times 10^3 - 1\right\} = 10^4 u^2$$

and
$$u^2 = \frac{5\ \text{MN/m}^2}{1\ \text{Mg/m}^3} \times \frac{1}{10^4}\left[\frac{\text{kg m}}{\text{N s}^2}\right] = 0.5\ \text{m}^2/\text{s}^2$$

∴
$$u = 0.707\ \text{m/s}$$

Thus, the quantity delivered by three pipes is

$$Q = 3 \times \frac{\pi}{4} \times 0.15^2\ \text{m}^2 \times 0.707\ \frac{\text{m}}{\text{s}}\left[\frac{60\ \text{s}}{\text{min}}\right] = 2.25\ \text{m}^3/\text{min}$$

Also,
$$H_1 = \frac{p_1}{w} + \frac{u^2}{2g} = \frac{5\ \text{MN/m}^2}{9.807\ \text{kN/m}^3} + \frac{0.5\ \text{m}^2/\text{s}^2}{19.61\ \text{m}^2/\text{s}^2} = 510\ \text{m}$$

Hence the input of power to the pipe-lines is:

$$P_i = wQH_1 = 9.807\ \frac{\text{kN}}{\text{m}^3} \times \frac{2.25\ \text{m}^3}{60\ \text{s}} \times 510\ \text{m} = 188\ \text{kW}$$

and the power delivered to the factory is $0.94 \times 188\ \text{kW} = \textbf{177 kW}$.
From (1) the pressure head of the water delivered is:

$$\frac{p_2}{w} = \frac{p_1}{w} - \frac{4fl}{d}\frac{u^2}{2g} = 510\ \text{m} - 30.6\ \text{m} = 479.4\ \text{m}$$

If the same quantity is transmitted through two pipes instead of three then the velocity in each of the two pipes is:

$$\frac{1}{2} \times \frac{2.25\ \text{m}^3}{60\ \text{s}} \times \frac{4}{\pi} \times \frac{10^4}{225\ \text{m}^2} = 1.06\ \text{m/s}$$

and the head lost in friction is:

$$\frac{4 \times 0.007\ 5 \times 6\ \text{km}}{0.15\ \text{m}} \times \frac{1.06^2\ \text{m}}{19.61} = 69.5\ \text{m} = h_L$$

Hence, the head at entry to each pipe is:

$$H_1 = H_2 + h_L = \frac{p_2}{w} + \frac{u^2}{2g} + h_L = (479.4 + 0.01 + 69.5)\ \text{m} = 548.9\ \text{m}$$

and
$$p_1 = w\left(H_1 - \frac{u^2}{2g}\right) = 9.807\ \text{kN/m}^3 \times 548.9\ \text{m} = 5.38\ \text{MN/m}^2$$

Thus, the increase in pressure at the power station would need to be

$$(5.38 - 5.05)\ \text{MN/m}^2 = \textbf{330 kN/m}^2$$

and the efficiency of transmission would be

$$\epsilon = \frac{H_2}{H_1} = \frac{479 \cdot 41}{548 \cdot 9} = 87 \cdot 4 \ per \ cent$$

## 12.11. Nozzles

Nozzles are specially shaped contractions or short streamlined exits fitted to the ends of pipes in order to increase the velocity of the fluid—the kinetic head being increased at the expense of the pressure head by a stable process with low losses. Fig. 12.11(a) illustrates such a nozzle

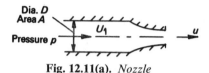

**Fig. 12.11(a).** *Nozzle*

formed by a converging channel for an incompressible fluid through which the rate of volumetric flow of incompressible fluid is $Q = A U_1 = au$.

The coefficient of velocity $C_u$ of the nozzle is the ratio of the actual and theoretical issuing velocities, i.e. $C_u = u/\sqrt{(2gH)}$ or

$$C_u = \frac{u}{\sqrt{\left[2\left(\dfrac{p}{\rho} + \dfrac{U_1^2}{2}\right)\right]}}$$

whose value is usually about 0·975.

Thus, using the energy equation for the nozzle, namely:

$$\text{Head at entrance} = \text{Head at exit} + \text{Loss of head}$$

the loss of head in the nozzle is

$$h_L = H - \frac{u^2}{2g} = \left(\frac{1 - C_u^2}{C_u^2}\right)\frac{u^2}{2g} = k\frac{u^2}{2g}$$

in which $k$ is usually about 0·05

Hence,
$$\frac{p}{\rho} + \frac{U_1^2}{2} = \frac{u^2}{2}(1+k)$$

or
$$\frac{p}{\rho} = \frac{u^2}{2}\left(1+k - \frac{a^2}{A^2}\right)$$

and
$$Q = au = a\sqrt{\left(\frac{2(p/\rho)}{1+k-(a^2/A^2)}\right)}$$

where
$$k = \frac{1 - C_u^2}{C_u^2}$$

From the latter formula we can deduce the value of $k$ (and hence of

$C_u$) by measuring the gauge pressure $p$ just behind the nozzle and the rate of volumetric flow $Q$. Afterwards, such a nozzle can be used for measuring discharges merely by observing $p$ for use in the formula.

EXAMPLE 12.11 (i)

*Find the coefficient of velocity of a nozzle of exit diameter 50 mm when fitted to a pipe of 100 mm diameter if 3 m³/min of water flows through when the gauge pressure just behind the nozzle is 318 kN/m². Also, calculate the discharge rate when the gauge pressure is 275 kN/m², and the power of the jet.*

$$1+k-\frac{a^2}{A^2} = 2\frac{p}{\rho}\frac{a^2}{Q^2}$$

$$1+k-\tfrac{1}{16} = \frac{2\ \text{m}^3}{10^3\ \text{kg}}\times 318\ \frac{\text{kN}}{\text{m}^2}\times\left(\frac{\pi}{4}\cdot\frac{25}{10^4}\ \text{m}^2\right)^2\times\frac{3\,600\ \text{s}^2}{9\ \text{m}^6}\left[\frac{\text{kg m}}{\text{N s}^2}\right]$$

i.e.
$$k+\tfrac{15}{16} = 0{\cdot}981$$

Hence $k = 0{\cdot}981 - 0{\cdot}937 = 0{\cdot}044$ and, since $(1/C_u^2)-1 = k$, the coefficient of velocity

$$C_u = \sqrt{\left(\frac{1}{1{\cdot}044}\right)} = \mathbf{0{\cdot}977}$$

If the pressure gauge reads 275 kN/m², the rate of discharge is:

$$Q = a\sqrt{\left(\frac{2p/\rho}{1+k-(a^2/A^2)}\right)} = \frac{\pi}{4}\frac{25}{10^4}\ \text{m}^2\sqrt{\left\{\frac{2\times 275\times 10^3\ \text{N/m}^2}{(1+0{\cdot}044-\tfrac{1}{16})}\times\frac{\text{m}^3}{10^3\ \text{kg}}\left[\frac{\text{kg m}}{\text{N s}^2}\right]\right\}}$$

$$= 0{\cdot}046\ 43\ \text{m}^3/\text{s}\quad\text{or}\quad \mathbf{46{\cdot}43\ litre/s}$$

Power of jet is

$$P = \frac{\dot{m}u^2}{2} = \tfrac{1}{2}\rho a u^3 = \frac{\rho}{2}\frac{Q^3}{a^2} = \frac{10^3\ \text{kg}}{2\ \text{m}^3}\left(\frac{4{\cdot}643}{10^2}\right)^3\frac{\text{m}^9}{\text{s}^3}\left(\frac{4}{\pi}\frac{10^4}{25\ \text{m}^2}\right)^2$$

$$= 12{\cdot}97\times 10^3\ \frac{\text{kg m}^2}{\text{s}^3}\left[\frac{\text{Ns}^2}{\text{kg m}}\right]\left[\frac{\text{Ws}}{\text{Nm}}\right] = \mathbf{12{\cdot}97\ kW}$$

EXAMPLE 12.11 (ii)

*A pump feeds two hoses each of which is 45 m long and is fitted with a nozzle. Each nozzle has a coefficient of velocity of 0·97 and discharges a 40 mm diameter jet of water at 25 m/s, when the nozzle is at the same level as the pump.*

*If the power lost in overcoming friction in the hoses is not to exceed 20 per cent of the hydraulic power available at the inlet ends of the hoses, calculate:*
*(a) the diameter of the hoses taking $f = 0{\cdot}007$ in the formula*

$$h_f = \frac{4fl}{d}\frac{u^2}{2g}$$

*(b) the power required to drive the pump if its efficiency is 70 per cent and it draws its water supply from a level 3 m below that of the nozzle.*

The principles involved in this problem are those of:
(i) conservation of energy resulting in equations

$$H_1 = H_2 + h_f$$

for the hose and

$$H_2 = \frac{p}{w} + \frac{U^2}{2g} = \frac{u_t^2}{2g} = \frac{u^2}{2g} + h_L$$

for the nozzle

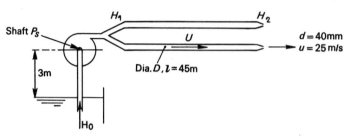

Fig. 12.11(b). *Hoses in parallel*

(ii) continuity of mass flow resulting in equation

$$Q = \frac{\pi}{4} D^2 U = \frac{\pi}{4} d^2 u$$

for one hose and nozzle
(iii) definitions

$$C_u = \frac{u}{u_t}$$

pump efficiency

$$\epsilon = \frac{wQ_{tot}(H_1 - H_0)}{P_s}$$

Thus for one hose

$$Q = \frac{\pi}{4} \frac{16 \text{ m}^2}{10^4} \times \frac{25 \text{ m}}{\text{s}} = \frac{\pi}{10^2} \text{ m}^3/\text{s}$$

The theoretical velocity of the jets is:

$$u_t = \frac{u}{C_u} = \frac{25}{0 \cdot 97} \text{ m/s} = 25 \cdot 78 \text{ m/s}$$

Hence, the head at entry to the nozzle is:

$$H_2 = \frac{u_t^2}{2g} = 33 \cdot 9 \text{ m}$$

The head lost in friction in each hose is:

$$h_f = H_1 - H_2 = \frac{20}{100} H_1$$

since the power in the water at inlet to each hose is $P_1 = wQH_1$.

Thus, $H_1 = \frac{5}{4}H_2 = \frac{5}{4} \times 33 \cdot 9$ m $= 42 \cdot 5$ m.

(a) For each hose

$$h_f = \frac{H_1}{5} = 8 \cdot 5 \text{ m} = \frac{4fl}{DA^2}\frac{Q^2}{2g}$$

$$\therefore \quad A^2 D = \left(\frac{\pi}{4}\right)^2 D^5 = \frac{4 \times 0 \cdot 007 \times 45 \text{ m}}{8 \cdot 5 \text{ m} \times 19 \cdot 61 \text{ m/s}^2} \times \left(\frac{\pi}{10^2}\frac{\text{m}^3}{\text{s}}\right)^2 = \frac{0 \cdot 076\pi^2}{10^5} \text{ m}^5$$

Hence $\quad D^5 = \dfrac{16 \times 0 \cdot 076 \text{ m}^5}{10^5} = 1 \cdot 21 \times 10^{10} \text{ mm}^5$ and $D = 104$ mm.

(b) $H_1 = 42 \cdot 5$ m and the quantity delivered by two nozzles is

$$Q_{\text{tot}} = \frac{2\pi}{10^2}\frac{\text{m}^3}{\text{s}}$$

Hence, the shaft power of the pump is:

$$P_s = \frac{wQ_{\text{tot}}(H_1 - H_0)}{\epsilon} = 9 \cdot 807 \frac{\text{kN}}{\text{m}^3} \times \frac{2\pi}{10^2}\frac{\text{m}^3}{\text{s}} \times \frac{(42 \cdot 5 + 3)}{0 \cdot 7} \text{ m} \left[\frac{\text{kW s}}{\text{kN m}}\right]$$

$$= 40 \text{ kW}$$

## 12.12. Capacity effects

The behaviour of hydraulic (or electric) circuits cannot be described completely in terms of their resistances, particularly under transient or changing conditions. For example, although a simple stand-pipe (or surge tank), such as that shown in Fig. 12.12(a) may be ignored for steady

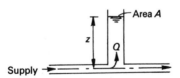

Fig. 12.12(a). *Stand-pipe or surge tank*

flows, it may have a profound effect on the pressures developed during surges. Unless a surge tank is provided near machinery, destructive pressures might be generated if, say, a governor of a turbine or Pelton wheel suddenly reduces the flow of water in a long pipeline. Water rises in the stand-pipe or surge tank and creates a pressure opposing flow; and water may flow over the top if the pressure becomes too high. If, owing to a surge, fluid enters it at a rate $Q$, the level will, during a time $\delta t$, rise by an amount:

$$\delta z = \frac{Q}{A}\delta t$$

*Ignoring, for the moment, inertia effects in the stand-pipe*, the corresponding static pressure in the line ($p = wz = \rho gz$) will rise at a rate which is proportional to $Q$:

$$\frac{dp}{dt} = w\frac{dz}{dt} = \frac{w}{A}Q = \frac{1}{C}Q$$

Recalling that $p$ is analogous to e.m.f. and $Q$ to current, electrical engineers (having in mind the definition of capacitance as $C = I/(dE/dt)$) will observe that $C (= A/w = A/\rho g)$ is a measure of the capacitance of the stand-pipe. This is a piece of descriptive terminology which hydraulic engineers can cheerfully endorse since, if the volume of fluid entering the pipe in a time $\delta t$ is denoted by $\delta V$:

$$C = \frac{Q\delta t}{\delta p} = \frac{\delta V}{\delta p}$$

*Hence C is a measure of the (small) volume required to produce a corresponding (small) unit rise in pressure.* Typical units for $C$ are thus $m^3/(N/m^2)$, a $1$ $m^2$ stand-pipe in a hydraulic system having a capacitance of $1$ $m^2/(9 \cdot 807$ $kN/m^3)$, i.e. $0 \cdot 102$ $m^3/(kN/m^2)$ or simply $0 \cdot 102$ $m^5/kN$.

Such a device or 'capacity column', when 'teed' into a line under pressure (which is equivalent to an earthed condenser) can be very valuable in limiting the surge pressures, due for example to sudden closure of a valve downstream. Equally well, when such a valve is suddenly opened, thus making a large transient demand on the line, the stand-pipe (or surge tank in the case of the large amount of water required when a turbine is started up) will discharge fluid into the line and so reduce the fall in pressure.

For high pressure, or 'suction' applications, the stand-pipe is sealed with a suitable volume of air trapped in it as in Fig. 12.12(b). The rate of

**Fig. 12.12(b).**  *Air vessel*

change of pressure with fluid volume $(dp/dV = 1/C)$ for such an air vessel is almost entirely due to the compressibility of the air—the gravitational changes in potential energy being relatively small. If the bulk modulus of the gas is $\kappa$ (see Section 2.3, on p. 30):

$$\kappa = -\frac{\delta p}{\delta V/V}$$

and we note that the *decrease* in volume of the gas $(-\delta V)$ is equal to the *increase* in the liquid volume $\delta V = Q\delta t$.

The capacitance of the air vessel in the sense defined above, may be expressed as:

$$C = \frac{\delta V}{\delta p} = -\frac{V}{\kappa}$$

in which $V$ denotes the volume of air trapped in the vessel. For rapid expansions and contractions conditions may be assumed approximately adiabatic and reversible. Hence, $pV^\gamma = $ constant so that $\kappa = \gamma p$; whereas. for very slow changes we may assume the isothermal law $pV = $ constant so that $\kappa$ may approach the value $p$. More generally, if the $p$, $V$ relationship for the air is: $pV^n = $ constant

$$C = \frac{\delta V}{\delta p} = -\frac{V}{np} = -\frac{V}{\kappa}, \quad \text{or} \quad \kappa = np$$

Thus, if 49 dm³ of air is trapped at 3·5 MN/m² abs. and $n = \gamma = 1·4$, $C = 0·049 \text{ m}^3/(1·4 \times 3·5 \text{ MPa}) = 10^{-2} \text{ m}^3/\text{MPa}$.

In oil hydraulic systems, dissolved air can be particularly troublesome, so that a flexible bag or a piston may be used as in Fig 12.12(c). This

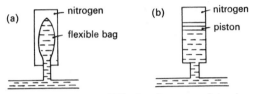

Fig. 12.12(c). *Small 'hydraulic accumulators'*

separates the gas (usually nitrogen to avoid the possibility of compression–ignition if leakage should occur) from the oil and such a device, which is extensively used in aircraft as a convenient form of energy storage, is referred to as a hydraulic accumulator. Its relatively low capacitance ($10^{-2}$ m³/MPa) in the case given above, implies that the pressure falls rapidly with demand. This can be overcome by using dead-weights to pressurize the piston on heavy land installations— Tower Bridge, London, provides the classical example of this. Since such a dead-weight accumulator receives and delivers fluid virtually at constant pressure whatever the volume of fluid contained within it (i.e. $dp/dV$ is zero), its 'capacitance' is infinite. The need for accumulators became prominent after the invention of hydraulic presses supplying heavy loads intermittently. As its volumetric capacity is limited, trips are fitted so that, when the accumulator is fully charged, the pump is stopped, and when discharged, the pump is restarted.

Occasionally, as in pneumatic or hydraulic servo-control systems, a capacitance may be connected into a line, so that both sides of a diaphragm, piston or bellows are under pressure as illustrated in Fig. 12.12(d).

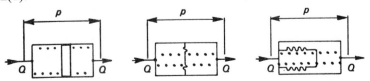

Fib. 12.12(d). *Capacitances in control systems*

## 12.13. Inertia effects

As mentioned in Section 6.3, if a plug of fluid moving through a pipe is being either speeded up say, by the opening of a valve in the pipe-line (or slowed down by gradual closing of a valve) there must be a resultant force on it, to produce the necessary mass acceleration (or deceleration). This must, neglecting viscosity or friction, be brought about by a pressure difference, $\delta p$ say, across the ends of the plug, see Fig. 12.13(a).

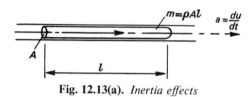

**Fig. 12.13(a).** *Inertia effects*

The mass of the plug is $\rho$ times its volume $Al$, so that, if the acceleration is $du/dt$:

$$F = \delta p A = \rho A l \frac{du}{dt}$$

i.e.
$$\delta p = \rho l \frac{du}{dt} \quad \text{and} \quad \frac{\delta p}{w} = \frac{l}{g} \frac{du}{dt}$$

In terms of the discharge rate $Q = uA$, i.e.

$$\delta p = \frac{\rho l}{A} \frac{dQ}{dt}$$

The change of head of fluid is:

$$h = \frac{\delta p}{w} = \frac{l}{g} \frac{du}{dt} = \frac{l}{g} a$$

i.e.
$$i = \frac{h}{l} = \frac{a}{g}$$

as in Section 6.7, p. 132.

Comparison with the corresponding electrical phenomenon $[E = L(di/dt)]$ suggests that we should write:

$$\delta p = L \frac{dQ}{dt}$$

and refer to $\rho l/A = L$ as the 'inductance' of the line. Its numerical value is, in hydraulic problems, equal to the pressure difference required to cause unit rate of change in the flow rate $Q$. It is therefore a measure of the 'inertia' of the line, and suggests that its effect is most marked with dense fluids in lines which are either long ($l$ large) or restricted in area ($A$ small). Engineers with an electrical background will naturally represent such a line by the symbol shown in Fig. 12.13(b).

Thus, if the closing of a valve in a pipe-line is carried out such that the deceleration $a$ is constant then the inertia head is $h_i = (l/g)a$. This will remain constant until the valve is completely shut, after which the pressure in the pipe-line will reduce and settle down to the hydrostatic pres-

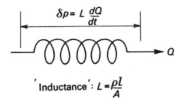

$$\delta p = L\,\frac{dQ}{dt}$$

$$\text{'Inductance'}: L = \frac{\rho l}{A}$$

**Fig. 12.13(b).** *Inductance of a pipe-line*

sure as indicated in Fig. 12.13(c). From this it will be seen that the hydraulic gradient when the valve is fully open is TV and that as the valve is gradually closed the inertia head $h_i$ (which remains constant if $a$ is constant) changes the pressure gradient from the position TA on commencement of closure to OB at the moment of complete closure (the kinetic head $u^2/2g$ having been reduced to zero) and finally to the steady hydrostatic pressure represented by the horizontal line TC.

*In the slow closing of a valve the whole mass of fluid in the pipe-line is decelerated and the inertia pressure* $\delta p = \rho l(du/dt)$ *depends on l and the*

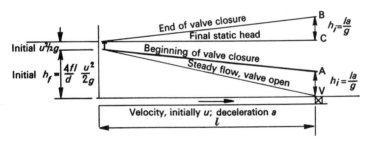

**Fig. 12.13(c).** *Hydraulic gradient during deceleration*

*time taken in closing the valve, whereas in the sudden closing of a valve the inertia pressure created is independent of l and t and depends only on the initial velocity of the fluid and the velocity of a sound wave in the fluid* $[\delta p = u\,\sqrt{(\rho k)}]$ *as will be seen in the next section.*

EXAMPLE 12.13 (i)

*The speed of flow of water in a pipe 60 m long is 3 m/s. Calculate the rise of pressure due to the inertia of the water if the stream be stopped in half a second by closing a valve at the outflow end such that the retardation of the water is constant. What would happen if the closing of the valve did not produce uniform retardation?*

The rise in pressure

$$\delta p = \frac{\rho l}{A} \frac{dQ}{d} = \rho l \frac{du}{dt}$$

$$= \frac{1 \text{ Mg}}{\text{m}^3} \times 60 \text{ m} \times \frac{3 \text{ m}}{0 \cdot 5 \text{ s}^2} \left[ \frac{\text{N s}^2}{\text{kg m}} \right] = 360 \text{ kPa}$$

If the closing of the valve did not produce uniform retardation, the rise in pressure would be greater according to the maximum value of the retardation

$$\frac{du}{dt} \quad \text{or} \quad \frac{dQ}{dt}$$

## 12.14. Water hammer

Hydraulic circuit analysis under transient conditions is, in general, difficult without the aid of a computer. There are two major reasons for this. The first is that the resistance law is not linear, as in electrics, but quadratic. In addition the resistive, elastic (capacitive) and inertia (inductive) effects are distributed throughout the lines and can be tackled in terms of lumped blocks only if the number of the latter is made large. In terms of the latter idea, we imagine a moving column of fluid to be analogous to a train of trucks, each of which possesses inertia, experiences

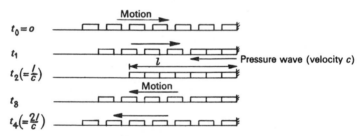

**Fig. 12.14(a).** *Water hammer and reflection of shock wave*

friction, and is separated from its neighbour by a spring which represents the fluid elasticity, or its effective value when expansion and contraction of the pipe is taken into account.

We can, with the aid of this analogy, discuss the pressure rise due to sudden and complete valve closure in a pipe-line. This corresponds to the trucks striking rigid buffers, see Fig. 12.14(a).

The buffers would experience an impulse each time a truck were stopped. If the number of the latter were to be increased indefinitely to correspond to the case of the sudden arresting of a column of fluid, say, by quickly closing a valve, a steady pressure would be experienced during the stoppage and a pressure wave or shock wave transmitted along the fluid in the pipe. The magnitude of the rise in pressure may be deduced quite simply by considering the energy changes of the system.

The kinetic energy per unit volume of moving fluid is, initially: $e = \frac{1}{2}\rho u^2$, and when the whole column of fluid has been stopped (as at time $t_2$ in Fig. 12.14(a)) this energy will be stored uniformly throughout the length of the fluid as strain energy if the pipe is assumed rigid. It is a simple exercise in strength of materials to show that if a material or fluid has a bulk modulus $\kappa$, the strain energy per unit volume, i.e. the resilience of the fluid corresponding to a pressure rise $\delta p$ is $e = \frac{1}{2}(\delta p)^2/\kappa$.

Hence, *assuming in the first instance that the pipe remains rigid* and neglecting the small amount of energy dissipated by friction while the pressure wave is advancing to the other end of the column, we may equate the two energies per unit volume to give the rise in pressure which is independent of the size and length of pipe, namely:

$$\delta p = u \sqrt{(\rho\kappa)} = \rho u \sqrt{\frac{\kappa}{\rho}} = \rho u \times \textbf{velocity of sound in the fluid}$$

EXAMPLE 12.14 (i)

*Find the rise in pressure caused by the sudden closing of a valve in a pipe-line in which water is flowing at the rate of 3 m/s, assuming water to have a bulk modulus of 2·07 GPa and a density of 1 Mg/m³.*

Rise in pressure

$$\delta p = u \sqrt{(\rho\kappa)} = 3 \,\frac{\text{m}}{\text{s}} \sqrt{\left(\frac{10^3 \text{ kg}}{\text{m}^3} \times 2{\cdot}07 \times \frac{10^9 \text{ N}}{\text{m}^2} \left[\frac{\text{N s}^2}{\text{kg m}}\right]\right)}$$

$$= 4{\cdot}32 \text{ MN/m}^2 \quad \text{or} \quad 4{\cdot}32 \text{ MPa}$$

which is the equivalent of a hammer blow.

EXAMPLE 12.14 (ii)

*If water flowing in a pipe is suddenly stopped by closure of a valve find the velocity of the interface between the moving and stationary fluid—i.e. of the pressure wave or shock wave in the column of water.*

Since the rise in pressure $\delta p$ on a fluid is connected with bulk modulus $\kappa$ and increase in volume $\delta V$ by the expression

$$\delta p = \kappa \left(-\frac{\delta V}{V}\right)$$

and density of fluid is $\rho = m/V$, then the increase in density ($\delta\rho$) of fluid is given by:

$$\delta\rho = -mV^{-2}\delta V = -\frac{m}{V}\frac{dV}{V} = \frac{\rho}{\kappa}\delta p \quad \text{and} \quad \frac{\delta\rho}{\rho} = \frac{\delta p}{\kappa}$$

From Fig. 12.14(b) we deduce that the mass of fluid flowing into and arrested in length $\delta s$ in time $\delta t$ and which increases the density by $\delta\rho$ is $\rho A u \delta t = A \delta s \delta\rho$. Hence, the velocity of the pressure wave receding from the valve is:

$$c = \frac{ds}{dt} = \frac{\rho u}{\delta\rho} = u\frac{\kappa}{\delta p} = \frac{u\kappa}{u\sqrt{(\rho\kappa)}} = \sqrt{\frac{\kappa}{\rho}}$$

which is the velocity of sound in the fluid. For water this is:

$$\sqrt{\left(\frac{2 \cdot 07 \times 10^9 \text{ N/m}^2}{10^3 \text{ kg/m}^3}\left[\frac{\text{kg m}}{\text{N s}^2}\right]\right)} = 1 \cdot 44 \text{ km/s}$$

From Figs. 12.14(a) and (b) we deduce that a pipe containing fluid whose flow is suddenly arrested by the closure of a valve will contain stationary fluid for an instant (pressing in the direction of the valve) at time $t_2 = l/c$ after sudden closure of the valve and that the rise in pressure $\delta p = u \sqrt{(\rho\kappa)}$ will continue to press on the valve for a time $t_4 = 2l/c$ after the instant of sudden closure of the valve. Then for a

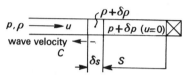

**Fig. 12.14(b).** *Velocity of shock wave*

time $2l/c$ the pressure on the valve is reduced while the shock wave travels to and is reflected from the other end of the pipe. Reflections continue until the pressure wave damps itself out by friction within the fluid.

In the case of water the time $2l/c$ in a pipe 72 m long is:

$$\frac{144 \text{ m}}{1\ 440 \text{ m/s}} = 0 \cdot 1 \text{ s}$$

and if the initial flow velocity is 3 m/s the rise in pressure is 4·32 MPa (see Example 12.14(i)) but only if the valve is closed in less than a time $2l/c$, i.e. 0·1 s in the case of a 72 m length of pipe.

*If the closing of the valve takes longer than a time of $2l/c$ any pressure-wave will have returned to the valve before its complete closure and partly escaped through the valve which will not suffer as great an increase of pressure as $\delta p = u \sqrt{(\rho\kappa)}$ but $\delta p = \rho l(du/dt)$ as proved in Section 12.13.*

EXAMPLE 12.14 (iii)

*Compare the rise in pressure caused by the instantaneous closure of a valve at the end of a pipe-line 60 m long in which water is flowing at 3 m/s with the rise in pressure when the valve is closed in 0·5 second.*

Example 12.14 (i) shows the rise in pressure for instantaneous closure to be 4 320 kPa. This same pressure is generated for any closure time less than

$$\frac{2 \times 60}{1\ 440} = \frac{1}{12} \text{ s}$$

But if the closure is slow, taking 0·5 second, then Example 12.13 (i) shows the rise in pressure to be 360 kPa. Thus, if the time of closure is increased from 0·083 s to 0·5 s the rise in pressure caused is reduced by 3.96 MPa. Hence, the need to avoid sudden stoppages of fluid flow.

The results so far in this section are based on the assumption that the pipe remains rigid. In the more general case, however, increase in hoop stress, $\Delta f$, in the pipe implies that the pipe has an increased resilience amounting to $(\Delta f)^2/2E$ per unit volume of pipe material where $E$ is the

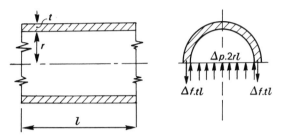

**Fig. 12.14(c).** *Stress induced by inertia pressure or water hammer*

modulus of elasticity. Thus, the energy balance allowing for both the resilience of the pipe and fluid becomes

$$\tfrac{1}{2}\rho u^2(\pi r^2 l) = \frac{1}{2}\frac{(\Delta p)^2}{\kappa}(\pi r^2 l) + \frac{1}{2}\frac{(\Delta f)^2}{E}(2\pi rtl)$$

referring to Fig. 12.14(c).

Thus,    $$\tfrac{1}{2}\rho u^2 = \frac{(\Delta p)^2}{2\kappa} + \frac{(\Delta f)^2}{2E}\frac{2t}{r} = \frac{(\Delta p)^2}{2}\left(\frac{1}{\kappa} + \frac{2r}{Et}\right)$$

and hence the bulk modulus $\kappa$ for fluid in a rigid pipe may be replaced by an equivalent bulk modulus $\kappa'$ in the expression $\tfrac{1}{2}\rho u^2 = (\Delta p)^2/2\kappa'$ in which:

$$\frac{1}{\kappa'} = \frac{1}{\kappa} + \frac{2r}{Et}$$

Since $\kappa'$ is thus less than $\kappa$ the pressure rise due to instantaneous valve closure:

$$\Delta p = u \sqrt{(\rho\kappa')}$$

is reduced below that for a rigid pipe in the ratio of:

$$\sqrt{\frac{\kappa'}{\kappa}} = \frac{1}{\sqrt{\left(1 + \frac{\kappa}{E}\frac{2r}{t}\right)}}$$

The above pressure rise, which is referred to as 'water hammer', is seen to be limited only by the elasticity of the system. The greater the rigidity of the latter, the more intense will be the water hammer.

EXAMPLE 12.14 (iv)

*Develop a formula for the rise in pressure of a pipe through which water is flowing at a constant rate, due to a sudden closing of a valve, allowing both for the expansion of the pipe and the compressibility of the water.*

*A cast-iron pipe is* 150 mm *bore and* 15 mm *thick. Calculate the maximum permissible rate of flow if a sudden stoppage is not to stress the pipe to more than* 46·5 MPa.

*Modulus of compressibility of water,* 2·07 GPa.

*Young's modulus for the pipe material,* 124 GPa.

Using the formula developed in this section, namely

$$\tfrac{1}{2}\rho u^2 = \frac{1}{2}\frac{p^2}{\kappa} + \frac{1}{2}\frac{p^2}{E}\frac{2r}{t}$$

or

$$\tfrac{1}{2}\rho u^2 = \tfrac{1}{2}f^2\frac{t}{r}\left(\frac{t}{r\kappa} + \frac{2}{E}\right)$$

we deduce that

$$u^2 = \frac{(46\cdot5 \text{ MPa})^2}{10^3 \text{ kg/m}^3} \times \frac{1}{5} \times \left(\frac{1}{5 \times 2\cdot07} + \frac{2}{124}\right)\left(\frac{1}{\text{GPa}}\right)$$

$$= \frac{2\,163 \times 10^{12} \text{ Pa}^2}{5 \times 10^3 \text{ kg/m}^3} \times \left(\frac{1}{8\cdot87 \times 10^9 \text{ Pa}}\right)\left[\frac{\text{kg}}{\text{Pa m s}^2}\right] = 48\cdot6 \text{ m}^2/\text{s}^2$$

Hence, $u = 6\cdot97$ m/s.

$\therefore$    Max permissible $Q = au = \dfrac{\pi}{4} \times \dfrac{225}{10^4}$ m$^2 \times 6\cdot97$ m/s $= \mathbf{0\cdot123}$ m$^3$/s

# Exercises on Chapter 12

1. Derive an expression for the loss of energy of fluid when it passes from one pipe to a second pipe by a sudden enlargement. Explain the assumptions made with the reasons for making them.

A horizontal pipe, 75 mm diameter, is joined by a sudden enlargement to a 150 mm diameter pipe. Water is flowing through it at a rate of 0·84 m³/min. Calculate the dissipated power at the enlargement.

If the pressure in the smaller pipe just before the junction is 6 m head, what would be the pressure in the larger pipe well clear of the junction if pipe friction is neglected?

2. In order to regulate the gravitational flow between two reservoirs, a sluice-valve or full-way valve is interposed in the pipe 10 m from the outlet end. This pipe is 1 510 m long and 400 mm diameter, its friction coefficient is 0·004, and the total head drop between the reservoirs is 18 m.

As the valve is progressively closed, from fully-open to fully-closed, the rate of flow will decline. Show this relationship by means of a curve between (a) area of valve opening, and (b) discharge through the system in cubic metres per second. Only three points need be plotted, from estimations of rates of volumetric flow when: (i) valve fully open, (ii) area of valve opening = 0·04 m², (iii) valve fully closed.

Take the coefficient of contraction of partly-open valve as 0·7.

3. A tank containing water and having a constant plan area of 1 m² is emptied to atmosphere through two pipes A of length 9 m and diameter 60 mm and B of length 6 m and diameter 40 mm connected to the bottom of the tank.

The inlets to both pipes are rounded so that the entrance losses are small while both outlets are 4 m below the original level in the tank.

Find the time required for the level to fall 1 m, given that $f = 0·006$ in the formula

$$h_L = \frac{4fl}{d}\frac{u^2}{2g}$$

4. Two sharp-ended pipes of 50 mm and 100 mm diameter, each 30 m long, are connected in parallel between two reservoir tanks whose difference of level is 7·5 m. Find

(a) the flow in m³/min for each pipe and draw the corresponding hydraulic gradients;

(b) the diameter of a single pipe, 30 m long, which will give the same flow as the two actual pipes.

Assume that the entrance loss is $0·5\ u^2/2g$ and take $\lambda = 0·032$ in the formula

$$h_f = \frac{\lambda l}{d}\frac{u^2}{2g}$$

in each case.

5. Two reservoirs whose difference of level is 15 m are connected by a pipe ABC whose highest point B is 2 m below the level in the upper reservoir A. The portion AB has a diameter of 200 mm and the portion BC a diameter of 150 mm, the friction coefficient $\lambda = 4f$ being the same for both portions.

The total length of the pipe is 3 km.

Find the maximum allowable length of the portion AB if the pressure head at B is not to be more than 2 m below atmospheric pressure. Neglect the secondary losses—velocity head in the pipe, loss of head at pipe entry and loss of head at change of diameter.

6. A hydraulic machine is operated by water pressure acting on a ram in a horizontal cylinder of diameter 120 mm. Water is supplied to the cylinder from a hydraulic main maintained at a pressure of 5 MN/m² through a pipe 100 m long and 40 mm bore. Find what resisting force can be overcome by the ram when moving at a uniform velocity of 750 mm/s. Neglect the friction between the ram and the cylinder and assume that the friction coefficient for the pipe is $\lambda = 4f = 0·032$.

If the force resisting the motion of the ram, found above, is assumed to be independent of velocity, and the total mass at the ram is 8 Mg, determine the acceleration of the ram at the instant of starting.

7. To improve the water supply between two points A and B, 3 km apart, two different schemes have been proposed.

The existing 150 mm pipe may be replaced entirely by a 200 mm pipe over the whole distance, or, alternatively, a new 200 mm pipe may be installed over the first half distance, and the old pipes relaid in parallel over the remaining half. The permitted overall loss of head for maximum flow is 15 m in each case.

Determine which scheme will give the better supply, stating the maximum discharge in each case. Friction losses only need be considered.

$$\left.\begin{array}{l} f = 0\!\cdot\!01 \text{ for old pipes} \\ f = 0\!\cdot\!008 \text{ for new pipes} \end{array}\right\} \text{in the formula} \frac{4fl}{d}\frac{u^2}{2g}$$

8. In a long pipe-line carrying crude oil, it is usual to install several pumping stations spaced at intervals along the line, rather than to rely on a single station at the pipe inlet. Why is this? What are the advantages of multiple stations? Illustrate your answer by sketching comparative hydraulic gradients.

A horizontal pipe-line, 400 km long and 0·6 m diameter, is to carry 1 500 m³/h of crude oil of relative density 0·89. Its viscosity is such that in the formula $h = (4fl/d) \times$ velocity head, the value of the friction coefficient $f$ is 0·007. It is stipulated that, at each pumping station, the pressure-difference generated is not to exceed 4 MN/m². How many identical stations would be needed, and what should be the power delivered to the oil in each one?

9. A horizontal pipe, 150 mm diameter and 1·2 km long, has a friction coefficient $\lambda = 4f = 0\!\cdot\!020$.

Water entering the pipe is all drawn off at a uniform rate per unit of length along the pipe. Neglecting losses other than pipe friction, find the quantity of water that enters the pipe when the drop in pressure along the pipe is 250 kN/m².

Also draw the hydraulic gradient assuming the pressure at the pipe entry is 275 kN/m², and from the diagram obtain the pressure half-way along the pipe.

10. A pipe, 600 mm diameter and 6 km long, connects two reservoirs, A and B, whose constant difference of level is 12 m. The friction coefficient $\lambda$ of the pipe is 0·016 = 4f.

A branch pipe 600 mm diameter is taken from a point distant 1·5 km from reservoir A leads to a third reservoir C. A valve on this branch pipe is adjusted so that the quantity of water entering reservoir C is twice the quantity entering reservoir B.

Determine the rate at which water enters the reservoirs B and C.

11. Water is supplied to a double-jet Pelton wheel through a 450 mm pipe 1·8 km long for which the coefficient of friction $f$ is 0·0075.

Given that the gross head is 200 m, calculate (a) the diameter of the jets if the friction loss in the pipe-line is limited to 8 per cent of the gross head and other losses are neglected, (b) the power of the two jets under these conditions.

12. Water from a reservoir flows through a 300 mm diameter pipe, 3 km long, and is discharged through a nozzle which gives a 65 mm diameter jet.

The total head from reservoir level to nozzle outlet is 200 m. The velocity coefficient for the nozzle is 0·975 and the friction coefficient for the pipe is $f = 0·0045$.

Find (a) the velocity of the jet;

(b) the loss of head in the pipe;

and (c) the loss of head on the nozzle.

13. A hose-pipe of diameter $D$ and length $L$ is fitted with a nozzle of diameter $d$ at the end. If the losses in the nozzle and pipe entry, etc., are 10 per cent of the friction loss in the pipe, show that, for any fixed supply head and pipe, the force of the jet from the nozzle is a maximum when

$$\frac{d}{D} = \left(\frac{D}{4·4fL}\right)^{1/4}$$

where $f$ is the friction coefficient of the pipe.

A hose-pipe, 75 mm diameter and 180 m long, discharges water from a nozzle fitted to the end of the pipe. The head at entry to the pipe is 40 m measured above the nozzle, which has a coefficient of velocity of 0·97. The friction coefficient $f$ is 0·009.

Find the rate of volumetric flow if the useful energy of the jet is 70 per cent of the supply head, and all losses except pipe friction and nozzle loss are neglected.

14. A water turbine is supplied from a reservoir 120 m above its own level through a pipe of 300 mm diameter. A Venturi contraction in the pipe, with a throat area equal to half of the pipe area, shows a pressure drop at the throat of 230 mm of water when the turbine is developing 18 kW. At this condition the loss of head in the pipe-line is 10·5 m of water and the turbine discharge to atmosphere is at 6 m/s. Determine the overall efficiency of the plant and find the losses in the turbine and in the discharge as percentages of the total head.

15. A small trailer fire pump delivers 0·55 m³/min of water with overall efficiency of 60 per cent when the pressure heads at the delivery and suction flanges are respectively 70 m above and 3 m below atmospheric pressure.

The water is discharged through a nozzle at the end of 150 m of horizontal 70 mm diameter hose pipe whose coefficient of friction $\lambda$ is 0·024. The suction pipe is also 70 mm diameter. Find

(a) The pressure head at the nozzle in metres of water.

(b) The maximum height to which the jet would rise, assuming this to be 70 per cent of the theoretical due to air resistance and a nozzle velocity coefficient of 0·97.

(c) The diameter of the nozzle assuming its discharge coefficient is unity.

(d) The power required to drive the pump.

16. (a) A hydraulic machine is connected to a constant pressure supply through lines in which the losses are proportional to the velocity squared. Under what conditions is the power available at the machine a maximum?

(b) A hydraulic ram has a diameter of 150 mm and is connected to a 7 MPa supply. Tests show that on no load the ram has a maximum run-out speed of

2·5 m/s all the available pressure then being destroyed in pipe losses. Estimate the maximum power available at the ram.

17. A hydraulic crane has a ram 300 mm diameter and with rope gearing giving a velocity ratio of 8:1. The ram friction may be taken as 12·5 kN and the efficiency of the gearing may be taken as 72 per cent. The crane raises a load of 30 kN through a height of 15 m once every two minutes at a speed of 1·5 m/s.

Calculate the minimum capacity of the accumulator if the pump runs at constant speed. Find also the load on the accumulator ram if its diameter is 350 mm and the packing friction is equivalent to 170 kPa of ram area. Disregard any other losses such as pipe friction, etc.

18. Deduce an expression for the rise of pressure head in a pipe-line of length $L$ when the normal flow velocity $u$ is uniformly decelerated to zero in a time $t$. Hence find the rise in pressure head in a penstock pipe 1 km long and 1·5 m diameter when the flow velocity of 5 m/s is stopped with uniform deceleration in a time of 25 s.

Indicate the usual position and purpose of a surge tank in a long supply pipe from a storage reservoir to a hydraulic turbine.

19. Derive a formula for the pressure rise in a fluid flowing in a pipe when the valve at the end from which the fluid escapes is closed (*a*) slowly; (*b*) very quickly.

Water flows through a steel pipe 200 mm inside diameter and 5 mm thick with a velocity of 2·5 m/s. The length is 100 m.

Calculate the maximum pressure rise if: (*a*) the flow is reduced uniformly to zero in 5 s; (*b*) the valve is closed in a time which may be treated as instantaneous.

Bulk modulus of water, 2·07 GN/m².
Young's modulus of steel, 207 GN/m².

20. Describe briefly the pressure phenomena produced in a pipe by sudden valve closure when the elasticity of the water is taken into account.

The pressure change produced in a water pipe by a sudden decrease of velocity $du$ is given by

$$dp = \rho c\,du$$

where $\rho$ is the density of the water and $c$ is the velocity of sound transmission through water.

(*a*) A valve is placed at the outlet end of a pipe-line 1 km long through which water flows at 2·5 m/s. Find the pressure produced by complete closure of the valve in 1 s and show that this pressure is not affected by the rate of valve closure. How long will the maximum pressure continue to act on the valve?

Take the value for $c$ equal to 1 460 m/s.

(*b*) Also, neglecting effects of shock waves, find the pressures produced by valve closure in 5 and 10 seconds assuming deceleration occurs at uniform rates as the valve closes.

(*c*) From the relationship between pressure rise and time of closure plot a graph of pressure rise against time of closure over the range 2 to 10 s.

## Chapter 13

# Compressibility effects in fluids

'*The atmosphere is an element capable of being compressed within itself when it is struck by something moving at a greater rate of speed than that of its own velocity and it then forms a cloud within the rest of the air.*'  LEONARDO

## 13.1. Equilibrium of a compressible fluid

Although it is convenient and often reasonably accurate, when the pressure differences are not large, to treat air as if it were an incompressible fluid, there are many instances when neglect of compressibility is not justified. When we come to consider the earth's atmosphere for example,

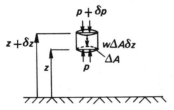

Fig. 13.1. *Equilibrium of compressible fluid*

and realize that each layer must support the considerable weight of air above it, it becomes obvious that those strata nearer to the earth's surface are relatively more compressed, and hence denser. Referring to Fig. 13.1 it is thus *not* permissible to assume that the vertical cylinder of air above the level $z$ is of constant density; it is therefore necessary to restrict ourselves to a small slice, say of thickness $\delta z$, in order to formulate an equilibrium equation. In doing this it is neither desirable nor necessary, to recognize that the pressure at $(z + \delta z)$ is in fact less than that at a height $z$. We merely assume that it is slightly different from the value $p$ at $z$. If $\delta p$ is in fact negative, the mathematics will sort this out for us. Resolving vertically, noting that the weight of the elementary cylinder is the product of its weight per unit volume ($\rho g = w$) and its volume ($\Delta A \delta z$), and that the unbalanced downward force due to the pressure across its ends is $\delta p \Delta A$:

$$\delta p \Delta A + w \Delta A \delta z = 0$$

i.e.
$$\frac{\delta p}{\delta z} = -w = -\rho g$$

The implied assumption of a constant value for $w$ through the slice becomes increasingly accurate as $\delta z$ is decreased indefinitely so that, for an element at rest the expression is exact in the limit, as $\delta z$ tends to zero, when we should write:

$$\frac{dp}{dz} = -w = -\rho g \tag{13.1}$$

This expression cannot be integrated, i.e. summed over a number of slices, until we provide some further assumption which will determine how $\rho g$ varies with height $z$.

## 13.2. Equilibrium of an isothermal atmosphere

Over a limited range of height it may be plausible to assume that the temperature is sensibly constant. Although there are occasions when this is reasonably true near the earth's surface, it is in general valid only above a height of about 10 000 m, for some distance into what is termed the *stratosphere*. For such a region, which is said to be in 'isothermal equilibrium', Boyle's law may be applied:

$$\frac{p}{\rho} = \text{constant}$$

Denoting the corresponding values of the pressure and density at the lower limit, corresponding to the beginning of the stratosphere for example, by $p_0$ and $\rho_0$:

i.e.
$$\frac{p}{\rho} = \frac{p_0}{\rho_0}$$

This expression enables Equation 13.1 to be integrated:

$$\frac{dp}{dz} = -\frac{\rho_0}{p_0} p g$$

i.e.
$$\int_{p_0}^{p} \frac{dp}{p} = -\frac{\rho_0 g}{p_0} \int_{z_0}^{z} dz$$

Hence the pressure $p$ at any point in the stratosphere at a height $z$, is related to that ($p_0$) at the point where the stratosphere begins ($z_0$), by:

$$-\log_e \frac{p_0}{p} = -\frac{\rho_0 g}{p_0}(z-z_0)$$

i.e.

$$\frac{p_0}{p} = e^{\rho_0 g(z-z_0)/p_0}$$

or

$$\frac{p}{p_0} = e^{-\rho_0 g(z-z_0)/p_0}$$

As the atmosphere may be treated as a near perfect gas, for which the equation:

$$p = \rho RT$$

applies, $(p_0/\rho_0)$ may alternatively be written as:

$$\frac{p_0}{\rho_0} = RT_0$$

where $T_0$ is the assumed constant temperature throughout the region. Hence, in these terms, the expression for the pressure variation with height becomes:

$$\frac{p}{p_0} = e^{-g(z-z_0)/RT_0} \tag{13.2}$$

EXAMPLE 13.2 (i)

*If the temperature may be assumed to remain constant up to a height z from the surface of the earth, deduce the corresponding formula relating pressure to height. Find the height at which the pressure would be 735 mm of mercury in air at 15°C if the height of the barometer on the earth's surface is 760 mm of mercury.*

In this problem we are asked to assume that the temperature remains constant from ground level, so that $(z-z_0)$ is simply replaced by $z$ and $p_0$, $\rho_0$, $T_0$, etc., refer to ground-level conditions.

Thus:

$$gz = -RT_0 \log_e \left(\frac{p}{p_0}\right)$$

The molar ideal-gas constant is 8·314 3 J/mol K and the molar mass of air is 29 g/mol. Hence, the specific gas constant of air is

$$\frac{8\cdot314\ 3}{29} \frac{\text{kJ}}{\text{kg K}} = 287 \text{ Nm/kg K}$$

and, since each pressure is proportional to the corresponding barometer height:

$$gz = 0\cdot287 \text{ kJ/kg K} \times (273\cdot1 + 15)\text{K} \log_e \left(\frac{760}{735}\right) = 2\cdot840 \text{ kJ/kg}$$

$$\therefore \quad z = 2\cdot84 \frac{\text{kJ}}{\text{kg}} \times \frac{\text{s}^2}{9\cdot807 \text{ m}} \left[\frac{\text{kg m}^2}{\text{J s}^2}\right]$$

$$= \mathbf{289\cdot5 \text{ m}}$$

### 13.3. Convective equilibrium and standard atmospheres

To obtain a better picture of normal atmospheric conditions near the earth's surface it is necessary to make some more realistic assumption than that of constant temperature. Lord Kelvin introduced the concept of an atmosphere which was in convective equilibrium. By this he implied that if we imagine isolating an element of air in a weightless sac, and then removing it to a new position, it would automatically expand or compress to the appropriate local conditions, without any heat being added to (or removed from) it. If the movement is made slowly and gradually the air will always be supported by the local buoyancy force. In addition, such a gradual change would not involve any friction, i.e. in thermal terms, it would be reversible. The crucial point concerning the assumption of convective equilibrium is that the gas would always be in thermal equilibrium with its new surroundings. Thus, if it were moved nearer the earth's surface the gas would, as it compressed, adopt the higher local pressure and temperature without any heat exchange, i.e. simply by reversible adiabatic compression. Similarly if the air were lifted, the cooling associated with reversible adiabatic expansion would automatically bring it to the same state as the surrounding air. The equation for such reversible adiabatic changes is:

$$\frac{p}{\rho^\gamma} = \text{constant}$$

or, since $p = \rho RT$, as a law relating temperature to pressure:

$$\frac{p}{T^{\gamma/(\gamma-1)}} = \text{constant}$$

Thus, on taking logs, the law becomes:

$$\log p = \frac{\gamma}{\gamma-1} \log T + \log \text{constant}$$

so that differentiation results in:

$$\frac{dp}{p} = \frac{\gamma}{\gamma-1}\frac{dT}{T}$$

Hence the temperature variation with height may be deduced for an atmosphere in convective equilibrium, by using this expression to eliminate $p$ from the equilibrium (Section 13.1, p. 362) equation:

$$\frac{dp}{dz} = -\rho g = -\frac{pg}{RT}$$

$$\frac{\gamma}{\gamma-1}\frac{dT}{T} = -\frac{gdz}{RT}$$

i.e.
$$\frac{dT}{dz} = -\frac{\gamma-1}{\gamma}\frac{g}{R}$$

Since the right-hand side of this equation is a constant the temperature falls uniformly with height at a rate:

$$\frac{dT}{dz} = -\frac{0.4}{1.4} \times 9.807 \frac{m}{s^2} \times \frac{kg\,K}{287\,J}\left[\frac{J\,s^2}{kg\,m^2}\right] = -0.01\ K/m$$

This value of the lapse rate, as $(dT/dz)$ is called, corresponds to air which is in neutral equilibrium, in the sense that a mass of air, if moved, merely stays in its new position. Were the rate of fall of temperature with height greater than this, a mass of air, if lifted, would by adiabatic expansion, arrive at its new position warmer than its surroundings. Thus, as with hot gas from a flue, it will continue to rise; i.e. a sharper lapse rate than the adiabatic, corresponds to unstable conditions. Conversely, a smaller fall of temperature than this implies stability, and a figure of 0.005 K/m may be taken to be representative of average stable atmospheric conditions. Just as the adiabatic law is associated with a constant value of the lapse rate $g(\gamma-1)/\gamma R$, so the polytropic law:

$$\frac{p}{\rho^n} = \text{constant}$$

corresponds to a lapse rate, say:

$$\frac{dT}{dz} = -\frac{g}{R}\frac{(n-1)}{n} = -\lambda$$

Thus a lapse rate of 0.005 K/m implies a polytropic index:

$$n = \frac{1}{1-\lambda R/g} = \frac{1}{1-0.146} = 1.172$$

Such values define a 'standard atmosphere' which, like S.T.P., serve as a convenient set of reference conditions that are reasonably representative of those encountered in practice. Thus, aircraft instruments and performance are related to such a standard and, in so far as local conditions depart from the standard, so the instrument readings or predicted performance will be in error.

EXAMPLE 13.3 (i)

*At sea level the pressure of the air is 101·5 kN/m² and the temperature is 15°C. The temperature decreases uniformly with increase in altitude and is found to be −25°C at an altitude where the pressure is 45·5 kN/m². Estimate the temperature lapse rate and the pressure and density of the air at an altitude of 3000 m.*

The temperature drops from 288 K to 248 K while the pressure drops from 101·5 to 45·5 kPa. Hence, using laws $p/\rho^n = \text{constant}$ and $p = \rho RT$, we deduce:

$$\frac{T_0}{T} = \left(\frac{p_0}{p}\right)^{(n-1)/n}$$

and

$$\frac{n-1}{n} = \frac{\log\left(\frac{288}{248}\right)}{\log\left(\frac{101\cdot5}{45\cdot5}\right)} = \frac{0\cdot064\ 9}{0\cdot347\ 6} = 0\cdot187$$

i.e.                        $n = 1\cdot23$

Since $p/T^{n/(n-1)} = \text{constant}$ $(c)$, then $\log p = n/(n-1)\log T + \log c$ and $dp/p = \{n/(n-1)\}\{dT/T\}$. Also, from static equilibrium of columns of air

$$\frac{dp}{dz} = -w = -\rho g = -\frac{pg}{RT}$$

Hence,

$$dp = \frac{n}{n-1}\frac{pdT}{T} = -\frac{g}{R}\frac{p}{T}dz$$

or        $-\lambda = \frac{dT}{dz} = -\left(\frac{n-1}{n}\right)\frac{g}{R} = -\frac{0\cdot23}{1\cdot23} \times 9\cdot807\frac{m}{s^2} \times \frac{kg\ K}{287\ Nm}$

i.e.                        $\lambda = 6\cdot39\ K\ per\ 1\ 000\ m$

By integration, $T - T_0 = -\lambda(z - z_0)$; hence at 3 000 m

$$T = 288\ K - 6\cdot39 \times 3\ K = 268\cdot8\ K \quad \text{or} \quad -4\cdot3°C$$

Also        $\frac{p_0}{p} = \left(\frac{T_0}{T}\right)^{n/(n-1)} = \left(\frac{288}{268\cdot8}\right)^{1\cdot23/0\cdot23} = 1\cdot459$

hence $p = 69\cdot5\ kN/m^2$, and

$$\rho = \frac{p}{RT} = 69\cdot5\frac{kN}{m^2} \times \frac{1\ kg\ K}{287\ N\ m} \times \frac{1}{268\cdot8\ K} = 0\cdot882kg/m^3$$

## 13.4. Total energy of a compressible fluid at rest

If a region of fluid is in convective equilibrium we may imagine inter-changing two equal elements without disturbing the equilibrium of the system. Further, these movements may be made without having to add energy to either element. Since this is true of any element in the system, the latter is, in the widest sense of the word, one of constant energy, i.e. all parts of it have the same capacity for doing work. Thus we may formulate a new equation for gases, which corresponds to the hydrostatic equation:

$$\frac{p_1}{\rho} + gz_1 = \frac{p_2}{\rho} + gz_2$$

which applies to liquids. The corresponding equation for static or stagnant gases may be obtained from the convective equilibrium equation:

$$\frac{dT}{dz} = -\frac{\gamma-1}{\gamma}\frac{g}{R}$$

simply by integration between any two heights $z_1$ and $z_2$:

$$\frac{\gamma}{\gamma-1}R(T_2-T_1) = -g(z_2-z_1)$$

i.e.
$$\frac{\gamma}{\gamma-1}RT_1+gz_1 = \frac{\gamma}{\gamma-1}RT_2+gz_2$$

Noting that $(RT_1)$ is $(p_1/\rho_1)$ this equation may be more closely related to the hydrostatic equation by writing it as:

$$\left(\frac{\gamma-1+1}{\gamma-1}\right)RT_1+gz_1 = \left(\frac{\gamma-1+1}{\gamma-1}\right)RT_2+gz_2$$

i.e.
$$\frac{1}{\gamma-1}RT_1+RT_1+gz_1 = \frac{1}{\gamma-1}RT_2+RT_2+gz_2$$

$$\therefore \qquad \frac{R}{\gamma-1}T_1+\frac{p_1}{\rho_1}+gz_1 = \frac{R}{\gamma-1}T_2+\frac{p_2}{\rho_2}+gz_2$$

Hence in considering the total energy of a compressible fluid (such as a gas) at rest, it is not sufficient to say that it is simply the sum of its pressure and height energies. There is another factor involved, which depends on its temperature. If we recall that the specific heat capacities of a gas are related by:

$$c_p - c_v = R$$

in which $(c_p/c_v) = \gamma$, then

$$c_v = \frac{R}{\gamma-1}$$

and the additional term $R/(\gamma-1)T_1$ is seen to be $c_vT_1$ which is the specific internal energy $e_1$

i.e.
$$e_1+\frac{p_1}{\rho_1}+gz_1 = e_2+\frac{p_2}{\rho_2}+gz_2 \textbf{ for gases at rest}$$

This equation may be abbreviated by writing

$$h = e+\frac{p}{\rho}$$

$h$ being known as the specific 'enthalpy'. For a gas, for which $e = c_vT$ and $p/\rho = RT$:

$$h = (c_v+R)T = c_pT$$

so that the equation:

$$h_1+gz_1 = h_2+gz_2$$

becomes, for a gas at rest:

$$c_p T_1 + g z_1 = c_p T_2 + g z_2$$

This expression may alternatively be deduced directly from:

$$\frac{\gamma}{\gamma - 1} R T_1 + g z_1 = \frac{\gamma}{\gamma - 1} R T_2 + g z_2$$

by noting that:

$$\frac{\gamma R}{\gamma - 1} = c_p$$

To sum up, for a region of constant energy at rest, the sum of the internal, pressure and height energies is a constant. Alternatively the sum of the enthalpy and the height energy is the same at all points.

## 13.5. Total energy of a compressible fluid in motion

If a compressible fluid be in motion, its kinetic energy must be taken into account in drawing up a balance sheet. *Thus if it flows between two sections* 1 *and* 2 *without work being done on* (or by) *it, and if, in addition, no heat is transferred to* (or from) *it, then the sum of its static and kinetic energies will remain constant for steady flow:*

$$h_1 + g z_1 + \frac{u_1^2}{2} = h_2 + g z_2 + \frac{u_2^2}{2} \tag{13.5}$$

In this equation energy is expressed per unit mass.

This equation would not apply if sections 1 and 2 enclosed a machine doing work, any more than would Bernoulli's equation for an incompressible fluid. In addition it could not be applied if the gas were being heated or cooled from outside, or even by burning fuel in the stream. *Thus Equation* 13.5 *is appropriate only to adiabatic flows, that is, to flows along a stream-tube the boundaries of which are heat-insulated, and transfer no energy to or from the stream. Although this would appear to restrict the range of its validity quite severely it is in fact more widely used than any other in high-speed flows. The temperature gradients across such streams are generally small, and, if the gas is moving fast, very little heat transfer will in fact take place in many cases.*

Thus the equation (13.5) is fundamental to studies of the internal aerodynamics of, say, a gas turbine, or of the external aerodynamics of the aircraft which it powers. In each case it is important to choose the frame of reference so that the boundaries do not move, as this would invalidate the assumption that no work is done on the fluid between the sections considered. Thus it is necessary to consider the compressor, turbine or aerofoil to be at rest and to consider the motion of the fluid relative to it, see Fig. 13.5(a).

This artifice has been used earlier in connection with Bernoulli's equation and in fact Equation 13.5 may be considered to be an extension of the latter which allows, in addition, for the specific internal energy $e$ of the fluid:

$$e_1 + \frac{p_1}{\rho_1} + gz_1 + \frac{u_1^2}{2} = e_2 + \frac{p_2}{\rho_2} + gz_2 + \frac{u_2^2}{2}$$

This implies that, in one respect, the range of its validity is considerably greater than that of Bernoulli's equation. Any mechanical 'losses', due

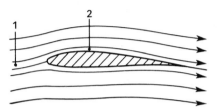

**Fig. 13.5(a).** *Motion of fluid relative to aerofoil*

to friction for example, simply generate an equivalent amount of heat which, since the tube is assumed to be heat insulated, must be conserved in the stream. Thus, *as the energy balance or equation of energy conservation takes account of the dissipation 'losses' which always generate heat, the equation is equally valid for adiabatic flows involving friction, or other irreversible changes.*

## 13.6. Temperature equivalent of velocity

As a simple example, if a compressible fluid, moving with negligible velocity through a pipe (the walls of which are insulated) encounters some form of throttle, such as a valve (see Fig. 13.6(a)) then, neglecting

**Fig. 13.6(a).** *Effect of valve on flow*

changes in height and kinetic energy between sections 1 and 2:

$$h_1 + gz_1 + \frac{u_1^2}{2} = h_2 + gz_2 + \frac{u_2^2}{2}$$

simplifies to:

$$h_1 = h_2$$

i.e. the enthalpy is unchanged. For a perfect gas this implies that:

$$c_p T_1 = c_p T_2$$

i.e.

$$T_1 = T_2$$

Thus although its pressure and density may be much reduced, the temperature remains unchanged. At the throttle itself, where the velocity ($u$) through the restriction may be quite appreciable, the corresponding specific enthalpy ($h$) is (for an approach velocity $u_1$) given by:

$$h + \frac{u^2}{2} = h_1 + \frac{u_1^2}{2}$$

i.e.

$$h = h_1 - \left(\frac{u^2}{2} - \frac{u_1^2}{2}\right)$$

Thus the temperature at the throttle ($T$) is reduced below that of the mainstream ($T_1$) by an amount given by:

$$T = T_1 - \frac{u^2 - u_1^2}{2c_p}$$

The subsequent eddying downstream, as the kinetic energy of the throttled gas is converted into heat, performs Joule's experiment, and restores the gas to its original temperature $T_1$. This explains the 'frosted' appearance of partly closed valves in high pressure air-lines. The high velocity through the throttle can be acquired only at the expense of the enthalpy, and hence the temperature, of the air; as soon as the latter slows down again, the enthalpy, and hence the temperature, begin to increase in accordance with the equation $T_2 = T + [(u^2 - u_2^2)/2c_p]$.

It is only when exceptionally large differences of elevation are considered, as in studies of atmospheric conditions, that the changes in the gravitational potential energy are of any great importance for gases and vapours. Thus the equation of energy conservation for steady compressible flow along a stationary, heat-insulated, stream-tube will from now on be simplified to:

$$h + \frac{u^2}{2} = \text{constant}$$

For *a near-perfect gas* this may be written:

$$c_p T + \frac{u^2}{2} = \text{constant}$$

and the right-hand side represents the value the enthalpy would reach if the stream were brought to rest, no matter how, so long as no heat were added from outside and no external work were done. The corresponding temperature $T_0$ is thus given by:

$$c_p T + \frac{u^2}{2} = c_p T_0$$

and $T_0$ **is known as the stagnation (or total head) temperature** of the stream.

Thus when an airstream impinges on an object as in Fig. 13.6(b) there is some point P where the stream divides and in which region the stream is virtually brought to rest. This point experiences a corresponding temperature rise:

$$(T_0 - T) = \frac{u^2}{2c_p}$$

which, it may be observed, depends only on speed, and on nothing else except the specific heat of the gas. If a body is to undergo sustained flight

**Fig. 13.6(b).** *Stagnation point*

at high speeds, it must therefore be specifically designed to withstand temperatures of this order, particularly in the region of the nose. Unlike the so-called sonic-barrier, which presents certain flight problems which are generally most acute in a limited speed range, the misnamed 'heat barrier' has been more aptly re-christened the 'thermal-thicket', in so far as the faster we attempt to move, the more acutely do we become entangled with heating problems.

EXAMPLE 13.6 (i)

*Calculate the velocity and Mach number of a stream of air in which the static temperature is 15°C and the total head (or stagnation) temperature is 40°C. Assume $\gamma = 1\cdot4$ for air. If the compression is assumed to be reversible adiabatic (isentropic) estimate the percentage rise in pressure between the stagnation and static pressures.*

$$T_0 - T = \frac{u^2}{2c_p}$$

but since $c_p - c_v = R$ then

$$\frac{\gamma - 1}{\gamma} = \frac{R}{c_p}$$

Hence

$$\frac{(\gamma - 1)u^2}{2\gamma R} = T_0 - T$$

or

$$\frac{T_0}{T} = 1 + \frac{(\gamma - 1)}{2} Ma^2$$

where Mach number

$$Ma = \frac{u}{a} = \frac{u}{\sqrt{(\gamma RT)}}$$

(see also Section 13.10, on p. 385).

Thus

$$Ma^2 = \frac{2}{\gamma-1}\left(\frac{T_0}{T}-1\right) = \frac{2}{0\cdot4}\left(\frac{313}{288}-1\right) = \frac{5\times25}{288} = 0\cdot434$$

i.e.
$$Ma = \mathbf{0\cdot658} = \frac{u}{a}$$

The velocity of sound in the free stream is:

$$a = \sqrt{(\gamma RT)} = \sqrt{(1\cdot4\times287 \text{ Nm/kg K}\times288 \text{ K})} = 340 \text{ m/s}$$

and the velocity of the free stream of air is:

$$u = a(Ma) = 340\times0\cdot658 = \mathbf{224} \text{ m/s}$$

For reversible adiabatic compression:

$$\frac{T_0}{T} = \left(\frac{p_0}{p}\right)^{(\gamma-1)/\gamma}$$

Hence,
$$\frac{p_0}{p} = (1\cdot087)^{3\cdot5} = 1\cdot339$$

and
$$\frac{p_0-p}{p} = 0\cdot339 \quad \text{or} \quad \mathbf{33\cdot9} \textit{ per cent}$$

## 13.7. Adiabatic and isentropic flow equations

It is worthwhile pausing at this stage to compare two similar, but in general different, equations for steady flows along a stream-tube. If there is no heat exchange with the outside—i.e. if the flow is, in thermodynamic terms, *adiabatic* (but not necessarily reversible) we may write, as in Section 13.6:

$$h+\frac{u^2}{2} = \text{constant}$$

If, however, the flow involves no loss of available energy—i.e. if the flow is thermodynamically *reversible* (but not necessarily adiabatic) the appropriate equation is that derived from Newton's second law, namely:

$$\int\frac{dp}{\rho}+\frac{u^2}{2} = \text{constant}$$

(see Section 6.3 p. 122).

The basic assumptions on which this result depended were that the flow was steady, frictionless and that there were no discontinuous changes along the stream-tube, i.e. there were no thermodynamic qualifications regarding the process by which the fluid changed its state, other than the fact that frictionless continuous changes imply thermodynamic reversibility. Thus the flow could nominally be adiabatic or equally well, isothermal, although in practice the latter is unlikely, except for very slow flows as may be in gas mains.

*Thus; for flows which may be considered both reversible and adiabatic—(these two conditions implying isentropic flow)—either equation may be used.*

That the equation of motion for *all reversible steady flows*, namely,

$$\int \frac{dp}{\rho} + \frac{u^2}{2} = \text{constant}$$

applies in the important particular case of *reversible adiabatic* flows may be proved by thermodynamic argument. For if the reversible flow is also

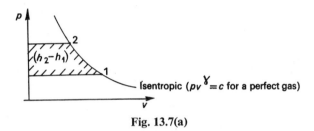

Fig. 13.7(a)

heat-insulated (adiabatic) then by the first law of thermodynamics: $\delta Q = \delta e + p \delta v = 0$; and from the definition of enthalpy:

$$\delta h = \delta(e + pv) = (\delta e + p \delta v) + v dp$$

Hence,                                   $\delta h = v dp$

*for reversible adiabatic (i.e. isentropic) changes.* Therefore, for such flows of (perfect) gas the law is $pv^\gamma = $ constant, and between sections 1 and 2, the change of enthalpy may be evaluated from:

$$h_2 - h_1 = \int_1^2 v dp = \int_1^2 \frac{dp}{\rho}$$

—a result expressed graphically in Fig. 13.7(a).

EXAMPLE 13.7 (i)

*Taking the ratio of the specific heat capacities of air as $\gamma = 1\cdot4$ deduce a formula relating velocity and pressure in a stream-tube 1–2 shown in Fig. 13.5(a), assuming adiabatic flow.*

For heat-insulated (i.e. adiabatic) steady flow

$$c_p T + \frac{u^2}{2} = \text{constant} = \frac{c_p}{R} \frac{p}{\rho} + \frac{u^2}{2}$$

assuming $p = \rho RT$ for a perfect gas.

Hence, using the fact $c_p - c_v = R$ and $\gamma = c_p/c_v$ we deduce:

$$\frac{\gamma}{\gamma-1}\frac{p}{\rho}+\frac{u^2}{2} = \text{constant}$$

and, therefore

$$\frac{p}{\rho}+\frac{u^2}{7} = \text{constant for air.}$$

## EXAMPLE 13.7 (ii)

*At a particular section of a duct air with a static temperature of 32°C and static pressure 80 kN/m² flows with a velocity of 365 m/s. Assuming reversible adiabatic flow calculate the velocity and temperature at a section where the static pressure is 120 kN/m² and estimate the Mach number at both sections. The specific gas constant for air may be taken as 287 Nm/kg K.*

For adiabatic flow $h+(u^2/2) = \text{constant}$,
hence,

$$\frac{u_1^2-u_2^2}{2} = \frac{\gamma}{\gamma-1}\left\{\frac{p_2}{\rho_2}-\frac{p_1}{\rho_1}\right\}$$

and

$$u_2^2-u_1^2 = 7\left\{\frac{p_1}{\rho_1}-\frac{p_2}{\rho_2}\right\} \text{ for air.}$$

Since $p/\rho^\gamma = \text{constant}$ for reversible adiabatic changes

$$u_2^2-u_1^2 = 7\frac{p_1}{\rho_1}\left\{1-\left(\frac{p_2}{p_1}\right)^{(\gamma-1/\gamma)}\right\} = 7RT_1\left\{1-\left(\frac{p_2}{p_1}\right)^{1/3\cdot5}\right\}$$

Hence,

$$u_2^2 = (365 \text{ m/s})^2 + 7\times287\times305 \text{ Nm/kg}\left\{1-\left(\frac{12}{8}\right)^{1/3\cdot5}\right\}$$

$$= 13\cdot325\times10^4 \text{ m}^2/\text{s}^2 - 7\cdot52\times10^4 \frac{\text{Nm}}{\text{kg}}\left(\frac{\text{kg m}}{\text{N s}^2}\right) = 5\cdot8\times10^4 \text{ m}^2/\text{s}^2$$

and    $u_2 = \mathbf{241}$ m/s.

Also,

$$\frac{T_2}{T_1} = \left(\frac{p_2}{p_1}\right)^{(\gamma-1)/\gamma} = 1\cdot122\,8$$

hence $T_2 = 1\cdot122\,8 \times 305$ K $= 344$ K and $t_2 = \mathbf{71°C}$, and

$$Ma_2 = \frac{u_2}{\sqrt{(\gamma RT_2)}} = \frac{241}{\sqrt{(1\cdot4\times287\times344)}} = \frac{241}{372} = \mathbf{0\cdot63}$$

whereas

$$Ma_1 = \frac{u_1}{\sqrt{(\gamma RT_1)}} = \frac{365}{\sqrt{(1\cdot4\times287\times305)}} = \frac{365}{350} = \mathbf{1\cdot033}$$

## 13.8. Isothermal gas flow with pipe friction

In fast-moving flows it has been argued that there is generally insufficient time for appreciable heat transfer to take place, so that they may usually be considered adiabatic. The other extreme is approached by, for example, flow through a gas main, where the movement is slow enough for the gas to take on the temperature of its surroundings. Conditions will often therefore approach the isothermal case, but unlike the previous example, friction may well cause irreversible changes, so that the reversible flow equation:

$$\frac{\delta p}{\rho} + \delta \left( \frac{u^2}{2} \right) = 0$$

i.e.
$$-\delta p = \rho u \delta u$$

does not apply. When friction is present, the pressure drop (represented by $-\delta p$) is not so much expended in accelerating the fluid (for that is

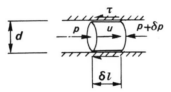

**Fig. 13.8(a).** *Gas flow with friction*

what the right-hand side of the last equation represents (see Section 6.3, p. 121)), as in overcoming the frictional resistance on the slice, see Fig. 13.8(a). If the gas were incompressible there would be no fluid accelera-tion. It is only because of its changing density that the fluid may acceler-ate. If the density changes only slowly, as in the case when gas flows slowly in a pipe, the force required to accelerate the fluid is small, and if this may be neglected, we deduce that:

$$-\delta p \frac{\pi d^2}{4} = \tau \pi d \delta l$$

i.e.
$$-\delta p = \frac{4 \delta l}{d} \tau$$

In an earlier section on pipe resistance (see Chapter 11, on p. 290) it was found convenient to express $\tau$ in terms of a non-dimensional friction factor $f$, defined by:

$$\tau = f \tfrac{1}{2} \rho u^2$$

In terms of this, the equation of motion becomes:

$$-\delta p = \frac{4 f \delta l}{d} \tfrac{1}{2} \rho u^2$$

which is another version of Darcy's formula

$$h_f = \frac{4fl}{d}\frac{u^2}{2g}$$

As in Section 13.3, when dealing with differential changes, we again proceed as follows to express the basic equations in logarithmic terms and then differentiate. In the present problem for example, we have, as additional information, the continuity equation, which states that the mass flow across each section is the same:

$$\dot{m} = \rho Au = \text{constant}$$

and the isothermal equation for a gas:

$$\frac{p}{\rho} = RT = \text{constant}$$

For a pipe of constant area the continuity equation may alternatively be written:

$$\delta \left(\log \rho + \log u\right) = 0$$

i.e.    $$\delta \left(\log \rho\right) + \delta \left(\log u\right) = 0$$

or    $$\frac{\delta \rho}{\rho} + \frac{\delta u}{u} = 0$$

and the isothermal equation may alternatively be written:

$$\delta \left(\log p\right) - \delta \left(\log \rho\right) = 0$$

i.e.    $$\frac{\delta p}{p} - \frac{\delta \rho}{\rho} = 0$$

Hence, on combining the two equations, the pressure change between any two adjacent sections is:

$$\delta p = p\frac{\delta \rho}{\rho} = -p\frac{\delta u}{u}$$

This expression enables the equation of motion to be integrated quite simply if we make the assumption that $f$ remains sensibly constant. Thus, since:

$$-\delta p = p\frac{\delta u}{u} = \frac{4f\delta l}{d}\tfrac{1}{2}\rho u^2$$

and because $(p/\rho) = RT$ remains constant in isothermal flow, we may write:

$$\left(\frac{p}{\rho}\right)\int_1^2 \frac{2du}{u^3} = \frac{4f}{d}\int_0^l dl$$

i.e.    $$RT\left(\frac{1}{u_1^2} - \frac{1}{u_2^2}\right) = \frac{4fl}{d}$$

Each of the velocities may be expressed in terms of the mass flow:

$$u_1 = \frac{\dot{m}}{\rho_1 A} \quad \text{and} \quad u_2 = \frac{\dot{m}}{\rho_2 A}$$

and the densities, in their turn, can be expressed in terms of pressure by means of the gas equation:

$$p = \rho RT$$

∴
$$u_1 = \frac{\dot{m} RT}{p_1 A} \quad \text{and} \quad u_2 = \frac{\dot{m} RT}{p_2 A}$$

Hence, the flow equation becomes:

$$RT\left\{ \left(\frac{p_1 A}{\dot{m} RT}\right)^2 - \left(\frac{p_2 A}{\dot{m} RT}\right)^2 \right\} = \frac{4fl}{d}$$

i.e.
$$\frac{A^2}{\dot{m}^2 RT}(p_1^2 - p_2^2) = \frac{4fl}{d}$$

which gives, for the rate of mass flow:

$$\dot{m} = A \sqrt{\left( \frac{d}{4flRT}(p_1^2 - p_2^2) \right)}$$

EXAMPLE 13.8 (i)

*Gas having a density of 0·64 kg/m³ is delivered at atmospheric pressure from a 300 mm diameter main 30 m long. Estimate the maximum volumetric flow of free gas that may be drawn from the main if the latter has a friction factor f = 0·01, and is supplied at a pressure of 0·6 m water.*

Let $Q_2$ denote the volumetric flow rate at the delivery end; if $\rho_2$ is the corresponding density:

$$Q_2 = \frac{\dot{m}}{\rho_2} = A \sqrt{\left[ \frac{d}{4fl\rho_2\rho_2} (p_1^2 - p_2^2) \right]}$$

in which $p_2$ = atmospheric = 101·3 kN/m² and

$$p_1 - p_2 = wh = 9\text{·}807 \text{ kN/m}^3 \times 0\text{·}6 \text{ m} = 5\text{·}9 \text{ kPa}$$

and since

$$p_1 + p_2 = (5\text{·}9 + 2 \times 101\text{·}3) \text{ kPa} = 208\text{·}5 \text{ kPa}$$

$$p_1^2 - p_2^2 = 1\ 230 \text{ kPa}^2$$

Hence,

$$Q = \frac{\pi}{4} \times 0\text{·}09 \text{ m}^2 \sqrt{\left\{ \frac{0\text{·}3 \text{ m} \times 1\ 230 \text{ kPa}^2}{0\text{·}04 \times 30 \text{ m} \times 101\text{·}3 \text{ kPa} \times 0\text{·}64 \text{ kg/m}^3} \left[ \frac{\text{kg}}{\text{Pa m s}^2} \right] \right\}}$$

$$= 0\text{·}0707 \text{ m}^2 \sqrt{(4\ 720 \text{ m}^2/\text{s}^2)} = \textbf{4·9 m}^3\textbf{/s}$$

## 13.9. Velocity and area changes; choking of nozzles

Suppose that gas flows steadily through a stream-tube, the area of which varies from section to section as in Fig. 13.9(a). The continuity equation:

$$\dot{m} = \rho u A = \text{constant}$$

can be expressed (after taking logs) in differential terms as:

$$\frac{\delta\rho}{\rho} + \frac{\delta u}{u} + \frac{\delta A}{A} = 0$$

For incompressible flows this reduces to:

$$\frac{\delta A}{A} = -\frac{\delta u}{u}$$

i.e. a small fractional decrease in area implies a corresponding equal fractional increase in velocity and vice versa. Thus a liquid speeds up in

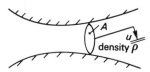

**Fig. 13.9(a).**  *Flow of gas*

a contracting nozzle, and slows down in an expanding diffuser. We cannot *a priori* make the same assumption for gas flows, in which the density changes. In general:

$$\frac{\delta A}{A} = -\left(\frac{\delta\rho}{\rho} + \frac{\delta u}{u}\right)$$

i.e.

$$\frac{\delta A}{A} = -\left[\frac{(\delta\rho/\rho)}{(\delta u/u)} + 1\right]\frac{\delta u}{u}$$

As density tends to fall with decreasing pressure [$\rho = p/(RT)$], and with increasing velocity [$\rho = \dot{m}/(uA)$], the first term within the square bracket is in fact, numerically, normally negative. So long as its numerical value is less than unity, fractional *decreases* in area [$-(\delta A/A)$] continue to cause some *increase* ($\delta u$) in speed. If however its negative value is great enough to reverse the sign of the right-hand side a fractional *increase* in area would be needed to cause a further fractional *increase* in speed. From a practical point of view it is obviously important to establish whether it is necessary to contract or expand a channel in order to speed up a fluid. To do this we need an expression for $(\delta\rho/\rho)/(\delta u/u)$ and this may in fact be quite simply obtained if we assume the reversible flow equation:

$$\frac{\delta p}{\rho} + u\delta u = 0$$

For, on dividing it by $(u^2\delta\rho)/\rho$, an expression for the reciprocal of the term we require results from:

$$\frac{\delta p}{u^2 \delta \rho} + \frac{u \delta u}{(u^2 \delta \rho)/\rho} = 0$$

hence:

$$\frac{(\delta \rho / \rho)}{(\delta u / u)} = -\frac{u^2}{(\delta p / \delta \rho)}$$

Since the left-hand side of this equation is a numerical ratio, so is the right-hand side. Hence $\sqrt{(dp/d\rho)}$ must have the dimensions of a velocity,

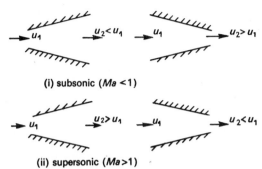

(i) subsonic (*Ma* <1)

(ii) supersonic (*Ma* >1)

**Fig. 13.9(b).** *Subsonic and supersonic flow of gas*

and is usually denoted by the symbol $a$. For isentropic changes which obey the law $p/\rho^\gamma = \text{constant} = pv^\gamma$:

$$a = \sqrt{\left(\frac{dp}{d\rho}\right)} = \sqrt{\left(\gamma \frac{p}{\rho}\right)} = \sqrt{(\gamma RT)}$$

Thus, the original equation relating area to velocity becomes:

$$\frac{\delta A}{A} = \left[\left(\frac{u}{a}\right)^2 - 1\right] \frac{\delta u}{u}$$

We shall later establish that $a$ is the velocity of sound in the fluid (see Section 13.10, p. 384) and it can be seen that the ratio which $u$ bears to $a$ is of fundamental importance in compressible flow. It is denoted here by the symbol $Ma$:

i.e.

$$Ma = \frac{u}{a}$$

and is referred to as the Mach number. If $Ma$ is less than one, the stream is said to be subsonic and if greater than one, supersonic. The equation:

$$\frac{\delta A}{A} = (Ma^2 - 1)\frac{\delta u}{u}$$

for reversible adiabatic (i.e. isentropic) flows establishes that, for subsonic streams, an *increase* in area $\delta A$ causes a *decrease* in velocity $\delta u$ and vice versa as is the case for incompressible flows. For supersonic flows, however, the velocity *increases* as the stream-tube *expands* in area, and vice versa, see Fig. 13.9(b).

*We may infer from this that if we fit a convergent nozzle to a gas reservoir, and allow the gas to escape through the nozzle, then no matter how high the pressure inside ($p_1$ say) nor how low the pressure outside ($p_2$ say), the velocity through the throat can never exceed the local speed of sound, see Fig. 13.9(c).*

This remarkable result may be explained as follows. The issuing jet is set in motion by the differences of pressure acting on it. It may be thought of as being thrust forward from upstream, and drawn forward from downstream. Let us suppose that $p_2$ is, initially, close enough to $p_1$ for the flow to be

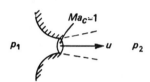

**Fig. 13.9(c).** *Flow of gas through a convergent nozzle*

subsonic throughout. The gas stream would then emerge in much the same way as would a liquid jet, except that it would possess considerable compressibility as well as inertia. From a dynamic viewpoint, therefore, it may be likened to an elastically coupled train. A pressure reduction downstream corresponds to an impulsive tension which would be propagated back through the fluid with the appropriate wave velocity. The latter is, for the gas, the same as the speed of sound which consists simply of repeated pressure disturbances in a certain frequency band. Provided that the fluid column is travelling to the right with a speed less than that at which the wave is advancing through it from the right, i.e. so long as the stream is subsonic, the flow will be increased by a reduction of pressure downstream. Once, however, the fluid reaches sonic velocity at the throat, any rarefaction impulses, i.e. any further pressure reduction downstream cannot be propagated past this section. The flow through the nozzle then becomes independent of $p_2$, and it is said to be *choked*.

We have already shown (see Example 12.7(i), p. 374) that integration of the reversible flow equation:

$$\frac{\delta p}{\rho} + u\delta u = 0$$

for isentropic expansion from reservoir conditions $p_1$, $\rho_1$ and $u_1 = 0$, leads to an expression for the velocity at any section $u$, where the pressure is $p$:

$$u = \sqrt{\left\{\frac{2\gamma}{\gamma-1}\frac{p_1}{\rho_1}\left[1-\left(\frac{p}{p_1}\right)^{(\gamma-1)/\gamma}\right]\right\}}$$

In particular it gives the throat velocity for subsonic conditions, and hence the corresponding mass flow for a convergent nozzle:

$$\dot{m} = \rho A u = \rho_1 \left(\frac{p}{p_1}\right)^{1/\gamma} A u$$

in which $\rho = \rho_1(p/p_1)^{1/\gamma}$ for reversible adiabatic expansion,

i.e. $$\dot{m} = A \sqrt{\left\{\frac{2\gamma}{\gamma-1}p_1\rho_1\left(\frac{p}{p_1}\right)^{2/\gamma}\left[1-\left(\frac{p}{p_1}\right)^{(\gamma-1)/\gamma}\right]\right\}}$$ (i)

In these expressions, the pressure at the throat is $p = p_2$ so long as conditions are sub-critical. As soon as the local Mach number reaches

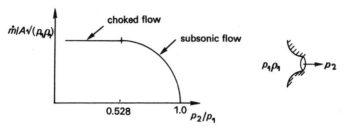

Fig. 13.9(d). *Rate of mass flow through a convergent nozzle*

unity at the exit of the convergent nozzle the throat velocity pressure and density assume their 'critical' values $u_c$, $p_c$, $\rho_c$ for choked flows.

Hence, $$u_c = \sqrt{\left\{\frac{2\gamma}{\gamma-1}\frac{p_1}{\rho_1}\left[1-\left(\frac{p_c}{p_1}\right)^{(\gamma-1)/\gamma}\right]\right\}}$$

and $$\dot{m}_c = A \sqrt{\left\{\frac{2\gamma}{\gamma-1}p_1\rho_1\left(\frac{p_c}{p_1}\right)^{2/\gamma}\left[1-\left(\frac{p_c}{p_1}\right)^{(\gamma-1)/\gamma}\right]\right\}}$$ (ii)

In order to apply these expressions (i) and (ii) it is necessary to relate local conditions to Mach number, and in particular to deduce the value of $p = p_c$ corresponding to $Ma = 1$.

In the next section it will be shown that

$$\frac{p_c}{p_1} = \left(\frac{2}{\gamma+1}\right)^{\gamma/(\gamma-1)}$$

so that for air, for which $\gamma = 1\cdot4$, the critical pressure ratio:

$$r_c = \frac{p_c}{p_1} = 0\cdot528$$

Thus, the mass flow through a nozzle is given by Equation (i) so long as $p_2 > 0\cdot528\,p_1$ and by Equation (ii) when $p_2 < 0\cdot528\,p_1$, see Fig. 13.9(d), plotted non-dimensionally.

EXAMPLE 13.9 (i)

*Show that the mass flow through an air-choked nozzle is:*

$$\dot{m} = \frac{40\cdot4}{10^6}\frac{aP}{\sqrt{T}} \text{ kg/s}$$

*for a nozzle area of a* mm², *a reservoir pressure P kPa and a reservoir temperature of T K.*

For a choked nozzle:

$$\dot{m} = A \sqrt{\left( \frac{2\gamma}{\gamma-1} p_1 \frac{p_1}{RT_1} r_c^{2/\gamma}[1 - r_c^{(\gamma-1)/\gamma}] \right)}$$

in which $r_c$ is the critical pressure ratio = 0·528, $\gamma$ is the ratio $c_p/c_v$ = 1·4, $R$ is the gas constant = 287 J/kg K for air.

If we express $A$, $p_1$ and $T_1$ in particular units by writing:

$$A = a \text{ mm}^2; \quad p_1 = Pk \text{ N/m}^2; \quad T_1 = T \text{ K}$$

we reduce the above equation to the working rule:

$$\dot{m} = a \text{ mm}^2 \times \frac{P \text{ kN}}{\text{m}^2} \sqrt{\left( \frac{2 \times 1\cdot4}{0\cdot4} \times \frac{\text{kg K}}{287 \text{ Nm}} \times \frac{(0\cdot528)^{2/1\cdot4}}{T \text{ K}} \times (1 - 0\cdot528^{0\cdot4/1\cdot4}) \right)}$$

On collecting together the units:

$$\dot{m} = \frac{aP}{\sqrt{T}} \sqrt{\left( \frac{7}{287} \times 0\cdot528^{10/7}(1 - 0\cdot528^{2/7}) \frac{\text{Nkg}}{10^6 \text{ m}} \right)}$$

it can be seen that we need to use the identity:

$$1 \text{ N} \equiv 1 \text{ kg m/s}^2$$

in order to obtain an expression in kg/s.

$$\dot{m} = \frac{aP}{\sqrt{T}} \sqrt{\left( 0\cdot009\ 75[1 - 0\cdot833] \frac{\text{kg}^2}{10^6 \text{ s}^2} \right)}$$

i.e.
$$\dot{m} = \frac{40\cdot4\ aP}{10^6 \sqrt{T}} \text{ kg/s} \quad or \quad 40\cdot4 \frac{aP}{\sqrt{T}} \text{ mg/s}$$

where $a$ is in mm², $P$ in kPa and $T$ in K.

### EXAMPLE 13.9 (ii)

*Carbon dioxide discharges through an orifice 10 mm diameter from a large container in which the gas has a temperature of 15°C and pressure 7 atmospheres. Calculate the speed of the jet and the rate of mass flow.*

*The atmospheric pressure is 101·3 kPa and $\gamma$ = 1·30 for CO₂.*

$$\text{The characteristic gas constant} = \frac{\text{molar ideal-gas constant}}{\text{molar mass}}$$

i.e., for $CO_2$ the specific gas constant

$$R = \frac{8\cdot314\ 3 \text{ J/mol K}}{44 \text{ g/mol}} = 189 \text{ Nm/kg K}$$

Hence the density of $CO_2$ in the container is:

$$\rho_1 = \frac{p_1}{RT_1} = \frac{709 \times 10^3 \text{ N/m}^2}{189 \text{ Nm/kg} \times 288\cdot1} = 13\cdot04 \text{ kg/m}^3$$

Critical pressure ratio for choked flow is:

$$\frac{p_c}{p_1} = \left(\frac{2}{\gamma+1}\right)^{\gamma/(\gamma-1)} = \left(\frac{2}{2\cdot3}\right)^{1\cdot3/0\cdot3} = 0\cdot545$$

Hence, since the ratio of the pressures on the two sides of the orifice is $\frac{1}{7} = 0\cdot143$, the flow is choked and the speed is that of sound in $CO_2$ at temperature $T_c$ given by:

$$\frac{T_c}{T_1} = \left(\frac{p_c}{p_1}\right)^{(\gamma-1)/\gamma} = \frac{2}{\gamma+1} = \frac{2}{2\cdot3}$$

i.e.
$$T_c = \frac{2}{2\cdot3} \times 288 \text{ K} = 250 \text{ K}$$

and the speed is

$$u_c = \sqrt{(\gamma R T_c)} = \sqrt{\left(1\cdot3 \times 189 \frac{\text{Nm}}{\text{kg K}} \times 250 \text{ K} \left[\frac{\text{kg m}}{\text{N s}^2}\right]\right)}$$

$$= \textbf{248} \text{ m/s}$$

Density of $CO_2$ at critical conditions obtaining at the orifice is:

$$\rho_c = \frac{p_c}{RT_c} = \frac{0\cdot545 \times 7 \times 101\cdot3 \text{ kN/m}^2}{189 \times 250 \text{ Nm/kg}}$$

$$= 8\cdot18 \text{ kg/m}^3$$

Hence, the rate of mass flow is $\rho_c a u_c$

$$= 8\cdot18 \frac{\text{kg}}{\text{m}^3} \times \frac{\pi}{4} \frac{\text{m}2}{10^4} \times 248 \frac{\text{m}}{\text{s}}$$

$$= 0\cdot1595 \text{ kg/s} \quad \text{or} \quad \textbf{9\cdot57} \text{ kg/min}$$

## 13.10. Speed of sound

In the previous Section we coupled the problem of compressible flow through a nozzle with that of the propagation of sound waves, i.e. small pressure disturbances. Fig. 13.10(a) represents such a wave, or rather

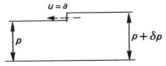

**Fig. 13.10(a).** *Velocity of sound or wavelet in a stationary gas*

wavelet (as the pressure rise across it is assumed infinitesimal) advancing through a stationary compressible medium. As each successive layer is jerked into motion impulsively, the problem cannot be analysed as it stands in terms of equations developed for steady flow conditions. If however we adopt the artifice of advancing with the wavelet, we reduce the problem to that of a uniform stream impinging on a stationary wave front with a velocity equal to *a* in the opposite direction, see Fig. 13.10(b).

Across the wavelet, the pressure rise $\delta p$ causes a velocity change, to some different value $(u+\delta u)$ and these are related, if the change is reversible, by:

$$\frac{\delta p}{\rho} + u\delta u = 0$$

As all stream-tubes must behave in the same way they cannot contract

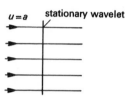

Fig. 13.10(b). *Velocity of small disturbance or wavelet*

or expand, and the mass flow per unit area must be the same, for continuity, each side of the wave. Hence, from the previous section, setting $\delta A/A = 0$:

$$\frac{\delta \rho}{\rho} + \frac{\delta u}{u} = 0$$

Dividing the two equations of this section:

$$\frac{(\delta p/\rho)}{(\delta\rho/\rho)} = \frac{-(u\delta u)}{-(\delta u/u)}$$

$$\frac{dp}{d\rho} = u^2$$

i.e.
$$u = a = \sqrt{\left(\frac{dp}{d\rho}\right)}$$

This confirms that for reversible flows, the limiting velocity through the throat of a stream-tube, which was shown in Section 13.9 to be equal to $\sqrt{(dp/d\rho)}$, is the same as the velocity of propagation of weak pressure waves. *Thus a steam or air nozzle 'chokes' when the throat velocity equals the velocity of sound at that point in the medium.* It is important to note that it is the *local value* of the velocity of sound which matters, as *a* depends on the 'state' of the fluid, and this, in general, varies from point to point.

In order to estimate *a* from the state of the fluid, we must make some assumption concerning the law by which fluid expands or contracts. Knowing that these changes are rapid, we assume a reversible adiabatic expansion or compression $p/\rho^\gamma = C$ (constant). Therefore,

$$\frac{dp}{d\rho} = C\rho^{\gamma-1} = \gamma\frac{p}{\rho} = \gamma RT$$

Hence
$$a = \sqrt{\left(\frac{dp}{d\rho}\right)} = \sqrt{\left(\gamma\frac{p}{\rho}\right)} = \sqrt{(\gamma RT)}$$

Thus, inserting numerical values for, say, air at 15°C, we obtain:

$$a = \sqrt{\left(1\cdot4 \times 287\frac{\text{Nm}}{\text{kg K}} \times 288 \text{ K}\left[\frac{\text{kg m}}{\text{N s}^2}\right]\right)}$$
$$= 340 \text{ m/s}$$

*The parameter for assessing the importance of compressibility in a gas flow is the Mach number:*

$$Ma = \frac{u}{a} = \frac{u}{\sqrt{(dp/d\rho)}} = \frac{u}{\sqrt{(\gamma p/\rho)}} = \frac{u}{\sqrt{(\gamma RT)}}$$

Because of its fundamental importance, it is often convenient to re-express the flow equations we have developed, in terms of $Ma$ and so relate the state of the fluid to the local Mach number.

Thus the basic equation of energy conservation used in Section 13.6 for adiabatic flow relating local temperature and velocity, to the stagnation temperature $T_0$, namely:

$$c_p T + \frac{u^2}{2} = c_p T_0$$

becomes, after dividing through by $c_p T$ and recalling that:

$$c_p/R = \gamma/(\gamma-1)$$

$$1 + \frac{u^2}{2c_p T}\left[\frac{c_p(\gamma-1)}{R\gamma}\right] = \frac{T_0}{T}$$

i.e.
$$1 + \frac{\gamma-1}{2}Ma^2 = \frac{T_0}{T}$$

If, in addition to being adiabatic, the changes are reversible, so that the corresponding stagnation pressure ($p_0$) and density ($\rho_0$) are related by:

$$\frac{p_0}{p} = \left(\frac{T_0}{T}\right)^{\gamma/(\gamma-1)}, \quad \frac{\rho_0}{\rho} = \left(\frac{T_0}{T}\right)^{1/(\gamma-1)}$$

we may obtain expressions for temperature, pressure and density in terms of local Mach number:

$$\frac{T}{T_0} = \frac{1}{1 + \frac{\gamma-1}{2}Ma^2}$$

$$\frac{p}{p_0} = \frac{1}{\left(1 + \frac{\gamma-1}{2}Ma^2\right)^{\gamma/(\gamma-1)}}$$

$$\frac{\rho}{\rho_0} = \frac{1}{\left(1 + \frac{\gamma-1}{2}Ma^2\right)^{1/(\gamma-1)}}$$

The values with a suffix 0 correspond to the state that would be reached by the fluid, were it to be brought to or from rest by a reversible adiabatic change.

If applied to the nozzle problem of the previous section, for example, $p_0$, $\rho_0$ and $T_0$ would be the same as those in the reservoir. In particular, the critical pressure ratio in that problem can thus be obtained directly by substituting $p_1$ for $p_0$, $p_c$ for $p$, with $Ma = 1$:

$$\therefore \qquad \frac{p_c}{p_1} = \left(\frac{2}{\gamma+1}\right)^{\gamma/(\gamma-1)}$$

## EXAMPLE 13.10 (i)

*Estimate the corrections needed to Pitot-static tube readings when the velocity of air relative to the tube takes on Mach numbers between 0·1 and 0·5.*

Denoting the pressure of the air moving relative to the tube by $p$, and the pressure at the stagnation point at entry to the tube by $p_0$, we have seen that for reversible (frictionless) adiabatic compression

$$\frac{p_0}{p} = \left(1+\frac{\gamma-1}{2}Ma^2\right)^{\gamma/(\gamma-1)}$$

which, when expanded according to the binomial theorem becomes:

$$\frac{p_0}{p} = 1+\frac{\gamma}{2}Ma^2+\frac{\gamma}{8}Ma^4+\frac{\gamma(2-\gamma)}{48}Ma^6+\ldots$$

or

$$\frac{p_0-p}{p} = \frac{\gamma}{2}Ma^2\left(1+\frac{Ma^2}{4}+\frac{2-\gamma}{24}Ma^4+\ldots\right)$$

and since

$$p\frac{\gamma}{2}Ma^2 = p\frac{\gamma}{2}\frac{u^2}{a^2} = p\frac{\gamma}{2}\frac{u^2}{\gamma(p/\rho)} = \tfrac{1}{2}\rho u^2$$

then

$$\frac{p_0-p}{\tfrac{1}{2}\rho u^2} = 1+\frac{Ma^2}{4}+\frac{2-\gamma}{24}Ma^4+\ldots$$

For air $\gamma = 1\cdot4$ and the following table shows that if the Mach number does not exceed about 0·2 then compressibility has a negligible effect on Pitot-static tube readings.

| $Ma$ | 0·1 | 0·2 | 0·3 | 0·4 | 0·5 |
|---|---|---|---|---|---|
| $\dfrac{p_0-p}{\tfrac{1}{2}\rho u^2}$ | 1·003 | 1·010 | 1·023 | 1·041 | 1·064 |

## EXAMPLE 13.10 (ii)

*Using the expression for compressibility error of Pitot-static tube readings estimate the speed which would be recorded on a Pitot-static tube of normal calibration and calculate the percentage error when the tube is indicating the speed of an aeroplane which is actually flying at 270 m/s or 525 knot at sea level where the velocity of sound in air is 340 m/s.*

Neglecting $Ma^4$ and higher powers we have

$$\frac{p_0 - p}{\frac{1}{2}\rho u^2} = 1 + \frac{Ma^2}{4}$$

and an actual speed given by

$$u_{\text{act.}}^2 = \frac{2(p_0 - p)}{\rho[1 + (Ma^2/4)]}$$

as compared with a recorded speed given by a normally calibrated tube of

$$u^2 = \frac{2(p_0 - p)}{\rho}$$

which, of course, overestimates in the ratio

$$\frac{u}{u_{\text{act.}}} = \sqrt{\left(1 + \frac{Ma^2}{4}\right)}$$

When flying at 270 m/s at sea level

$$Ma = \frac{270}{340} = 0.795$$

Hence $$\frac{u}{u_{\text{act.}}} = \sqrt{1.157} = 1.075$$

and $u = $ **291** m/s which is an error of **7·8** *per cent*.

## EXAMPLE 13.10 (iii)

*Calculate the actual pressure, density and temperature of air at stagnation points relative to an aeroplane flying at 270 m/s or 525 knot at s.t.p. conditions and the percentage rises, respectively above s.t.p. conditions of 101·3 kN/m², 1·225 kg/m³ and 15°C.*

From Example 13.10 (ii):

$$Ma = 0.795$$

Hence, at stagnation points:

$$p_0 = p\left(1 + \frac{\gamma - 1}{2} Ma^2\right)^{\gamma/(\gamma - 1)} = p\left(1 + \frac{Ma^2}{5}\right)^{3.5}$$

$$= 101\cdot3 \text{ kPa} \left(1 + \frac{0\cdot795^2}{5}\right)^{3.5} = 153\cdot4 \text{ kN/m}^2$$

and $$\frac{p_0 - p}{p} = \frac{52\cdot1}{101\cdot3} \quad \text{or} \quad \textbf{51·4} \textit{ per cent}$$

Density

$$\rho_0 = \rho\left(1 + \frac{\gamma - 1}{2} Ma^2\right)^{1/(\gamma - 1)} = 1\cdot225 \text{ kg/m}^3 \left(1 + \frac{0\cdot795^2}{5}\right)^{2.5} = 1\cdot648 \text{ kg/m}^3$$

and $$\frac{\rho_0 - \rho}{\rho} = \frac{0\cdot423}{1\cdot225} \quad \text{or} \quad \textbf{34·5} \textit{ per cent}$$

Temperature

$$T_0 = T\left(1+\frac{\gamma-1}{2}Ma^2\right) = 288\ \text{K}\left(1+\frac{0{\cdot}795^2}{5}\right) = 324\ \text{K}$$

and $\qquad \dfrac{T_0-T}{t} = \dfrac{36}{15}$  or  **240** *per cent on Celsius scale*

## EXAMPLE 13.10 (iv)

*An aeroplane is flying at 225 m/s at a low altitude where the velocity of sound is 335 m/s. At a certain position just outside the boundary layer of the wings the velocity of the air relative to the aeroplane is 315 m/s. Determine the pressure drop on the wing surface near this position. Assume frictionless adiabatic flow and* $\gamma = 1{\cdot}4$*, and the pressure of ambient air is* $0{\cdot}103$ *MPa.*

From Example 13.7 (i):

$$\frac{\gamma}{\gamma-1}\frac{p}{\rho}+\frac{u^2}{2} = \text{constant}$$

Hence, referring to Fig. 13.5(a),

$$\tfrac{1}{2}(u^2-u_1^2) = \frac{\gamma}{\gamma-1}\left(\frac{p_1}{\rho_1}\right)\left\{1-\frac{p}{p_1}\frac{\rho_1}{\rho}\right\}$$

$$= \frac{\gamma}{\gamma-1}\frac{a_1^2}{\gamma}\left\{1-\left(\frac{p}{p_1}\right)^{(\gamma-1)/\gamma}\right\}$$

since $\qquad a_1^2 = \gamma\dfrac{p_1}{\rho_1}$  and  $\dfrac{\rho_1}{\rho} = \left(\dfrac{p_1}{p}\right)^{1/\gamma}$

Thus, $\qquad \left(\dfrac{p}{p_1}\right)^{(\gamma-1)/\gamma} = 1-\dfrac{\gamma-1}{2}\left(\dfrac{u^2-u_1^2}{a_1^2}\right)$

$$= 1-\frac{0{\cdot}4}{2}\left(\frac{315^2-225^2}{335^2}\right)$$

$$= 1-\frac{1}{5}\times\frac{540}{335}\times\frac{90}{335} = 0{\cdot}913\ 5$$

and $\qquad \dfrac{p}{p_1} = (0{\cdot}913\ 5)^{3{\cdot}5} = 0{\cdot}728\ 63$

Hence

$$p = 103\ \text{kPa}\times0{\cdot}729 = \textbf{75}\ \textbf{kN/m}^2$$

and the pressure drop on the wing is $(103-75)\ \text{kN/m}^2 = \textbf{28 kPa}$.

## EXAMPLE 13.10 (v)

*Calculate the static pressure of air flowing in a duct at 270 m/s if the stagnation temperature is 57°C and the stagnation pressure is 72 kN/m².*

$$\frac{T_0{}'}{T_1} = 1+\frac{\gamma-1}{2}Ma_1^2 = \left(\frac{p_0}{p_1}\right)^{(\gamma-1)/\gamma}$$

and

$$a_1 = \sqrt{(\gamma R T_1)} = \sqrt{\left(1{\cdot}4 \times 287 \frac{\text{Nm}}{\text{kg K}} \times T_1\right)} = 20 \text{ m/s } \sqrt{(T_1/\text{K})}$$

First find $Ma_1$ by trial (see table).

| $Ma_1$ | 0·8 | 0·77 | 0·79 |
|---|---|---|---|
| $\dfrac{T_1}{\text{K}} = \dfrac{330}{1+0{\cdot}2\ Ma_1^2}$ | 292·3 | 295·1 | 293·5 |
| $\dfrac{a_1}{\text{m/s}} = 20 \sqrt{\dfrac{T_1}{\text{K}}}$ | 342·2 | 343·4 | 342·6 |
| $Ma_1 = \dfrac{u_1}{a_1} = \dfrac{270 \text{ m/s}}{a_1}$ | 0·788 | 0·786 | 0·79 |

Hence, using $Ma_1 = 0{\cdot}79$ and $T_1 = 293{\cdot}5$ K we have:

$$\frac{p_0}{p_1} = \left(\frac{T_0}{T_1}\right)^{\gamma/(\gamma-1)} = \left(\frac{330}{293{\cdot}5}\right)^{3\cdot5} = 1{\cdot}507$$

Hence,

$$p_1 = \frac{72 \text{ kPa}}{1{\cdot}507} = 47{\cdot}8 \text{ kN/m}^2$$

## 13.11. Oblique Mach waves; Prandtl–Mayer expansions

In the previous Section we showed that a stream may undergo an infinitesimal compression (or, conversely, expansion) through a pressure (or

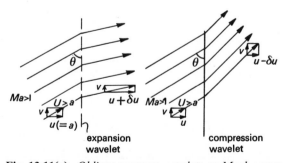

Fig. 13.11(a). *Oblique pressure wavelets or Mach waves*

rarefaction) wavelet, provided that the component of velocity normal to it is equal to the local speed of sound. If we now imagine observing such a Mach wave, as it is often called, from a vehicle travelling with uniform velocity $v$ along the line of the wavelet, we should describe conditions as follows. Fluid travelling with a normal component of velocity $u\ (= a)$

relative to a stationary plane wave, crosses the latter obliquely and, in doing so, undergoes a small change in its component velocity normal to the wave, although its velocity component along the wave remains unchanged. Thus the fluid direction changes in crossing the wave, and

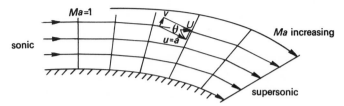

**Fig. 13.11(b).** *Mach waves and Mach angle*

the sense of this deflection depends on whether the wave is one of compression or expansion, see Fig. 13.11(a).

We can infer from this the mechanism by which a supersonic stream can expand through a series of wavelets to a higher speed. As any streamline may, neglecting viscosity effects, be replaced by a solid boundary so we may think of Mach waves springing from each change in direction of the latter and being propagated outwards in straight lines at an angle $\theta$ to the oncoming flow as in Fig. 13.11(b). The Mach angle, as $\theta$ is called, is such that:

$$\sin \theta = \frac{a}{u} = \frac{1}{Ma}$$

i.e.

$$\theta = \sin^{-1}\left(\frac{1}{Ma}\right)$$

Because of the refraction of light in regions where there is a change in density in a gas, such waves can be made visible. Thus by passing light through the divergent part of a supersonic nozzle, the existence of stationary waves inclined to the stream at the appropriate Mach angle may be confirmed on a shadow-graph, see Fig. 13.11(c). By the use of more elaborate optical techniques, known as a schlieren arrangement, density changes as well as shock patterns may be observed.

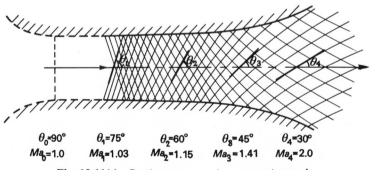

| $\theta_0=90°$ | $\theta_1=75°$ | $\theta_2=60°$ | $\theta_3=45°$ | $\theta_4=30°$ |
|---|---|---|---|---|
| $Ma_0=1.0$ | $Ma_1=1.03$ | $Ma_2=1.15$ | $Ma_3=1.41$ | $Ma_4=2.0$ |

**Fig. 13.11(c).** *Stationary waves in supersonic nozzle*

If the gradual convexity of the boundaries shown in Fig. 13.11(c) is replaced by a much more sudden change of direction, the pattern of

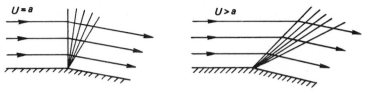

**Fig. 13.11(d).** *Mach waves at sudden change of direction or Prandtl–Mayer expansion*

Mach waves will tend towards a pencil of rays focused on the point at which the boundary changes direction, see Fig. 13.11(d). Although the velocity change across each Mach wave is infinitesimal, the overall

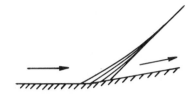

**Fig. 13.11(e).** *Mach waves at sudden concavity*

velocity increase across such a fan, or Prandtl–Mayer expansion, as it is called, is finite.

If a concavity is sharpened, however, the rays soon coalesce, see Fig. 13.11(e), and this mutual interaction causes a finite jump in pressure. Such changes, which take place virtually discontinuously, need special consideration and are referred to as shock waves. They will be examined from a related point of view in the next section.

## 13.12. Shock waves

A small pressure disturbance in an elastic medium will be propagated forward with the speed of sound, and if the medium is a gas [$a = \sqrt{(\gamma RT)}$],

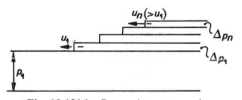

**Fig. 13.12(a).** *Successive compressions*

the latter will be proportional to the square root of the absolute temperature. Let us suppose that we cause a large impulse by applying, in rapid succession, a large number of small ones. Each successive compression will cause a corresponding temperature rise, and hence an increase in the velocity of propagation of the wavelets generated at a higher pressure, see Fig. 13.12(a).

It follows therefore that instead of moving forward, unchanged in

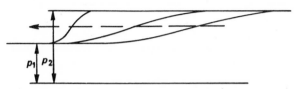

**Fig. 13.12(b).** *Generation of compression wave*

form, the wave-front will steepen as it advances. Eventually our assumption that the process of compression is continuous, ceases to be valid, and we are left to examine the changes which may take place discontinuously across a *sudden* wave of compression. Before doing so however,

**Fig. 13.12(c).** *Generation of expansion wave*

we may observe that when the above argument is applied to a similarly generated expansion wave, we infer that no matter however intense, or sudden, the initial rarefaction, the wave-form becomes less and less steep as it is propagated forward, see Fig. 13.12(c). We conclude therefore that the equations for continuous changes of state apply to the expansion of a gas through a series of what are termed Mach waves, but that the cumulative effect of the superposed compression waves causes a jump phenomenon, in which the pressure rises virtually discontinuously, from

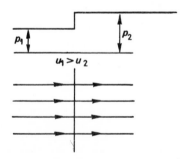

**Fig. 13.12(d).** *Steady flow relative to wave front*

an initial value $p_1$ to some higher value $p_2$. We shall again adopt the artifice of studying conditions relative to the wave-front so that the conditions for steady flow apply, see Fig. 13.12(d). Again there can be no contraction of the stream-tube so that continuity requires:

$$\dot{m} = \rho A u = \text{constant}$$

i.e.
$$\rho_1 u_1 = \rho_2 u_2 \tag{1}$$

The equation of motion is:

$$(p_1 - p_2)A = \dot{m}(u_2 - u_1)$$
$$= A(\rho_2 u_2^2 - \rho_1 u_1^2)$$

i.e.
$$p_1 + \rho_1 u_1^2 = p_2 + \rho_2 u_2^2 \tag{2}$$

from which, since

$$a = \sqrt{\left(\frac{\gamma p}{\rho}\right)} = \frac{u}{Ma}$$

we deduce that:

$$\rho_1(a_1^2 + \gamma u_1^2) = \rho_2(a_2^2 + \gamma u_2^2) \quad \text{or} \quad \rho_1 u_1^2 \left(\frac{1}{Ma_1^2} + \gamma\right) = \rho_2 u_2^2 \left(\frac{1}{Ma_2^2} + \gamma\right) \tag{2a}$$

Also, energy conservation for adiabatic flow implies that:

$$\tfrac{1}{2}u_1^2 + h_1 = \tfrac{1}{2}u_2^2 + h_2 \tag{3}$$

from which, since

$$h = c_p T = \frac{\gamma R T}{\gamma - 1} = \frac{a^2}{\gamma - 1}$$

we deduce:

$$\left(\gamma - 1 + \frac{2}{Ma_1^2}\right)\frac{u_1^2}{u_2^2} = \left(\gamma - 1 + \frac{2}{Ma^2}\right) \tag{3a}$$

By combining Equations (1) and (2a) we can obtain:

$$\frac{u_1}{u_2}\left(\gamma - 1 + \frac{2}{Ma_1^2}\right) + \left(\frac{u_1}{u_2} - 1\right)(\gamma + 1) = \left(\gamma - 1 + \frac{2}{Ma_2^2}\right)$$

and when this is subtracted from (3a) we get:

$$\left(\gamma - 1 + \frac{2}{Ma_1^2}\right)\left(\frac{u_1^2}{u_2^2} - \frac{u_1}{u_2}\right) - (\gamma + 1)\left(\frac{u_1}{u_2} - 1\right) = 0$$

i.e.
$$\left(\frac{u_1}{u_2} - 1\right)\left\{\left(\gamma - 1 + \frac{2}{Ma_1^2}\right)\frac{u_1}{u_2} - (\gamma + 1)\right\} = 0$$

Hence, either $[(u_1/u_2) - 1] = 0$, i.e. $u_1 = u_2$ and there is no shock wave or

$$\frac{u_1}{u_2}\left(\frac{2}{Ma_1^2} + \gamma - 1\right) - (\gamma + 1) = 0$$

which is the condition for a normal shock wave. This may alternatively be expressed as:

$$\frac{u_2}{u_1} = \frac{(\gamma - 1)Ma_1^2 + 2}{(\gamma + 1)Ma_1^2} = \frac{Ma_1^2 + 5}{6 Ma_1^2}$$

for air and which, from Equation (1), is also the ratio of the densities $\rho_1/\rho_2$.

Inserting this ratio for $u_2/u_1$ in Equation (3a) we deduce, after some algebraic manipulation that:

$$Ma_2^2 = \frac{(\gamma-1)\,Ma_1^2+2}{2\gamma\,Ma_1^2-(\gamma-1)} = \frac{Ma_1^2+5}{7\,Ma_1^2-1}$$

as the most convenient equation relating the upstream and downstream Mach numbers for adiabatic air flow.

Similarly, since

$$a = \sqrt{\left(\frac{\gamma p}{\rho}\right)} = \frac{u}{Ma}$$

then the ratio of the pressures

$$\frac{p_2}{p_1} = \frac{\rho_2\,u_2^2\,Ma_1^2}{\rho_1\,u_1^2\,Ma_2^2} = \frac{u_2\,Ma_1^2}{u_1\,Ma_2^2}$$

which reduces to:

$$\frac{p_2}{p_1} = \frac{2\gamma\,Ma_1^2-(\gamma-1)}{\gamma+1} = 1+\frac{2\gamma}{\gamma+1}(Ma_1^2-1) = \frac{7\,Ma_1^2-1}{6}$$

for air.

Also, since

$$\frac{T_2}{T_1} = \frac{p_2\,\rho_1}{p_1\,\rho_2} = \frac{u_2^2\,Ma_1^2}{u_1^2\,Ma_2^2} = \frac{Ma_1^2}{Ma_2^2}\left[\frac{\gamma-1+(2/Ma_1^2)}{\gamma-1+(2/Ma_2^2)}\right]$$

we deduce that

$$\frac{T_2}{T_1} = \frac{(\gamma-1)\,Ma_1^2+2}{(\gamma-1)\,Ma_2^2+2} = \frac{Ma_1^2+5}{Ma_2^2+5}$$

for air.

Alternatively, the three fundamental equations regarding shock waves may be written:

$$\rho u = \text{constant} \tag{1}$$

i.e.

$$\left(\frac{p}{RT}\right)(Ma\times a) = \text{constant} \quad\text{or}\quad \frac{p^2\,Ma^2}{T} = \text{constant} \tag{1a}$$

$$p+\rho u^2 = \text{constant} \tag{2}$$

i.e.

$$p+\rho u^2\left[\frac{\gamma p}{\rho u^2}\,Ma^2\right] = p(1+\gamma\,Ma^2) = \text{constant} \tag{2a}$$

$$h+\tfrac{1}{2}u^2 = \text{constant} \tag{3}$$

i.e.

$$T+\tfrac{1}{2}\frac{u^2}{c_p} = T+\frac{(\gamma-1)\,u^2 T}{2\gamma\,RT} = T[1+\frac{(\gamma-1)}{2}\,Ma^2] = \text{constant} \tag{3a}$$

From these we may deduce, by eliminating $p$ from (1a) and (2a) that

$$\frac{T(1+\gamma\,Ma^2)^2}{Ma^2} = \text{constant} \tag{A}$$

and, using (3a) to eliminate $T$ from the latter equation, that

$$\frac{(1+\gamma\, Ma^2)^2}{Ma^2\,[1+\{(\gamma-1)\,Ma^2\}/2]}\ \text{constant} \tag{B}$$

Also, using (2a) and (A), we may deduce from the gas law:

$$\frac{\rho T}{p}=\frac{1}{R}=\text{constant}$$

that

$$\frac{\rho\, Ma^2}{1+\gamma\, Ma^2}=\textbf{constant} \tag{C}$$

EXAMPLE 13.12 (i)

*If the air density is doubled on passing through a stationary plane shock wave which is perpendicular to the direction of the flow, determine (i) the speed of the incident stream if $\gamma = 1\cdot4$ at sea level, (ii) the percentage increase in pressure.*

(i)

$$\frac{\rho_1}{\rho_2}=\frac{u_2}{u_1}=\frac{(\gamma-1)\,Ma_1^2+2}{(\gamma+1)\,Ma_1^2}=\frac{Ma_1^2+5}{6\,Ma_1^2}=\frac{1}{2}$$

Hence, $\qquad 4Ma_1^2 = 10 \quad\text{or}\quad Ma_1 = 1\cdot58 = u_1/a_1$

and $\qquad u_1 = 1\cdot58\times340\text{ m/s} = 538\text{ m/s}$

(ii)

$$\frac{p_2}{p_1}=\frac{2\gamma Ma_1^2-(\gamma-1)}{\gamma+1}=\frac{7Ma_1^2-1}{6}=\frac{7\times2\cdot5-1}{6}=2\cdot75$$

Hence,

$$\frac{p_2-p_1}{p_1}=1\cdot75\quad\text{or}\quad\textbf{175}\ \textit{per cent}$$

The expressions for pressure ratio temperature ratio and velocity ratio are shown graphically on Fig. 13.12(e) with the incident Mach number $Ma_1$ as the abscissa according to the following table of values:

| $Ma_1$ | 1·0 | 1·5 | 2·0 | 3·0 | 4·0 | 5·0 |
|---|---|---|---|---|---|---|
| $Ma_2$ | 1·0 | 0·70 | 0·577 | 0·475 | 0·435 | 0·415 |
| $p_2/p_1$ | 1·0 | 2·46 | 4·50 | 10·33 | 18·50 | 29·00 |
| $T_2/T_1$ | 1·0 | 1·32 | 1·688 | 2·679 | 4·047 | 5·800 |
| $u_2/u_1$ | 1·0 | 0·537 | 0·375 | 0·259 | 0·218 | 0·200 |

It can be seen that there is a large pressure rise across a shock wave for high incident Mach numbers (e.g. when $Ma_1 = 5\cdot0, p_2/p_1 = 29\cdot0$) but the pressure attained falls considerably below that which would be attainable by reversible adiabatic (isentropic) compression through the same temperature rise. For since for reversible adiabatic compression:

$$r\frac{p_2}{p_1} = \left(\frac{T_2}{T_1}\right)^{\gamma/(\gamma-1)} = \left(\frac{T_2}{T_1}\right)^{3\cdot5}$$

we should, if such compression were possible achieve the following ideal pressure ratios and the ratio actual/ideal gives a measure of efficiency of compression by shock wave.

| Ideal $r\dfrac{p_2}{p_1}$ | 1·0 | 2·64 | 6·24 | 31·62 | 133·35 | 462·38 |
|---|---|---|---|---|---|---|
| Ratio $\dfrac{\text{actual}}{\text{ideal}}$ | 1·0 | 0·933 | 0·723 | 0·327 | 0·138 | 0·063 |

If these results are applied to the intake, say, of a ram-jet we note that shock compression cannot provide an adequate pressure ratio $p_2/p_1$ at low Mach numbers. Hence, the pressure rise must be attained primarily

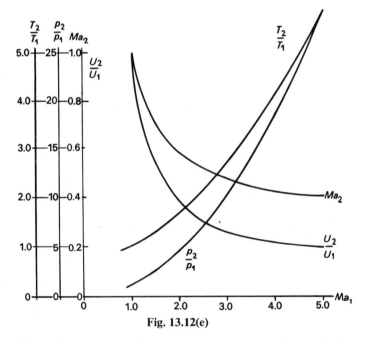

Fig. 13.12(e)

in the diffuser. At high Mach numbers a normal shock across the entry would cause a considerable compression, but the poor efficiency of compression would be intolerable. In order to increase the efficiency the nose must be so designed that the compression occurs through oblique shock waves which are considered in the next section.

Gas flow through a shock wave is only possible from a lower to a higher pressure and from a supersonic to a subsonic velocity, accompanied by an increase in entropy inevitable in all real processes in isolated systems. Thus, it is impossible to have a sudden 'rarefaction wave' or a sudden discontinuity leading from a higher to a lower pressure since this would involve a decrease in entropy. Such a spontaneous change or process in an isolated stream of fluid would contravene the second law of thermodynamics.

EXAMPLE 13.12 (ii)

*Calculate the ratio of rise in pressure to the pressure in the incident stream of air before a normal shock wave for a range of incident Mach numbers between 1·0 and 5·0. Make a similar calculation for density and plot graphs against the ratio of pressure rise to show the discrepancy between the actual and ideal ratio*

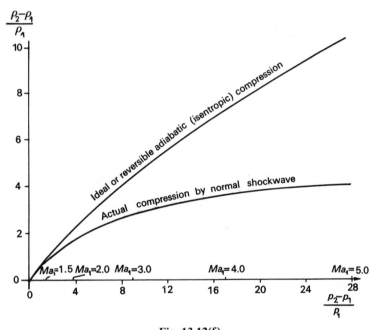

Fig. 13.12(f)

*of density rises—the ideal compression being according to the reversible adiabatic (isentropic) $p/\rho^\gamma = constant$, where $\gamma = 1·4$.*

From the formulae of this section:

$$\frac{p_2}{\rho_1} = \frac{u_1}{u_2} = \frac{6Ma_1^2}{Ma_1^2 + 5} \quad \text{and} \quad \frac{p_2}{p_1} = \frac{7Ma_1^2 - 1}{6}$$

and for reversible adiabatic (isentropic) compression

$$\frac{\rho_2}{\rho_1} = \left(\frac{p_2}{p_1}\right)^{1/\gamma}$$

we may draw up the following table:

| $Ma_1$ | 1·5 | 2·0 | 3·0 | 4·0 | 5·0 |
|---|---|---|---|---|---|
| $\dfrac{p_2-p_1}{p_1}$ | 1·46 | 3·50 | 9·33 | 17·50 | 28·0 |
| $\dfrac{\rho_2-\rho_1}{\rho_1}$ | 0·86 | 1·67 | 2·86 | 3·58 | 4·0 |
| Ideal $\dfrac{\rho_2-\rho_1}{\rho_1}$ | 0·90 | 1·92 | 4·56 | 7·51 | 10·1 |

From these numbers Fig. 13.12(f) may be drawn showing that it can by no means be assumed that the reversible adiabatic law $p/\rho^\gamma = $ constant applies to the change across a normal shock wave.

EXAMPLE 13.12 (iii)

(a) *Deduce expressions for the pressure rise* $(p_2-p_1)$ *across a shock wave in terms of mass flow per unit area* $(\dot{m}')$ *and the velocity change* $(u_2-u_1)$ *across it from:*
   (i) *the momentum equation,*
   (ii) *the energy equation.*
(b) *Hence, show that across a normal shock wave, the upstream and down-stream velocities* $u_1$ *and* $u_2$ *are related to the stagnation temperature* $T_0$ *by the equation:*

$$u_1 u_2 = \frac{2\gamma}{\gamma+1}\, RT_0$$

*and give a physical interpretation to the right-hand side of this equation.*

(a) By equating (per unit area) the pressure difference across the wave to the momentum flux, we obtain the equation of motion:

$$(p_2-p_1) = \dot{m}'(u_1-u_2) \tag{i}$$

Energy conservation requires that:

$$\frac{u_1^2}{2}+h_1 = h_0$$

the stagnation enthalpy per unit mass which, since

$$h = c_p T = \frac{\gamma RT}{\gamma-1} = \frac{\gamma}{\gamma-1}\frac{p}{\rho}$$

may be written:

$$\frac{u_1^2}{2}+\frac{\gamma}{\gamma-1}\frac{p_1}{\rho_1} = \frac{\gamma}{\gamma-1}\frac{p_0}{\rho_0} = \frac{\gamma}{\gamma-1}RT_0 = \frac{u_2^2}{2}+\frac{\gamma}{\gamma-1}\frac{p_2}{\rho_2}$$

Thus, on multiplying one by $[(\gamma-1)/\gamma]\rho_1$ and the other by $[(\gamma-1)/\gamma]\rho_2$ and subtracting, we obtain:

$$\frac{\gamma-1}{2\gamma}(\rho_2 u_2^2 - \rho_1 u_1^2) + (p_2 - p_1) = RT_0(\rho_2 - \rho_1)$$

and eliminating density by means of the equation for the mass flow per unit area, namely, $\dot{m}' = \rho_1 u_1 = \rho_2 u_2$, we get:

$$\frac{(\gamma-1)}{2\gamma}\dot{m}'(u_2 - u_1) + (p_2 - p_1) = \dot{m}' RT_0 \left(\frac{1}{u_2} - \frac{1}{u_1}\right)$$

Hence,

$$(p_2 - p_1) = \dot{m}'(u_1 - u_2)\left\{\frac{RT_0}{u_1 u_2} + \frac{\gamma-1}{2\gamma}\right\} \qquad \text{(ii)}$$

(b) Comparison of Equations (i) and (ii) indicates that, in general, the term within the last bracket must equal unity, i.e.

$$u_1 u_2 = \frac{2\gamma}{\gamma+1} RT_0$$

The right-hand side of this equation has the dimensions of a velocity squared, and is sometimes denoted by $a^{*2}$.

i.e. $\qquad\qquad u_1 u_2 = a^{*2}$

in which

$$a^* = \sqrt{\left(\frac{2\gamma}{\gamma+1} RT_0\right)}$$

a relationship due to Prandtl.

If $u_1$ were to approach $u_2$, $a^*$ would correspond to the fluid velocity $u_1 = u_2 = a^*$, across an infinitely weak wave. Conversely, this is of course the speed at which a weak wave would propagate into stationary fluid. This particular case may be compared with earlier work on Mach waves.

Recalling (p. 385) that:

$$\frac{T}{T_0} = \frac{1}{1 + [(\gamma-1) Ma^2/2]}$$

the temperature at the section where the local Mach number $Ma$ equals unity, $T = T^*$ say, is thus,

$$\frac{T^*}{T_0} = \frac{2}{\gamma+1}$$

The corresponding velocity at this section is equal to the local speed of sound, i.e.

$$u = a^* = \sqrt{(\gamma RT^*)} = \sqrt{\left(\frac{2\gamma}{\gamma+1} RT_0\right)}$$

This confirms that a vanishingly weak shock wave is the same as a Mach wave. Prandtl's equation indicates that if $u_1$ is greater than the local speed of sound $a^*$, $u_2$ must be less. In other words, the velocity downstream of a normal shock across a supersonic stream is always subsonic.

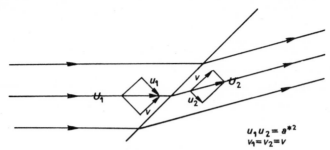

**Fig. 13.12(g).** *Oblique shock wave*

Just as conditions across an oblique Mach wave may be inferred from those for a wave normal to the stream, so may we deduce the changes across an oblique shock by superimposing a velocity component $v$ along the wave which remains unchanged before and after the shock, see Fig. 13.12(g).

EXAMPLE 13.12 (iv)

*Calculate the percentage error in using the normal formula*

$$u = \sqrt{\left[\frac{2(p_0 - p)}{\rho}\right]}$$

*for estimating or recording the speed of air from the observation $(p_0 - p)$ on a Pitot-static tube when flow is subsonic and rises to supersonic flow in which*

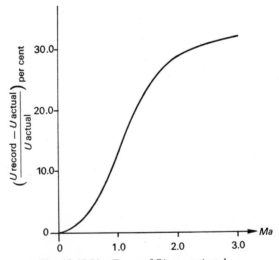

**Fig. 13.12(h).** *Error of Pitot-static tube*

*shock waves occur. What would be the error in recording speed of air having a Mach number of 0·3 and with a Mach number of 2·0?*

For adiabatic flow, the energy equation is:

$$h + \frac{u^2}{2} = \text{constant}$$

and if changes be, also, reversible then $\delta h + u\,du = 0 = (\delta p/\rho) + u\,du$ (p. 372).

(1) If compressibility be neglected so that the density of air $\rho$ in the free stream be assumed not sensibly to change up to the stagnation point at entry to the Pitot tube we have:

$$\frac{p_0 - p}{\rho} = \frac{1}{2}u^2 \quad \text{or} \quad \frac{p_0 - p}{p} = \frac{1}{2}\frac{\rho}{p}u^2 = \frac{1}{2}\gamma Ma^2$$

where $Ma$ denotes the Mach number of the incident free stream of velocity

$$u = \sqrt{\left[\frac{2(p_0 - p)}{\rho}\right]}$$

which is the speed recorded ($u_{\text{rec.}}$) by a Pitot tube.

(2) If compressibility cannot be neglected and the density rises from $\rho$ in the free stream to $\rho_0$ at the stagnation point according to the reversible adiabatic (isentropic) law $p/\rho^\gamma = \text{constant}$, then the energy equation $h + (u^2/2) = h_0$ which may be written:

$$\frac{h_0 - h}{h} = \frac{T_0}{T} - 1 = \frac{u^2}{2c_pT} = \frac{\gamma-1}{2}\frac{u^2\rho}{\gamma p} = \frac{\gamma-1}{2}Ma^2$$

shows that:
$$\frac{T_0}{T} = \left(\frac{p_0}{p}\right)^{(\gamma-1)/\gamma} = 1 + \frac{\gamma-1}{2}Ma^2$$

Alternatively

$$\frac{p_0}{p} = \left(1 + \frac{\gamma-1}{2}Ma^2\right)^{\gamma/(\gamma-1)} = 1 + \frac{\gamma}{2}Ma^2 + \frac{\gamma}{8}Ma^4 + \frac{\gamma(2-\gamma)}{48}Ma^6 + \ldots$$

or
$$\frac{p_0 - p}{(\gamma/2)pMa^2} = \frac{p_0 - p}{\frac{1}{2}\rho u^2} = 1 + \frac{Ma^2}{4} + \frac{Ma^4}{40} + \ldots = f(Ma)$$

Thus, for subsonic flow in which compression to the stagnation point may be assumed reversible adiabatic (isentropic)—since no shock wave can form in subsonic flow, we have the actual speed of air given by:

$$\tfrac{1}{2}\rho u_{\text{actual}}^2 = \frac{p_0 - p}{f(Ma)} = \frac{p_0 - p}{[1 + (Ma^2/4) + (Ma^4/40) + \ldots]}$$

whereas the speed estimated or recorded on the normal Pitot equation would be given by:

$$\tfrac{1}{2}\rho u_{\text{rec.}}^2 = p_0 - p$$

Therefore,

$$\frac{u_{\text{rec.}}^2}{u_{\text{act.}}^2} = f(Ma) = 1 + \frac{Ma^2}{4} + \frac{Ma^4}{40} + \ldots$$

for air.

| $Ma$ | 0·1 | 0·2 | 0·3 | 0·5 | 0·7 | 0·9 |
|---|---|---|---|---|---|---|
| $\dfrac{u_{\text{rec.}}^2}{u_{\text{act.}}^2} = f(Ma)$ | 1·003 | 1·010 | 1·023 | 1·064 | 1·129 | 1·203 |
| $\dfrac{u_{\text{rec.}} - u_{\text{act.}}}{u_{\text{act.}}}\%$ | 0·2 | 0·5 | 1·15 | 3·2 | 6·2 | 9·7 |

Thus, up to a Mach number of 0·3 the error in estimating or recording the speed of air by means of an ordinary Pitot-static tube without correction is a maximum of only one per cent.

(3) If there is a shock wave upstream of the tube reducing the Mach number $Ma$ of the free stream to $Ma_s$ immediately after the shock wave, the ratio of pressures on the two sides will be:

$$\frac{p_s}{p} = 1 + \frac{2\gamma}{\gamma+1}(Ma^2 - 1) \quad \text{and} \quad Ma_s^2 = \frac{(\gamma-1)Ma^2 + 2}{2\gamma Ma^2 - (\gamma-1)}$$

After the shock wave the flow will be subsonic and the compression from $p_s$ to the stagnation pressure $p_0$ at entry to the Pitot tube may be assumed reversible adiabatic (isentropic) as in (2). Hence,

$$\frac{p_0}{p_s} = \left(1 + \frac{\gamma-1}{2}Ma_s^2\right)^{\gamma/(\gamma-1)} = \left(\frac{\frac{1}{2}Ma^2(\gamma+1)^2}{2\gamma Ma^2 - (\gamma-1)}\right)^{\gamma/(\gamma-1)}$$

Therefore

$$\frac{p_0}{p} = \frac{p_0}{p_s} \times \frac{p_s}{p} = \left(\frac{\frac{1}{2}Ma^2(\gamma+1)^2}{2\gamma Ma^2 - (\gamma-1)}\right)^{\gamma/(\gamma-1)}\left(1 + \frac{2\gamma}{\gamma+1}(Ma^2 - 1)\right)$$

which reduces to

$$\frac{p_0}{p} = \left\{\frac{(\gamma+1)\left[(\gamma+1)\dfrac{Ma^2}{2}\right]^\gamma}{2(\gamma Ma^2 + 1) - (\gamma+1)}\right\}^{1/(\gamma-1)} = \varphi(Ma) = \left(\frac{7·7454 Ma^{2·8}}{7Ma^2 - 1}\right)^{2·5}$$

for air, and

$$\frac{p_0 - p}{\frac{1}{2}\gamma Ma^2 p} = \frac{p_0 - p}{\frac{1}{2}\rho u^2} = \frac{\varphi(Ma) - 1}{\frac{1}{2}\gamma Ma^2}$$

or

$$\frac{1}{2}\rho u_{\text{actual}}^2 = (p_0 - p)\left(\frac{\frac{1}{2}\gamma Ma^2}{\varphi(Ma) - 1}\right)$$

Thus, for Pitot-tube measurements of supersonic flow in which there is a shock wave upstream of the tube

$$\frac{u_{\text{rec.}}^2}{u_{\text{act.}}^2} = \frac{\varphi(Ma) - 1}{\frac{1}{2}\gamma Ma^2}$$

| $Ma$ | 1·25 | 1·5 | 2·0 | 2·5 | 3·0 |
|---|---|---|---|---|---|
| $\dfrac{u^2_{\text{rec.}}}{u^2_{\text{act.}}} = \dfrac{\varphi(Ma)-1}{\frac{1}{2}\gamma Ma^2}$ | 1·41 | 1·523 | 1·662 | 1·703 | 1·745 |
| $\dfrac{u_{\text{rec.}} - u_{\text{act.}}}{u_{\text{act.}}}$ % | 19 | 23·5 | 29 | 31 | 32 |

Thus, an ordinary Pitot-static tube used without correction to estimate or record the speed of air having a Mach number of 2·0 would be in error by 29 per cent.

## 13.13. Effect of back pressure on nozzles

The object of a convergent-divergent nozzle of the type first used by de Laval in impulse turbines is to generate, as efficiently as possible, a stream of kinetic energy by expansion of the fluid from the pressure

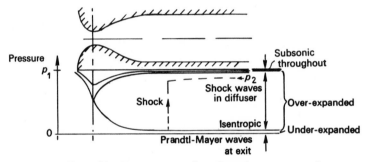

**Fig. 13.13(a).** *Effect of back pressure on flow through convergent-divergent nozzle*

upstream to that at exit. Suppose that the back pressure is dropped progressively, whilst that in the reservoir remains constant, see Fig. 13.13(a).

For small pressure drops, the flow is virtually incompressible, and the ideal performance would be in accordance with Bernoulli's equation, with a minimum pressure at the throat, where the velocity is a maximum. Qualitatively the picture remains unchanged as the pressure is dropped still further, so long as the flow remains subsonic throughout. Once the pressure at the throat has dropped to its critical value, however, i.e. to:

$$p_c = p_1 \left(\frac{2}{\gamma+1}\right)^{\gamma/(\gamma-1)}$$

(see Section 13.10, p. 386) the nozzle chokes (assuming reversible adiabatic flow) and further reduction in back pressure causes no change in the mass flow (see Section 13.9, p. 381). In Example 13.9(i) it was shown for the typical case of air at $p_1 = P$ kN/m² and $TK$ in the reservoir, the choked mass flow for throat of area $a$ mm² is then:

$$\dot{m}_c = \frac{40 \cdot 4}{10^6} \frac{aP}{\sqrt{T}} \text{kg/s}$$

For a given exit area $A_2$, there is only one value of the pressure ratio $(p_2/p_1)$ which allows isentropic expansion throughout the nozzle, and this is such that:

$$\dot{m}_c = \rho_2 A_2 u_2 = \rho_1 \left(\frac{p_2}{p_1}\right)^{1/\gamma} A_2 u_2$$

i.e. the value of $r = p_2/p_1$ is such that:

$$\dot{m}_c = A_2 \sqrt{\left(\frac{2\gamma}{\gamma - 1} p_1 \rho_1 r^{2/\gamma} [1 - r^{(\gamma - 1)/\gamma}]\right)}$$

because the exit velocity for isentropic expansion is:

$$u_2 = \sqrt{\left(\frac{2\gamma}{\gamma - 1} \frac{p_1}{\rho_1} [1 - r^{(\gamma - 1)/\gamma}]\right)}$$

(see Section 13.9, p. 380).

This corresponds to the lower branch of the curves shown in Fig. 13.13(a). If the back pressure is maintained at a lower value than this, the fluid must, if it has flowed through the nozzle isentropically, expand subsequently downstream of the nozzle exit. Observations confirm that this takes place initially through Prandtl–Mayer expansion waves (see Fig. 13.13(b), and Section 13.11, p. 391).

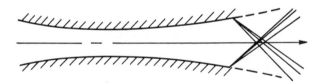

**Fig. 13.13(b).** *Under-expanded nozzle*

Similarly if the back pressure is above that to which the fluid expands in the divergent portion of the nozzle it must recompress subsequently through a system of shock waves, see Fig. 13.13(c). Such processes imply

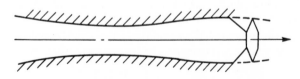

**Fig. 13.13(c).** *Over-expanded nozzle*

degradation of available energy in heat generated, i.e. they are mechanically inefficient. The actual shock pattern depends on the magnitude of the recompression. If the latter is large a normal shock is formed on the axis and this may, if the back pressure is increased, travel upstream into the divergent part of the nozzle. It decays in intensity, ultimately reaching the throat as a Mach wave, as the back pressure approaches that for isentropic recompression. Further increase in back pressure causes the wave to die out completely and the flow reverts to subsonic.

From a practical point of view, if a uniform supersonic flow is required at exit, it is necessary to operate a supersonic nozzle at its designed pressure ratio and at no other.

Whilst these phenomena require an adequate supply of compressed air or other fluid, and optical apparatus for their observation, analogous effects may be observed on the surface of water flowing down a similarly shaped open channel. These, and other surface phenomena, will be discussed in the next chapter.

## Exercises on Chapter 13

1. Deduce expressions for change of density with height and the ratio of the pressures over a height $h$ in the stratosphere where the temperature remains constant. Use the facts that the velocity of sound is $a = \sqrt{(\gamma RT)}$ and $p \propto \rho$ at constant temperature.

2. Assuming a constant lapse rate $\lambda$ and the law $p/\rho^n = $ constant show that $n = 1/[1-(\lambda R/g)]$.

Hence, calculate the value of $n$ and estimate the height, the relative density and relative pressure in the atmosphere where the temperature is $-33°C$ if the temperature at sea level is $15°C$.

3. In the stratosphere, where the temperature is assumed constant at $-56.5°C$, deduce an expression for the pressure ratio at two heights and calculate this ratio for a change in height of 3 km.

4. At zero altitude the pressure is $0.101$ MPa, the temperature is $15°C$ and the air density is $1.23$ kg/m³. Find (*a*) the temperature and (*b*) the density, at an altitude of 3 000 m assuming that the relation between pressure and density is $p = k\rho^n$ where $k$ is constant and $n$ is $1.25$.

5. Define the term 'convective equilibrium' used in connection with the stability of the atmosphere.

Working from first principles, estimate the temperature lapse rate when the atmosphere is in a state of convective equilibrium. Assume the adiabatic law applies and take $\gamma = 1.4$ and the gas constant $R = 288$ J/kg K for air.

6. The pressure recorded through a small hole in the wall of a pipe conveying air at $38°C$ with a velocity of 300 m/s is 145 kN/m². Calculate the Mach number of the air stream and the pressure measured by a Pitot tube facing directly upstream. Assume that, for air, $\gamma = 1.4$ and the characteristic constant is 288 J/kg K. Take account of compressibility.

7. The variation in the cross-sectional area $S$ of a stream-tube is determined by the equation of continuity $\rho u S = $ constant, in which $\rho$ denotes density of fluid and $u$ is its velocity at the same section.

406     *Compressibility effects in fluids*

Apply this equation in conjunction with the Bernoulli equation for compressible flow, $(\delta p/\rho) + u\delta u = 0$, to show that, for adiabatic conditions defined by $p = k\rho^\gamma$, $dS/du = S/u(Ma^2 - 1)$ in which $Ma$ denotes Mach number.

Explain the significance of this equation with reference to the flow pattern round a body as the velocity of flow approaches and exceeds the velocity of sound given by $a = \sqrt{(\gamma p/\rho)}$.

8. Prove that if $p$, $\rho$, $u$ denote pressure, density and velocity at any point in the steady adiabatic (not necessarily frictionless) flow of a compressible gas then along a horizontal streamline:

$$\tfrac{1}{2}u^2 + \frac{\gamma}{\gamma - 1}\frac{p}{\rho} = \text{constant}$$

where $\gamma$ is the ratio of the specific heat-capacities $c_p/c_v$.

A steady stream of air is flowing past a fixed aerofoil with initial velocity $v_0$, the corresponding speed of sound being $a_0$. Deduce a formula for the speed at a point just outside the boundary layer where the local Mach number is unity. Take $\gamma = 1.4$.

9. Show how the equation $[\gamma/(\gamma - 1)][p/\rho] + (u^2/2) = \text{constant}$ for adiabatic flow of compressible gas is obtained.

Assuming reversible adiabatic (i.e. isentropic) flow from entry to the throat of a Venturi meter, calculate the flow of air in kg/min if the inlet diameter is 400 mm and the throat diameter is 125 mm. The temperature and pressure at inlet were 17°C and 140 kPa abs. when the pressure in the throat was 119 kPa abs. Assume 0.96 as the coefficient of discharge for the meter.

10. Taking the ratio of specific heat-capacities $c_p/c_v$ as 1.4, develop the formula

$$u^2 + 5a^2 = \text{constant}$$

for steady adiabatic flow of air along a streamline where $p$, $\rho$ and $u$ denote pressure, density and velocity of the fluid and $a$ is the local speed of sound.

A steady stream of air is flowing past a fixed aerofoil at initial speed $u_1$, the corresponding speed of sound being $a_1$ and air density $\rho_1$. Determine, in terms of $\rho_1$, $u_1$ and $a_1$, the speed and density of the air just outside the boundary layer at a point where the local Mach number is $\tfrac{3}{4}$.

11. Deduce and use an appropriate formula (see Section 13·8) to calculate the rate of mass flow of air through a horizontal pipe 150 mm diameter and 150 m long assuming isothermal condition ($p/\rho = \text{constant}$) when the inlet conditions are 1 MN/m² abs. and 67°C and the discharge pressure is 0.9 MN/m² abs. Assume $f = 0.005$—the analogous incompressible flow formula being

$$h = 4flu^2/2gd \quad \text{(see pp. 375 and 284)}$$

12. Estimate the dynamic pressure of air for a speed of 965 km/h or 522 knot at standard pressure of 0.101 3 MPa abs. assuming (i) incompressible flow, (ii) isothermal compressible flow, (iii) frictionless adiabatic compressible flow.

13. Compressed air flows through a long pipe of uniform diameter which is buried in the ground so that conditions may be considered isothermal. Derive a formula for the fall in pressure due to friction and state any assumptions made.

A pipe is 50 mm diameter and 1 200 m long. Calculate the flow in m³/min of free air at 15°C, and 101·3 kPa abs. if the initial pressure is 1 MPa abs. and the final pressure 0·7 MPa abs. The flow temperature is 5°C and the friction coefficient $f = 0·004$ where the drag stress referred to the pipe area in contact with air is $f \times \frac{1}{2}\rho u^2$.

14. A convergent-divergent nozzle is fitted into the side of a large vessel containing air under constant pressure. Prove that, when the rate of discharge is a maximum, the speed through the throat is equal to the speed of sound at the throat (i.e. $\sqrt{(dp/d\rho)}$ where $p$ and $\rho$ are respectively the pressure and density of the air).

Determine also the percentage change in the air density between the reservoir and the throat under these conditions.

Take the ratio of specific heat-capacities, $\gamma = 1·40$.

15. Write down Bernoulli's equation for the steady flow of an inviscid compressible fluid satisfying the isentropic law $p/\rho^\gamma$ = constant. Hence show that for flow at subsonic Mach numbers the difference between the Pitot ($p_0$) and static ($p$) pressures of a Pitot-static tube in the fluid is given by:

$$p_0 - p = \tfrac{1}{2}\rho u^2 \left\{ 1 + \frac{Ma_1^2}{4} + \frac{Ma_1^4}{40} + \ldots \right\}$$

where $u$ is the velocity and $Ma_1$ the Mach number of the undisturbed stream. Assume $\gamma$ is 1·4.

16. In streamline flow near the upper surface of an aeroplane wing, the velocity just outside the boundary layer changes from 258 km/h at a point A near the leading edge to 466 km/h at a point B to the rear of A. If the temperature at A is 8°C, calculate the temperature at B. Assume the reversible adiabatic law of compression $p/\rho^\gamma$ = constant and $\gamma = 1·4$ for air. Also find the value of the local Mach number at point B.

17. An aerofoil is set in a wind-stream operating at a Mach number $Ma_0$ at pressure $p_0$. Calculate, from first principles, the pressure on the upper surface of the aerofoil at a point where the Mach number is unity. Take $\gamma = 1·4$.

18. Estimate the percentage error in calculating the stagnation pressure of air at 0·1013 MPa abs. and 15°C flowing at 92 m/s when compressibility is ignored. Also, estimate the percentage error in the calculation for difference of pressure recorded on a Pitot-static tube. For air $\gamma = 1·4$ and $R = 288$ J/kg K.

19. When an aeroplane is flying at a Mach number of 0·7 at 10 km, where the relative density of the air $\rho_1/\rho_0$ is 0·375, the relative airflow attains a maximum local Mach number of 0·95. Determine the local density of the air where this maximum occurs, taking $\gamma = 1·4$ and $\rho_0 = 1·23$ kg/m³.

20. A Pitot-static tube records a difference between the stagnation and

static pressures of 19·3 MPa. If the static pressure and static temperature of the air-stream are 0·1 MPa abs. and 15°C, calculate the speed of the air:

    (i) assuming the density of the air-stream remains constant,

    (ii) taking compressibility of air into account, and

    (iii) state by what percentage does method (i) overestimate the speed.

21. Write down the three fundamental equations governing the flow of a gas through a plane normal shock wave.

A normal shock wave forms in an air-stream at a static temperature of 222 K, the total temperature being 400 K. Estimate the Mach number and static temperature behind the shock wave, given that the pressure rise through the shock wave, $p_2/p_1$, is $(7Ma_1^2 - 1)/6$. Prove the latter formula.

22. Air flows through a parallel passage in which a shock wave is formed. If the subscripts (1) and (2) refer to conditions first before and after the wave, combine the energy and momentum equations to prove that

$$\frac{u_1^2}{2} + \frac{\gamma}{\gamma-1}\frac{p_1}{\rho_1} = \frac{u_2^2}{2} + \frac{\gamma}{\gamma-1}\left(\frac{u_2}{u_1}\right)\left\{u_1^2 - u_1 u_2 + \frac{p_1}{\rho_1}\right\}$$

At a position (1) in a pipe $p_1 = 690$ kN/m² abs., $\rho_1 = 5·45$ kg/m³ and $u_1 = 450$ m/s. Calculate the values of $u_2$ and $p_2$ immediately after the shock wave, given that $\gamma = 1·4$. Also calculate the Mach numbers before and after the wave.

# Varying flow in open channels

'*In order to understand better what I mean,
watch the blades of straw that because of their
lightness are floating on the water, and observe
how they do not depart from their original
position in spite of the waves underneath
them. . . .*'
                                                    LEONARDO

## 14.1. Shallow water wavelets caused by gravity

At the close of the previous chapter it was remarked that there are striking
similarities between the surface wave phenomena observed in open
channels, and the Mach and shock waves encountered at supersonic

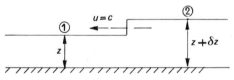

**Fig. 14.1(a).** *Surface wavelet*

speeds in compressible media. This suggests that we should examine the
way in which surface wavelets travel forward along a channel, in order
to establish *their* velocity of propagation. Fig. 14.1(a) shows such a
plane wavelet, which we shall, as in Section 13.10, imagine being observed
from a vehicle travelling forward with it. If the wave speed $u$ be denoted
by $c$, then the oncoming fluid, as we now imagine it (on passing through
the now apparently stationary wave), undergoes an increase in its height
energy which, if it involves no losses, is given by Bernoulli's equation:

$$\frac{p_1}{\rho} + gz_1 + \frac{u_1^2}{2} = \frac{p_2}{\rho} + gz_2 + \frac{u_2^2}{2}$$

or
$$\frac{p}{\rho} + gz + \frac{u^2}{2} = \text{constant}$$

Noting that the free surface pressures are both atmospheric and that the amplitude is assumed small then, for the free surface:

$$g\delta z + u\delta u = 0$$

If the wave is travelling up a channel of uniform width, as in Fig. 14.1(b), continuity requires that the oncoming flow relative to the wave

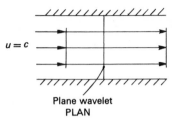

Plane wavelet
PLAN
**Fig. 14.1(b).** *Wavelet in uniform channel*

shall equal that downstream of it, i.e.:

$$uz = \text{constant}$$

or

$$\frac{\delta u}{u} + \frac{\delta z}{z} = 0$$

Thus we have two expressions for $(\delta z/\delta u)$, namely:

$$\frac{\delta z}{\delta u} = -\frac{z}{u} = -\frac{u}{g}$$

i.e.

$$u^2 = c^2 = gz$$

and

$$c = \sqrt{(gz)}$$

The speed of propagation of shallow surface waves plays the same role as does the speed of sound in aerodynamics, for in assessing regions of flow, we use as a parameter, the ratio which the fluid speed bears to this wave speed:

$$\boldsymbol{Fr} = \frac{u}{c} = \frac{u}{\sqrt{(gz)}}$$

and refer to this ratio as the **Froude number.**

If friction is present, we may extend the validity of the above equations to the case of a stationary wavelet, provided that the oncoming flow is of uniform depth in the channel. That is, *if the slope of the channel is such that it just compensates for the friction*, i.e. $h_f = d$ in Fig. 14.1(c), then

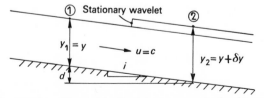

**Fig. 14.1(c).** *Channel inclined to compensate for friction*

the energy equations

$$\frac{p_1}{w}+\frac{u_1^2}{2g}+(d+y_1) = \frac{p_2}{w}+\frac{u_2^2}{2g}+y_2+h_f$$

or

$$\frac{u_1^2}{2g}+y_1 = \frac{u_2^2}{2g}+y_2+(h_f-d)$$

may be written:

$$gy+\frac{u^2}{2} = \text{constant}$$

i.e.

$$g\delta y+u\delta u = 0$$

applies, $y$ being measured from the channel bed, which is no longer a horizontal datum. The continuity equation also remains unchanged:

$$\frac{\delta u}{u}+\frac{\delta y}{y} = 0$$

so that the previous analysis applies, giving:

$$u = c = \sqrt{(gy)}$$

That is, *if the wavelet is to remain stationary* and normal to the stream, $u$ and $y$ must have corresponding values given by this equation *and*

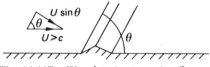

**Fig. 14.1(d).** *Wavelets in streaming flow*

*conditions in the stream are then said to be 'critical'* and $Fr = 1$. For Froude numbers less than one, i.e. **for $u < \sqrt{(gy)}$, the flow is said to be 'tranquil' or 'streaming', and for $u > \sqrt{(gy)}$ the flow is referred to as 'shooting' or 'rapid'.**

The analogy with compressible flow suggests that streaming (i.e. sub-critical) flows are analogous to subsonic conditions and shooting flows to supersonic. Referring back to the previous chapter it will be recalled that any small disturbances cause wavelets inclined at the appropriate Mach angle in supersonic flows, see Fig. 13.11(c) (p. 390).

Similar wavelets occur in shooting flows, although the pattern may be complicated (especially on small-scale models) by capillary waves. Just as in Section 13.11, the wavelets set themselves at such an angle that they are held stationary by fluid crossing it normally with a component velocity $u \sin \theta$ equal to the rate at which the waves would propagate into still water, i.e.

$$u \sin \theta = c = \sqrt{(gz)} = \frac{u}{Fr}$$

∴

$$\theta = \sin^{-1}\left(\frac{1}{Fr}\right)$$

## 14.2.  Broad-crested weirs; shooting and streaming flows

If water flows down a channel with a mean velocity $u$ then, neglecting friction, the surface streamline lies below a datum corresponding to the total energy line by an amount $(u^2/2g)$ which can be measured experimentally by putting a total head tube into the flow (see Fig. 14.2(a)).

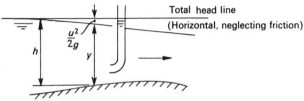

**Fig. 14.2(a).** *'Specific energy'*

The elevation of the total head line *above the bed* is known as the

**'specific energy'**
$$h = y + \frac{u^2}{2g}$$

*It will be noted that in this particular context, the word 'specific' does not imply energy per unit mass.*

There exists a unique relationship between the Froude number $Fr$ of the stream at any section and the ratio $(h/y)$, which is quite simply obtained from this equation:
$$\frac{h}{y} = 1 + \frac{u^2}{2gy}$$

i.e.
$$\frac{h}{y} = 1 + \tfrac{1}{2}Fr^2$$

Thus if we plot $h$ against $y$, lines of constant Froude number are straight and have a slope $(1 + \tfrac{1}{2} Fr^2)$, see Fig. 14.2(b). In particular the

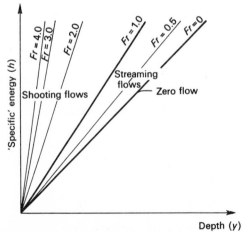

**Fig. 14.2(b).** *'Specific' energy and Froude number*

line for no flow ($Fr = 0$) is of unit slope and that for critical conditions ($Fr = 1$) has a slope of 3/2. These two lines define the zone in which all streaming flows lie (compare subsonic flows) and above the line $Fr = 1$ lies the zone of shooting flows (cf. supersonic flows). *When $Fr = 1$ $h = \frac{3}{2} y$ and the critical depth is 2/3 of the elevation of the still-water surface above the bed*; so long as the depth is greater than this, the flow is a streaming one, and if less than this it is shooting, see Fig. 14.2(c).

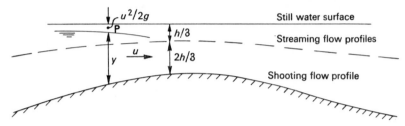

Fig. 14.2(c). *Streaming and shooting flows*

If there is no sideways contraction of the stream the flow *per unit width*, $Q'$ say, will be constant, i.e.

$$Q' = uy = \text{constant}$$

or, considering rates of change along the length of the channel, i.e.:

$$\frac{\delta u}{u} + \frac{\delta y}{y} = 0$$

Hence any small fractional increase in depth causes a corresponding fractional decrease in velocity and vice versa (cf. Section 13.9, p. 378).

The 'specific' energy expression:

$$h = \frac{u^2}{2g} + y$$

yields a second expression relating $u$ and $y$, which in differential terms is:

$$g\delta h = u\delta u + g\delta y$$

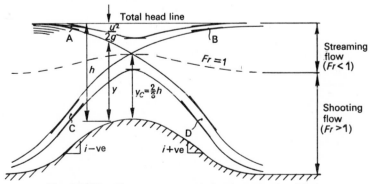

Fig. 14.2(d). *Flow over a rounded weir neglecting friction*

For example, *over the crest* (Fig. 14.2(d)), $\delta h = 0$ so that we have the two relationships:

$$u\delta u + g\delta y = 0$$

$$\frac{\delta u}{u} + \frac{\delta y}{y} = 0$$

These are identical in form to a similar pair of equations developed for the speed of a shallow water wave in the previous section. Thus, one solution is:

$$u = c = \sqrt{(gy)}$$

i.e. *conditions over the crest are critical* and the corresponding depth is then:

$$y = y_c = \tfrac{2}{3}h = \textbf{critical depth}$$

This principle is exploited at a broad-crested weir, for a single reading at the crest is nominally sufficient to establish the discharge, so long as conditions are critical:

$$Q' = uy_c$$

$$Q' = y\sqrt{(gy)} = g^{1/2}y^{3/2}$$

In terms of the elevation of the total energy line:

$$Q' = \tfrac{2}{3}h\sqrt{(g\tfrac{2}{3}h)} = \frac{2\sqrt{2}}{3\sqrt{3}}g^{1/2}h^{3/2} = \left(1\cdot71\frac{m^{1/2}}{s}\right)h^{3/2}$$

The other solution and the only alternative to conditions changing from streaming to shooting flow, or vice versa, is for $\delta u$ and $\delta y$ to be zero over the crest. This possibility also satisfies the equations:

$$u\delta u + g\delta y = 0$$

$$\frac{\delta u}{u} + \frac{\delta y}{y} = 0$$

See Fig. 14.2(d) profiles AB and CD.

At sections *other than above the crest*:

$$\frac{u\delta u}{g} + \delta y = \delta h$$

from the equation for 'specific' energy.

Eliminating $\delta u$ by means of the continuity equation:

$$Q' = uy \quad \text{or} \quad \frac{\delta u}{u} + \frac{\delta y}{y} = 0$$

we get:

$$-\frac{u^2\delta y}{gy} + \delta y = \delta h$$

i.e.

$$\delta y = \frac{1}{1 - Fr^2}\delta h$$

If the total head line is very nearly horizontal, as it would be for frictionless flows, we may infer the nature of the surface profiles. For if *Fr* is less than unity, an adverse bed slope (i.e. a decrease in *h*) causes an even greater decrease in *y*, i.e. a fall in level, as at A. Similarly if the bed level falls in a streaming flow, the surface level rises, as at B. On the other hand, if the flow is shooting (*Fr* > 1), the stream thickens as it approaches

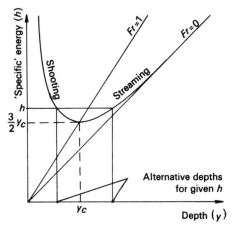

**Fig. 14.2(e).** *Curve for a particular discharge over a broad-crested weir*

the hump, as at C, and decreases in thickness as the bed falls again as at D.

The nature of the possible surface profiles over, for example, a broad-crested weir may thus be inferred, and leads to four possibilities:

(i) a streaming flow throughout—curve AB,
(ii) a streaming flow which passes through the critical depth to shooting flow—curve AD,
(iii) a shooting flow throughout—curve CD,
(iv) a shooting flow which passes through the critical depth to streaming flow—curve CB.

These changes may conveniently be followed on a plot of 'specific' energy against depth. Thus, since $u = Q'/y$, 'specific' energy, in terms of the discharge and depth, is:

$$h = \frac{Q'^2}{2gy^2} + y$$

Fig. 14.2(e) shows a typical curve for a particular discharge.

Case (i) corresponds to an excursion down the right-hand branch of the curve and back up again.

Case (ii) corresponds to an excursion from the right-hand branch to the left-hand one, passing through the critical depth $Fr = 1$.

Case (iii) corresponds to an excursion down the left-hand branch followed by a return along the same branch.

Case (iv) traverses the curve from left to right, passing through the point of minimum 'specific' energy.

The particular path followed in any given case depends on the surface levels upstream and downstream.

### EXAMPLE 14.2 (i)

*A block* 40 mm *thick is placed across the bed of a horizontal channel* 380 mm *wide to form a broad-crested weir. The depth of water just upstream of the weir is* 70 mm. *Find the rate of flow on the assumption that conditions over the block are critical and make one correction for velocity of approach.*

If conditions over the block are critical $y = \frac{2}{3}h$ and $u = \sqrt{(gy)}$. Hence

$$Q = buy = b(\tfrac{2}{3}h)^{3/2}\sqrt{g} = \left(1\cdot71\frac{\mathrm{m}^{1/2}}{\mathrm{s}}\right)bh^{3/2}$$

Therefore, as a first approximation

$$Q = \left(1\cdot71\frac{\mathrm{m}^{1/2}}{\mathrm{s}}\right)\times 0\cdot38 \text{ m}\times(0\cdot03 \text{ m})^{3/2}$$

$$= 0\cdot00338 \text{ m}^3/\text{s} \quad \text{or} \quad 3\cdot38 \text{ litre/s}$$

Using this to find the velocity of approach, namely

$$u_1 = \frac{Q}{A} = \frac{3\,380\times10^3\,\mathrm{mm}^3/\mathrm{s}}{38\times7\times10^2\,\mathrm{mm}^2} = 127 \text{ mm/s}$$

the kinetic head upstream is

$$\frac{u_1^2}{2g} = \frac{161\cdot5}{1\,961\cdot3} = 0\cdot0823 \text{ mm}$$

and the corrected $h = 30+0\cdot082 = 30\cdot082$ mm.

This gives

$$Q = \left(1\cdot71\frac{\mathrm{m}^{1/2}}{\mathrm{s}}\right)\times\frac{38 \text{ m}}{10^2}\left(\frac{3\cdot0082 \text{ m}}{10^2}\right)^{3/2}$$

$$= 0\cdot00339 \text{ m}^3/\text{s} \quad \text{or} \quad 3\cdot39 \text{ litre/s}$$

## 14.3. Effect of friction on channel flow

In the previous Section it was shown that under varying conditions, any increase in total head $(\delta h)$ was related to the corresponding increase in depth by the equation:

$$\delta y = \frac{1}{1-Fr^2}\delta h$$

In that section the total head line was taken to be horizontal. If friction is present, and the rate of *loss* of head is $k$, then the total head loss over a length $\delta x$ of channel is $k\delta x$ for small slopes over a length $\delta x$

which may be taken to be either horizontal or along the bed. Similarly, if the *downward* slope of the bed is *i*, see Fig. 14.3(a):

$$h + i\delta x = h + \delta h + k\delta x$$

$$\delta h = (i - k)\delta x$$

Thus, eliminating the specific energy term $\delta h$ from the two equations,

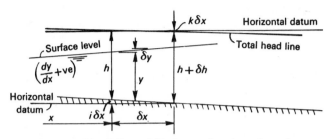

**Fig. 14.3(a).** *Effect of friction on flow in a channel*

we obtain a differential equation for the depth of water *y* in channel of slope *i* under slowly varying conditions:

$$\frac{dy}{dx} = \frac{i - k}{1 - Fr^2}$$

in which the denominator of the right-hand side represents the rate of change of 'specific' energy with depth:

$$\frac{dh}{dy} = 1 - Fr^2 = 1 - \frac{u^2}{gy}$$

This latter expression, which springs from the definition of 'specific' energy and continuity for a channel of uniform width, *is equally true whether or not friction is present.*

EXAMPLE 14.3 (i)

*A wide rectangular channel has a slope of 0·000 3, the normal depth of flow in it being 0·6 m. A weir is placed across the channel increasing the local depth to 0·8 m. Find approximately how far back from the weir that depths are 0·75 and 0·725 m given that the Chezy constant is 55·5 m$^{1/2}$/s.*

$$u = C \sqrt{(mi)} = 55.5 \text{ m}^{1/2}/\text{s} \sqrt{(0.6 \text{ m} \times 0.000\ 3)} = 0.745 \text{ m/s}$$

$$Fr^2 = \frac{u^2}{(gy)} = \frac{0.555}{9.807 \times 0.6} = 0.093\ 7$$

Neglecting friction (i.e. $k = 0$)

$$\frac{dy}{dx} = \frac{i}{1 - Fr^2} = \frac{0.000\ 3}{0.906\ 3} = \frac{1}{3\ 021}$$

When *y* is 0·75 m

$\delta x = 3\,021\ \delta y = (0{\cdot}8 - 0{\cdot}75)3\,021$ m $= \mathbf{151}$ m back from the weir

and $y = 0{\cdot}725$ m at a distance of $\mathbf{227}$ m back from the weir.

## 14.4. Hydraulic jumps

Suppose that water in flowing down a channel of uniform width from a reservoir encounters an obstacle such as a weir. There are a number of possible surface profiles, see Fig. 14.4(a), depending on the level downstream.

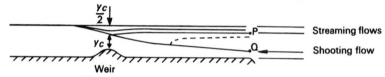

**Fig. 14.4(a).** *Effect of a weir on channel flow*

As the latter is progressively reduced, the velocity over the crest of the weir eventually reaches the critical value $u_c = \sqrt{(gy_c)}$. The flow through the channel is then dictated by the critical conditions at this section (cf. a choked nozzle, see Section 13.9, p. 380). There remain only two possible discharge levels, P and Q, which can be reached by the fluid, the former corresponding to streaming, and the latter to shooting flow, downstream of the weir. Any surface level below Q can ostensibly be reached by a fall at exit, but we have yet to determine how a shooting flow downstream of the weir can be made compatible with any discharge level lying between points P and Q. Analogy with convergent-divergent nozzles which are over-expanded suggests that we should examine whether there is not a similar hydraulic jump phenomenon, analogous to the shock waves which can occur in supersonic gas flows. This is found to be so, and we shall see that hydraulic jumps occur when the flow at a depth less than the critical ($Fr > 1$ or $y_c < \tfrac{2}{3}h$) suddenly changes to flow with a depth greater than the critical, and hence with a reduced mean velocity.

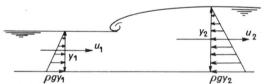

**Fig. 14.4(b).** *Standing wave and hydraulic jump*

Fig. 14.4(b) illustrates a more or less sudden jump in level across such a standing wave for which, assuming the vertical accelerations are negligible at sections 1 and 2, the equation of motion is:

Force $=$ Rate of mass flow $\times$ Change in velocity

$$\tfrac{1}{2}\rho g y_1^2 - \tfrac{1}{2}\rho g y_2^2 = \rho Q'(u_2 - u_1)$$

The left-hand side of this equation is the difference between the mean pressure times depth upstream and downstream of the wave and as this is the force per unit width it equals the mass flow per unit width $\rho Q'$ times its velocity change where:

$$Q' = u_1 y_1 = u_2 y_2$$

Solving for the change in levels for a given upstream velocity:

$$\tfrac{1}{2}g(y_1^2 - y_2^2) = u_1 y_1 \left( u_1 \frac{y_1}{y_2} - u_1 \right)$$

i.e.

$$\tfrac{1}{2}g(y_1 - y_2)(y_1 + y_2) = \frac{u_1^2 y_1}{y_2}(y_1 - y_2)$$

$\therefore$

$$\tfrac{1}{2}g(y_1 + y_2) = \frac{u_1^2 y_1}{y_2}$$

i.e.

$$u_1 = \sqrt{\left[ \frac{g y_2 (y_1 + y_2)}{2 y_1} \right]}$$

If we impress (say by an oar) a velocity of $u_1$ to the left in the above diagram, the expression may be interpreted as one for the velocity with

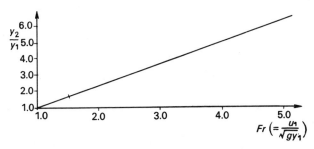

Fig. 14.4(c). *Ratio of depths across a standing wave*

which such a wave invades still water and is then referred to as a 'bore'. In the limiting case of a weak wave, $y_2$ approaches the value $y_1$ and we rediscover the expression:

$$u_1 = \sqrt{(g y_1)}$$

For stronger waves, $y_2 > y_1$ and so:

$$u_1 > \sqrt{(g y_1)}$$

Thus, the stronger the wave the greater its velocity of propagation.

Returning to the standing wave, an expression for the ratio of the depths across it may be related to the Froude number

$$Fr_1 = \frac{u_1}{\sqrt{(g y_1)}}$$

of the stream by writing:

$$\tfrac{1}{2}g\left(1+\frac{y_2}{y_1}\right) = \frac{u_1^2 y_1}{y_2 y_1}$$

i.e.    $1+\left(\dfrac{y_2}{y_1}\right) = 2Fr_1^2\left(\dfrac{y_1}{y_2}\right)$   or   $\left(\dfrac{y_2}{y_1}\right)^2+\dfrac{y_2}{y_1}-2Fr_1^2 = 0$

Solving this quadratic results in:

$$\frac{y_2}{y_1} = \tfrac{1}{2}(\sqrt{(1+8Fr_1^2)}-1)$$

see Fig. 14.4(c).

E<small>XAMPLE</small> 14.4 (i)

(a) *Find a relationship for critical depth of flow in an open channel in terms of depths before and after a standing wave.*

(b) *The depth of flow prior to a standing wave in a rectangular channel 130 mm wide is 100 mm. If the rate of flow is 14 litre/s, find the depth after the standing wave and the critical depth.*

(a) Referring to Fig. 14.4(b) we deduce, since $Q' = u_1 y_1 = u_c y_c = g^{1/2} y_c^{3/2}$ that

$$\tfrac{1}{2}g(y_1+y_2) = \frac{u_1^2 y_1}{y_2} = \frac{(Q')^2}{y_1 y_2} = \frac{g y_c^3}{y_1 y_2}$$

Hence    $2y_c^3 = y_1 y_2 (y_1+y_2)$

(b) Since $Q_1 = b y_1 u_1 = 14 \times 10^{-3}$ m³/s

$$u_1 = \frac{14 \times 10^6 \text{ mm}^3/\text{s}}{13 \times 10^3 \text{ mm}^2} = 1\,077 \text{ mm/s}$$

Also    $\sqrt{(g y_1)} = \sqrt{(9\,807 \text{ mm/s}^2 \; 100 \text{ mm})} = 990$ mm/s

Hence,    $Fr_1 = \dfrac{u_1}{\sqrt{(g y_1)}} = \dfrac{1\,077}{990} = 1\cdot088$

and    $\dfrac{y_2}{y_1} = \tfrac{1}{2}(\sqrt{(1+8Fr_1^2)}-1) = \tfrac{1}{2}(\sqrt{(1+8\times1\cdot185)}-1) = 1\cdot14$

Thus,    $y_2 = 1\cdot14 \times 100$ mm $= 114$ mm

and $2y_c^3 = 100 \times 114(100+114)$ mm³ $= 2\cdot44 \times 10^6$ mm³, i.e. the critical depth $y_c = \sqrt[3]{(1\cdot22 \times 10^6 \text{ mm}^3)} = 106\cdot8$ mm which lies between $y_1$ and $y_2$ necessarily since a standing wave can only be created from shooting flows.

E<small>XAMPLE</small> 14.4 (ii)

(a) *Develop a formula for the dissipation of energy or loss of total head due to a hydraulic jump.*

(b) *In a rectangular channel 0·6 m wide a jump occurs where the Froude number is* 3. *The depth after the jump is* 0·6 m. *Estimate the loss of total head and the power dissipated at the jump.*

(*a*) The loss of total head due to a jump is:

$$\left(y_1+\frac{u_1^2}{2g}\right)-\left(y_2+\frac{u_2^2}{2g}\right) = (y_1-y_2)+\frac{u_1^2}{2g}\left(1-\frac{y_1^2}{y_2^2}\right)$$

$$= (y_2-y_1)\left\{\frac{u_1^2}{2gy_2^2}(y_2+y_1)-1\right\}$$

Also, since from Fig. 14.4(b), we deduce

$$\frac{u_1^2 y_1}{y_2} = \frac{g}{2}(y_1+y_2)$$

the loss of head is:

$$h_L = (y_2-y_1)\left\{\frac{(y_2+y_1)^2}{4y_1y_2}-1\right\}$$

i.e.

$$h_L = \frac{(y_2-y_1)^3}{4y_1y_2}$$

(*b*) Using the formula already proved:

$$\frac{y_2}{y_1} = \tfrac{1}{2}(\sqrt{(1+8Fr_1^2)}-1) = \tfrac{1}{2}(\sqrt{(73)}-1) = \frac{7\cdot55}{2} = 3\cdot775$$

then

$$y_1 = \frac{0\cdot6}{3\cdot775}\,\text{m} = 0\cdot159\text{ m} \quad\text{and}\quad y_2-y_1 = 0\cdot441\text{ m}$$

Hence, the loss of head due to hydraulic jump is

$$h_L = \frac{(0\cdot441)^3\text{ m}}{4\times0\cdot159\times0\cdot6} = 0\cdot225\text{ m}$$

Also,

$$Fr_1 = \frac{u_1}{\sqrt{(gy_1)}} = 3$$

Hence $u_1 = 3\sqrt{(9\cdot807\times0\cdot159)} = 3\cdot75$ m/s and the mass flowing per unit time is:

$$\rho Q_1 = \rho b_1 y_1 u_1 = 1\text{ Mg/m}^3\times0\cdot6\text{ m}\times0\cdot159\text{ m}\times3\cdot75\text{ m/s} = 356\text{ kg/s}$$

Hence, the power lost due to the jump is: $P = h_L w Q_1 = \rho Q_1 g h_L$

$$= 356\frac{\text{kg}}{\text{s}}\times9\cdot807\frac{\text{m}}{\text{s}^2}\times0\cdot225\text{ m} = 786\text{ kg}\frac{\text{m}^2}{\text{s}^3} = 0\cdot786\text{ kW}$$

**Fig. 14.4(d).** *Energy dissipation in hydraulic jump*

Fluid in crossing the jump cannot undergo a spontaneous increase in energy; the energy dissipation which occurs for all but the weakest waves occurs in a violent belt of eddies as suggested by Fig. 14.4(d). Experiments show that if the surface level downstream is higher than that to which a shooting flow tends (established, e.g., by a weir upstream), then a standing wave system will so form that the two levels are made compatible.

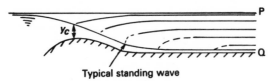

Typical standing wave

**Fig. 14.4(e).** *Typical standing wave in a channel*

Fig. 14.4(e) is thus complementary to the earlier diagram Fig. 14.4(a) and indicates how the fluid may discharge at the intermediate range of levels between P and Q.

## 14.5. Jump conditions downstream of a sluice

Hydraulic jumps may occur downstream of any obstacle which causes a shooting flow, and a sluice gate is a typical example. If the stream emerges from under the sluice gate with shooting flow and conditions of head downstream require streaming flow a jump will occur as indicated in Fig. 14.5(a). The effective head to be allowed across such a gate depends on

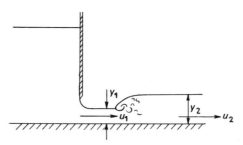

**Fig. 14.5(a).** *Jump in shooting flow caused by a sluice gate*

whether the difference in levels is sufficient to establish critical conditions across the vena contracta. If so, then this will control the discharge, which will then tend to become independent of the downstream level. The latter will subsequently be reached by a hydraulic jump, somewhat as shown in Fig. 14.5(a).

Since

$$\frac{y_2}{y_1} = \tfrac{1}{2}[\sqrt{(1+8Fr_1^2)}-1]$$

then for a jump to occur, the value must be greater than unity, i.e.

$$\sqrt{(1+8\ Fr_1^2)} > 3 \quad \text{or} \quad Fr_1 > 1$$

which means that there must be shooting flow at the vena contracta.

EXAMPLE 14.5 (i)

*The stream issuing from beneath a vertical sluice gate is 300 mm deep at the vena contracta. Its mean velocity is 6 m/s. A standing wave is created on the level bed below the sluice gate. Find the height of the jump, the loss of energy and the power dissipated per unit width of sluice.*

Referring to Fig. 14.5(a), the Froude number at the vena contracta is:

$$Fr_1 = \frac{6}{\sqrt{(9\cdot807 \times 0\cdot3)}} = 3\cdot5$$

Hence
$$\frac{y_2}{y_1} = \tfrac{1}{2}(\sqrt{(1+8Fr_1^2)} - 1) = \tfrac{1}{2}(\sqrt{(1+98)} - 1) = 4\cdot47$$

and $y_2 = 4\cdot47 \times 0\cdot30$ m $= 1\cdot34$ m.

Thus the height of the jump is $y_2 - y_1 = 1\cdot04$ m, and the head lost is:

$$h_L = \frac{(y_2-y_1)^3}{4y_1y_2} = \frac{(1\cdot04 \text{ m})^3}{4 \times 0\cdot3 \times 1\cdot34 \text{ m}^2} = 0\cdot7 \text{ m}$$

The rate of flow per metre width is

$$\rho a u = \dot{m} = 1\frac{\text{Mg}}{\text{m}^3} \times 0\cdot3 \text{ m}^2 \times 6\frac{\text{m}}{\text{s}} = 1\cdot8 \text{ Mg/s}$$

Hence the power dissipated per m width is:

$$P = \dot{m}gh_L = 1\cdot8\frac{\text{Mg}}{\text{s}} \times 9\cdot807\frac{\text{m}}{\text{s}^2} \times 0\cdot7 \text{ m} \left[\frac{\text{W s}^3}{\text{kg m}^2}\right] = 12\cdot36 \text{ kW}$$

## 14.6. Venturi flumes

The sides of a channel may be contracted to form a Venturi-shaped passage in plan as shown in Fig. 14.6(a). Such Venturi flumes can be

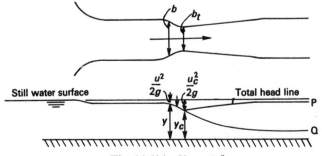

**Fig. 14.6(a).** *Venturi flume*

made of concrete and are not liable to damage in floods by debris as are sharp-crested weirs, and do not silt up.

Flumes are used when little head is available for measuring purposes or when it is impracticable to use weirs. No damming is necessary as with weirs and there is only a small loss of head as compared with the loss of kinetic head over weirs.

At the throat, where $b$ is constant ($b_t$), at least over a short length, the continuity equation:

$$Q = uby = \text{constant}$$

reduces to:

$$\frac{\delta u}{u} + \frac{\delta y}{y} = 0$$

Further, if energy losses may be neglected in this region:

$$gy + \frac{u^2}{2} = \text{constant}$$

i.e.

$$g\delta y + u\delta u = 0$$

These two differential relationships were first met in the first section of this chapter, in connection with the propagation of shallow water wavelets, and again for the flow over a broad-crested weir, see Section 14.2, p. 413. There it was shown that either both $\delta u$ and $\delta y$ are zero, as shown by branch P of the curve, or alternatively conditions at the throat are critical, i.e. at this section:

$$u_c = c = \sqrt{(gy_c)}$$

in which:

$$y_c = \tfrac{2}{3}h$$

where $h$ is the height of the total head line above the bed:

$$h = y + \frac{u^2}{2g}$$

The discharge under these conditions:

$$Q_c = u_c y_c b_t = y_c b_t \sqrt{(gy_c)} = b_t g^{1/2} y_c^{3/2}$$

i.e. the actual

$$Q = C_d \frac{2\sqrt{2}}{3\sqrt{3}} b_t g^{1/2} h^{3/2} = \left(1\cdot71 \frac{\text{m}^{1/2}}{\text{s}}\right) C_d b_t h^{3/2}$$

where $C_d$ is usually about 0·94.

As it is difficult to measure $y_c$, which is located at a section where the depth is varying rapidly it is generally preferable to estimate the discharge from $h$. A first approximation to the latter is the depth $y$ in the approach channel which may subsequently be corrected by the addition of the velocity of approach term $u^2/2g$.

EXAMPLE 14.6 (i)

*A Venturi flume with a level bed is 12 m wide and 1·5 m deep upstream and has a throat width of 6 m. Calculate the rate of flow of water if the discharge coefficient is 0·94. Correct for velocity of approach.*

If a standing wave forms downstream we take:

$$Q = 1·71 \ m^{1/2}/s \ C_d b_t h^{3/2}$$
$$= 1·71 \times 0·94 \times 6 \times (1·5)^{3/2} \ m^3/s = 17·37 \ m^3/s$$

as a first approximation. Using this we estimate the velocity of approach in the main channel to be:

$$u = \frac{Q}{A} = \frac{17·37}{12 \times 1·5} \ m/s = 0·963 \ m/s$$

Hence, using the corrected value for the head upstream of

$$1·5 \ m + \frac{u^2}{2g} = 1·5 \ m + 0·047 \ m$$

we estimate:

$$Q = 1·71 \times 0·94 \times 6 \times (1·547)^{3/2} = \mathbf{18·56} \ m^3/s$$

## Exercises on Chapter 14

1. The bed of a river 18 m wide has a uniform slope of 1 in 15 000, and the Chezy constant $C$ is 50 $m^{1/2}/s$. Find the flow along the river 1·25 m deep.

A dam across the river has a spillway along the top, in the form of a broad-crested weir 17 m wide, the height of the sill being 3·5 m from the river bed.

Assuming that the whole of the river flow passes over the spillway, which has a coefficient of discharge of 0·90, find the efflux at the dam (i.e. the increase in the river depth caused by the dam) when the depth of the river well upstream is 1·25 m. Neglect the effect of the end spillway, and assume the banks of the river are vertical.

2. Sketch curves showing how depth changes with specific energy of a fixed quantity of water flowing in three rectangular channels of different widths. Thus, explain why it is not necessary to measure the depth of water at the throat of a Venturi flume if a standing wave is formed just downstream of this section.

Prove that at the throat the speed and depth are related by $u^2 = gy$ and hence develop the expression $Q = 1·71 \ m^{1/2}/s \ BH^{3/2}$ where $B$ is the breadth at the throat and $H$ the depth at a point upstream where the velocity is small.

3. Water flows across a broad-crested weir having length $B$. Show that the rate of flow is given theoretically by $Q = 1·71 \ BH^{1·5} \ m^3/s$, where $B$ and $H$ are measured in metres—$H$ being from the crest of the weir to the level of undisturbed water in the reservoir.

Water from a reservoir flows over a broad-crested weir 30·5 m long. Determine the time taken from the head above the weir to fall from 300 mm to

200 mm if, during this time, (*a*) no water enters the reservoir, and (*b*) water enters the reservoir from the catchment area at the rate of 170 m³/min. The surface area of the reservoir is $93 \times 10^3$ m².

4. Develop the formula

$$\frac{dy}{dx} = (i-k) - \frac{u}{g}\frac{du}{dx} = \frac{i-k}{1-Fr^2}$$

for the slope of the water surface in a channel of constant rectangular section and distinguish clearly between the slopes of $dy/dx$, $i$ and $k$.

5. Show that the critical depth $y_c$ in a channel of uniform width in which the flow is non-uniform is given by:

$$y_c^3 = Q^2/b^2g$$

where $Q$ is the rate of volumetric flow and $b$ is the width of the channel. In a channel 75 mm wide the depth of the stream increases from 12·5 mm to 22·5 mm by a hydraulic jump.

Estimate the rate of flow.

6. Calculate the Froude number before and after the jump, the height of the jump, the loss of head, and the power dissipated when a flow of 25·5 m³/min in an open channel of rectangular section 1 m wide passes through a hydraulic jump. The depth of the stream just before the jump is 100 mm.

7. Water flows under a sluice gate in a rectangular channel with a velocity $u_1$ and uniform depth $d_1$ at the vena contracta. Show that provided that $u_1$ is greater than $gd_1$ a 'standing wave' or 'hydraulic jump' will occur downstream. Deduce an expression for the depth $d_2$ after the jump in terms of $u_1$ and $d_1$.

8. Describe the action of a Venturi flume as used for determining the discharge of a stream—in particular when a hydraulic jump is formed. Obtain from first principles the expression $Q = 1 \cdot 71\ BC_d H^{3/2}$ and state what quantities are represented by the symbols. A Venturi flume with a level bed is 12 m wide and 1·5 m upstream and has a throat width of 6 m. Calculate the discharge if the discharge coefficient is 0·94. Make one correction for velocity of approach.

9. Explain the terms 'normal depth' and 'critical depth' of flow of water in an open channel of rectangular cross-section. Develop an expression for the critical depth.

The cross-section of a river 30 m wide is rectangular. At a point where the bed is approximately horizontal, the width is restricted to 24 m by the piers of a bridge. If a flood of 455 m³/s is to pass the bridge with the minimum upstream depth, describe the flow past the piers, and calculate the upstream depth. Prove any formula used and state any assumptions made.

10. A Venturi flume having a horizontal floor and throat width 725 mm is fitted in a channel 1 m wide. When the flow is 17 m³/min the upstream depth is 340 mm and the downstream depth 290 mm. A standing wave and backward roll develop and the minimum depth is 175 mm near the outlet from the flume where the width is 1 m.

Determine:

    (a) the total head upstream and downstream and express the loss of head as a percentage of that upstream,

    (b) the theoretical velocity at the position of minimum depth,

    (c) the actual mean velocity at this position,

    (d) the 'critical velocity' for this minimum depth to maintain conditions of 'shooting flow'.

11. A river of depth $H_1$ is flowing with a velocity $u$ towards the sea. Tide conditions cause a sudden increase in depth downstream to $H_2$ and this depth is maintained constant by the tide. Show that a 'bore' will travel upstream with a velocity

$$U = -u + \sqrt{\left[\frac{g\,H_2}{H_1}\left(\frac{H_1+H_2}{2}\right)\right]}$$

# Hydro-kinetic machines

'*When you put together the science of the motions of water, remember to include under each proposition its application and use, in order that this science may not be useless.*'

LEONARDO

## 15.1. Euler's turbine equation and hydraulic efficiencies

A hydro-kinetic machine is one in which forces are exerted on a moving rotor, through a change in the momentum of the fluid moving across it.

**Fig. 15.1(a).** *Hero's turbine*

If the rotor drives the shaft at the expense of the energy of the fluid, the machine is operating as a turbine and, conversely, if shaft work is supplied to increase the energy of the fluid, the machine is said to be a pump, or compressor. The turbine has a much longer history than the turbo-pump, or compressor. It comes down to us, associated with the name of Hero of Alexandria, and was known in Ancient Egypt, if only as an ingenious toy in the Temple, devouring the liquid oblations offered it, see Fig. 15.1(a).

It was not until the time of Newton that we have any record of a hydro-kinetic machine being visualized as a pump. Papin, who not only foreshadowed the steam-engine, also invented the centrifugal pump.

To interpret their behaviour we use the same fundamental scientific laws which we have recognized are at the base of most of the work in

Mechanics of Fluids, namely (i) *the law of conservation of matter*—the rate of mass flow of fluid through a rotor or impeller is constant, (ii) *Newton's second law of motion*—torque is equal to the time rate of change of angular momentum, and (iii) *the law of conservation of energy*—for any energy transformer the rate of reception of energy is equal to the

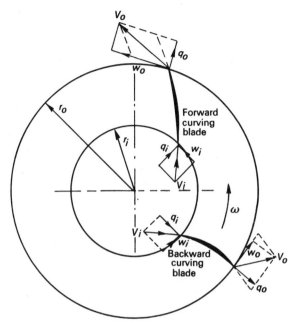

**Fig. 15.1(b).** *Velocity components at entry and exit of rotor*

rate of rejection of energy. The energy involved is usually in one or other of five forms as a result of pressure, velocity, height, mechanical energy or heat resulting from the dissipation of energy by friction. Subsequent analysis is merely a statement of these fundamental laws in symbols and a merging or reduction of them into formulae useful to the engineer.

As has so often been the case, man's inventiveness provided the stimulus for theory, and it is to the great mathematician Euler that we owe 'the turbine equation' which bears his name. *It can, in fact, be applied to turbines or pumps, and holds equally well for axial, radial or mixed flows.* Fig. 15.1(b) shows an axial view of a rotor in which fluid enters at a radius $r_i$ and leaves at some other radius $r_o$, which may be greater or less than $r_i$ according to the type of machine.

The absolute velocity of the fluid at inlet is denoted by $V_i$, of which the radial and whirl components are shown as $q_i$ and $w_i$ respectively. If, as a result of deflection caused by the rotor blades, the fluid leaves with an absolute velocity $V_o$, having components $q_o$ and $w_o$ as shown, corresponding momentum forces will be experienced as reactions acting inwards on the rotor, as shown in Fig. 15.1(c) (see Section 7.7, p. 153). It should be noted that although the pressures at inlet and exit control the flow, they exert no torque. Similarly, although the radial velocity components determine the mass flow, they too do not give rise, in themselves, to forces

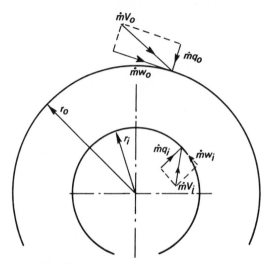

Fig. 15.1(c). *Momentum forces on rotor*

which exert a moment about the axis of rotation. The resultant moment or the torque experienced by the rotor is simply (Euler's equation):

$$M = \dot{m}(w_i r_i - w_o r_o)$$

If $M$ is positive $w_i r_i > w_o r_o$ and the rotor is rotating in the same direction as $w_i$; the fluid does work on the rotor, i.e. the machine operates as a turbine. If the fluid torque resists rotation of the rotor or impeller then $M$ is negative, i.e. $w_o r_o > w_i r_i$ and hence the machine is operating as a pump. If the angular speed of the rotor is $\omega$, then the magnitude of the power developed (turbine) or imparted (pump) by the blading of the rotor is given by:

$$P = M\omega$$

—sometimes called the 'water power'—in either case. If the blade tip (i.e. tangential) velocities $u$ of the wheel at inlet and outlet are denoted by $u_i = r_i\omega$ and $u_o = r_o\omega$, respectively, then the rate of energy transfer from the water in the case of a turbine and to the water in the case of a pump is:

$$P = \dot{m}[w_i r_i \omega - w_o r_o \omega]$$

or
$$P = \dot{m}(w_i u_i - w_o u_o)$$

In terms of the volumetric flow rate $Q$, the torque and power expressions may be written:

$$M = \rho Q(w_i r_i - w_o r_o)$$

and
$$P = M\omega = \rho Q(w_i u_i - w_o u_o) = \rho Q W$$

The work per unit mass done by or imparted to fluid in passing through a runner or impeller, namely $(w_i u_i - w_o u_o)$ may be estimated from velocity diagrams—and hence, $W$ is often referred to as the diagram work or specific diagram work.

The above equations may often be simplified both for turbines and for pumps. The object of a turbine is to convert as much as possible of the available energy in the fluid into shaft work. It is obviously wasteful therefore to reject fluid with a velocity of whirl; therefore, under the most efficient conditions the blading should be so arranged that $w_o = 0$. Then the above equations when applied to a turbine reduce to:

$$M = \rho \, Q w_i r_i$$

$$P_o = \rho \, Q w_i u_i = \rho Q W$$

in which $P_o$ distinguishes the power output of the water from the available power supplied, namely,

$$P_i = \rho Q g H$$

$H$ denoting the available head across the turbine. Thus the hydraulic efficiency $(P_o/P_i)$ of a turbine is, in general:

$$_t\epsilon_h = \frac{w_i u_i - w_o u_o}{g H}$$

and reduces, in the absence of swirl at exit, to:

$$_t\epsilon_h = \frac{w_i u_i}{g H} = \frac{W}{g H}$$

The mechanical efficiency of a turbine is the ratio of the power delivered along the impeller shaft to the power output of the water in flowing through the impeller,

i.e.
$$\epsilon_m = \frac{P_s}{P_o}$$

Hence,    $P_s = \epsilon_m \rho Q w_i u_i$   or   $\epsilon_m \rho Q(\epsilon_h g H)$   or   $\rho Q g H \epsilon_o$

in which the overall efficiency is

$$\epsilon_0 = \frac{P_s}{P_i} = \frac{P_s}{P_o} \times \frac{P_o}{P_i} = \epsilon_m \times \epsilon_h$$

Similarly, under ideal or most efficient conditions, there should be no swirl at entry to the eye of a centrifugal pump (i.e. $w_i = 0$) if, as is usually the case, there are no inlet guide blades. In this case the equations applied to a centrifugal pump reduce to:

$$M = \rho \, Q w_o r_o$$

$$P_i = \rho \, Q w_o u_o = \rho Q W$$

$P_i$ denotes the power supplied to the fluid by the rotating impeller, and if the pump delivers against a head $H$, so that its power output is:

$$P_o = \rho Q g H$$

then its hydraulic efficiency will be $P_o/P_i$ or

$$_p\epsilon_h = \frac{gH}{w_o u_o} = \frac{gH}{W}$$

The total gain in head by fluid in passing through a centrifugal pump is due solely to transference of energy from the impeller. The rest contributes nothing to the head but causes losses by friction in eddies and shock whirls if entry of fluid to the impeller (and diffuser) is not smooth, by skin friction in passage through channels of blades, and by mechanical friction between metals of bearings.

The mechanical efficiency is $\epsilon_m = P_i/P_s$, and the overall efficiency of a pump is:

$$\epsilon_0 = \frac{P_o}{P_s} = \frac{P_o}{P_i} \times \frac{P_i}{P_s} = \epsilon_h \times \epsilon_m$$

Hence, the shaft power to drive a pump is:

$$P_s = \frac{P_i}{\epsilon_m} = \frac{\rho Q w_o u_o}{\epsilon_m} = \frac{\rho Q}{\epsilon_m}\left(\frac{gH}{\epsilon_h}\right) \quad \text{or} \quad \frac{\rho Q g H}{\epsilon_0}$$

## EXAMPLE 15.1 (i)

*An impulse turbine is required to give* 45 kW *shaft power using water with an available head across the turbine of* 37 m. *The energy transferred from the water to the runner is* 320 Nm/kg. *Assuming a mechanical efficiency of* 93 per cent *calculate the overall efficiency and the rate of flow of water through the turbine.*

Energy abstracted from water

$$= w_i u_i - w_o u_o = 320 \text{ Nm/kg} = \text{runner output}$$

or diagram work from an available head $H = 37$ m.

Hence, hydraulic efficiency is:

$$\epsilon_h = \frac{w_i u_i - w_o u_o}{gH} = \frac{320 \text{ Nm}}{37 \text{ m kg}} \times \frac{s^2}{9\cdot807 \text{ m}}\left[\frac{\text{kg m}}{\text{N s}^2}\right] = 0\cdot88$$

and the overall efficiency

$$\epsilon_0 = \epsilon_m \times \epsilon_h = 0\cdot93 \times 0\cdot88 = \mathbf{0\cdot82}$$

Shaft power

$$P_s = \rho g Q H \epsilon_0$$

Hence

$$Q = \frac{P_s}{\rho g H \epsilon_0} = \frac{45 \text{ kNm/s}}{9\cdot807 \text{ kN/m}^3 \times 37 \text{ m} \times 0\cdot82} = \mathbf{0\cdot15 \text{ m}^3/\text{s}}$$

## 15.2. Velocity triangles for single-stage axial machines

From the previous section it may be seen that *the mechanical performance of a hydro-kinetic machine is, for a given rotor speed, determined primarily by the circumferential or whirl velocity of the fluid as it enters ($w_i$) and leaves ($w_o$) the runner or impeller.* The simplest assumption is that fluid enters without whirl or, if there are inlet guide vanes, at an angle dictated by the latter. Similarly we may *assume that the relative velocity of the fluid leaving the rotor is tangential to the exit angle of the rotor blades.*

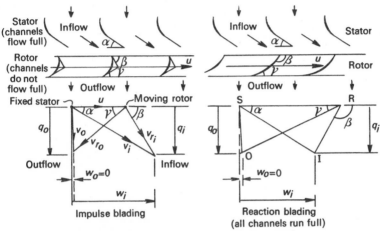

Fig. 15.2(a).  *Velocity triangles for turbines*

The inlet angle of the blades should be so arranged that, when operating under the design conditions, the fluid enters without shock, i.e. it is able to slide on to the blades without leaving eddying regions of separation.

These ideas have already been applied to rows of blading (see Section 7.4, p. 147) and may be applied without modification to the mean radius of an axial-flow machine (pump or turbine) *provided that the flow follows the direction of the blades*, as assumed. If the fluid is to leave the machine without swirl, as considered for a turbine at the end of the previous section, the angular momentum introduced by passage of the fluid across the first set of blades (stator or guide blades for a turbine) must be removed during its passage across the second set of blades (runner of a turbine). Figure 15.2(a) illustrates the application of this idea to impulse and reaction blading for an axial-flow turbine (an axial-flow reaction turbine is usually referred to as a Kaplan or propeller turbine)—the outlet velocity of the fluid having no whirl component (i.e. $w_o = 0$).

*The distinct difference between the blading for impulse and reaction machines is caused by the fluid speed remaining virtually constant relative to the blade in the former case* (i.e. RO $\leqslant$ RI according to friction—there being no pressure drop over the blade since the channels do not run full), but *increasing progressively at the expense of pressure during its passage across the blading of a reaction turbine.* For in the latter case the channels run full and a pressure drop over the moving blades causes a corresponding increase in velocity relative to the moving blades—i.e. RO > RI in a reaction turbine.

It can be seen that for a given rotor speed, fluid velocities thus tend to be smaller in reaction machines, and, therefore, reaction turbines are better suited to low head applications. Also, a reaction turbine can be placed at any height within, say, 8·5 m above the tail water level—the

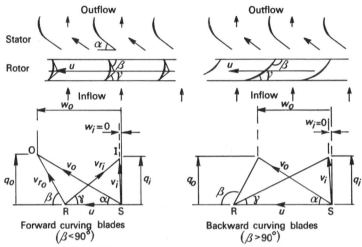

**Fig. 15.2(b).** *Velocity triangles for axial-flow pumps*

discharge being into a draught tube which produces a negative pressure at the outlet of the turbine rotor and has its lower end below the tail water level.

The diagrams of Fig. 15.2(a) refer to turbines, whereas those of 15.2(b) refer to pumps. Note that the tip angles of the rotor are measured from tangents in the direction of motion in the case of turbines and backwards to the direction of motion for pumps (Fig. 15.2(b)). As the purpose of a pump is to cause a rise in pressure, the rotor must be 'drowned'—i.e. the channels must run full, as in a reaction turbine, so that there may be a change of pressure across the rotor or impeller.

For an axial-flow *pump, fan or compressor*, the flow direction of Fig. 15.2(a) is reversed, as is the direction of rotation, see Fig. 15.2(b). In this case the rotor receives fluid without initial whirl at a relative angle $\gamma$ and by deflecting it rejects it with a whirl velocity $w_o$ which is taken out on the stator blades which make the final ejection axial. Also $u_i = u_o = u$.

From the previous section we may quote the hydraulic efficiencies:

$$_t\epsilon_h = \frac{w_i u}{gH} = \frac{W_t}{gH}$$

for the turbine and

$$_p\epsilon_h = \frac{gH}{w_o u} = \frac{gH}{W_p}$$

for the pump, *provided that there is no velocity of whirl at exit from the turbine, or at inlet to the pump.* Also, if the fluid follows the directions set by the blades, then the change in the whirl velocity across the rotor is, in either case, simply:

$$w = (u + q \cot \beta)$$

where $q$ is the axial fluid velocity, *assumed to remain constant.* For a given flow area this is related to the delivery through the machine and thus enables the rotor speed ($u$) to be related to the corresponding head ($H$) across it when the machine is operating at its best point of maximum $\epsilon_h$— i.e. when there is no shock because the velocity triangles in Fig. 15.2($a$) or ($b$) are being complied with,

$$_t\epsilon_h = \frac{(u + q \cot \beta)u}{gH} = \frac{1 + (q/u) \cot \beta}{gH/u^2}$$

for a turbine, and

$$\frac{1}{_p\epsilon_h} = \frac{1 + (q/u) \cot \beta}{gH/u^2}$$

for a pump.

Thus, in order to relate $gH$ to $u^2$, at the best point, it is seen that the important (non-dimensional) parameters are $\epsilon_h$ which may be of the order of 0·85, $\beta$ which may be less than 90° in forward-curving or greater than 90° in backward-curving blades, and ($q/u$). From the triangles it may be seen that ($q/u$) must be such that:

$$\frac{q}{u} = \tan \gamma$$

If this condition is not satisfied, the fluid will be rejected from the turbine with a velocity of whirl, whereas for a pump this would imply that the fluid would not flow tangentially on to the moving blades. The corresponding condition for *shock-free entry* on to a turbine rotor or pump stator is that:

$$u + q \cot \beta = q \cot \alpha$$

i.e.

$$\frac{q}{u} = \frac{1}{\cot \alpha - \cot \beta}$$

Thus for such optimum conditions to be realized for axial-flow machines it is necessary that the blade angles be related by:

$$\frac{q}{u} = \frac{1}{\cot \gamma} = \frac{1}{\cot \alpha - \cot \beta}$$

i.e.

$$\cot \gamma = \cot \alpha - \cot \beta$$

## EXAMPLE 15.2 (i)

*A nozzle, having a coefficient of velocity of 0·96, operates under a head of 90 m of water and directs a 50 mm diameter jet of water on to a ring of axial-flow impulse blades, which have an inlet angle of 40° measured relative to the direction of blade motion and which turn the water through an angle of 105° during its passage over the blades (i.e. blade angle at outlet = 35°). Because of friction the velocity of the water relative to the blades at outlet is only 0·85 of that at inlet. The blade speed is to be 18 m/s and the water is to flow on to the blades without shock.*

*Draw the velocity diagram and determine:*
*(a) the diagram work and the angle which the line of the jet should make with the direction of motion of the blades;*
*(b) the power developed by the blade ring, the shaft horse-power assuming a mechanical efficiency of 95 per cent;*
*(c) the hydraulic and overall efficiencies.*

The velocity diagram will be similar to that shown in Fig. 15.2(a) for impulse blading. The velocity of the jet is:

$$V_1 = C \sqrt{(2gH_1)} = 0{\cdot}96 \sqrt{(19{\cdot}613 \text{ m/s}^2 \times 90 \text{ m})} = 40{\cdot}5 \text{ m/s}$$

The quantities $u$, $\beta$ and $\gamma$ are given as 18 m/s, 40° and 35°, respectively. Hence the velocity triangles can be drawn and the angle $\alpha$ measured at **23°** and the change in velocity of whirl $(w_i - w_o)$ measured as 36 m/s.

Thus, the (specific) diagram work is:

$$W = u(w_i - w_o) = 18 \text{ m/s} \times 36 \text{ m/s} = \textbf{648} \text{ Nm/kg}$$

and the output of power from the water while flowing through the rotor is:

$$P_o = \rho Q W = 10^3 \frac{\text{kg}}{\text{m}^3} \left( \frac{\pi}{4} \frac{25}{10^4} \text{ m}^2 \right) 40{\cdot}5 \frac{\text{m}}{\text{s}} \times 648 \frac{\text{Nm}}{\text{kg}} \left[ \frac{\text{W s}}{\text{N m}} \right]$$

Hence the shaft power is $0{\cdot}95 \times 51{\cdot}5 = \textbf{49}$ kW; the hydraulic efficiency

$$\epsilon_h = \frac{W}{gH} = \frac{648 \text{ m}^2/\text{s}^2}{9{\cdot}807 \times 90 \text{ m}^2/\text{s}^2} = 0{\cdot}734 \quad or \quad \textbf{73·4} \textit{ per cent}$$

and the overall efficiency is

$$\epsilon_0 = \frac{P_s}{\rho Q W} = \epsilon_m \frac{W}{gH} = \epsilon_m \epsilon_h = 0{\cdot}95 \times 0{\cdot}734 = 0{\cdot}697 \quad or \quad \textbf{69·7} \text{ per cent}$$

## EXAMPLE 15.2 (ii)

*In an axial-flow or propeller pump, the rotor has an outer diameter of 0·74 m and an inner diameter of 0·4 m: it revolves at 500 rev/min. At the mean blade radius, the inlet blade angle is 12 deg. and the outlet blade angle is 165 deg. both measured from rear of the blade. Sketch the corresponding velocity diagrams at inlet and outlet, and estimate from them (i) the head the pump will generate,*

*(ii) the discharge or rate of flow, (iii) the shaft power input required to drive the pump. Assume a manometric or hydraulic efficiency of 88 per cent and a gross or overall efficiency of 81 per cent.*

It will be seen that the velocity diagrams at inlet and outlet in Fig. 15.2(c) can be superimposed on the common vector $u$ as in Fig. 15.2(b). The inlet velocity $V_i$ is axial and the mean velocity of flow $q$ is constant.

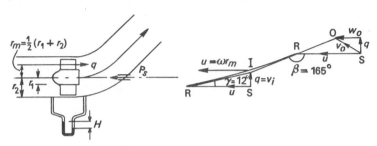

**Fig. 15.2(c).** *Axial-flow pump*

The velocity of the blades at mean radius is:

$$u = \omega r_m = \omega \left(\frac{r_1+r_2}{2}\right) = \frac{500 \text{ rev}}{60 \text{ s}} \times 0\cdot285 \text{ m} \left[\frac{2\pi}{\text{rev}}\right] = 14\cdot93 \text{ m/s}$$

and the velocity of flow

$$q = u \tan \gamma = 14\cdot93 \text{ m/s} \times 0\cdot212\ 6 = 3\cdot17 \text{ m/s}$$

The velocity of whirl at outlet is:

$$w_o = u + q \cot \beta = (14\cdot93 - 3\cdot17 \times 3\cdot732) \text{ m/s} = 3\cdot1 \text{ m/s}$$

Hence, the diagram work or head the pump will generate (*if $w_i = 0$*) theoretically is:

$$W = uw_o = 14\cdot93 \times 3\cdot1 \ \frac{\text{m}^2}{\text{s}^2}\left[\frac{\text{N s}^2}{\text{kg m}}\right] = 46\cdot3 \text{ Nm/kg}$$

(i) The hydraulic efficiency is the ratio of the head actually to that calculated (theoretically generated) from velocity diagrams, i.e.

$$\epsilon_h = \frac{gH}{W}$$

Hence, the head generated is $H = 0\cdot88 \times \dfrac{46\cdot3 \text{ m}^2/\text{s}^2}{9\cdot807 \text{ m/s}^2} = 4\cdot15 \text{ m.}$

(ii) The discharge rate is

$$\pi(r_2^2 - r_1^2)q = \pi(0\cdot37^2 - 0\cdot2^2) \text{ m}^2 \times 3\cdot17 \text{ m/s} = 0\cdot976 \text{ m}^3/\text{s} = \textbf{976 litre/s}$$

(iii) The overall efficiency is $\epsilon_0 = P_o/P_s$. Hence the shaft power is

$$P_s = \frac{P_o}{\epsilon_0} = \frac{\rho g Q H}{\epsilon_0} = 9\cdot807 \ \frac{\text{kN}}{\text{m}^3} \times \frac{0\cdot976 \text{ m}^3}{0\cdot81 \ \ \text{s}} \times 4\cdot15 \text{ m} \left[\frac{\text{W s}}{\text{N m}}\right] = \textbf{49 kW}$$

## 15.3.  Velocity triangles for single-stage radial-flow machines

Fig. 15.3(a) shows the corresponding inflow and outflow triangles for radial-flow machines. In order not to 'fight the centrifugal action', the flow through a pump is outward radially and for a turbine it is normally inward as in Francis turbines—which are of the reaction type since the channels run full.

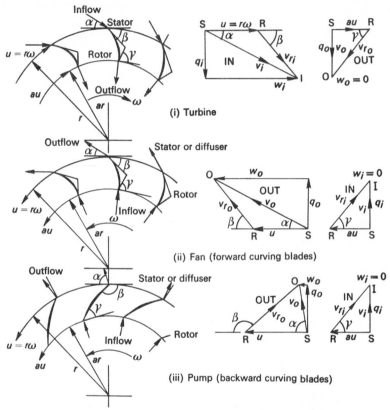

**Fig. 15.3(a).** *Velocity triangles for radial-flow machines*

The pump impeller (ii) has been drawn to illustrate forward-curving blades ($\beta < 90°$) as may be fitted in a ventilating fan, and impeller (iii) with backward-curving blades ($\beta > 90°$) as in the case in centrifugal pumps dealing with liquids.

A set of stator blades has also been shown in each case to correspond with the turbine (i). Such stator or diffuser blades are not always fitted in practice unless the velocity of fluid leaving the impeller is high. By suitable design, the casing of pumps into which the fluid is rejected from the impeller can be so shaped as a spiral or volute so as to remove angular momentum of the discharged fluid. It is customary in radial as in axial machines to make the velocities $q_i$ and $q_o$ approximately equal which, referring to Fig. 15.3(b), requires that the inner and outer cylindrical areas shall be the same for an incompressible fluid.

If the inner radius is some fraction $a$ of the outer radius then:

$$\pi\, Db D = \pi a D\; cD$$

i.e.
$$c = \frac{b}{a}$$

Thus if the inner radius is $\frac{1}{5}$ that of the outside ($a = \frac{1}{5}$) the channel width at the inside must be five times that at the outside ($c = 5b$).

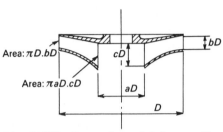

**Fig. 15.3(b).** *Rotor of a radial-flow machine*

By comparison with the previous section (15.2) it may be seen that for radial flow (i.e. velocity of whirl $w = 0$) at the inner (exit) radius the hydraulic efficiency of a Francis turbine is, as for a single-stage axial machine:

$$_t\epsilon_h = \frac{1 + (q/u)\cot\beta}{gH/u^2}$$

*provided that the fluid follows the directions set by the blades.* Similarly, for a centrifugal pump or fan:

$$\frac{1}{_p\epsilon_h} = \frac{1 + (q/u)\cot\beta}{gH/u^2}$$

The condition that there is no swirl ($w_o = 0$) at the inner radius is more critical for a turbine, and requires that:

$$\frac{q}{u} = \alpha\tan\gamma$$

For a pump this same condition ensures shock-free entry. The condition for shock-free flow between rotor and stator or diffuser blades (if fitted) is in either case the same as for an axial-flow machine:

$$\frac{q}{u} = \frac{1}{\cot\alpha - \cot\beta}$$

Hence *at the value of* $(q/u)$ *dictated by the rotor geometry in this way,* the corresponding value of $(gH/u^2)$ may be inferred, *and this defines the best (max $\epsilon_h$) condition for operating the machine.*

Departure from this condition does not destroy the validity of the analysis for a pump so long as there is no swirl ($w_1 = 0$) at entry; it merely reduces the value of $\epsilon_h$. Hence we may draw a set of straight lines representing constant efficiency on a plot of $(gH/u^2)$ against $(q/u)$ and this has been done on Fig. 15.3(c).

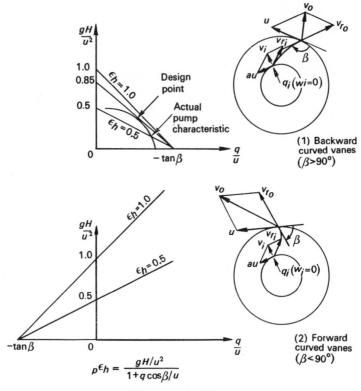

**Fig. 15.3(c).** *Pump and fan characteristics*

$$p\epsilon_h = \frac{gH/u^2}{1+q\cos\beta/u}$$

It is therefore possible to evaluate an existing pump by plotting its measured performance to such (non-dimensional) co-ordinates. This is suggested on the diagram for a centrifugal pump with backward curving vanes. Conversely, it is possible, with experience, to deduce the likely shape of the characteristic for a conventional machine from its expected variation of hydraulic efficiency with $(q/u)$. Departure from the 100 per cent efficiency straight line is, of course, due to friction and shock—there being only one $q/u$ according to the exit velocity triangle for which there can be no shock at entry to diffuser blades (if fitted), and the same applies at entry to the impeller. This particular value of $q/u$ is the best point at which to operate, and gives maximum efficiency—the losses at this design point being due to friction only. A widely used law for the characteristic curve of a pump is:

$$\frac{gH}{u^2} = a+b\left(\frac{q}{u}\right)-c\left(\frac{q}{u}\right)^2$$

where $a$, $b$ and $c$ are constants.

The ordinates may, with a change of scale, be changed to ones of:

$$\left(\frac{gH}{\omega^2 D^2}\right) = \frac{1}{4}\left(\frac{gH}{u^2}\right)$$

since the outer-tip velocity of the impeller blades is $u = r\omega = D\omega/2$.
Similarly, since the radial velocity is related to the flow rate by:

$$Q = \pi Db Dq$$

$$\frac{q}{u} = \frac{Q}{\pi b D^2}\frac{2}{\omega D}$$

i.e.

$$\left(\frac{Q}{\omega D^3}\right) = \frac{\pi b}{2}\left(\frac{q}{u}\right)$$

Hence, the $x$ axis may be recalibrated in terms of the delivery rate $Q$ divided by $\omega D^3$, if preferred. Yet another alternative is to give a physical meaning to the $y$ axis by plotting $H$ divided by $(u^2/2g)$ rather than $(u^2/g)$. $H/(u^2/2g)$ is then a measure of the head rise across the machine expressed as so many times the kinetic head which would be recorded by a Pitot-static tube, rotating in stationary fluid at the same speed as the peripheral speed of the rotor tips.

Also, we may note that the empirical law:

$$\frac{gH}{u^2} = a + b\left(\frac{q}{u}\right) - c\left(\frac{q}{u}\right)^2$$

for centrifugal pumps may be written:

$$\frac{gH}{\omega^2 D^2} = a' + b'\frac{Q}{\omega D^3} - c'\left(\frac{Q}{\omega D^3}\right)^2$$

i.e.

$$H = A'\omega^2 D^2 + B'\omega\frac{Q}{D} - C'\frac{Q^2}{D^4}$$

Hence, for one particular pump at one particular speed the head-delivery law is: $H = A + BQ - CQ^2$, where $A$, $B$ and $C$ are constants.

## EXAMPLE 15.3 (i)

*An inward-flow reaction turbine wheel has radial tips at the outer periphery, and at the inner periphery the blades make an angle of 30° with the forward tangent. If the total head of the supply is 20 m and the outer radius is twice the inner radius, find the velocity of the rim of the wheel if the water discharges radially, and the hydraulic efficiency of the turbine. Assume a constant radial velocity of flow and that friction in the blading accounts for a dissipation of energy equivalent to 10 per cent of the kinetic flow of energy.*

Referring to Fig. 15.3(a), both the inlet and outlet velocity diagrams are right-angle triangles, with $q_i = q_o = V_o$ and the ratio of the inner and outer radii and tip speeds $u_o/u_i = a = \frac{1}{2}$. Also $\beta = 90°$, $\gamma = 30°$, and $w_o = 0$ for radial exit.

Thus, the specific diagram work—i.e. the energy transmitted from the water to the wheel or rotor per unit mass of fluid flowing through it—is:

$$W = w_i u_i = u_i^2 = 4u_0^2$$

The specific energy equation is:

$$gH = W + h_f + \frac{V_0^2}{2}$$

$$gH = 8\frac{u_0^2}{2} + \frac{1}{10}\frac{q_0^2}{2} + \frac{q_0^2}{2} = 8\frac{u_0^2}{2} + \frac{11}{20}\left(\frac{u_0}{\cot 30}\right)^2 = \frac{83 \cdot 67\, u_0^2}{20}$$

Hence,

$$u_0 = \left(\frac{9 \cdot 807 \times 20}{4 \cdot 184}\right)^{1/2} \text{m/s} = 6 \cdot 84 \text{ m/s}$$

and $u_i = \mathbf{13 \cdot 68}$ m/s = *speed of rim.*
Hydraulic efficiency is:

$$\epsilon_h = \frac{W}{gH} = \frac{4 \times 20}{83 \cdot 67} = 0 \cdot 955 \quad \text{or} \quad \mathbf{95 \cdot 5} \text{ per cent}$$

EXAMPLE 15.3 (ii)

(a) *The impeller of a centrifugal pump is* 300 mm *diameter and* 50 mm *wide at the periphery and has blades whose tip angles incline backwards* 60° *from the radius (i.e.* $\beta = 150°$). *Calculate the speed and direction of water as it leaves the impeller when the pump delivers* 18 m³/min *and the impeller rotates at* 1 000 rev/min.

(b) *Assuming the pump is designed to admit radially, calculate the torque exerted by the impeller on the water and the shaft power required and the lift of the pump, taking the mechanical efficiency as* 95 *per cent and the hydraulic efficiency as* 75 *per cent.*

(a) The velocity diagrams will be similar to the ones for liquid in Fig. 15.3(a). The values of vectors in the outlet diagram will be:

$$q_0 = \frac{Q}{\pi D b_0} = \left(\frac{18 \text{ m}^3}{60 \text{ s}}\right)\frac{10^4}{\pi \times 30 \times 5 \text{ m}^2} = 6 \cdot 37 \text{ m/s}$$

$$u = r\omega = 0 \cdot 15 \text{ m} \times \frac{1\ 000 \text{ rev}}{60 \text{ s}}\left[\frac{2\pi}{\text{rev}}\right] = 15 \cdot 73 \text{ m/s}$$

$$w_0 = u + q \cot \beta = 15 \cdot 73 + 6 \cdot 37 \cot 150$$
$$= 15 \cdot 73 - 6 \cdot 37 \sqrt{3} = 4 \cdot 7 \text{ m/s}$$

Hence, $V_0 = \sqrt{(q_0^2 + w_0^2)} = 7 \cdot 92$ m/s and $\alpha = \tan^{-1}(q_0/w_0) = 53 \cdot 6°$.

(b) Torque $M = \dot{m}(w_0 r_0 - w_i r_i) = \dot{m} w_0 r_0$ for radial inlet ($w_i = 0$)

$$= \rho Q\, w_0 r_0 = 10^3\,\frac{\text{kg}}{\text{m}^3} \times 0 \cdot 3\,\frac{\text{m}^3}{\text{s}} \times 4 \cdot 7\,\frac{\text{m}}{\text{s}} \times 0 \cdot 15 \text{ m}\left[\frac{\text{N s}^2}{\text{kg m}}\right]$$

$$= \mathbf{211 \cdot 5} \text{ Nm}$$

Power exerted on water by the impeller is:

$$P_i = M\omega = 211 \cdot 5 \text{ Nm} \times \frac{1\,000\,\text{rev}}{60\ \text{s}}\left[\frac{2\pi}{\text{rev}}\right] = 22 \cdot 15 \text{ kW}$$

Shaft power

$$P_s = \frac{P_i}{\epsilon_m} = \frac{22 \cdot 15}{0 \cdot 95}\text{ kW} = \mathbf{23 \cdot 3 \text{ kW}}$$

Hydraulic efficiency

$$\epsilon_h = \frac{P_o}{P_i} \quad \text{or} \quad \epsilon_h P_i = P_o = \rho Q g H$$

Hence, the lift of the pump

$$H = \frac{\epsilon_h P_i}{\rho g Q} = \frac{0 \cdot 75 \times 22 \cdot 15 \text{ kN m/s}}{9 \cdot 807 \text{ kN/m}^3 \times 0 \cdot 3 \text{ m}^3/\text{s}} = \mathbf{5 \cdot 64 \text{ m}}$$

EXAMPLE 15.3 (iii)

*Develop a formula for the pressure head required across the impeller of a centrifugal pump to maintain flow, and the minimum head required to start flow. If the radii of the impeller tips are 100 mm and 300 mm calculate the minimum speed at which delivery will commence against a static head of 12 m.*

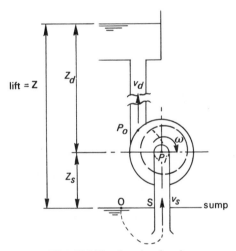

**Fig. 15.3(d).** *Starting head*

Assuming the pump to be working then the energy equations for flow in suction and delivery pipes according to Fig. 15.3(d) may be written (using $w = \rho g$) in terms of head as:

$$0 = \frac{P_s}{w} + \frac{v_s^2}{2g} = \frac{P_i}{w} + \frac{v_s^2}{2g} + Z_s + h_{fs}$$

i.e.

$$-\frac{P_i}{w} = \frac{v_s^2}{2g} + Z_s + h_{fs}$$

and
$$\frac{P_o}{w}+\frac{v_d^2}{2g} = 0+0+Z_d+h_{fd}$$

i.e.
$$\frac{P_o}{w}+\frac{v_d^2}{2g} = Z_d+h_{fd}$$

Hence,

$$\frac{P_o-P_i}{w} = Z+\frac{v_s^2}{2g}-\frac{v_d^2}{2g}+(h_{fs}+h_{fd})$$

and before delivery will commence a minimum pressure difference of $P_o-P_i = wZ$ must be created by the centrifugal action of the impeller which rotates the water in a forced vortex.

Considering a mass of water $\delta m = \rho\delta a\delta r$ being rotated at radius $r$ and angular velocity $\omega$ with a pressure difference $\delta P$ between two layers $\delta r$ apart, then the centripetal force on the effective area $\delta a$ is $\delta F = \delta P\delta a = (\rho\delta a\delta r)r\omega^2$ i.e.

$$\delta P = \rho\omega^2 r\delta r$$

or
$$\frac{P_o-P_i}{\rho} = \frac{\omega^2}{2}(r_o^2-r_i^2) = \frac{u_o^2-u_i^2}{2}$$

Delivery will not commence until

$$\frac{P_o-P_i}{w} = Z = \frac{\omega^2}{2g}(r_o^2-r_i^2)$$

i.e. at speed

$$\omega = \left(\frac{2gZ}{r_o^2-r_i^2}\right)^{1/2} = \left(\frac{19{\cdot}613 \text{ m/s}^2 \times 12 \text{ m}}{(9-1)\times 10^{-2} \text{ m}^2}\right)^{1/2}$$

$$= \frac{54{\cdot}2}{\text{s}}\left[\frac{\text{rev}}{2\pi}\right]\left[\frac{60 \text{ s}}{\text{min}}\right] = \textbf{518 rev/min}$$

EXAMPLE 15.3 (iv)

(a) *Assuming that the radial component of flow of fluid through a centrifugal pump remains constant and that the fluid enters the impeller radially, deduce formulae for the energy imparted by the impeller to increase the pressure and velocity of the fluid, neglecting losses. Plot non-dimensional graphs to show the effect of blade angle $\beta$ on these and on the total energy imparted to the fluid due to the action of the impeller. Comment on the effect of curvature of impeller blades.*

(b) *Calculate the total energy imparted per unit mass of fluid and also that to increase the pressure and kinetic energy of the fluid being forced through a ventilating fan of impeller diameter 600 mm with peripheral width 150 mm and blades curving forward with $\beta = 35°$. Assuming 25 per cent of the ideal total energy is dissipated (causing 20 per cent reduction of kinetic energy and 5 per cent reduction of pressure) calculate the increase in pressure when 60 m³/min of air flows through the impeller which rotates at 1 000 rev/min.*

*Assume the mean density of air is* 1·12 kg/m³.

(a) The total energy imparted per unit mass of fluid entering radially ($w_i = 0$) into an impeller is the diagram work

$$W_T = w_o u_o = u_o^2 \left(1 + \frac{q_o}{u_o} \cot \beta \right)$$

which, neglecting shock and friction losses, may be divided into $W_P$ and $W_V$, where the latter (referring to Fig. 15.3(a)) is:

$$W_V = \frac{V_o^2 - V_i^2}{2} = \frac{(w_o^2 + q_o^2) - q_i^2}{2} = \frac{w_o^2}{2} = \frac{(u_o + q_o \cot \beta)^2}{2}$$

i.e.

$$W_V = \frac{u_o^2}{2} \left(1 + \frac{q_o}{u_o} \cot \beta \right)^2$$

∴

$$W_P = W_T - W_V = \frac{u_o^2}{2} \left\{ 1 - \left(\frac{q_o}{u_o} \cot \beta \right)^2 \right\}$$

Hence

$$\frac{W_P}{W_V} = \frac{1 - (q_o/u_o) \cot \beta}{1 + (q_o/u_o) \cot \beta} = \frac{\text{specific energy to increase pressure}}{\text{gain in specific kinetic energy}}$$

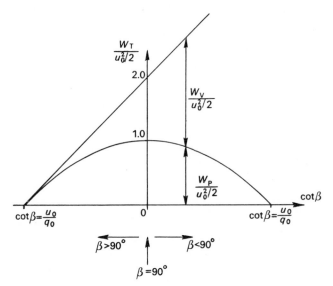

**Fig. 15.3(e).** *Specific energies imparted in centrifugal pumps*

Thus, if $q_o/u_o$ is constant for outlet velocity triangles with different angles $\beta$ as shown in Fig. 15.3(a) the distribution of energy imparted to fluid by the impeller (in the ideal case neglecting losses) can be shown by means of graphs as on Fig. 15.3(e). It will thus be seen that if a pump is required to deliver the greater portion of its energy in the form of pressure energy (as in a boiler feed pump) the angle $\beta$ must be greater than 90° and the greater the backward curving of the impeller blades the greater will be the ratio of pressure/kinetic energies. Conversely, if kinetic head or high exit velocity of fluid is required as in a ventilating fan, $\beta$ is less than $90°$.

(b) The tip speed

$$u_0 = r_0\omega = 300 \text{ mm} \times \frac{1\ 000 \text{ rev}}{60 \text{ s}} \left[ \frac{2\pi}{\text{rev}} \right] = 31 \cdot 4 \text{ m/s}$$

and the radial velocity of flow

$$q_0 = \frac{Q}{A_0} = \frac{1 \text{ m}^3/\text{s}}{\pi \times 0 \cdot 6 \times 0 \cdot 15 \text{ m}^2} = 3 \cdot 54 \text{ m/s}$$

Total energy (ideal) imparted by impeller per unit mass of air is:

$$W_T = u_0^2 \left( 1 + \frac{q_0}{u_0} \cot \beta \right) = (31 \cdot 4)^2 \frac{\text{m}^2}{\text{s}^2} \left( 1 + \frac{3 \cdot 54}{31 \cdot 4} \cot 35 \right) \left[ \frac{\text{N s}^2}{\text{kg m}} \right]$$

$$= 979(1 \cdot 161\ 8)\ \text{Nm/kg}$$

$$= \mathbf{1\ 138\ J/kg.}$$

Dissipation $= \frac{1}{4} \times 1\ 138$ J/kg of which $\frac{1}{5}$ is loss of $W_P$, i.e. $\frac{1}{5} \times 285$ J/kg $= 57$ J/kg and 228 J/kg loss of specific kinetic energy $W_V$. Hence, the actual gain in total specific energy is

$$(W_T)_{\text{act}} = (1\ 138 - 285) \text{ J/kg} = 853 \text{ J/kg}$$

The ideal gain in specific kinetic energy is:

$$W_V = \frac{u_0^2}{2} \left( 1 + \frac{q_0}{u_0} \cot \beta \right)^2 = \frac{(31 \cdot 4)^2}{2} \frac{\text{m}^2}{\text{s}^2} (1 \cdot 161\ 8)^2 = 683 \text{ J/kg}$$

Hence the actual gain in specific kinetic energy is $(W_V)_{\text{act}} = 683 - 228 = 455$ J/kg, and the actual gain in energy increasing the pressure is $(W_P)_{\text{act}} = 853 - 455 = 398$ J/kg.

$$\therefore \qquad \text{Gain in pressure } \Delta P = \rho_{\text{air}} \times (W_P)_{\text{act}}$$

which may be recorded on a water manometer gauge as:

$$\frac{\Delta P}{w_w} = \left( \frac{\rho_{\text{air}}}{\rho_w} \right) \times \frac{(W_P)_{\text{act}}}{g} = \left( \frac{1 \cdot 12}{10^3} \right) \left( \frac{398 \text{ Nm/kg}}{9 \cdot 807 \text{ m/s}^2} \right) \left[ \frac{\text{kg m}}{\text{N s}^2} \right]$$

$$= 0 \cdot 045\ 4 \text{ m} \quad \text{or} \quad \mathbf{45 \cdot 4} \text{ mm of water}$$

EXAMPLE 15.3 (v)

*Tests on a centrifugal pump running at a constant speed gave the following results:*

| Output (l/min) | 0 | 225 | 455 | 680 | 910 | 1135 |
|---|---|---|---|---|---|---|
| Head (m) | 12·2 | 12·5 | 11·9 | 10·4 | 7·3 | 3·7 |
| Overall efficiency (per cent) | 0 | 48 | 68 | 76 | 70 | 50 |

*Two such pumps are installed to run in parallel with common suction and delivery pipes with a useful lift of 6·4 m. The friction and other losses external to the pumps are calculated as $1·47\ Q^2 \times 10^{-6}$ m (where Q is the flow in l/min). Plot graphs and use information from them to estimate the delivery in gal/min and the driving horse-power to the pumps when:*

*(a) only one pump is connected:*

*(b) the two pumps are working in parallel.*

One pump works with a useful lift of 6·4 m represented by AA on Fig. 15.3(f) but overcomes a total resistance $H_T = 7·6$ m when delivering **890** l/min with an overall efficiency $\epsilon_0 = 71$ per cent.

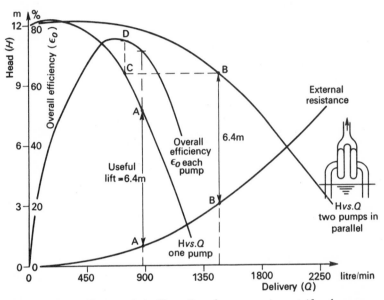

**Fig. 15.3(f).** *Characteristic H vs. Q and $\eta_0$ curves in centrifugal pumps*

Thus, the power output of the pump is $P_0 = wQH_T$ and the shaft power supplied is:

$$P_s = \frac{wQH_T}{\epsilon_0} = 9·807\ \frac{kN}{m^3} \times \frac{890\ litre}{60\ s} \times 7·6\ m \left[\frac{m^3}{10^3\ litre}\right]\left[\frac{W\ s}{N\ m}\right] = \mathbf{1·11\ kW}$$

*Two pumps in parallel* work with a useful lift of 6·4 m represented by BB and have to overcome a total lift $H_T = 9·6$ m. Each pump will be working at the point C on the *H vs. Q* graph of one pump and at an efficiency represented by D where $\epsilon_0 = 75$ per cent. The delivery of the two pumps in parallel will be **1 460** l/min. Hence, the total power output of the pumps is: $P_0 = wQH_T$ and the shaft power is:

$$P_s = \frac{wQH_T}{\epsilon_0} = 9·807\ \frac{kN}{m^3} \times \frac{1·46\ m^3}{60\ s} \times \frac{9·6\ m}{0·75} = \mathbf{3·06\ kW}$$

## 15.4. Non-dimensional characteristics and specific speeds

The previous sections of this chapter were concerned with an analysis of the performance of hydro-kinetic machines with particular reference to the geometry of the guide vanes and blades which deflect the fluid, the best performance being seen to be achieved when the fluid enters the machine without shock and leaves a turbine (or enters a pump) without whirl. We shall now consider a machine of a given shape (defined, say, by a set of undimensioned drawings or blue prints) and whose size is characterized by some convenient linear dimension, such as the outside diameter, $D$, of the rotor. Let us suppose that the Reynolds number is sufficiently high for viscous actions to exert little influence on the flow pattern and that, in addition, there are no other factors, such as compressibility or cavitation to be taken into account. If we imagine testing such a machine over a range of loads we shall conclude that the volumetric flow rate $Q$ depends on size ($D$), the pressure difference ($\Delta p$) across the machine and on the rotor speed ($\omega$) for a fluid of given density ($\rho$):

i.e.    $$Q = f(\Delta p, \omega, \rho, D)$$

Rearranging this expression into a non-dimensional form we find that its performance is, on the basis of the above hypothesis, capable of being expressed as a unique relationship between two parameters (see Section 10.8, p. 267), such as:

$$\frac{Q}{\omega D^3} = \psi\left(\frac{\Delta p}{\rho \omega^2 D^2}\right) = \psi\left(\frac{gH}{\omega^2 D^2}\right)$$

Fig. 15.4(a) illustrates experimental tests of a centrifugal or turbo-pump and demonstrates the validity of the above expression. By varying the throttle valve setting, the resistance of the circuit is varied and with

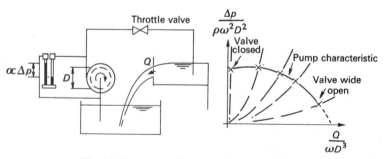

**Fig. 15.4(a).** *Test of a centrifugal or turbo pump*

it, the pump delivery $Q$. The curve shown is primarily one of pressure rise ($\Delta p$) as a function of discharge ($Q$) for a given fluid, impeller size and speed. However, by plotting non-dimensionally as shown, the same curve appears whatever the values of $\rho$, $D$ and $\omega$. That the curve takes into account or absorbs any variations of $\omega$ is a particularly valuable feature as it is *not* usually possible to arrange a constant speed drive. The above analysis suggests just how the rotor speed modifies the characteristic curve. To fix ideas, *if the speed of a given pump is doubled the maximum pressure rise is increased four times, and the corresponding discharge is doubled, if the machine is operated at its corresponding point* see Fig. 15.4(b).

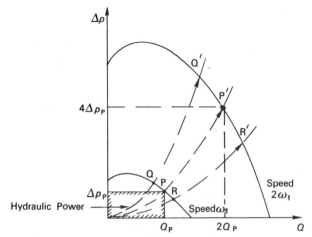

**Fig. 15.4(b).** *Effect of doubling the speed of a turbo-pump*

The work done on the fluid is closely related to the hydraulic power $\Delta p \times Q$, which at any point P on the $\Delta p - Q$ characteristic is represented by the shaded rectangle. This varies from zero to a maximum and down to zero again as we work along the characteristic. The efficiency of the machine must therefore also vary in a similar fashion, see Figs. 15.3(*f*)

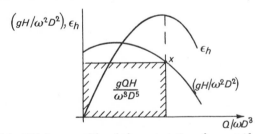

**Fig. 15.4(c).** *Efficiency and head characteristics of a centrifugal or turbo-pump*

and 15.4(c). On this diagram the pressure rise has been re-expressed in terms of head, using the relationship:

$$\Delta p = \rho g H$$

$$\therefore \quad \frac{\Delta p}{\rho \omega^2 D^2} = \frac{gH}{\omega^2 D^2}$$

thus leading to a rediscovery of a parameter introduced in Section 15.3, p. 441.

*Thus a turbo-pump, whatever its type, when operating at its best point (max. $\epsilon_h$ corresponding to X on the characteristic), works with definite values of:*

$$\text{the head coefficient} = \frac{gH}{\omega^2 D^2} = C_H$$

and

$$\text{the flow coefficient} = \frac{Q}{\omega D^3} = C_Q$$

As each of these non-dimensional groups has *a unique value at the best point, for a given geometrical arrangement* (i.e. is the same value for geometrically similar machines), so does any other group obtained through multiplying or dividing them when raised to any power. In particular if we cube the first of them and divide it by the square of the second we obtain *a constant for a particular machine, which is independent of size* (D), namely:

$$\left(\frac{gH}{\omega^2 D^2}\right)^3 \left(\frac{\omega D^3}{Q}\right)^2 = \frac{g^3 H^3}{\omega^4 Q^2} = \text{constant}$$

Thus, for any given *type* of machine, no matter what its size:

$$\omega^4 \propto \frac{g^3 H^3}{Q^2}$$

i.e.

$$\frac{\omega}{g^{3/4} H^{3/4}/Q^{1/2}} = \text{constant}$$

*This constant has a unique non-dimensional value for any given geometry, independent of size—i.e. for geometrically similar machines.* It became customary to assimilate the factor $g^{3/4}$ into the constant and to express $\omega$, $H$ and $Q$ in particular units. The 'constant' thus obtained, although more convenient to evaluate, *depends on the units chosen*; thus we may write:

$$\omega = N \frac{\text{rev}}{\text{min}}$$

$$H = H' \text{ m}$$

$$Q = Q' \frac{\text{m}^3}{\text{min}}$$

and refer to the resulting constant as the **specific speed** of the machine (turbine or pump):

$$N_s = \frac{N}{(H')^{3/4}/(Q')^{1/2}} \left(\frac{\text{rev. m}^{3/4}}{\text{min}^{3/2}}\right)$$

Thus for centrifugal and turbo-pumps:

$$N_s = \frac{N\sqrt{Q'}}{(H')^{3/4}}$$

and $N_s$ is constant for geometrically similar pumps working at maximum efficiency.

Fig. 15.4(d) shows the range of specific speeds covered by each machine, and enables the most appropriate type to be selected for a given duty.

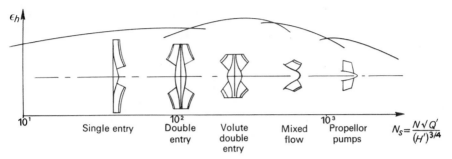

$\epsilon_h$

$10^1$        $10^2$        $10^3$    $N_s = \dfrac{N\sqrt{Q'}}{(H')^{3/4}}$

     Single entry    Double    Volute    Mixed    Propellor

                entry    double    flow     pumps

                        entry

**Fig. 15.4(d).** *Specific speeds of the several types of turbo-pump*

Similarly each type of turbine, *when working at its best ($\epsilon_h$ max) point*, runs at particular values of:

$$\frac{gH}{\omega^2 D^2} \quad \text{and} \quad \frac{Q}{\omega D^3}$$

and in terms of potential power available ($wQH$), which is of more interest for turbines, this implies that:

$$\frac{wH}{\rho\omega^2 D^2}\frac{Q}{\omega D^3} = \frac{P}{\rho\omega^3 D^5} = \text{constant}$$

*Thus by doubling the size of a turbine, the theoretical power available increases $2^5$, i.e. thirty-two times.* Thus such machines can achieve spectacular increases in power output by relatively modest increases in size.

As was shown for a pump or for a turbine, $g^3 H^3/\omega^4 Q^2 = \text{constant}$ so that one non-dimensional group, independent of $D$, is:

$$\frac{\omega Q^{1/2}}{g^{3/4} H^{3/4}} = \text{constant}$$

*This has a distinguishing value for each type of turbine when operating at its best point.* As the purpose of a turbine is to deliver power it is more convenient to express this group in terms of:

$$P = wQH$$

rather than $Q$, so obtaining the more useful result that:

$$\frac{\omega Q^{1/2}}{g^{3/4} H^{3/4}} \left[ \frac{P^{1/2}}{w^{1/2} Q^{1/2} H^{1/2}} \right] = \frac{\omega P^{1/2}}{\rho^{1/2} g^{5/4} H^{5/4}} = \text{constant}$$

This leads to an alternative definition of **specific speed**, more appropriate to turbines:

$$N_s = \frac{N\sqrt{P'}}{(H')^{5/4}} \left( \frac{\text{rev}}{\text{min}} \frac{(\text{kW})^{1/2}}{\text{m}^{5/4}} \right)$$

*in which $N_s$ is a constant for each type of turbine at its best point*, obtained from the above expression. $N$ denotes the *number* of rev/min, $P'$ denotes *the number* of kW of power and $H'$ denotes the *number* of metres head available across the turbine. The specific speeds of pumps range from less than 100 for single-entry centrifugal pumps to 1 000 for axial flow or propeller pumps, whereas for turbines the range is from less than 20 for Pelton wheels to 800 for axial flow or Kaplan turbines.

Fig. 15.4(e) shows the variation of turbine efficiency with specific speed for the more commonly used types of machine. Impulse turbines

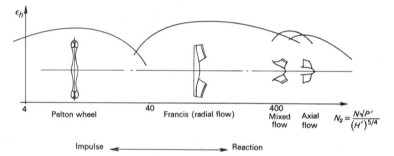

Fig. **15.4(e)**. *Specific speeds of the several types of turbine*

perform best with relatively high heads and low flow, whereas reaction turbines perform best with low heads and large rates of flow.

The non-dimensional quantities or coefficients $C_H = gH/\omega^2 D^2$ and $C_Q = Q/\omega D^3$ have constant values for geometrically similar machines (turbines and pumps) when working at the best point (i.e. in accordance with the velocity triangles determined by the fixed blade angles). Consequently the results of laboratory tests on *scale models* can be used to predict the probable output of projected full-scale designs of turbines or pumps by use of similarity equations deduced from $C_H$ and $C_Q$:

Since
$$H \propto \omega^2 D^2$$

therefore
$$\frac{h}{H} = \left( \frac{\omega}{\Omega} \right)^2 \left( \frac{d}{D} \right)^2$$

and
$$Q \propto \omega D^3 \propto \frac{H^{3/2}}{\omega^2}$$

therefore
$$\frac{q}{Q} = \frac{\omega}{\Omega} \left( \frac{d}{D} \right)^3 = \left( \frac{\Omega}{\omega} \right)^2 \left( \frac{h}{H} \right)^{3/2}$$

and
$$P \propto QH \propto D^2 H^{3/2} \propto \frac{H^{5/2}}{\omega^2}$$

therefore
$$\frac{p}{P} = \left(\frac{d}{D}\right)^2 \left(\frac{h}{H}\right)^{3/2} = \left(\frac{\Omega}{\omega}\right)^2 \left(\frac{h}{H}\right)^{5/2}$$

where the small letters $h$, $q$, $\omega$, $d$ refer to the scale model and the large letters $H$, $Q$, $\Omega$, $D$ refer to the full-size machine.

EXAMPLE 15.4 (i)

*If two geometrically similar centrifugal pumps are to run at the same speed, estimate the head and size of a pump to deliver half the quantity of one which runs at 1 450 rev/min and delivers 2·3 m³/min through a head of 18 m with an impeller of diameter 0·26 m.*

The specific speed
$$N_s = \frac{N\sqrt{Q}}{H^{3/4}} = \frac{1\ 450\ \sqrt{2\cdot3}}{(18)^{3/4}} = 252 = \frac{n\sqrt{q}}{h^{3/4}}$$

for the geometrically similar smaller pump.

Hence,
$$h^{3/4} = \frac{1\ 450 \times \sqrt{1\cdot15}}{252} = 6\cdot17 \quad \text{and} \quad h = \mathbf{11\cdot32}\ \text{m}$$

Using similarity equations
$$\frac{h}{H} = \left(\frac{\omega}{\Omega}\right)^2 \left(\frac{d}{D}\right)^2$$

we have
$$\frac{d}{D} = \sqrt{\left(\frac{11\cdot32}{18}\right)} = 0\cdot796$$

or $d = 0\cdot26\ \text{m} \times 0\cdot796 = \mathbf{0\cdot206}$ m or alternatively $q/Q = (d/D)^3 = \frac{1}{2}$, hence all linear dimensions will be reduced in the ratio $\sqrt[3]{\frac{1}{2}}$.

EXAMPLE 15.4 (ii)

*A double-jet Pelton wheel required to develop 5 400 kW has a specific speed of 20 and is supplied through a pipe-line 800 m long from a reservoir, the level of which is 350 m above the nozzles.*

*Allowing for 5 per cent friction loss in the pipe-line, calculate:*

*(a) the speed in rev/min; (b) diameter of the jets; (c) mean diameter of bucket circle; (d) diameter of the supply pipe.*

*Assume $C_v$ for the jets is 0·98, bucket speed 0·46 × jet speed, overall efficiency of wheel 85 per cent and f for pipe 0·006.*

Specific speed
$$N_s = 20 = \frac{N\sqrt{P}}{H^{5/4}}$$

Head at jets = $0\cdot95 \times 350$ m = 333 m = $H$.

Therefore, speed of Pelton wheel is:
$$N = \frac{20 \times (333)^{5/4}}{\sqrt{5\ 400}} = 400\ \text{rev/min}$$

Velocity of jets

$$v_j = C_v \sqrt{(2gH)} = 0.98 \sqrt{(19.613 \text{ m/s}^2 \times 333 \text{ m})} = 80.8 \text{ m/s}$$

Velocity of buckets $= 0.46 \times v_j = 37.2 \text{ m/s} = \omega(D/2)$.
Hence, mean diameter of bucket circle is:

$$D = \frac{2 \times 37.2 \text{ m/s}}{2\pi \times \dfrac{400}{60\text{s}}} = \mathbf{1.77} \text{ m}$$

Power of two jets:

$$\frac{P_o}{\epsilon_0} = 2\left(\frac{\dot{m}}{2}v_j^2\right) = \rho a v_j^3 = \rho \frac{\pi}{4} d_j^2 v_j^3$$

Hence (for each jet)

$$d_j^2 = \frac{4}{\pi} \frac{P_o}{\epsilon_0 \rho v_j^3} = \frac{4}{\pi} \times \frac{5.4 \times 10^6 \text{ W}}{0.85 \times 10^3 \dfrac{\text{kg}}{\text{m}^3} \times (80.8)^3 \dfrac{\text{m}^3}{\text{s}^3}} \left[\frac{\text{N m}}{\text{W s}}\right]\left[\frac{\text{kg m}}{\text{N s}^2}\right]\left[\frac{10^6 \text{ mm}^2}{\text{m}^2}\right]$$

$$= 15450 \text{ mm}^2$$

i.e. diameter of each jet is $d_j = \mathbf{124.5}$ mm.
Rate of flow through the two jets is:

$$Q = 2av_j = 2\left(\frac{\pi}{4} \times \frac{154.5 \text{ m}^2}{10^4}\right)\left(80.8 \frac{\text{m}}{\text{s}}\right)$$

Hence, flow through main supply pipe is:

$$Q = 1.96 \text{ m}^3/\text{s}$$

Head lost in pipe-line is:

$$h_f = \frac{5}{100} \times 350 \text{ m} = 17.5 \text{ m} = \frac{4fl}{d}\frac{v^2}{2g} = \frac{32fl}{\pi^2 g}\left(\frac{Q^2}{d^5}\right)$$

Hence for the pipe-line

$$d^5 = \frac{32fl}{\pi^2 g}\frac{Q^2}{h_f} = \frac{32}{9.87} \times \frac{0.006 \times 800 \text{ m}}{9.807 \text{ m/s}^2} \times \frac{3.84 \text{ m}^6/\text{s}^2}{17.5 \text{ m}}$$

$$= 0.352 \text{ m}^5$$

i.e. the diameter of the supply pipe is $d = \mathbf{0.812}$ m.

EXAMPLE 15.4 (iii)

*A constant speed test of a centrifugal pump resulted in the law*

$$H = 43.5 + 258Q - 3\,770Q^2$$

*where H is the total head generated measured in m and Q is the rate of flow
in m³/s. The pump is to be used to deliver water through a pipe-line 800 m long
and 300 mm diameter, the static lift being 24·5 m. Neglecting velocity heads and*

*taking a pipe friction coefficient $f = 0.007\,5$ calculate the operating head and discharge rate of the pump. Also estimate the power required to drive the pump assuming an overall efficiency of 70 per cent for the particular H vs. Q point at which the pump operates.*

Using the energy principle: Head created by pump = static lift + head lost in pipe friction, we may write, using metre and second units:

$$43\cdot5 + 258Q - 3\,770Q^2 = H = 24\cdot5 + \frac{4fl}{d2g}\left(\frac{4}{\pi}\frac{Q}{d^2}\right)^2$$

$$= 24\cdot5 + \frac{4\times0\cdot007\,5\times800\times16Q^2}{19\cdot163\times243\times10^{-5}\times9\cdot87}$$

$$= 24\cdot5 + 816Q^2$$

Thus,     $4\,586Q^2 - 258Q - 19 = 0 = 241\cdot2Q^2 - 13\cdot57Q - 1$

Hence,     $Q = \dfrac{13\cdot57 \pm \sqrt{(184 + 964\cdot8)}}{482\cdot4} = \mathbf{0\cdot098\,2}\ \mathbf{m^3/s}$

and     $H = 24\cdot5 + 816Q^2 = 24\cdot5 + 8 = \mathbf{32\cdot5\ m}$

The power output of the pump is $P_o = wQH$ and the shaft power to drive the pump is:

$$P_s = \frac{P_o}{\epsilon_0} = \frac{9\cdot807\ \text{kN}}{0\cdot70\ \text{m}^3}\times0\cdot098\,2\ \frac{\text{m}^3}{\text{s}}\times32\cdot5\ \text{m} = \mathbf{44\cdot7\ kW}$$

## EXAMPLE 15.4 (iv)

*A Kaplan turbine is to develop 20 000 kW when running at 240 rev/min under a net head of 45 m. In order to predict its performance, a model to a scale of $\frac{1}{7}$ is tested under a net head of 20 m. At what speed should the model run, and what power would it develop?*

*If the overall efficiency of the model were found to be 0.84, at the best operating point, what quantity of water would the model and the full-scale turbine require?*

For the model $h = 20$ m and $d = D/7$ and $\epsilon_0 = 0\cdot84$.

For the full-scale turbine $H = 45$ m, $P = 20\,000$ kW and $\Omega = 240$ rev/min.

When operating at the best point the head and flow coefficients for the full-scale machine, namely:

$$C_H = \frac{gH}{\Omega^2 D^2} \quad \text{and} \quad C_Q = \frac{Q}{\Omega D^3}$$

are the same as for the model.

Hence

$$\frac{\omega}{\Omega} = \left(\frac{h}{H}\right)^{1/2}\left(\frac{D}{d}\right) = \sqrt{\left(\frac{20}{45}\right)}\times\frac{7}{1} = \frac{14}{3}$$

i.e. the speed of the model $= \omega = \dfrac{14}{3}\times240 = \mathbf{1\,120}$ rev/min.

Since power output $P \propto QH \propto \Omega D^3 H$ then

$$\frac{p}{P} = \frac{\omega}{\Omega}\left(\frac{d}{D}\right)^3 \left(\frac{h}{H}\right) = \frac{14}{3}\left(\frac{1}{7}\right)^3 \left(\frac{20}{45}\right) = \frac{1}{165 \cdot 4}$$

Hence, the power developed by the model would be

$$p = \frac{P}{165 \cdot 5} = \frac{20\,000}{165 \cdot 5} = \textbf{121 kW}$$

Since the overall efficiency $\epsilon_0 = P_0/wQH$, the quantity of water required by the model would be:

$$q = \frac{p_0}{\epsilon_0 wh} = \frac{121 \text{ kW}}{0 \cdot 84 \times 9 \cdot 807 \, \dfrac{\text{kN}}{\text{m}^3} \times 20 \text{ m}}$$

$$= 0 \cdot 734 \, \frac{\text{m}^3}{\text{s}}\left[\frac{10^3 \text{ 1}}{\text{m}^3}\right] = \textbf{734 litre/s}$$

From $C_Q$ = constant for model and full-scale turbine we deduce:

$$\frac{Q}{q} = \frac{\Omega}{\omega}\left(\frac{D}{d}\right)^3 = \frac{3}{14}(7)^3 = 73 \cdot 5$$

Hence the rate of flow through the full-scale turbine would be **53·9** m³/s.

EXAMPLE 15.4 (v)

*A multi-stage boiler feed pump is required to pump 120 Mg of water per hour against a pressure difference of 3 200 kN/m² when running at a speed of 2 900 rev/min. The density of the preheated feed water is 960 kg/m³.*

*If all the impellers are identical and the specific speed per stage is not to be less than 170, find*

(a) *the least number of stages and the head h per stage;*

(b) *the diameter of the impeller assuming the peripheral velocity* $0 \cdot 96 \sqrt{(2gh)}$;

(c) *the shaft power required to drive the pump, assuming an overall efficiency of 78 per cent.*

Rate of volumetric flow of hot water through each stage of the pump is:

$$Q = \frac{\dot{m}}{\rho} = \frac{120 \times 10^3 \text{ kg}}{3\,600 \text{ s}} \times \frac{\text{m}^3}{960 \text{ kg}} = \frac{1}{28 \cdot 8} \frac{\text{m}^3}{\text{s}}\left[\frac{10^3 \text{ 1}}{\text{m}^3}\right]$$

$$= 34 \cdot 75 \text{ l/s} \quad \text{or} \quad 2 \cdot 083 \text{ m}^3/\text{min}$$

and the total head created by the pump is:

$$H = \frac{\Delta P}{w} = \frac{\Delta P}{\rho g} = 3\,200 \times 10^3 \, \frac{\text{N}}{\text{m}^2} \times \frac{\text{m}^3}{960 \text{ kg}} \, \frac{\text{s}^2}{9 \cdot 807 \text{ m}} = 340 \text{ m}$$

(a) Thus, if there are $x$ stages, the head per stage is $h = (340/x)$ m. Specific speed is:

$$N_s = \frac{N \sqrt{Q}}{(h)^{3/4}} = \frac{2\,900 \sqrt{(2 \cdot 083)}}{\left(\dfrac{340}{x}\right)^{3/4}} = 170$$

$$\therefore \qquad x^{3/4} = \frac{170 \times (340)^{3/4}}{2\,900 \times (2 \cdot 083)^{1/2}} = 3 \cdot 21$$

Hence, the number of stages $x = 4 \cdot 7$, i.e. **5** *stages* and

$$h = \frac{340 \text{ m}}{5} = \textbf{68 m}$$

(*b*) Peripheral or tip speed of impeller is

$$u = 0 \cdot 96 \sqrt{(2gh)} = 0 \cdot 96 \sqrt{(19 \cdot 6 \times 68)} \text{ m/s} = 35 \text{ m/s}$$

Hence, since $n = \omega \, (D/2)$ then

$$D = \frac{2u}{\omega} = \left(2 \times 35 \frac{\text{m}}{\text{s}}\right) \left(\frac{60 \text{ s}}{2\,900 \text{ rev}}\right) \left[\frac{\text{rev}}{2\pi}\right] \left[\frac{10^3 \text{ mm}}{\text{m}}\right] = \textbf{230 mm}$$

(*c*) Power output of pump is $P_o = g(\rho Q)H$ and the shaft power is

$$\frac{P_o}{\epsilon_0} = \frac{9 \cdot 807 \text{ m}}{\text{s}^2} \left(\frac{120 \times 10^3 \text{ kg}}{3\,600 \text{ s}}\right) \frac{340 \text{ m}}{0 \cdot 78}$$

$$= \textbf{142·2 kW}$$

## 15.5. Unit power and unit speed

In the previous Section it was shown that if we neglect such complicating factors as viscosity, compressibility and cavitation, then each type of hydro-kinetic machine has a unique characteristic when we plot

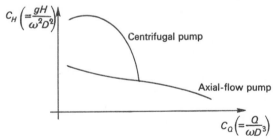

**Fig. 15.5(a).** *Characteristic curves for centrifugal and axial-flow pumps*

$(gH/\omega^2 D^2)$ against $(Q/\omega D^3)$. If the machine is a pump, these parameters conveniently correspond to head and delivery in non-dimensional terms and may be denoted by $C_H$ and $C_Q$:

$$C_H = \frac{gH}{\omega^2 D^2} \quad . \quad . \quad . \quad \text{head coefficient}$$

$$C_Q = \frac{Q}{\omega D^3} \quad . \quad . \quad . \quad \text{flow coefficient}$$

Fig. 15.5(a) shows typical curves for two commonly used types, the centrifugal pump (for high heads and low flows) and the axial-flow pump (for low heads and high flows).

*For negative values of $C_Q$ the machines are operating as turbines*, and when specifically designed for this duty, are said to be of the Francis type if the flow is substantially radial, and of the Kaplan or propeller type if the flow is axial. Again, the radial-flow machine is best suited to high heads and the axial machine to low heads. For turbines the emphasis is primarily on *power as a function of speed* for a given available head and we may derive corresponding power and speed coefficients $C_P$ and $C_S$ as follows: (see also the relationships for scale models in Section 15.4, p. 452).

Let
$$C_P = \frac{C_Q}{\sqrt{C_H}} = \frac{Q}{\omega D^3} \frac{\omega D}{\sqrt{(gH)}} = \frac{Q}{D^2\sqrt{(gH)}}$$

That this is a measure of power in terms of size and head may be seen by noting that:

$$P = \rho g Q H$$

$\therefore$
$$C_P = \frac{Q}{D^2\sqrt{(gH)}}\left[\frac{P}{\rho g Q H}\right]$$

i.e.
$$C_P = \frac{P}{\rho D^2(gH)^{3/2}} \cdots \textbf{power coefficient}$$

a parameter which is independent of speed. A corresponding non-dimensional measure of the latter (independent of $Q$) is obtained simply by taking the reciprocal of the square root of $C_H$:

$$C_S = \frac{1}{\sqrt{C_H}} = \frac{\omega}{\sqrt{(gH)}/D} \cdots \textbf{speed coefficient}$$

Thus if $C_H$ is a unique function of $C_Q$, as indicated in Figs. 15.4(c) and 15.5(a), then $C_P$ or $(C_Q/\sqrt{C_H})$ is similarly uniquely related to $C_S$ or $(1/\sqrt{C_H})$, see Fig. 15.5(b).

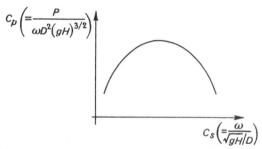

**Fig. 15.5(b).** *Plot of power coefficient against speed coefficient for a turbine*

*A curve of $C_P$ against $C_S$, obtained on any size of hydro-kinetic machine with any convenient range of driving heads (say, available in a laboratory) may thus be used to predict with fair accuracy the behaviour of any geometrically similar machine working under what may possibly be a different range of heads altogether.*

If such curves are merely needed to show the performance of *one particular machine*, the factor $D$ as well as such constants as $\rho$ and $g$ may conveniently be merged in with the values of $C_P$ and $C_S$.

The characteristic head-delivery of a pump (Fig. 15.5(a)) then reduces to a curve of $(H/\omega^2)$ as a function of $(Q/\omega)$; the power-speed curve of a turbine to one of $(P/H^{3/2})$ as a function of $\omega/H^{1/2}$ as shown in Fig. 15.5(c). Each point on the characteristic corresponds to a certain flow

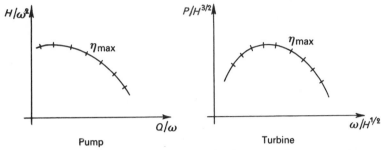

**Fig. 15.5(c).** *Head-delivery curve of a centrifugal pump and power-speed curve of a turbine*

pattern, and the corresponding efficiency may conveniently be marked on it. If power, head and speed are expressed in specific units, namely:

$$P = P'\,\mathrm{kW}$$

$$H = H'\,\mathrm{m}$$

$$\omega = N\,\mathrm{rev/min}$$

i.e.
$$\frac{P}{H^{3/2}} = \frac{P'}{(H')^{3/2}}\left(\frac{\mathrm{kW}}{\mathrm{m}^{3/2}}\right)$$

and
$$\frac{\omega}{H^{1/2}} = \frac{N}{(H')^{1/2}}\left(\frac{\mathrm{rev/min}}{\mathrm{m}^{1/2}}\right)$$

then $P_1 = P'/(H')^{3/2}$ **is referred to as 'unit power'**, and $N_1 = N/(H')^{1/2}$ **as 'unit speed'** for the machine. The unit power, unit speed curves shows that, if the driving head is doubled for example, then the best speed (i.e. for best point or maximum efficiency) increases $\sqrt{2}$ times and the corresponding power output $2\sqrt{2}$ times, and similarly for every other point on the characteristic, see Fig. 15.5(d). In the same way, if the speed of a pump is doubled, the corresponding head size goes up four times, with twice the delivery through the pump, see Fig. 15.4(b).

EXAMPLE 15.5 (i)

*The following data refer to a Pelton wheel: 4 nozzles each 50 mm in diameter with $C_V = 0\cdot97$; reservoir head 300 m; head lost in friction 30 m on 360 m of pipe-line with $f = 0\cdot006$; bucket pitch circle diameter 0·83 m; bucket speed*

0·46 × *jet speed; bucket friction reduces the relative velocity by* 15 *per cent; angle through which the buckets deflect the jet* 165°; *mechanical efficiency* 0·94.
    Determine:

(a) *the diameter of the pipe-line; the hydraulic and overall efficiencies of the machine;*

(b) *the unit power, unit speed and specific speed.*

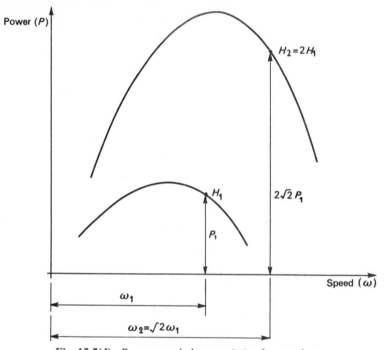

**Fig. 15.5(d).** *Power-speed characteristics for a turbine*

From a velocity diagram similar to Fig. 7.2(b) we measure or calculate that the change in velocity of water in the direction of motion is $(f_1 f_2) = \Delta V = 70$ m/s having used an available head of 270 m at the nozzles to calculate the velocity of the jets $(ef_1)$ as $v_j = C_V \sqrt{(2gH)} = 0.97 \sqrt{(19.613 \times 270)}$ m/s $= 70.5$ m/s and the bucket speed $(eb)$ as $v_B = 0.46 v_j = 32.5$ m/s and allowed for 15 per cent reduction of relative speed of water as it passes over the bucket surfaces.

Thus, the quantity of water flowing through four nozzles is:

$$Q = 4av_j = 4 \times \frac{\pi}{4} \frac{25}{10^4} \text{ m}^2 \times 70.5 \frac{\text{m}}{\text{s}} = 0.55 \text{ m}^3/\text{s}$$

which also flows through the main supply line in which:

$$h_f = \frac{4fl}{2gD} \left( \frac{Q}{(\pi/4)D^2} \right)^2 = \frac{32}{\pi g} \frac{fl}{D^5} Q^2$$

Hence,

$$D^5 = \frac{32}{\pi \times 9.807}\frac{s^2}{m} \times \frac{0.006 \times 360\ m}{30\ m} \times (0.55)^2\frac{m^6}{s^2} = 0.022\ 8\ m^5$$

and the diameter of the supply pipe-line is **0·47 m**  or  **470 mm**.
The power transferred from the four jets to the bucket wheel is:

$$P_o = \dot{m}\Delta V \times v_B = \rho Q \times \Delta V \times v_B$$

$$= 10^3\frac{kg}{m^3} \times 0.55\frac{m^3}{s} \times 70\frac{m}{s} \times 32.5\frac{m}{s}\left[\frac{W\ s^3}{kg\ m^2}\right] = 1\ 250\ kW$$

The hydraulic efficiency

$$\epsilon_h = \frac{P_o}{P_i} = \frac{\dot{m}\Delta V \times v_B}{\frac{1}{2}\dot{m}v_j^2} = \frac{2\ \Delta V \times v_B}{v_j^2} = \frac{2 \times 70 \times 32.5}{(70.5)^2} = 0.915 \ \text{ or } \ \textbf{91·5}\ \textit{per cent}$$

Hence the shaft power available is:

$$P_s = \epsilon_m P_o = 0.94 \times 1\ 250\ kW = 1\ 175\ kW$$

The overall efficiency of the machine

$$\epsilon_0 = \epsilon_h \times \epsilon_m = 0.915 \times 0.94 = 0.86 \ \text{ or } \ \textbf{86}\ \textit{per cent}$$

The speed of the wheel is:

$$\omega = \frac{v_B}{r} = 32.5\frac{m}{s}\frac{2}{0.83\ m}\left[\frac{rev}{2\pi}\right]\left[\frac{60\ s}{min}\right] = \textbf{747}\ \text{rev/min}$$

Unit power

$$P_1 = \frac{P'}{(H')^{3/2}} = \frac{1\ 175}{(270)^{3/2}} = \textbf{0·265}$$

Unit speed

$$N_1 = \frac{N}{(H')^{1/2}} = \frac{747}{(270)^{1/2}} = \textbf{45·4}$$

Specific speed

$$N_s = \frac{N\ \sqrt{P'}}{(H')^{5/4}} = \frac{747\ \sqrt{1\ 175}}{(270)^{5/4}} = \textbf{23·4}$$

EXAMPLE 15.5 (ii)

*A centrifugal pump when tested at constant speed of 1 500 rev/min gave the following results:*

| Delivery m³/min Q | 9·09 | 11·36 | 13·65 | 15·91 |
|---|---|---|---|---|
| Total head m *water* H | 29·6 | 28·0 | 24·7 | 19·8 |
| Shaft kW *power* $P_s$ | 57·4 | 62·7 | 67·1 | 70·1 |

*Find the efficiency values and plot curves showing how the head, shaft power and efficiency vary with the delivery. Find the specific speed corresponding to the best efficiency, assuming dynamically similar conditions of operation. Also find (a) the speed at which the pump should run to give a total head of 36·6 m at the best efficiency and (b) the corresponding discharge from the pump and shaft power required to drive it.*

Overall efficiency of pump is $\epsilon_0 = (wQH)/P_s = $ **76·4** *per cent*; **83·1** *per cent*, **81·9** *per cent*; **73·3** *per cent* for the four cases which are plotted on Fig. 15.5(e).

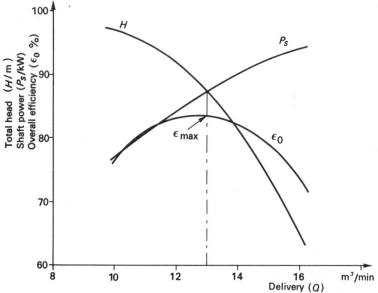

**Fig. 15.5(e).** *Head, power and efficiency curves for a centrifugal pump*

Maximum $\epsilon_0 = 83·5$ per cent when $P_s = 64·9$ kW, $Q = 12·5$ m³/min and $H = 26·5$ m.

Specific speed
$$N_s = \frac{N\sqrt{Q'}}{(H')^{3/4}} = \frac{1\,500\sqrt{12·5}}{(26·5)^{3/4}} = 454$$

Since we are only concerned with one machine the general non-dimensional coefficients (e.g. $(gH/\omega^2 D^2) = C_H$ and $C_Q = (Q/\omega D^3)$) for the point of best efficiency can be simplified to:

$$\frac{H}{\omega^2} = \text{constant} \quad \text{or} \quad \frac{\omega}{H^{1/2}} = \text{constant}, \quad \frac{Q}{\omega} = \text{constant}, \quad \frac{P}{H^{3/2}} = \text{constant}$$

i.e. we may use the formulae for unit speed, unit quantity and unit power. Thus, when $H_2 = 36·6$ m,

$$\frac{\omega_2}{\omega_1} = \sqrt{\left(\frac{H_2}{H_1}\right)} \quad \text{or} \quad \omega_2 = 1500\sqrt{\left(\frac{36·6}{26·5}\right)} = \textbf{1 752} \text{ rev/min}$$

$$\frac{Q_2}{Q_1} = \frac{\omega_2}{\omega_1} \quad \text{or} \quad Q_2 = 12·5\left(\frac{1\,752}{1\,500}\right) = \textbf{14·6} \text{ m}^3/\text{min}$$

and $\qquad \dfrac{P_2}{P_1} = \left(\dfrac{H_2}{H_1}\right)^{3/2} \quad$ or $\quad P_2 = 64\cdot9\left(\dfrac{36\cdot6}{26\cdot5}\right)^{3/2} = \mathbf{103}$ **kW**

EXAMPLE 15.5 (iii)

*The figures given below relate to a test of a water turbine working under its designed head of* 8·5 m.

| Unit power $P_1$ | 8·86 | 9·26 | 9·53 | 9·53 | 9·35 | 9·04 |
|---|---|---|---|---|---|---|
| Unit speed $N_1$ | 56·2 | 65·2 | 74·2 | 83·4 | 92·4 | 101·5 |
| Mass flow $\dot{M}$ kg/s | 3 590 | 3 530 | 3 460 | 3 380 | 3 295 | 3 195 |

*Plot a curve of overall efficiency against unit speed. Find the turbine speed at maximum efficiency, and hence the specific speed of the machine.*

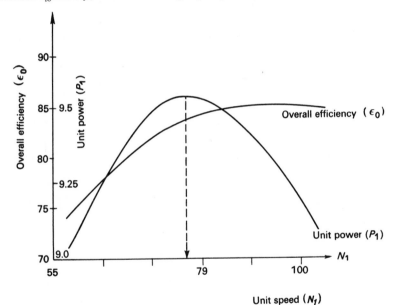

**Fig. 15.5(f).** *Overall efficiency and unit power curves for a turbine*

*If the head is changed to* 10 m, *estimate the power developed and the efficiency at a turbine speed of* 250 rev/min.

Unit power $\quad P_1 = \dfrac{P'}{(H')^{3/2}} = \dfrac{P/\text{kW}}{(8\cdot5)^{3/2}} \quad$ or $\quad P = 24\cdot8P_1$ kW

Overall efficiency

$$\epsilon_0 = \frac{P}{wQH} = \frac{P}{(\dot{M}g)H} = \frac{24\cdot8P_1 \text{ kW}}{9\cdot807 \times 8\cdot5 \ \dot{M} \text{ kgm}^2/\text{s}^3} \left[\frac{\text{kg m}^2}{\text{W s}^3}\right]$$

$$= 297\frac{P_1}{\dot{M}}$$

i.e. $\epsilon_0 = 297(P_1/\dot{M}) = 0\cdot716; 0\cdot778; 0\cdot807; 0\cdot829; 0\cdot84; 0\cdot839$ in the six cases
Maximum $\epsilon_0 = 0\cdot84$ at unit speed

$$N_1 = 92\cdot4 = \frac{N}{(8\cdot5)^{1/2}} = \frac{N}{2\cdot92}$$

and unit power

$$P_1 = 2\cdot11 = \frac{P'}{(8\cdot5)^{3/2}} = \frac{P'}{24\cdot8}$$

Hence, speed at maximum efficiency is:

$$N = 92\cdot4 \times 2\cdot92 = 270 \quad \text{or} \quad \omega = 270 \text{ rev/min}$$

and $\qquad P' = 9\cdot35 \times 24\cdot8 = 232 \quad \text{or} \quad P = 232 \text{ kW}.$

Specific speed

$$N_s = \frac{N\sqrt{P'}}{(H')^{5/4}} = \frac{270 \sqrt{(232)}}{(8\cdot5)^{5/4}} = 283$$

If head is changed to $H = 10$ m and $\omega = 250$ rev/min, i.e. $H' = 10$ and $N = 250$, unit speed is $N_1 = N/(H')^{1/2} = 250 \sqrt{(10)} = 79$ at which point we see from the curves that the efficiency $\epsilon_0 = 83\cdot5$ *per cent* and unit power $P_1 = 9\cdot55$.

Hence, $P = P_1(H')^{3/2} \text{ kW} = 9\cdot55(10)^{3/2} \text{ kW} = 302 \text{ kW}.$

## 15.6. Theoretical predictions of coefficients and specific speeds

In Sections 15.2 and 15.3 it was shown from the velocity triangles that, when working at its best point, the performance of a turbine was given by:

$$_t\epsilon_h = \frac{1 + (q/u)\cot\beta}{gH/u^2}$$

and for a pump by:

$$\frac{1}{_p\epsilon_h} = \frac{1 + (q/u)\cot\beta}{gH/u^2}$$

The value of $(q/u)$ in these expressions was, in terms of the blade geometry, such that:

$$\frac{q}{u} = a\tan\gamma = \frac{1}{\cot\alpha - \cot\beta}$$

where $a$ has the value $1\cdot0$ for an axial-flow machine.

From these expressions a theoretical value of $C_H = (gH/u^2)$ may be derived, as may the corresponding value of $C_Q$. This follows because:

$$C_Q = \left(\frac{Q}{\omega D^3}\right) = \frac{\pi b}{2}\left(\frac{q}{u}\right)$$

for a radial-flow machine. For an axial-flow machine $Q = qA$ where $A$ is the area through which the fluid moves with a velocity $q$, and if $u$ is based on the tip radius, i.e.

$$u = \frac{\omega D}{2}$$

we have, for an axial machine:

$$C_Q = \left(\frac{Q}{\omega D^3}\right) = \frac{Aq}{\omega D^3}\left[\frac{\omega D}{2u}\right] = \frac{A}{2D^2}\left(\frac{q}{u}\right)$$

The optimum theoretical values of $C_H$ and $C_Q$ so obtained enable the corresponding theoretical value of the specific speed to be found by re-forming the fundamental parameter which has a constant value (different constant according to type of turbine or pump). For any type of hydro-kinetic machine working at its best point, the parameter (see Section 15.4, p. 450) is:

$$\frac{\omega\sqrt{Q}}{(gH)^{3/4}} = \left(\frac{\omega^2 D^2}{gH}\right)^{3/4}\left(\frac{Q}{\omega D^3}\right)^{1/2} = \frac{C_Q^{1/2}}{C_H^{3/4}}$$

The specific speed:

$$N_s = \frac{N\sqrt{Q}}{H^{3/4}}$$

is thus proportional to $(C_Q^{1/2}/C_H^{3/4})$.

For a turbine it is also useful to be able to relate the 'unit power' and 'unit speed' values to the geometry of the machine. The former, which is derived from the power coefficient:

$$C_P = \frac{C_Q}{C_H^{1/2}} = \frac{P}{\rho D^2(gH)^{3/2}}$$

enables $(P/H^{3/2})$ to be found, and the latter, which is related to the speed coefficient:

$$C_S = \frac{1}{C_H^{1/2}}$$

gives $(\omega/H^{1/2})$ in any given case.

Example 15.6 (i)

*Show that for a single Pelton wheel having jet diameter d and mean bucket circle diameter D the specific speed may be expressed as K(d/D).*

*Calculate the constant K assuming the coefficient of velocity for the nozzle 0·98, ratio of bucket velocity to jet velocity 0·46 and overall efficiency 88 per cent.*

Specific speed

$$N_s = \frac{N\sqrt{P'}}{(H')^{5/4}}$$

Velocity of jet

$$v_j = C_V\sqrt{(2gH)} = 0.98\sqrt{\left(\frac{19\cdot613\ \text{m}\ H'\ \text{m}}{\text{s}^2}\right)}$$

$$= 4\cdot34(H')^{1/2}\ \text{m/s}$$

Velocity of buckets

$$\omega \frac{D}{2} = 0 \cdot 46 v_j = \frac{D}{2} N \frac{\text{rev}}{\text{min}}$$

Hence,

$$N = \frac{0 \cdot 92}{D} \times 4 \cdot 34 (H')^{1/2} \frac{\text{m}}{\text{s}} \times \frac{60 \text{ s}}{\text{rev}} \left[ \frac{\text{rev}}{2\pi} \right]$$

$$= 38 \cdot 1 (H')^{1/2} \left( \frac{\text{m}}{D} \right)$$

Quantity flowing from jet

$$Q = a v_j = \frac{\pi}{4} d^2 \times 4 \cdot 34 (H')^{1/2} \text{ m/s}$$

$$= 3 \cdot 41 (H')^{1/2} d^2 \text{ m/s}$$

Power output

$$P_o = \epsilon_0 w Q H = 0 \cdot 88 \times 9 \cdot 807 \text{ kN/m}^3 \, (3 \cdot 14 (H')^{1/2} d^2 \text{ m/s}) H' \text{ m}$$

$$= 29 \cdot 4 (H')^{3/2} \left( \frac{d}{\text{m}} \right)^2 \text{ kW} = P' \text{ kW}$$

Specific speed

$$N_s = \frac{N \sqrt{P'}}{(H')^{5/4}} = 38 \cdot 1 (H')^{1/2} \left( \frac{\text{m}}{D} \right) \left\{ 29 \cdot 4 (H')^{3/2} \left( \frac{d}{\text{m}} \right)^2 \right\}^{1/2} \bigg/ (H')^{5/4}$$

i.e.
$$N_s = 206 \cdot 8 \frac{d}{D} = K \frac{d}{D}$$

where        $K = \mathbf{206 \cdot 8}$

## EXAMPLE 15.6 (ii)

*If viscosity of fluid cannot be neglected, deduce non-dimensional expressions to show how it appears in relationships for lift, efficiency and power required to drive geometrically similar centrifugal pumps. Indicate how you would test a pump and plot curves to show its effect.*

Using dimensional analysis we need not introduce $g$ separately since it is always associated with the lift or head $H$ developed. Hence taking $gH$ as the dependent variable we may assume $gH$ is a function of $(\rho, \mu, \omega, D, Q)$ from which we deduce the non-dimensional relationship:

$$\frac{gH}{\omega^2 D^2} = \varphi_1 \left( \frac{\omega D^2}{\nu}, \frac{Q}{\omega D^2} \right) \quad \text{or} \quad C_H = \varphi_1 (Re, C_Q)$$

and similarly for power $P$ we deduce:

$$\frac{P}{\rho \omega^3 D^5} = \varphi_2 \left( \frac{\omega D^2}{\nu}, \frac{Q}{\omega D^3} \right) \quad \text{or} \quad C_P = \varphi_2 (Re, C_Q)$$

and efficiency

$$\epsilon = \frac{w Q H}{P} = \frac{\rho Q g H}{P} = \frac{C_Q C_H}{C_P}$$

If viscosity can be neglected we get the relationships previously deduced in Sections 15.4 and 15.5 and the unique curves for geometrically similar pumps, see Figs. 15.4(c) and 15.5 (b).

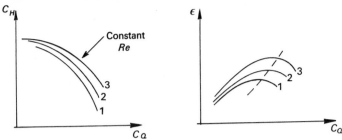

**Fig. 15.6(a).** *Effect of viscosity on pump performance*

If viscosity cannot be neglected, then its effect can be shown by testing with different Reynolds numbers

$$\frac{\omega D^2}{\nu} \quad \text{or} \quad \frac{\rho u D}{\eta}$$

and plotting as shown in Fig. 15.6(a).

## 15.7. Draught tubes and cavitation

The function of a draught tube shown diagrammatically in Fig. 15.7(a) is to reduce the loss of energy finally rejected into the tail race by reducing

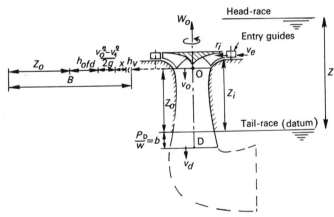

**Fig. 15.7(a).** *Inward-flow turbine with draught tube*

the velocity of the water ($v_o$) leaving the turbine and entering the gradually widening tube. It enables the pressure ($P_o$) at exit from a reaction turbine to be less than atmospheric, thus utilizing the height between the turbine and tail race to increase the effective head to $Z$.

·Fig. 15.7(a) is diagrammatic of a Francis turbine from which, using $h_f$ to denote head loss due to friction and $H$ in general to denote

$$\frac{P}{w}+\frac{v^2}{2g}+z$$

we may deduce the following energy equations.

For the guides: $\qquad\qquad Z = H_e = H_i + h_{efi}$ $\qquad\qquad$ (1)

For the impeller:

$$H_i = W_i + H_o + h_{ifo} = \frac{W_o}{\epsilon_m} + H_o + h_{ifo} \qquad\qquad (2)$$

For the draught tube:

$$H_o = H_d + h_{ofd} = \frac{v_d^2}{2g} + h_{ofd} \qquad\qquad (3)$$

Hence,

$$W_i = H_i - H_o - h_{ifo} = H_e - H_o - (h_{efi} + h_{ifo})$$

$$= H_e - \frac{v_d^2}{2g} - (h_{efi} + h_{ifo} + h_{ofd})$$

i.e.

$$W_i = Z - \frac{v_d^2}{2g} - \sum_e^d h_f$$

The hydraulic efficiency is:

$$\epsilon_h = \frac{W_i}{Z} = 1 - \frac{\left(\dfrac{v_d^2}{2g} + \sum h_f\right)}{Z}$$

and the overall efficiency is:

$$\epsilon_0 = \frac{W_o}{Z} = \frac{W_o}{W_i} \times \frac{W_i}{Z} = \epsilon_m \times \epsilon_h$$

For the draught tube Equation (3) may be written:

$$\frac{P_o}{w} + \frac{v_o^2}{2g} + Z_o = \frac{v_d^2}{2g} + h_{ofd}$$

or

$$\frac{P_o}{w} = -\left\{Z_o + \frac{v_o^2 - v_d^2}{2g}\right\} + h_{ofd}$$

Hence, if $h_{ofd} \to 0$, the minimum pressure head $(P_o/w)$ at entry to the draught tube is:

$$\frac{P_o}{w} = -\left\{Z_o + \left(\frac{v_o^2 - v_d^2}{2g}\right)\right\}$$

When water of temperature $T$ and of corresponding vapour pressure $h_v$, is used in at atmosphere of barometric head $B$, the 'ideal' minimum pressure head (i.e. with zero friction and just avoiding boiling) at entrance to the draught tube is $-(B - h_v)$ as indicated by the horizontal length on Fig. 15.7(a). In this case the appearance of cavities depends on the differences between the ambient pressure $p$ and the vapour pressure $p_v$, i.e. $p - p_v$ is the relevant pressure unit. However, dissolved air usually separates out of water before such a low pressure as $p_v$ is reached and the performance of the turbine begins to fall off; so that a margin, as indicated by head $x$, is necessary in practice. Hence, as cavities may form in aerated fluid at an absolute pressure $p_c$ which is greater than $p_v$, a 'cavitation number' $c = (p - p_c)/\frac{1}{2}\rho u^2$ may be used—the factor $\frac{1}{2}$ being introduced to give the denominator a physical meaning, namely, dynamic pressure.

If a turbine is run with a draught tube head $(z_0)$ which is gradually increased from zero to a maximum value, it is found that the performance of the turbine begins to deteriorate before the limiting head at the entrance to the draught tube is reached. The reason is that cavitation occurs in those regions or pockets of depression in the runner blades where the absolute pressure is less than the absolute pressure at the entrance to the draught tube. Cavitation is not the same thing as flow separation of stream lines, and can occur without flow separation. If the pressure anywhere reduces to such a value that liquid boils—say on the convex side of curved blades, then vapour pockets form and remain until the liquid has moved to a locality of higher pressure where the vapour bubbles collapse suddenly and cause noise, vibration and pitting of the blades and may cause failure of metals. Separation of air or 'air cavitation' usually occurs before 'vapour cavitation' the latter being detected by an increase in noise and vibration and both by a falling-off of lift, thrust and efficiency in pumps, turbines, propellers, venturi meters, etc.

Noise and accompanying vibration occur to some extent in all pumps when operating at points other than the point of maximum efficiency, since at other points the blade angle at entry to the impeller will not be correct and smooth entry of water cannot occur. Therefore shock with accompanying noise is inevitable. Small amounts of air admitted to the suction flow will cushion the collapsing of vapour bubbles and eliminate some of the noise due to shock.

In the case of centrifugal pumps cavitation usually occurs first at the eye of the impeller when the absolute pressure is equivalent to about 1·8 m of water head, at which pressure dissolved air begins to separate out of the water.

For the suction pipe of Fig. 15.7(b):

$$0 = \frac{P_s}{w} + \frac{v_s^2}{2g} = \frac{P_e}{w} + \frac{v_s^2}{2g} + Z_s + h_{fs}$$

Hence, the absolute pressure at the eye is

$$B + \frac{P_e}{w} = B - \left(Z_s + \frac{v_s^2}{2g} + h_{fs}\right)$$

Thus, since vapour cavitation will occur when the absolute pressure at the eye is $h_v$ and air cavitation before that when the absolute pressure is

$(h_v + x)$ where $x$ depends upon the amount of aeration, we deduce that to avoid cavitation at the eye of the impeller $(Z_s + (v_s^2/2g) + h_{fs}]$ must be less than $B - (h_v + x)$ or $B - h_v - (Z_s + h_{fs})$ must be greater than $(v_s^2/2g) + x$.

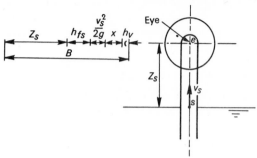

**Fig. 15.7(b).** *Suction pipe of a centrifugal pump*

The ratio $\sigma$ of the head $B - h_v - (Z_s + h_{fs})$ to $H_T$ the total operating or static head (i.e. lift of a rotary pump or height from head to tail races of a turbine) has become known as the 'Thoma cavitation number'

i.e.   $$\sigma H_T = B - h_v - (Z_s + h_{fs}) = \left(\text{abs.}\frac{P_e}{w} + \frac{v_s^2}{2g}\right) - h_v = \frac{v_s^2}{2g} + x$$

Observations and data obtained when efficiency begins to fall and noise due to cavitation begins to appear enable the critical (i.e. lowest before cavitation) $\sigma_c$ to be estimated from

$$\sigma_c H_T = \left(\text{abs.}\frac{P_e}{w} + \frac{v_s^2}{2g}\right)_{\text{crit}} - h_v$$

Its value is usually about $0\cdot1$.

Should cavitation occur, it may be suppressed by partially closing the delivery valve. This reduces $v_s$ and $h_{fs}$ (which is proportional to $v_s^2$) will also be reduced. The pressure at the eye below atmospheric pressure will thus be reduced, i.e. the absolute pressure at the eye will be increased and cavitation will be reduced or eliminated. An alternative to reducing the flow by partially closing the valve, is to reduce $v_s$ by lowering the speed of the pump without altering the setting of the valve.

Thus, referring to Fig. 15.7(c), if a pump begins to cavitate when

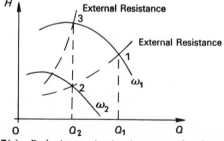

**Fig. 15.7(c).** *Reducing cavitation in a centrifugal pump*

delivering $Q_1$ at speed $\omega_1$, i.e. when operating at point (1), cavitation can be eliminated by reducing the speed of flow to $Q_2$ either by reducing the speed from $\omega_1$ to $\omega_2$ with the same external resistance and thus operating at point (2) or by continuing to run at the same speed $\omega_1$ and increasing the external resistance, say by partially closing the delivery valve so that the pump operates at point (3).

EXAMPLE 15.7 (i)

*The axis of a reaction machine and its draught tube is vertical, the head in the spiral casing at inlet being 38 m above atmospheric and the speed of water 5 m/s. The flow through the turbine is 1·7 m³/s, the hydraulic and overall efficiencies being 0·82 and 0·79 respectively. The top of the draught tube is 1 m below the horizontal centre-line of the spiral casing while the tail-race surface is 3 m below the top of the draught tube. At inlet and outlet of the draught tube the velocities are 4 and 2 m/s, respectively, and there is no whirl at either position. Find:*
(a) *the total head across the machine,*
(b) *the power output,*
(c) *the head lost in friction in turbine and draught tube,*
(d) *the power lost in mechanical friction.*

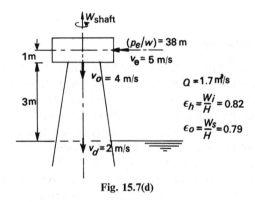

**Fig. 15.7(d)**

Referring to Fig. 15.7(d).
(a) Total head across the machine is

$$\frac{P_e}{w}+\frac{v_e^2}{2g}+4 \text{ m} = (38+1\cdot3+4) \text{ m} = \textbf{43·3 m} = H$$

(b) The shaft work $W_s = \epsilon_0 H$, hence the shaft-power is:

$$P_s = wQH\epsilon_0 = 9\cdot807 \text{ kN/m}^3 \times 1\cdot7 \text{ m}^3/\text{s} \times 43\cdot3 \text{ m} \times 0\cdot79 \left[\frac{\text{W s}}{\text{N m}}\right] = \textbf{572 kW}$$

(c) Considering the energy entering and leaving the impeller casing per unit weight of water:

$$\frac{P_e}{w}+\frac{v_e^2}{2g}+1 \text{ m} = \frac{P_o}{w}+\frac{v_o^2}{2g}+h_{ft}+W$$

Also for the draught tube:

$$\frac{P_o}{w}+\frac{v_o^2}{2g}+3\text{ m} = \frac{v_d^2}{2g}+h_{fd}$$

Hence

$$W = (38+1\cdot3+1)\text{ m}-\left(h_{ft}+h_{fd}+\frac{(2)^2\text{ m}}{19\cdot61}-3\text{ m}\right)$$

$$= 43\cdot1\text{ m}-(h_{ft}+h_{fd})$$

Since hydraulic efficiency $\epsilon_h = W/H = 0\cdot82$ then

$$W = 43\cdot1\text{ m}-(h_{ft}+h_{fd}) = 0\cdot82\times43\cdot3\text{ m} = 35\cdot5\text{ m}$$

Hence $\qquad h_{ft}+h_{fd} = (43\cdot1-35\cdot5)\text{ m} = \textbf{7·6 m}$

(*d*) Mechanical efficiency

$$\epsilon_m = \frac{\epsilon_0}{\epsilon_h} = \frac{0\cdot79}{0\cdot82} = 0\cdot964$$

and since

$$\epsilon_m = \frac{P_s}{P_s+P_f}$$

i.e. $\quad \dfrac{1}{\epsilon_m} = 1+\dfrac{P_f}{P_s}$ then $P_f = P_s\left(\dfrac{1}{\epsilon_m}-1\right) = 572\text{ kW}\left(\dfrac{1}{0\cdot964}-1\right) = \textbf{21·4 kW}$

## 15.8. Fluid couplings

We have seen in Section 15.1 that torque and angular momentum of a fluid are inter-related (torque $T = \rho Q\varDelta(wr)$ (and power $P = T\omega = \dot m\varDelta(wu)$). If the rotor performs work on the fluid, the machine operates as a pump, whereas if the fluid does work on the rotor, the machine is operating as a turbine. Pumps connected by supply lines to turbines or motors are the essential components in hydraulic transmission systems. The interconnecting lines and, hence, the resistance and inertia are reduced to a minimum in fluid couplings and torque converters which consist essentially of a centrifugal pump and a turbine in a single housing and with no solid contact or mechanical link between them. Oil is the usual fluid as it also lubricates.

A fluid coupling has an impeller or primary (driver) wheel which delivers fluid directly into an inward-flow turbine runner or secondary (driven) wheel as in Fig. 15.8(a). Both have straight radial blades (i.e. 90° to plane of rotation) which face each other. When the impeller is rotated, the product of the velocity of whirl and radius increases by $\varDelta(wr) = w_2r_2-w_1r_1$ and thus fluid is expelled with increased angular momentum. The bulk of the fluid enters the turbine and flows radially inwards against the lesser centrifugal pressure due to the slower speed of the turbine runner. The fluid gains energy from the impeller and delivers energy to the runner. The moment of momentum of the fluid decreases to create a torque on the secondary or driven wheel whilst flowing towards the eye of the impeller which, in turn, receives an inflow ($\dot{m} = \rho Q = \rho A u_{\text{axial}}$) at the inner radii (of mean value $r_1$). The vortex motion and rate of volumetric circulation ($Q$) continues so long as slip

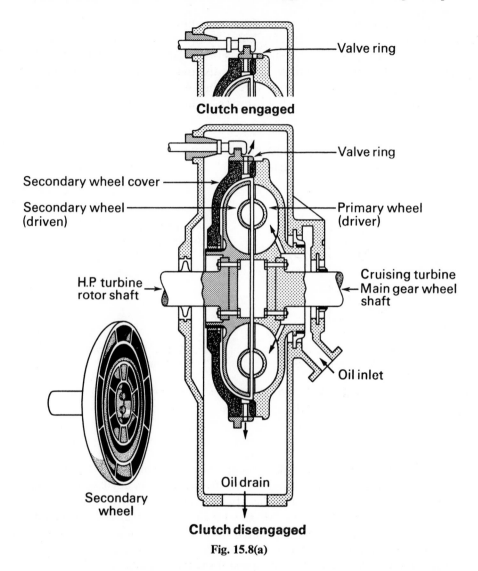

**Clutch engaged**

Valve ring

Secondary wheel cover

Secondary wheel (driven)

Primary wheel (driver)

H.P. turbine rotor shaft

Cruising turbine Main gear wheel shaft

Oil inlet

Secondary wheel

Oil drain

**Clutch disengaged**

**Fig. 15.8(a)**

$[S = (\omega_p - \omega_s)/\omega_p]$ occurs between the speeds of the primary ($\omega_p$) and secondary ($\omega_s$) shafts. If the speeds become equal ($S = 0$) the centrifugal effects or radial pressure gradients in the two halves (Fig. 15.8(a)) balance, the circulation of fluid ceases, and the torque falls to zero. The torque ($T_p$) on the input, or primary, and ($T_s$) output, or secondary, shafts must be equal under steady conditions (i.e. $\alpha = 0$ in $T = I\alpha$) since no other external moment acts on the system.

Thus, $T_p = T_s = T = \rho Q \Delta (wr)$ and the efficiency of transmission is

$$\epsilon = \frac{P_s}{P_p} = \frac{T_s \omega_s}{T_p \omega_p} = \frac{\omega_s}{\omega_p} = 1 - S = 1 - \frac{P_{lost}}{P_p}$$

and which, in the case of, say, a 3 per cent slip, is 97 per cent.

The eddies caused by the shock due to the abrupt change in velocity of whirl (owing to the different speeds of the two wheels) together with frictional drag on the fluid, results in a dissipation of mechanical energy and a consequent rise in temperature of the fluid. Fluid-friction losses are of more importance than shock losses at low slips, and vice versa at high slips.

Figs. 15.8(b) and (c) show superimposed velocity diagrams at the two mean radii in which it will be seen that

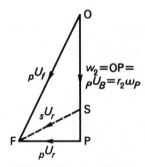

**Fig. 15.8(b).** *Velocity diagrams at the periphery* ($v_2$)

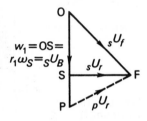

**Fig. 15.8(c).** *Velocity diagrams at the hub* ($v_1$)

(i) Slip, $S = \dfrac{SP}{OP} = \dfrac{OP - OS}{OP} = \dfrac{{}_pU_B - {}_sU_B}{{}_pU_B} = \dfrac{\omega_p - \omega_s}{\omega_p}$,

hence $\dfrac{\omega_s}{\omega_p} = 1 - S.$

(ii) There is a difference in the velocities of whirl, namely $OP - OS$ both as the oil leaves the primary wheel (with relative velocity ${}_pU_r$ at 90° to the plane of rotation) and the secondary wheel (with relative velocity ${}_sU_r$ also at 90°). This results in the directions (SF and PF in Figs. 15.8(b) and (c)) of the relative velocities ${}_sU_r$ and ${}_pU_r$ required for the fluid to enter the respective wheels smoothly, being different from the 90° receiving angle of the blades; hence there is shock and the consequent formation of eddies and dissipation of energy.

(iii) The axial velocity under steady conditions is constant, namely PF and SF in Figs. 15.8(b) and (c), respectively, and equal to

$$\frac{Q}{A} = \frac{\dot{m}}{\rho A}$$

where $A$ is the area of the annular rings at the periphery and at the hub;

hence the axial velocity PF in Fig. 15.8(a) is seen to be proportional to $\omega_p r_2$ or $u_{axial} = k\omega_p r_2$.

The power dissipated and causing a rise in temperature of the fluid is

$$P_p - P_s = T(\omega_p - \omega_s) = TS\omega_p$$

The $\Pi$ theorem may be applied to dimensional analysis of a fluid coupling in a steady state, assuming torque $T$ in a given design is a

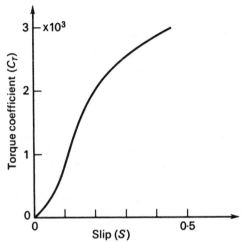

**Fig. 15.8(d).** *Torque coefficient of a fluid coupling*

function of $\rho$, $\eta$, $\omega_p$, $S$, $D$, $V$, where $D$ is the diameter of the runner wheels and $V$ the volume of fluid in the housing. Since $n - p = 7 - 3$, there are 4 non-dimensional groups giving a torque coefficient

$$C_T = \frac{T}{\rho\omega_p^2 D^5} = \varphi\left(S, \frac{\rho\omega_p D^2}{\eta}, \frac{V}{D^3}\right)$$

For a particular design $V/D^3$ is constant and so is the Reynolds number

$$Re = \frac{\rho\omega_p D^2}{\eta} = \frac{\omega_p D^2}{\nu} = \frac{2uD}{\nu}.$$

Tests show that the effect of slip is much more important than that of *Re* as shown in Fig. 15.8(d). Thus, under dynamically similar conditions fluid couplings will transmit torques proportional to $\rho\omega^2 D^5$—the chief conditions for dynamical similarity being that the slip ($S$) shall be the same and that the system shall circulate a volume of oil ($V$) proportional to the cube ($D^3$) of its size, i.e. $S$ and $V/D^3$ are, respectively, the same.

The principal advantages of fluid couplings lie in their unsteady-state characteristics. They enable prime movers which have poor starting characteristics to be directly coupled to heavy loads without fear of stalling the former, e.g. they provide low starting torques on electric motors, turbines and internal-combustion engines. Fluid couplings also tend to smooth out torsional oscillations originating either at the prime mover or at the load and, since all the blades are radially straight, they can operate and have the same efficiency in both directions of rotation. Under steady conditions they will provide a smoothly variable speed from a constant-speed prime mover by varying the amount of oil flow ($Q$) and, therefore, the slip(S). This flexibility is, however, obtained at the expense of efficiency since slip

$$S = 1 - \frac{\omega_s}{\omega_p} = 1 - \epsilon$$

hence, such speed control is unsuited to prolonged operation at reduced speed ($\omega_s$) if efficiency ($\omega_s/\omega_p$) is important. They are suited to variable speed drives in, e.g. automobile, railroad and marine power transmission systems.

EXAMPLE 15.8 (i)

*The torque coefficient of a particular fluid coupling is found to be*

$$C_T = \frac{T}{\rho \omega_p^2 D^5} = 0\cdot008 \sqrt{S}.$$

*If the efficiency of transmission is 98 per cent when the output is 11 kW from the coupling 300 mm diameter filled with oil of density 855 kg/m³, find the speed of the primary shaft.*

Power output $P_s = T\omega_s$ and power input $P_p = T\omega_p$.

Efficiency

$$\epsilon = P_s/P_p = \frac{\omega_s}{\omega_p}, \quad \text{and} \quad S = 1 - \frac{\omega_s}{\omega_p} = 1 - \epsilon$$

Hence, torque

$$T = \frac{P_s}{\epsilon \omega_p} = (0\cdot008 \sqrt{S})\rho \omega_p^2 D^5$$

and

$$\omega_p^3 = \frac{P_s}{\epsilon(0\cdot008 \sqrt{(1 - \epsilon)})\rho D^5}$$

$$= \frac{11 \text{ kN m/s}}{0\cdot98 \times 0\cdot008 \sqrt{0\cdot02} \times 855 \text{ kg/m}^3 \times (0\cdot3 \text{ m})^5}$$

$$= 4\cdot74 \times 10^6/\text{s}^3$$

and

$$\omega_p = \frac{147\cdot5}{\text{s}} \left[\frac{\text{rev}}{2\pi}\right] \left[\frac{60 \text{ s}}{\text{min}}\right] = 1\ 410 \text{ rev/min.}$$

EXAMPLE 15.8 (ii)

(a) *A fluid coupling in a ship transmits 4 MW at 300 rev/min when the fractional slip is 5 per cent. Estimate the power transmitted by a geometrically-similar clutch of twice the weight and working with the same slip and at the same speed.*

(b) *Also, estimate the rate of oil flow necessary to cool the larger clutch if oil enters at 15°C and leaves at 80°C and has a specific heat capacity of 2 kJ/kg K.*

The power coefficient is

$$\frac{P_{primary}}{\rho \omega_p^3 D^5} = \varphi\left(S, \frac{\omega_p D^2}{\nu}, \frac{V}{D^3}\right)$$

(a) Hence, for geometrically-similar clutches in which

$$S \quad \text{and} \quad \frac{V}{D^3}$$

are the same and

$$Re = \frac{\omega_p D^2}{\nu} = \frac{2uD}{\nu}$$

has negligible influence,

$$\frac{P_p}{\rho \omega_p^3 D^5} \text{ is constant.}$$

Also, the efficiency

$$\epsilon = \frac{P_s}{P_p} = 1 - S = 0\cdot95$$

Thus, the power of the primary wheel is $P_p = P_s/\epsilon$ and the speed of the primary wheel is $\omega_p = \omega_s/\epsilon$; hence, for geometrically-similar clutches with the same slip(S) and, therefore, the same efficiency $\epsilon = 1 - S$,

$$\left(\frac{P_s}{\omega_s^3 D^5}\right)_1 = \left(\frac{P_s}{\omega_s^3 D^5}\right)_2.$$

Also, since weight is proportional to $D^3$, then $D^5 \propto W^{5/3}$ and $P_s \propto \omega_s^3 W^{5/3}$.
Hence,

$$P_{s2} = P_{s1}\left(\frac{\omega_{s2}}{\omega_{s1}}\right)^3\left(\frac{W_2}{W_1}\right)^{5/3} = 4 \text{ MW } (1)^3(2)^{5/3} = \mathbf{12\cdot7} \text{ MW}$$

(b) Power dissipated

$$= P_p - P_s = P_s\left(\frac{1}{\epsilon} - 1\right) = \left(\frac{S}{1-S}\right)P_s$$

Hence, in the larger coupling the power dissipated

$$= \frac{5}{95} \times 12\cdot7 \text{ MW} = 0\cdot668 \text{ MW}$$

and the rate of oil flow

$$= \frac{668 \text{ kJ/s}}{2 \text{ kJ/kg K} \times (80-15) \text{ K}} = \mathbf{5\cdot13} \text{ kg/s}$$

EXAMPLE 15.8 (iii)

*A fluid clutch transmits* 3 MW *at* 600 rev/min. *It is cooled by oil from a pump supplying oil at* 350 kPa. *Estimate the ratio of the mean velocities of oil circulating and the pressure of the oil in a geometrically-similar clutch transmitting* 7·5 MW *at* 500 rev/min *working with the same slip and so that the temperature rise of oil cooling the larger clutch shall not exceed that of the smaller one. Assume that the flow in the ducts is laminar.*

For geometrically-similar fluid clutches working with the same slip,

$$\frac{P}{\rho\omega^3 L^5} = \text{constant.}$$

Hence,

$$\left(\frac{L_1}{L_2}\right)^5 = \frac{P_1}{P_2}\left(\frac{\omega_2}{\omega_1}\right)^3 = \frac{3}{7\cdot5}\left(\frac{5}{6}\right)^3 = \frac{1}{4\cdot32}$$

and the scale ratio $L_2/L_1 = 1\cdot34$.

Power dissipated

$$= P_p - P_s = \left(\frac{S}{1-S}\right)P_s = (\rho\dot{Q})C\delta\theta$$

Hence, for the same slip $(S)$, density $(\rho)$, specific heat capacity $(C)$ and temperature rise $(\delta\theta)$

$$\frac{P_{s1}}{P_{s2}} = \frac{\dot{Q}_1}{\dot{Q}_2} = \frac{\bar{u}_1 L_1^2}{\bar{u}_2 L_2^2}$$

and the ratio of the mean velocities

$$\frac{\bar{u}_2}{\bar{u}_1} = \frac{P_{s2}}{P_{s1}}\left(\frac{L_1}{L_2}\right)^2 = \frac{2\cdot5}{(1\cdot34)^2} = 1\cdot39$$

Assuming laminar flow,

$$\Delta p \propto \frac{\bar{u}}{L}$$

(by Poiseuille's law—$\eta$ being the same for the same temperature conditions).

Hence,

$$\frac{\Delta p_2}{\Delta p_1} = \frac{\bar{u}_2}{\bar{u}_1} \times \frac{L_1}{L_2} = \frac{1\cdot39}{1\cdot34} = 1\cdot04$$

i.e. the pump of the larger clutch would need to supply cooling oil at very much the same pressure (**360 kPa**) but in a quantity $\dot{Q}_2$ such that

$$\frac{\dot{Q}_2}{\dot{Q}_1} = \frac{u_2}{u_1}\left(\frac{L_2}{L_1}\right)^2 = 1\cdot39 \times (1\cdot34)^2 = \mathbf{2\cdot5}.$$

## 15.9. Torque converters

A torque converter may be obtained by designing for the introduction of stationary curved blading between the pump impeller and the turbine runner of a fluid coupling. It provides, in a single unit, a direct-drive fluid coupling with automatic torque conversion according to load. Torque converters were first developed by Föttinger in Hamburg for use with propulsion turbines of German warships prior to 1914. These

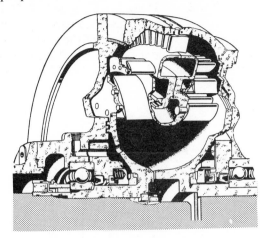

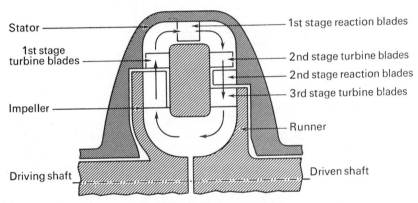

Stator

1st stage turbine blades

Impeller

Driving shaft

1st stage reaction blades

2nd stage turbine blades

2nd stage reaction blades

3rd stage turbine blades

Runner

Driven shaft

**Fig. 15.9(a).** *Torque converter*

were, however, superseded by helical gearing because of higher efficiency and lower cost. Modern torque converters are used in high-powered and heavy vehicles of various kinds to eliminate the necessity of gear changing.

If torque increase or multiplication (which is, of course, associated with reduction of speed $\omega_p - \omega_s$) is required, the fixed blades (or reaction elements) are arranged so as to redirect the fluid before it enters the next turbine stage. This gives rise to a torque ($T_R$) acting in the same sense as that ($T_p$) applied to the input or primary shaft (see Fig. 15.9(a)). Thus, $T_s = T_p + T_R$ and the efficiency of $\epsilon$ of the conversion is given by $T_s\omega_s = \epsilon T_p\omega_p$. The maximum efficiency is less than that attainable in a fluid coupling because of greater complexity—fluid couplings not needing reaction elements. The reaction elements exert a torque on the fluid on the same principle as operates in the impeller and runner, namely, by changing the angular momentum of the circulating fluid. The relatively large dissipation of energy due to shock that would take place with straight blades is minimized by forming or curving the blades and operating through a number of stages. These are designed so as to reduce shock as the fluid flows to or from the faster-moving ($\omega_p$) impeller to stationary reactor blades, but always via a set of the slower-moving ($\omega_s$) turbine blades rotating the secondary or output shaft. Fig. 15.9(a) shows such a development having 3 turbine stages and 2 reaction stages,

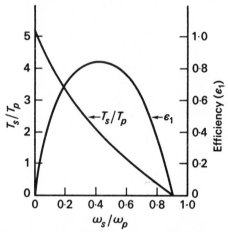

**Fig. 15.9(b).** *Torque multiplication and efficiency*

which can produce a torque multiplication of over 5 to 1 and a maximum efficiency of over 80 per cent. The drive of such a torque converter is uni-directional but only a simple gear is needed for reversing. Input or primary torque $T_p$ and output or secondary torque $T_s$ are, for geometrically-similar machines operating under dynamically similar conditions, both proportional to $\rho \omega_p D^5$, i.e. $T_p$ and $T_s \propto \rho \omega_p D^5$. Thus, so long as such hydro-kinetic machines circulate a volume of fluid proportional to the cube of their linear sizes (i.e. $V/D^3 = $ constant a header tank ensuring that the unit runs full), the major requirement for similarity is that $\omega_s \propto \omega_p$. Hence, torque multiplication, $T_s/T_p$, depends primarily on $\omega_s/\omega_p$—see Fig. 15.9(b) which also shows the corresponding efficiency variation with speed ratio.

The $\Pi$ theorem applied to dimensional analysis of a torque converter operating under steady conditions (assuming that $T_p$ and $T_s$ are functions of $\rho$, $\eta$, $\omega_p$, $\omega_s$, $D$, $V$) gives $n-\rho = 7-3 = 4$ dimensionless groups. The torque coefficients

$$C_p = \frac{T_p}{\rho \omega_p^2 D^5} = \varphi_1 \left( \frac{\omega_s}{\omega_p}, \frac{\omega_p D^2}{\nu}, \frac{V}{D^3} \right)$$

and
$$C_s = \frac{T_s}{\rho \omega_p^2 D^5} = \varphi_2 \left( \frac{\omega_s}{\omega_p}, \frac{\omega_p D^2}{\nu}, \frac{V}{D^3} \right),$$

The torque multiplication ratio

$$\frac{T_s}{T_p} = \varphi_3 \left( \frac{\omega_s}{\omega_p}, \frac{\omega_p D^2}{\nu}, \frac{V}{D^3} \right)$$

and the efficiency

$$\epsilon = \varphi_4 \left( \frac{\omega_s}{\omega_p}, \frac{\omega_p D^2}{\nu}, \frac{V}{D^3} \right)$$

As in fluid couplings, tests on torque converters show that the effect of $Re = (\omega_p D^2)/\nu$ can be neglected in comparison with that of speed ratio $\omega_s/\omega_p$.

## Exercises on Chapter 15

1. The mean blade circle diameter of an axial-flow impulse turbine is 1·37 m. The guide blade angle is 24°, the receiving and discharging angles of the runner blades are 48° and 23° respectively, and the breadth of the moving blades at inlet is 100 mm. Calculate the speed of the turbine so that the water may pass smoothly on to the blades when the turbine is working under a head of 85·5 m, and find the diagram power developed if, when there is full circumferential admission, the passages are 85 per cent full at inlet. Neglect friction.

2. Water is supplied to an axial-flow impulse turbine, having a mean diameter of runner 1·83 m revolving at 144 rev/min under a head of 30·5 m. The angle of the guide blade at entrance is 30° and the angle the vane makes with the direction of motion at exit is 30°. Eight per cent of the head is lost in the supply pipe and guide. Determine the relative velocity of water and wheel at entrance, and the blade angle at inlet. If 10 per cent of this relative velocity is lost in friction in the wheel, determine the velocity with which the water leaves the wheel. Find the hydraulic efficiency of the turbine.

3. The diameter of the central openings of an inward flow reaction turbine is 610 mm, the depth between the cover plates is 610 mm, the quantity of water discharged is 53·4 m³/min, and the wheel makes 150 rev/min. Find the best inclinations of the inner ends of the vanes.

4. The inner and outer diameters of an inward flow reaction turbine are 305 mm and 610 mm respectively. The water enters the outer circumference at 10° with the tangent, and leaves the inner circumference radially. The velocity of flow is 1·52 m/min at both circumferences, and the wheel makes 192 rev/min. Determine the angles of the vanes at both circumferences, and the theoretical hydraulic efficiency of the turbine.

5. In an outward flow reaction turbine the wheel has an internal radius of 763 mm, and an external radius of 915 mm. The speed is 250 rev/min. The wheel is 254 mm wide at inlet and outlet, and the net area for radial flow at inlet and at outlet is 0·86 of the gross area. The head of water is 42·7 m and the quantity of water is 5·67 m³/s. Neglect all friction losses and determine the angles of the tips of the vanes at inlet and outlet so that the water shall leave radially.

6. An outward radial-flow impulse turbine has nozzles with a total effective area of 970 mm² and the guide vanes make an angle of 20° to the tangent at exit.

The inner and outer diameter of the runner are 508 mm and 712 mm respectively, and the moving vanes have an outlet angle making 20° to the tangent.

When running at 620 rev/min under a head of 61 m the discharge is 1·8 m³/min and the power output is 11·9 kW.

The water at discharge is observed to leave the runner in a forward direction and inclined at 15° to the radius.

Assuming the water leaves tangential to the moving vanes, calculate the head lost in (*a*) the nozzles, (*b*) the moving vanes, (*c*) bearing friction and windage.

7. In a Francis-type turbine the guide-vane angle is 8° the inlet angle ($\beta$) of the moving blade is 100° and the outlet angle $\gamma$ is 20°. Both fixed and moving vanes reduce the flow area by 15 per cent. The runner is 610 mm outside diameter and 407 mm inside diameter, and the widths at entrance and exit are 50 mm and 86 mm respectively.

The pressure at entry to the guides is +26·5 m head and the kinetic energy there can be neglected. The pressure at discharge is −1·83 m head.

If the losses in the guides and moving vanes are $8f^2/2g$ where $f$ is the radial component of flow, calculate:

(*a*) the speed of the runner in rev/min for tangential flow on to the runner vanes,

(*b*) the power given to the runner by the water.

8. The peripheral speed of the impeller of a centrifugal pump is 915 mm/s. The blades are curved backwards at an angle of 55° from the radius and the water leaves the wheel with a component radial velocity of 152·5 mm/s. If 3·4 m³/min of water flows through the pump estimate the torque required to rotate the impeller of 1·25 m diameter.

9. Find the speed in rev/min at which a centrifugal pump must run in order to commence delivery against a head of 4, 9, 25, 49 m respectively. The inner and outer diameters of the impeller are 0·5 m and 1 m.

10. Find the tip speed of the impeller of a pump and the power required to lift 200 Mg of water per minute through a height of 1·5 m. The hydraulic efficiency of the pump is 0·6.

The constant velocity of flow through the impeller is 1·37 m/s. The blade tip at exit is bent back 70° from the radius. Water enters radially.

11. A boiler-room fan has an impeller diameter 2·25 m, blade width at periphery 350 mm and blades bent backwards 73° from the radius.

Assuming constant velocity of flow estimate the ideal change in pressure head of air through the fan when delivering 1 430 m³/min of air, the speed being 450 rev/min. Density of water 1 Mg/m³ and of air 1·12 kg/m³.

12. Two tanks A and B are connected by a main which may be assumed equivalent to a pipe 75 mm diameter and 300 m long. The cross-sectional areas of A and B are 45 m² and 63 m² respectively. A pump in the pipe-line

transfers oil from one tank to the other. If the levels in the two tanks are initially the same, estimate the rate of volumetric flow of oil at the beginning and end of the operation and the time necessary to lower the level in A by 6 m. Take $f = 0.009$ for the pipe and $H = 13.7 + 29.6Q - 204Q^2$ as the characteristic of the pump where $H$ is the total head in metres and $Q$ the rate of discharge in m³/min.

13. A boiler-room fan gave the following data on test:

| Speed (rev/min) $\omega$ | 500 | 490 | 477 | 455 | 495 |
|---|---|---|---|---|---|
| Total head (mm of water) $H$ | 121·8 | 120·8 | 108·9 | 81·2 | 62·0 |
| Discharge (m³/min) $Q$ | 735 | 1 435 | 2 120 | 2 690 | 3 570 |

Plot $H/\omega^2$ against $Q/\omega$ and find the optimum point. Hence, estimate the speed at which a geometrically similar fan with all linear dimensions reduced in the ratio of 1:0·8, should run to deliver 1135 m³/min with a total head equivalent to 101·5 mm of water.

14. A centrifugal pump has vanes curved backwards so that the tangent to the vanes at exit makes an angle of 30° with the tangent to the periphery. The impeller diameter is 250 mm and the width 19 mm. Vane thicknesses take up 15 per cent of the peripheral area.

Assuming 50 per cent of the kinetic energy at exit from the impeller is lost in eddies, calculate, on the usual assumptions, what will be the lift and manometric efficiency when discharging 1·8 m³/min at a speed of 1 100 rev/min.

15. Two identical pumps are connected to a main in such a way that they can be used in series or parallel. The characteristic of the pumps is

$$\Delta H = 91.5 + 3\,770Q - 22 \times 10^4 Q^2$$

where $Q$ is the flow in m³/s and $\Delta H$ the gain in total head in m. The resistance of the main is given by $\Delta H = 954 \times 10^2 Q^2$.

From graphs, estimate the maximum discharges that can be delivered when running in series and parallel and their ratio.

16. Find the specific speed of an inward-flow reaction turbine of runner diameter 2 m and blade width at entry 150 mm. When working under a head of 12 m and 120 rev/min the velocity of flow is 3·5 m/s and the efficiency is 0·91.

17. Calculate the diameter, speed and specific speed of a Kaplan or propeller turbine runner to develop 6 350 kW under a head of 4·8 m.

The speed ratio $\dfrac{u}{\sqrt{(2gH)}}$ is 2·1 on the outer diameter.

The flow ratio $\dfrac{q}{\sqrt{(2gH)}}$ is 0·65.

Diameter of boss = 0·35 times external diameter of runner.
Overall efficiency is 88 per cent.

18. A model of a Francis turbine built to a scale of one-fifth was found to develop 3·06 kW at a speed of 360 rev/min when tested under a head of 1·8 metres. Calculate the speed and power of the full-size turbine when working under a head of 5·8 m. Check that the specific speed is the same.

19. A centrifugal pump is intended to lift 42·5 m³/min against a head of 4·9 m. A model is made and delivers 1·7 m³/min against 4·9 m head when making 1 450 rev/min—the efficiency then being a maximum. Find the specific speed and estimate the speed and diameter of the impeller of the large pump if the diameter of the impeller of the model pump is 200 mm.

20. A centrifugal pump is to be designed to deliver 336 m³/min against a head of 135 m when running at a speed of 450 rev/min. In designing the model of this pump, the desired operating conditions include a rate of volumetric flow of 12 m³/min and a shaft power consumption of 225 kW. (*a*) If the model efficiency is assumed to be 89 per cent what will be (i) the speed of the model, (ii) the scale ratio.

(*b*) If the model is worked against full head find (i) the speed, (ii) the quantity pumped per unit time and (iii) the shaft power required.

21. Calculate the theoretical specific speed of a Pelton wheel which has a nozzle efficiency of 95 per cent, an overall efficiency of 85 per cent, a ratio of mean wheel speed to jet speed of 0·46 and a ratio of mean wheel diameter to jet diameter of 18.

22. A centrifugal pump, having 4 stages in *parallel*, delivers 12 m³/min of liquid against a head of 25 m, the diameter of the impeller being 220 mm and the speed 1 700 rev/min.

A pump is to be made up with a number of identical stages *in series*, of similar construction to those in the first pump, to run at 1 250 rev/min, and to deliver 15 m³/min against a head of 250 m.

Find the diameter of the impeller and the number of stages required.

23. Show that, for geometrically similar centrifugal pumps, the following relations hold between the external diameter of the impeller (*d*), the speed (*n*), the total head (*h*), the discharge (*Q*) and the power required to drive (*P*)

$$d \propto \frac{h^{1/2}}{n}; \quad Q \propto d^2 h^{1/2}; \quad n \propto \frac{h^{3/4}}{Q^{1/2}}; \quad P \propto \frac{h^{5/2}}{n^2}$$

It is required to predict the performance of a large centrifugal pump from that of a scale model one-quarter the diameter. The model absorbs 15 kW when pumping under the test head of 6 m at its best speed of 400 rev/min. The head for the full scale pump is 18 m. At what speed should it be driven, what power will be required to drive it and what will be the ratio of the quantities discharged by the large pump and the model?

24. Deduce a relation between the power transmitted, size and speed of geometrically-similar fluid couplings or clutches running under dynamically-similar conditions.

A fluid clutch transmitting 3 MW at 600 rev/min weighs 30 kN. Estimate the linear scale ratio and the weight of a geometrically-similar clutch to transmit 5 MW at 500 rev/min working with the same slip.

If the cooling oil in the smaller clutch is supplied with a pressure of 140 kPa, estimate the ratio of the volumetric flow rates and the supply pressure of the oil in the larger coupling to maintain the same temperature conditions. Assume laminar flow in both cases.

25. (*a*) A fluid coupling has a mean diameter $D$ and contains a volume of fluid $V$. The primary runner is driven with an angular speed $\omega_p$ and the secondary runner has an angular speed of $\omega_s$ so that the slip $S$ is $1 - (\omega_s/\omega_p)$. Apply the $\Pi$ theorem to show that the torque $T$ transmitted is given by:

$$T = \rho\omega_p^2 D^5 \varphi\left(S, \frac{\rho\omega_p D^2}{\eta}, \frac{V}{D^3}\right).$$

(*b*) Test shows that in the case of a particular coupling operating with $\omega_p = 1\,000$ rev/min and 50 per cent slip the torque is 14 times the normal full-load torque. If the speed of the primary runner is reduced to 667 rev/min without change in volume of the fluid, estimate the torque which will be transmitted with 50 per cent slip, assuming that the effect of the Reynolds number is relatively unimportant.

# Positive-displacement machines

## 16.1. Reciprocating pumps with seated valves

In Chapter 15 we were concerned with *hydro-kinetic* machines which can be traced back to Hero's turbine. The other important class of hydraulic machines comprises those of the *positive-displacement* type, and this too derives from ancient Egypt.

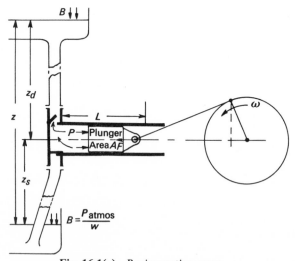

**Fig. 16.1(a).** *Reciprocating pump*

Fig. 16.1(a) shows, in diagrammatic form, a piston or plunger pump with self-acting lift-valves which, in a refined form, is extensively used today. In Fig. 16.1(b) the reciprocating plungers are actuated by a swash plate.

As their name suggests, positive-displacement pumps transform mechanical effort, such as the force $F$ on the piston, into fluid pressure $P$ which acts on the piston of area $A$ in Fig. 16.1(a). Such machines deliver fluid intermittently at a rate controlled by plunger displacement.

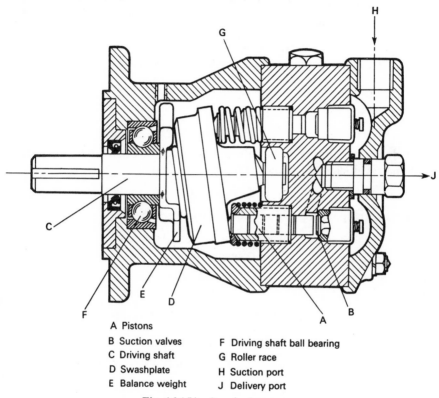

A Pistons

B Suction valves      F Driving shaft ball bearing
C Driving shaft        G Roller race
D Swashplate           H Suction port
E Balance weight       J Delivery port

**Fig. 16.1(b).** *Swash plate pump*

If the plunger moves sufficiently slowly, then the cylinder pressure during the suction stroke lies below atmospheric by an amount corresponding to the suction head $z_s$, the delivery valve being held closed automatically. When the plunger reverses its motion, the delivery valve opens and the suction valve closes, so that the pump delivers against the head $z_d$. Thus an ideal indicator diagram, corresponding to speeds sufficiently slow for the pressures to be distributed hydrostatically, is as shown in Fig. 16.1(c).

The area enclosed by the diagram represents $W$, the work done per cycle, during which a mean effective pressure $wz$ acts through a swept volume $AL$:

$$W = wzAL$$

Thus, if $C$ denotes the number of cycles completed in unit time (i.e. $C$ equals the cyclic speed of the crank for a single-acting pump) the

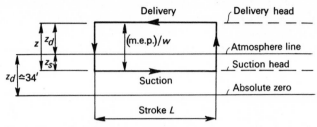

Fig. 16.1(c). *Hypothetical indicator diagram for slow-moving reciprocating pump*

indicated power can be seen to equal the weight ($wALC$) of fluid pumped in unit time, times the corresponding lift $z$, namely:

$$P = wzALC = wQz$$

If, due to mis-timing of the valves, leakage, or through other causes, the volumetric efficiency $\epsilon_v$ is not 100 per cent, then:

$$Q = \epsilon_v ALC$$

and the corresponding slip $s$ is defined by:

$$s = \frac{\text{swept volume} - \text{actual volume}}{\text{swept volume}}$$

$$s = 1 - \epsilon_v = 1 - \frac{Q}{ALC}$$

Lift valves, which are often spring-loaded on to their seats, generally

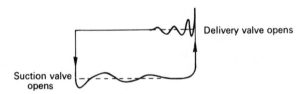

Fig. 16.1(d). *Effect of spring-loaded valves on indicator diagram*

tend to oscillate after opening, so causing a pressure fluctuation, which tends to be aggravated by the flow, see Fig. 16.1(d).

## 16.2. Reciprocating pump under dynamic conditions

Fig. 16.2(a) indicates conditions during the suction stroke of a plunger pump, drawing fluid initially at atmospheric pressure, through a suction line of length $l$ and area $A_1$. If no air vessel is fitted and the fluid column does not separate, then continuity of flow requires that:

$$Q = A_1 u_1 = A_0 u_0$$

that is, the velocity in the pipe is proportional to that of the plunger:

$$u_1 = \frac{A_0}{A_1} u_0$$

By differentiating this expression it may be seen that the accelerations are similarly related, namely:

$$a_1 = \frac{A_0}{A_1} a_0$$

so that the pressure drop required to overcome the friction and inertia effects in the suction pipe may be related to the plunger motion.

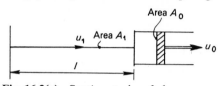

**Fig. 16.2(a).** *Suction stroke of plunger pump*

By expressing the former in terms of Darcy's formula (in which $\lambda = 4f$), we obtain for the pressure drop due to friction:

$$\Delta p_f = \frac{\lambda l}{d_1} \tfrac{1}{2} \rho u_1^2 = \frac{\lambda l}{d_1} \left(\frac{A_0}{A_1}\right)^2 \tfrac{1}{2} \rho u_0^2$$

and for the pressure drop due to the inertia of the fluid column of mass $\rho A_1 l$, moving with an acceleration $a_1$:

$$\Delta p_i = \frac{\rho A_1 l a_1}{A_1} = \frac{\rho l A_0 a_0}{A_1}$$

since $\Delta p_i$ acts over an area $A_1$. Hence, because of friction and inertia of fluid in the suction pipe, the pressure in the cylinder, where the kinetic and inertia effects will be relatively small if $A_0 \gg A_1$, lies below atmospheric by an amount at least equal to:

$$\Delta p = \Delta p_i + \Delta p_f = \rho l \frac{A_0}{A_1} \left[ a_0 + \frac{\lambda}{2d_1} \frac{A_0}{A_1} u_0^2 \right]$$

If, in addition, the cylinder is located at a height $z_s$ above the level in the sump, the cylinder pressure will be reduced still further below atmospheric pressure. In terms of head drops, the sum total is:

$$z = z_i + z_f + z_s$$

$$= \frac{1}{w} (\Delta p_i + \Delta p_f) + z_s$$

i.e.
$$z = \frac{l A_0}{g A_1} \left[ a_0 + \frac{\lambda}{2d_1} \frac{A_0}{A_1} u_0^2 \right] + z_s \qquad (1)$$

The fluid column will break due to separation of dissolved air if the cylinder pressure drops to within say 1·8 m above absolute zero, so that there is an upper limit of $z$ of about 8·5 m. If this limit to the running speed is exceeded, a cavity will first form and the subsequent fluid surge up the suction pipe will cause water hammer or 'knocking'.

If the crank throw is relatively small, the obliquity of the connecting rod may be neglected. The fundamental motion of the plunger is then

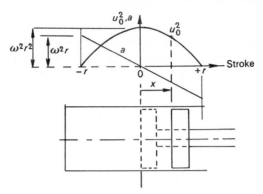

**Fig. 16.2(b).** *Plunger characteristics*

almost simple harmonic, with an amplitude equal to the crank radius, $r$, say, and of angular frequency ($\omega$) equal to the angular speed of the crank. If the plunger motion is referred to the mid-stroke position as datum, both for position and time, then:

$$x = r \sin \omega t$$

see Fig. 16.2(b).

The corresponding plunger velocity and accelerations are:

$$u_0 = \omega r \cos \omega t$$

$$a_0 = -\omega^2 r \sin \omega t$$

and the velocity squared, to which the frictional loss is assumed proportional:

$$u_0^2 = \omega^2 r^2 \cos^2 \omega t = \omega^2 r^2 (1 - \sin^2 \omega t)$$

**Fig. 16.2(c).** *Head in cylinder during suction stroke*

i.e.
$$u_0^2 = \omega^2 r^2 \left[ 1 - \left( \frac{x}{r} \right)^2 \right]$$

Thus the velocity squared varies parabolically with the stroke, reaching a maximum value of $\omega^2 r^2$ at mid-stroke and dropping to zero at the ends. The acceleration, which is proportional to the displacement from the mid-position, and is opposite in sign to it, reaches its maximum values of $\pm \omega^2 r$ at the ends of the stroke.

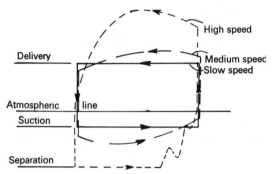

Fig. 16.2(d). *Effect of increasing speed*

Thus Equation (1) may be expressed graphically, it being noted that each of the terms represented there correspond to a *fall* in head, see Fig. 16.2(c) in which heads:

$$h_i(\text{max}) = \frac{lA_0}{g\,A_1}\omega^2 r \quad \text{and} \quad h_f(\text{max}) = \frac{\lambda l}{2gd_1}\left( \frac{A_0}{A_1}r\omega \right)^2$$

Fig. 16.2(d) illustrates the effect of increasing the pump speed, and indicates how a good static inlet head can inhibit separation and knock.

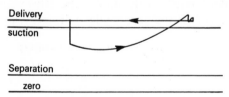

Fig. 16.2(e). *Negative slip*

If, however, the inlet head is sufficiently high, the inertia of the suction column may cause the delivery valve to open before the end of the suction stroke, as suggested by Fig. 16.2(e). The delivery is then greater than the nominal swept volume of the piston—a phenomenon known as 'negative slip'.

EXAMPLE 16.2 (i)

*The vertical suction pipe of a single-ram pump is 2 m long and 50 mm diameter. The stroke of the reciprocating plunger is 300 mm and its diameter is 150 mm.*

*Assuming simple harmonic motion, estimate the maximum speed at which the pump may run before cavitation may be expected at the suction valve when pumping water. The barometer stands at 10·3 m of water.*

Assuming cavitation occurs at the beginning of the stroke at 1·8 m of water above absolute zero, we have:

$$B - 1\cdot8 \text{ m} - Z_s = h_i(\text{max}) = \frac{l}{g}\frac{A_0}{A_1}\Omega^2 r$$

i.e.
$$(10\cdot3 - 1\cdot8 - 2)\text{ m} = \frac{2 \text{ m s}^2}{9\cdot807 \text{ m}} \times \frac{9}{1} \times 0\cdot15 \text{ m}\Omega^2$$

Hence,
$$\Omega^2 = \frac{6\cdot5 \times 9\cdot807}{2\cdot7 \text{ s}^2} = \frac{23\cdot6}{s^2}$$

or the maximum speed is

$$\Omega = \frac{4\cdot86}{s}\left[\frac{\text{rev}}{2\pi}\right]\left[\frac{60 \text{ s}}{\text{min}}\right] = \textbf{46·4 rev/min}$$

The maximum friction head (based on $h_f = (\lambda l/d)(u^2/2g)$ taking $\lambda = 0\cdot04$) is:

$$h_f(\text{max}) = \frac{\lambda l}{2gd_1}\left(\frac{A_0}{A_1}r\Omega\right)^2 = \frac{0\cdot04 \times 2 \text{ m}}{19\cdot613 \text{ m/s}^2 \times 0\cdot05 \text{ m}}\left(\frac{9}{1} \times 0\cdot15\text{m} \times \frac{4\cdot86}{s}\right)^2 = 3\cdot5 \text{ m}$$

which is only about half the maximum inertia head ($h_{i(\text{max})} = 6\cdot5$ m) at the beginning of the suction stroke where cavitation would begin.

## 16.3. The effect of air vessels

Fig. 16.3(a) illustrates the effect of fitting reservoirs to the suction and delivery lines. If these air vessels, or accumulators are of sufficient size, then the intermittent nature of the flow in the delivery and suction pipes will be much reduced. If they are made sufficiently large for the cyclic

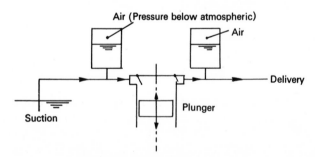

**Fig. 16.3(a).** *Air vessels on a pump*

changes in level in them to be neglected, constant flow conditions will be maintained in the suction and delivery pipes. In fact, however, as the pump makes a demand on the suction vessel in excess of the mean rate of flow during the charging stroke, and similarly delivers intermittently into the air vessel on the delivery side, pressure will vary cyclically in the air vessels. By making the area of the vessels considerably larger than those of the pipes, the fluctuations are kept small. Under these conditions the only fluid suffering appreciable acceleration is that contained between the air vessels, which should therefore be located as close to the pump as possible. *Not only is the fluid acceleration in the pipes minimized in this*

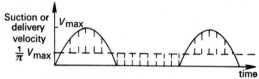

**Fig. 16.3(b).**  *Single-acting pump*

*way, but in addition, the friction losses correspond to the mean velocity in the suction and delivery pipes*, see Fig. 16.3(e). If the piston moves with *simple harmonic* motion then, without air vessels, the pipe velocity follows *this* law whenever the valve connecting it to the pump is open—see Figs. 16.3(b) and (c).

If air vessels of sufficient size are fitted, the pump continues to make a similar intermittent demand, but the pipe velocity tends to remain

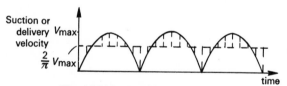

**Fig. 16.3(c).**  *Double-acting pump*

constant at the mean speed. The shaded areas in Figs. 16.3(b) and (c) then correspond to the fluctuating flows to and from the air vessels. Provided that the latter can accommodate these quantities without appreciable changes in air pressure, then for a single-cylinder pump the line velocities will be reduced to $1/\pi$ or $2/\pi$ of the corresponding maximum values with no air vessels, according as the pump is single or double acting.

Thus the friction work on the indicator diagram is changed in the ratio of the shaded areas of Figs. 16.3(d) and (e) by fitting adequate air vessels to a single-acting pump.

Since the area under a parabola is $\frac{2}{3}$ of the enclosing rectangle, *the work done against the pipe friction* is reduced by air vessels in the ratio:

$$\frac{W_1}{W_2} = \frac{\frac{2}{3}(V_{max})^2}{[(1/\pi)V_{max}]^2} = \frac{2\pi^2}{3} = 6.6{:}1$$

The corresponding figure for a double-acting pump is:

$$\frac{W_1}{W_2} = \frac{\frac{2}{3}}{(2/\pi)^2} = 1.6{:}1$$

This suggests that *the potential gain is much less spectacular when the number of working strokes per cycle is increased. Acceleration effects can similarly be reduced in this way and, by the use of an adequate number of cylinders, air vessels may be dispensed with.* This has some advantages,

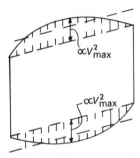

**Fig. 16.3(d).** *Indicator diagram—no air vessels*

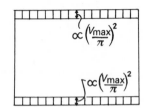

**Fig. 16.3(e).** *Indicator diagram— adequate air vessels*

as the latter néed constant attention, since air tends to dissolve in water under pressure.

EXAMPLE 16.3 (i)

*A single-acting reciprocating pump has a plunger diameter of* 250 mm *and a stroke of* 450 mm. *The delivery pipe is* 120 mm *diameter and* 50 m *long. If the motion of the plunger is simple harmonic, find the power saved in overcoming friction in the delivery pipe by the provision of a large air vessel, when the speed of the pump is* 60 rev/min. *Assume* $f = 0.01$ *when Darcy's formula is written*

$$h_f = \frac{4fl}{d}\frac{u^2}{2g}$$

Without air vessel but with simple harmonic motion

$$h_{f(\text{max})} = \frac{4fl}{2gd}\left(\frac{A_0}{A}r\omega\right)^2 = \frac{0.04 \times 50 \text{ m}}{19.61 \text{ m/s}^2 \times 0.12 \text{ m}}\left(\frac{625}{144}\times\frac{0.45 \text{ m}}{2}\times\frac{2\pi}{\text{s}}\right)^2 = 32 \text{ m}$$

Hence, the average over the stroke $= \frac{2}{3} \times 32$ m $= 21.3$ m.

With a large air vessel water may be assumed to be discharged at an even rate up the delivery pipe. The volume pumped per stroke is $A_0L$ in 1 second. Hence, the velocity of water in the delivery pipe is:

$$\frac{A_0L}{A \text{ s}} = \frac{625}{144}\times\frac{0.45 \text{ m}}{\text{s}} = 1.955 \text{ m/s}$$

and the friction head:

$$\frac{4fl}{d}\frac{u^2}{2g} = \frac{4 \times 0.01 \times 50 \text{ m}}{19.61 \text{ m/s}^2 \times 0.12 \text{ m}} \times (1.955)^2\frac{\text{m}^2}{\text{s}^2} = 3.3 \text{ m}$$

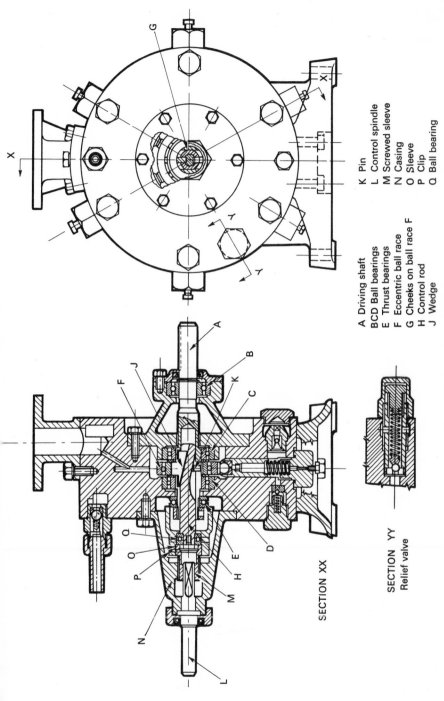

SECTION XX

SECTION YY
Relief valve

A  Driving shaft
BCD Ball bearings
E  Thrust bearings
F  Eccentric ball race
G  Cheeks on ball race F
H  Control rod
J  Wedge

K  Pin
L  Control spindle
M  Screwed sleeve
N  Casing
O  Sleeve
P  Clip
Q  Ball bearing

**Fig. 16.4(a).** *Radial piston pump*

Mass of water pumped without slip is:

$$\rho A_0 LC = 10^3 \, \frac{\text{kg}}{\text{m}^3} \times \frac{\pi}{4} \frac{625}{10^4} \times 0\cdot45 \text{ m} \times \frac{1}{\text{s}}$$

$$= 22\cdot1 \text{ kg/s}$$

Hence, the friction horse-power saved by fitting a large air vessel on the delivery side is:

$$22\cdot1 \, \frac{\text{kg}}{\text{s}} \times 9\cdot807 \, \frac{\text{m}}{\text{s}^2} \times (21\cdot3 - 3\cdot3) \text{ m} \left[ \frac{\text{W s}^3}{\text{kg m}^2} \right] = 3\cdot9 \text{ kW}$$

## 16.4. Multi-cylinder pumps

The majority of oil hydraulic pumps have a number of cylinders, so that the fluctuating delivery is reduced to an acceptable 'ripple', see Fig. 16.4(b). Fig. 16.4(a) illustrates a radial pump of this type, with seated

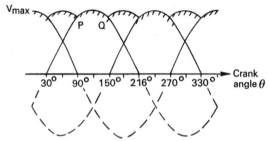

**Fig. 16.4(b).** *Delivery from a three-cylinder pump*

valves. It is in fact of the variable delivery type, as the eccentricity of the ball race controlling the stroke of the pistons may be adjusted by axial movement of the control rod. Each cylinder delivers fluid with a velocity which varies simple harmonically during 180° movement of the driving shaft and during the other half-turn is open to suction. Fig. 16.4(b) illustrates the delivery from three cylinders, each spaced 120° apart. It will be seen that, of the three cylinders considered, one delivers for 60° of crank angle whilst the other two are open to suction, as between P and Q. During the next 60° the falling delivery of this cylinder is supplemented by a growing delivery from a further cylinder, lagging it in phase by 120°. Hence their combined delivery is equivalent to that of a 'ghost' cylinder half-way between them (see Fig. 16.4(c)) and *the combined delivery of three cylinders at 120° introduces a ripple having a frequency which is six times the angular frequency of the crank.*

The average delivery of velocity $\bar{V}$ may be related to the peak speed of the pistons $V_{\text{max}}$, by integrating over the first 30° of crank angle:

$$\bar{V} \times \frac{\pi}{6} = V_{\text{max}} \int_0^{30°} \cos\theta \, d\theta = V_{\text{max}} \sin 30°$$

i.e.    $V = \dfrac{6}{2\pi} V_{\max} = \dfrac{3}{\pi} V_{\max} = 0.955\, V_{\max}$   or   $V_{\max} = 1.047 V$

The corresponding minimum speed, which occurs at 30° and thereafter at 60° intervals is:

$$V_{\min} = \frac{\sqrt{3}}{2} V_{\max} = \frac{\pi}{3}\frac{\sqrt{3}}{2} V = \frac{\pi}{2\sqrt{3}} V = 0.906 V$$

Hence the output velocity fluctuates between 0·906 and 1·047 of the mean value, with a ripple frequency equal to six times that of the crank.

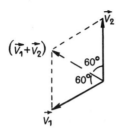

**Fig. 16.4(c).** *Combined delivery of two cylinders*

If, as in a six-cylinder pump, a further three cylinders set at 120° are introduced, shifted in phase by 60°, these merely contribute in phase with the 'ghost' cylinders and in their turn, double the output of the original three. Thus the ripple frequency remains unchanged, and is doubled in magnitude, as is the output of the pump. When an even number of pistons is used, there is no proportionate reduction in ripple. In contrast to the above figures, for example, a five-cylinder pump causes a maximum output which is only 1·015 of the mean, and a minimum which is 0·965 of the mean.

Hence, if ripple is of paramount importance, it is of advantage to use an odd number of cylinders.

In the above expressions, the peak velocity of the plungers is related to the angular speed of the shaft $\omega$ and the throw of the eccentric $r$. Taking the piston motion to be simple harmonic ($\omega r \cos \omega t$), then:

$$V_{\max} = r\omega$$

## EXAMPLE 16.4 (i)

*Calculate the shaft power required to drive a three-throw single-acting reciprocating pump having plungers 300 mm diameter and 600 mm stroke when lifting 5 m³ of water per minute through a static head of 115 m. Also, calculate the speed of the pump assuming that slip is 2 per cent and the mechanical efficiency is 90 per cent. Friction head in the suction and delivery pipes may be taken as 2 m and 17 m respectively; the kinetic head of water in the delivery pipe may be neglected. If the latter may not be neglected, write down an expression for shaft power.*

Slip

$$s = 1 - \frac{Q}{nALC} = 0.02$$

Hence, $Q = 0.98\, nALC$, where $n$ is the number of cylinders of cross-sectional area $A$.

Thus,

$$\frac{Q}{C} = 0.98 \times 3 \times \frac{\pi}{4}(0.3 \text{ m})^2 \times 0.6 \text{ m} = 0.1\,247 \text{ m}^3$$

Hence,

$$C = \frac{5 \text{ m}^3/\text{min}}{0.1\,247 \text{ m}^3} = \frac{40}{\text{min}}$$

and, since in a single-acting pump one cycle occurs each revolution of the crank, the speed of the pump is **40** rev/min.

Shaft power of motor driving the pump

$$= \frac{\text{Energy imparted to water to overcome resistance}}{\text{Mechanical efficiency of pump}} = \frac{\dot{M}g(z + \sum h_f)}{\epsilon_m}$$

$$= \left(\frac{5 \times 10^3}{60 \text{ s}} \text{ kg}\right) 9.807 \frac{\text{m}}{\text{s}^2}(115 + 2 + 17) \text{ m} \left[\frac{\text{W s}^3}{\text{kg m}^2}\right] = \textbf{109·5 kW}$$

In general, the **shaft power** $= \dfrac{\dot{M}g(1-s)VC}{\epsilon_m}\left\{z + \sum_{s}^{d} h_f + \dfrac{u_d^2}{2g}\right\}$

where $\dot{M} = \rho Q$ is the mass lifted/unit time, $V$ is the total swept volume, $z$ is the static lift and $u_d$ the velocity in delivery pipe.

## 16.5. Gear and vane pumps

As an alternative to using seated valves actuated by the pressure differences across them, there is an important class of pumps which has automatically-phased porting arrangements. In such machines, rotation of the pump opens and closes the suction and delivery ports mechanically at appropriate times. Fig. 16.5(a) illustrates a gear pump, which is of this type. As each tooth passes the point P or P' *a volume of oil is entrained and carried round between each gear and the casing*, to be delivered at points Q or Q'. Meshing of the gears serves both to transmit the drive and to maintain a seal between pressure and suction. Care must be taken to ensure that oil trapped between the successive lines of contact does not build up in pressure. Although mechanically simple, such pumps require precision manufacture for high-pressure duties. Otherwise a low volumetric efficiency results from internal leakage, particularly over the tips of the gears and between their end faces and the casing.

If, as is usually the case, the gear teeth occupy approximately one-half the area enclosed between the tip and root circles (diameters $D$ and $d$, respectively) the nominal delivery from a pair of gears of width $b$ is, for a running speed $\omega$:

$$Q_G = 2 \times \tfrac{1}{2} \times \frac{\pi}{4}(D^2 - d^2) \times b \times \omega$$

i.e.
$$Q_G = \frac{\pi b(D^2 - d^2)\omega}{4}$$

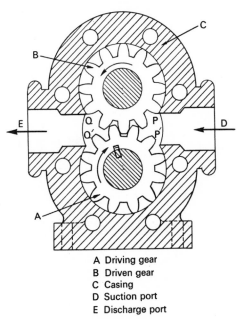

A Driving gear
B Driven gear
C Casing
D Suction port
E Discharge port

**Fig. 16.5(a).** *Gear pump*

This geometric estimate of the delivery must be multiplied by the volumetric efficiency $\epsilon_v$, which is generally greater than 80 per cent for oil hydraulic machines, as is the mechanical efficiency $\epsilon_m$ defined by:

$$\epsilon_m = \frac{pQ}{P}$$

In this expression $P$ is the input power (i.e. $T\omega$) and $pQ_G$ is the ideal hydraulic power available. As the actual hydraulic power is:

$$P_h = pQ = \epsilon_v p Q_G$$

the overall efficiency:

$$\epsilon_0 = \frac{P_h}{P} = \frac{\epsilon_v p Q_G}{P} = \epsilon_v \epsilon_m$$

The difficulties associated with wear are overcome in a **vane pump,** which is shown diagrammatically in Fig. 16.5(b), as the blades are forced against the liner when the rotor is driven. The latter is disposed eccentrically in the casing, and the ports are so arranged that when the pockets between the vanes tend to increase in volume, oil can be drawn from suction. When the point of maximum distance between the rotor and liner has been passed, the pocket is opened to delivery, so that fluid may be discharged as the space between the vanes again diminishes.

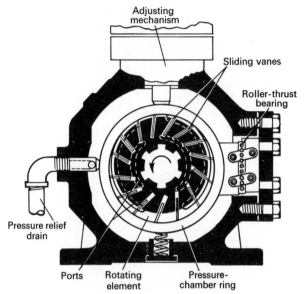

**Fig. 16.5(b).** *Racine vane pump*

The side thrust on the rotor due to the unbalanced hydraulic pressures acting across it may be avoided by arranging the suction and delivery ports in pairs, diagonally opposite each other. The pockets between the vanes then undergo two cycles of expansion and contraction during each revolution in the oval bore.

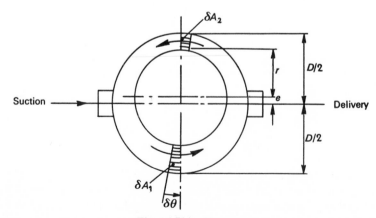

**Fig. 16.5(c).** *Vane pump*

EXAMPLE 16.5 (i)

*Show that the swept volume of a vane pump of width b rotating in a liner of bore D, with an eccentricity e, is $2\pi Deb$ per revolution, if the thickness of the vanes may be neglected.*

The swept volume per unit width is, for a small rotation $\delta\theta$, the difference between the two shaded areas, i.e. referring to Fig. 16.5(c):

$$\delta A_1 - \delta A_2 = \left[\frac{1}{2}\left\{\left(\frac{D}{2}+e\right)^2 - r^2\right\} - \frac{1}{2}\left\{\left(\frac{D}{2}-e\right)^2 - r^2\right\}\right]\delta\theta$$

Hence for a width $b$:

$$\delta V = bDe\delta\theta$$

so that the swept volume for one revolution is:

$$V = bDe\int_0^{2\pi}d\theta = 2\pi bDe$$

The rate of delivery for an angular speed $\omega$ ($= d\theta/dt$) is thus nominally

$$Q = \frac{dV}{dt} = bDe\omega$$

## 16.6. Reversible and regenerative machines

In the previous section it was shown that gear and vane pumps are ported automatically to pressure and suction, so eliminating the need for seated valves. It is similarly possible to design plunger-type machines to operate with automatic porting. They require extreme precision in manufacture

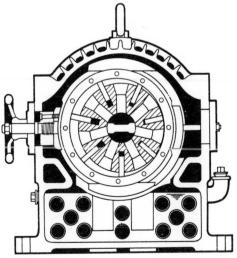

**Fig. 16.6(a).** *Churchill–Sturm transmission*

and tend to be noisier and/or less efficient than the corresponding type with seated valves. They do however offer important advantages in hydraulic drives where their *reversible* and *regenerative* characteristics are exploited. That they are reversible implies that, when operating as a pump they may be used to deliver fluid in either direction; being regenerative they are capable of operating as motors.

Fig. 16.6(a) shows a hydraulic drive which consists essentially of a vane pump driving a corresponding oil motor.

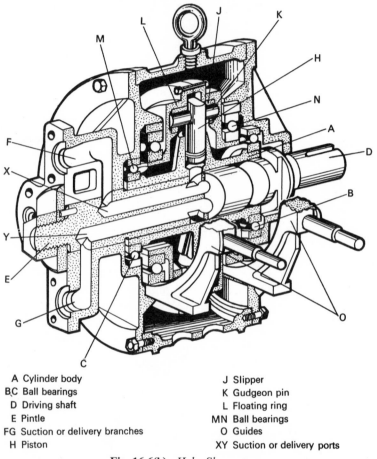

| | |
|---|---|
| A Cylinder body | J Slipper |
| B,C Ball bearings | K Gudgeon pin |
| D Driving shaft | L Floating ring |
| E Pintle | MN Ball bearings |
| FG Suction or delivery branches | O Guides |
| H Piston | XY Suction or delivery ports |

**Fig. 16.6(b).** *Hele–Shaw pump*

By adjusting the eccentricity of the pump, the speed of the motor may be continuously varied. By reversing the eccentricity, the direction of flow is changed, so reversing the motor. Since each pump setting corresponds to a certain rate of flow the motor speed ideally depends only on this, i.e. it is a speed control. Further provision is made for adjusting the cyclic capacity of the motor, so that the speed range is still further extended when the swept volume per revolution of the motor is reduced.

Figures 16.6(b) and 16.6(c) show corresponding plunger machines which can also be adapted for use as pumps or motors, or in combination as a continuously variable speed transmission.

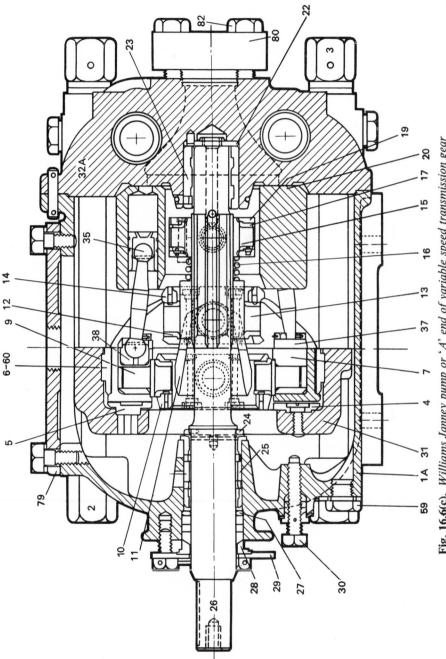

**Fig. 16.6(c).** *Williams Janney pump or 'A' end of variable speed transmission gear*

The cyclic capacity of the Hele-Shaw type of machine shown in Fig. 16.6(b) is controlled by adjusting the eccentricity of the floating ring, which, as shown in Fig. 16.6(d) carries slippers pinned to the plungers.

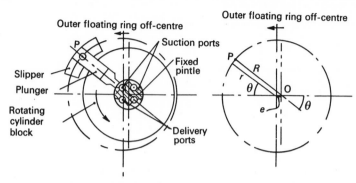

**Fig. 16.6(d).** *Principle of variable delivery radial piston pump*

The latter oscillate in the cylinder block, which in turn is rotated about the central fixed pintle containing the pressure and suction ports.

For small eccentricities, $e$, as shown in Fig. 16.6(d), the centre of the gudgeon pin P is at a distance from the fixed centre O given approximately by:

$$R = r + e \cos \theta$$

Thus for a uniform rotational speed $\omega$, the plungers move with an approximately simple harmonic motion having an amplitude equal to the eccentricity, and an angular frequency equal to the rotor speed.

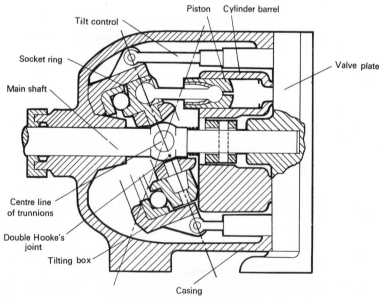

**Fig. 16.6(e).** *Axial piston pump*

This is analogous to the simple reciprocating pump, which is based on the engine mechanism, and is in fact simply another inversion of the slider crank chain. In each case the plunger movement may be represented by a fundamental simple harmonic motion, together with a number of harmonics, of small amplitude. Neglecting the latter, the delivery and the pump-ripple may be determined as in Section 16.4. This is also true of the Williams Janney design shown in Fig. 16.6(c) where each piston again executes one complete cycle for each shaft revolution. This is shown diagrammatically in Fig. 16.6(e) from which it may be seen that the stroke is determined by tilt of the socket ring, controlled by the tilting box trunnion mounted in the casing. The socket ring and main shaft are paired together by a double Hooke's joint which transmits the full driving torque. The double joint makes the valve timing independent of the tilt by transmitting a constant velocity ratio drive to the socket ring, which remains synchronized with the cylinder barrel when the whole assembly is rotated.

An alternative inversion of this arrangement, due to Thoma, avoids transmission of the driving torque through the universal joints. This is

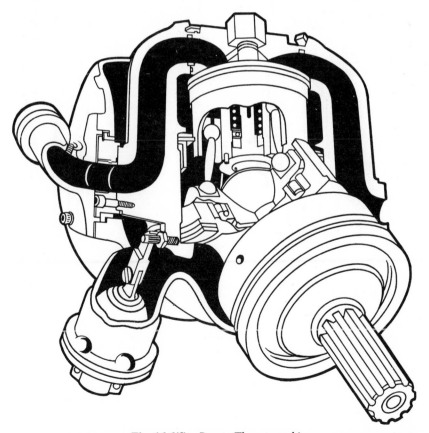

**Fig. 16.6(f).**  *Dowty Thoma machine*

achieved by making the socket ring integral with the driving shaft and by swinging the cylinder block on a yoke as shown in Fig. 16.6(f).

It thus substitutes the difficulty of carrying a large inertia on trunnions, through which the full delivery must pass to and from the cylinders.

## 16.7. Positive-displacement transmissions

The machines described in the previous section, when used in combination as a pump and a motor, constitute a hydraulic transmission of the positive-displacement type. In its simplest form, it consists of a simple

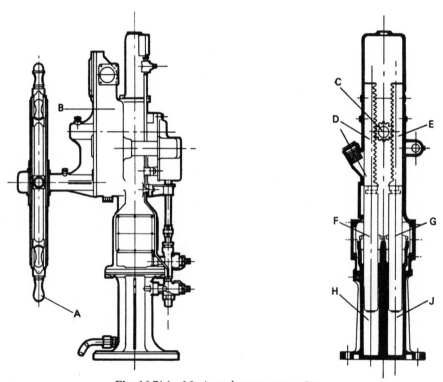

**Fig. 16.7(a).** *Marine telemotor transmitter*

pump-cylinder, actuating a motor or receiver, as in a car foot-brake or a marine telemotor, see Fig. 16.7(a).

Diagrammatically such systems are as shown in Fig. 16.7(b), where an effort $F_A$ applied to the pump, moves the pump piston at a speed $u_A$ displacing fluid at a rate $Q$ (which is less than the nominal (or geometric displaced) delivery $(Q_g)_A$ owing to leakage):

$$Q = (\epsilon_v u a)_A = (\epsilon_v Q_g)_A$$

This defines the volumetric efficiency of the pump, whereas its overall efficiency $\epsilon_A$ is the ratio of the input power $(Fu)_A$ to the hydraulic power delivered $(pQ)$:

i.e. $$\epsilon_A = \frac{pQ}{(Fu)_A}$$

The mechanical efficiency of the pump $(\epsilon_m)_A$ is so defined that:

$$\epsilon_A = (\epsilon_v \epsilon_m)_A$$

∴ $$(\epsilon_m)_A = \frac{pQ}{(Fu)_A} \frac{(Q_g)_A}{Q} = \frac{p(Q_g)_A}{(Fu)_A}$$

The latter is more conveniently, and more directly, expressed as the ratio:

$$(\epsilon_m)_A = \frac{p(Q_g)_A}{(Fu)_A} \left[ \frac{(ua)_A}{(Q_g)_A} \right] = \frac{pa_A}{F_A}$$

and thus allows for friction at the 'A'-end.

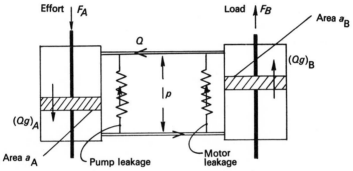

**Fig. 16.7(b).** *Diagram of positive displacement telemotor transmitter*

Similarly at the motor, or B-end, the geometric displaced volume may fall short of the corresponding oil volume delivered, so that in this case:

$$(ua)_B = (Q_g)_B = (\epsilon_v)_B Q$$

and the corresponding overall efficiency at the motor end is:

$$\epsilon_B = \frac{(Fu)_B}{pQ}$$

The mechanical efficiency of the B-end:

$$(\epsilon_m)_B = \frac{F_B}{pa_B}$$

is thus seen to be related to the other motor efficiencies by the same expression as for a pump:

$$\epsilon_B = \frac{(Fu)_B}{pQ} \left[ \frac{(\epsilon_m a)_B p}{F_B} \right] \left[ \frac{(\epsilon_v)_B Q}{(ua)_B} \right]$$

i.e. $$\epsilon_B = (\epsilon_v \epsilon_m)_B$$

A similar analysis may be made for rotary machines, see Fig. 16.7(c). If the geometric swept volume is $V$ per revolution the corresponding ideal delivery is, for an angular speed $\omega$:

$$Q_g = \frac{V}{2\pi}\omega$$

If the machine is operating as a pump, the actual delivery $Q$ will be some fraction $\epsilon_v$ of this:

$$Q = \epsilon_v Q_g = \frac{\epsilon_v V}{2\pi}\omega \tag{1}$$

When acting as a motor, the geometric swept volume will fall short of the

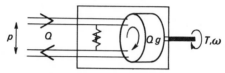

**Fig. 16.7(c).** *Diagram of motor end of rotary transmission*

corresponding oil volume supplied, i.e. the volumetric efficiency $\epsilon_v$ of a motor is given by:

$$Q_g = \epsilon_v Q$$

i.e.
$$\omega = \epsilon_v \frac{2\pi}{V} Q \tag{2}$$

In each case the mechanical efficiency $\epsilon_m$ is related to the overall efficiency $\epsilon$ by:

$$\epsilon = \epsilon_m \epsilon_v$$

in which for a pump:

$$\epsilon = \frac{pQ}{T\omega}$$

and for a motor:

$$\epsilon = \frac{T\omega}{pQ}$$

Thus the mechanical efficiency of the pump is, as in the previous section:

$$\epsilon_m = \frac{pQ}{T\omega}\frac{Q_g}{Q} = \frac{pQ_g}{T\omega}$$

and for a motor:

$$\epsilon_m = \frac{T\omega}{pQ}\frac{Q}{Q_g} = \frac{T\omega}{pQ_g}$$

i.e. in this case, it is the ratio of the power delivered by the shaft to the hydraulic power available, after allowing for leakage.

Just as the speed is related to the delivery and the cyclic swept volume by Equations (1) and (2) above, so the torque may be related to the pressure.

For a pump:

$$\epsilon_m = \frac{pQ_g}{T\omega}\left[\frac{V\omega}{Q_g 2\pi}\right] = \frac{pV}{2\pi T}$$

i.e.

$$p = \epsilon_m \frac{2\pi}{V} T$$

Similarly, the torque exerted by a motor is given in terms of $p$, $V$ and $\epsilon_m$ by:

$$\epsilon_m = \frac{T\omega}{pQ_g}\left[\frac{Q_g 2\pi}{V\omega}\right]$$

i.e.

$$T = \epsilon_m \frac{V}{2\pi} p$$

Fig. 16.7(d) illustrates a complete transmission for which:

$$T_B\omega_B = \epsilon T_A\omega_A = (\epsilon_m\epsilon_v)_A(\epsilon_m\epsilon_v)_B T_A\omega_A$$

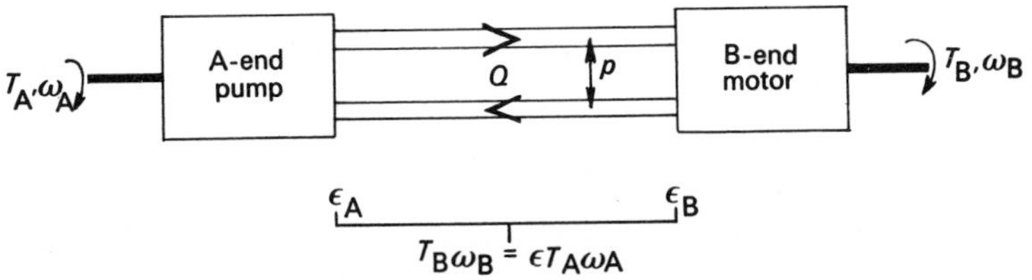

**Fig. 16.7(d).** *Diagram of rotary transmission*

Despite the formidable list of component efficiencies, precision manufacture enables 80 per cent of the driving power to the pump to be available at the output shaft of the motor.

Characteristics of fluids used both in positive displacement (or hydrostatic) and hydrokinetic (or hydrodynamic) transmissions are low compressibility and non-inflammability; adequate viscosity to enable good sealings and lubrication between moving parts is also needed and the fluid should be non-corrosive and chemically stable.

Where there is little risk of fire, petroleum oils are used but if there is a wide range of temperature during operation a fluid with a low change of viscosity with temperature and low fire risk is used, e.g. water in oil and synthetic emulsions.

## 16.8. Control characteristics of hydraulic drives

The essential components of a positive-displacement system are a pump, a motor (or actuator) and some means of controlling the motion of the latter. This may be effected by adjusting the relative cyclic capacities of the pump and the motor. If that of the motor is less than that of the pump, the drive is geared down; if greater, then geared up. Instead of controlling the pump or the motor, a third possibility is to control the flow in the pipes connecting them. The essential feature of line control is to absorb, or divert, the power not required by the motor. This has the advantage that it is possible to use relatively simple pumps and motors, as they need not be of variable capacity.

Almost any type of torque-speed characteristic may be obtained by using different arrangements of pump, motor or line control. Fig. 16.8(a) shows a fixed capacity motor connected to a variable delivery pump.

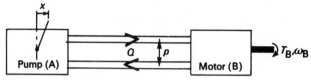

**Fig. 16.8(a).** *Fixed capacity motor and variable delivery pump*

If the cyclic swept volume of the latter is proportional to the control setting, i.e.:

$$V = kx$$

so that from Section 16.7:

$$Q = \frac{(\epsilon_v k \omega)_A}{2\pi} x$$

Hence the motor speed is controlled by $x$ according to the equation:

$$\omega_B = \frac{2\pi(\epsilon_v)_B}{V_B} Q$$

i.e.

$$\omega_B = (\epsilon_v)_A (\epsilon_v)_B \frac{k\omega_A}{V_B} x$$

Thus for a constant speed drive to the pump, the output speed would be independent of the load on the motor, provided that the volumetric efficiencies remained constant; this corresponds to the vertical characteristics on Fig. 16.8(b). In practice the leakage tends to be proportional

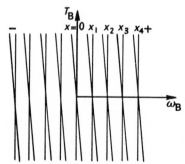

**Fig. 16.8(b).** *Effect of leakage on constant control setting*

to the load, so that the lines of constant control setting slope as indicated in Fig. 16.8(b).

Fig. 16.8(c) shows an alternative application for a variable delivery pump. Pressure in the delivery line acts as a spring to control the position of the capacity changing member. If the line pressure tends to fall, the delivery is increased; if it tends to rise, the delivery is reduced. Such a machine is known as an auto-pump, and ideally tends to maintain a constant pressure, independent of oil demand. If the system is made too

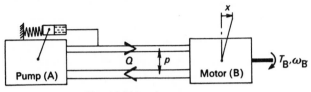

**Fig. 16.8(c).** *Auto-pump*

sensitive, self-sustained oscillations may occur. Consequently a certain drop of pressure with demand, known as regulation, is tolerated in practice.

Neglecting the latter, a variable capacity motor such that:

$$V = kx$$

when connected to a constant pressure supply, has a characteristic given by:

$$T_B = (\epsilon_m)_B \frac{kx}{2\pi} p$$

see Section 16.7. Thus it is essentially a system of torque control, see Fig. 16.8(d).

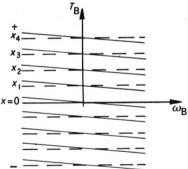

**Fig. 16.8(d).** *Torque control*

Due primarily to pump regulation, there is some fall-off in torque with speed, as shown.

Fig. 16.8(e) shows a system in which a suitable constant supply pressure is maintained by means of a relief valve, fitted to a constant delivery pump. The position of the control valve $x$ governs the connections between the pressure and exhaust lines to the pump, and the two lines to the motor or actuator. It is thus a four-way valve. When the latter is raised, pressure oil is throttled to the upper line to the pump, suffering a loss dependent on the valve opening, *a* say. This orifice loss ($\Delta p_1$) is given by:

$$Q = Ca\sqrt{\frac{2\Delta p_1}{\rho}}$$

so that, if the valve opening *a* is proportional to $x$:

$$\Delta p_1 \propto \left(\frac{Q}{x}\right)^2$$

A similar drop in pressure occurs through the exhaust port so that the pressure available across the motor falls below that of the pump delivery

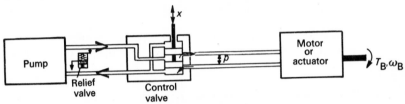

**Fig. 16.8(e).** *Valve control of an actuator*

by an amount proportional to $Q^2$, for a given valve setting. Thus, for any value of the latter, the motor pressure and hence its driving torque, which from Section 16.7 is given by:

$$T_B = \frac{(\epsilon_m V)_B}{2\pi}p$$

is reduced by an amount proportional to $Q^2$. For a given volumetric efficiency the latter is proportional to the square of the motor speed, since

$$\omega_B = 2\pi\left(\frac{\epsilon_v}{V}\right)_B Q$$

This results in a series of falling parabolic characteristics as suggested by Fig. 16.8(f) for upward movements of the valve. If the valve is

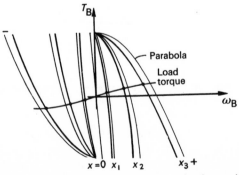

**Fig. 16.8(f).** *Torque-speed characteristics and valve position*

lowered the motor line connections are interchanged so that the direction of rotation and the sense of the driving torque are reversed. In practice, due to valve leakage, the characteristics tend to slope somewhat more, as

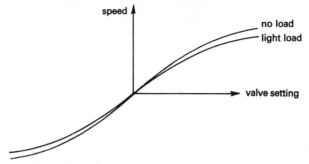

**Fig. 16.8(g).** *Control of motor speed by valve position*

shown. Provided that the load on the motor does not approach the stalled torque, the motor speed is substantially controlled by the valve position, as suggested by the hypothetical load characteristic shown, see Fig. 16.8(g).

Since the input effort required to move the valve is relatively small,

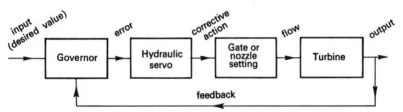

**Fig. 16.8(h).** *Servomechanism in speed governing system*

the system constitutes a controllable power amplifier, and as such, finds extensive use as a servomechanism. In such systems the output is compared with the desired value, and any error which exists is used to control a power amplifier in such a way as to correct the error. For example, the position of a governor sleeve indicates whether the speed is high or low. If this is connected to a control valve as indicated in Fig. 16.8(h), the

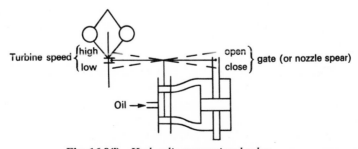

**Fig. 16.8(i).** *Hydraulic proportional relay*

servomotor may be used to change the gate setting of the turbine, in such a way as to wipe out the speed error.

Fig. 16.8(i) illustrates a simple proportional relay, which itself operates as a closed loop, serving simply to power-amplify the governor sleeve movement. In practice auxiliary devices are used to ensure a satisfactory compromise between accuracy and stability, without undue rise of pressure due to inertia of the water in the turbine supply pipes. Much has been written in recent years on the analysis of such control systems, which lies beyond the scope of this book.

## Exercises on Chapter 16

1. A double-acting reciprocating pump has a piston diameter of 200 mm and a stroke of 600 mm and runs at 20 rev/min. It discharges through a 150 mm main 76 m long ($f = 0.007\,5$) with a vertical lift of 46 m. Assuming the piston to have s.h.m. and that no air vessel is used, sketch the part of the hypothetical indicator diagram corresponding to discharge giving the heads in the cylinder at the ends and middle of the stroke. Neglect friction at the discharge valve.

2. The plunger diameter of a single-acting reciprocating pump is 120 mm diameter and stroke 240 mm. The suction pipe is 90 mm diameter and 4·3 m long ($f = 0.01$). If separation takes place at an absolute head of 1·5 m find the maximum speed at which the pump will run without separation taking place, if the barometer is 10·5 m of water, and the water level in the sump is 3 m below the pump cylinder axis. What power is expended in overcoming friction when running at this speed.

3. The plunger diameter of a single-acting reciprocating pump is 100 mm and the stroke 250 mm. The suction pipe is 75 mm diameter and 4·5 m in length, the water level in the sump being 2·5 m below the pump cylinder axis. There is no air vessel.

If separation takes place at the beginning of the stroke at an absolute head of 1·8 m of water, calculate the maximum speed at which the pump may be run without separation taking place. The height of the water barometer was 10·3 m.

4. Give a sketch of the theoretical pressure-volume diagram for the cylinder of a reciprocating pump, which is not fitted with air vessels. Show clearly the effects of acceleration and friction in both suction and delivery pipes. State the conditions under which 'separation' is likely to occur.

The following particulars relate to a reciprocating pump not fitted with air vessels: stroke 300 mm; plunger diameter 130 mm; diameter of suction pipe 75 mm; length of suction pipe 6 m; suction head 3 m.

Atmospheric pressure is equivalent to 10·3 m of water and separation may be assumed to occur when the absolute pressure head in the cylinder falls below 2·3 m of water.

Calculate the maximum speed at which the pump may be run if separation is to be avoided.

5. A reciprocating pump, working with simple harmonic motion may be fitted with large air vessels, near to the pump cylinder, in both the suction and delivery pipes.

Working from first principles show that, in the case of a single-acting pump, the ratio of the work done against pipe friction, if air vessels are fitted, compared with that done if there are no air vessels, is $3/2\pi^2$. It is assumed that the friction coefficient does not vary with velocity.

6. A double-acting reciprocating pump, having a bore and stroke of 200 mm and 400 mm, respectively, runs at 20 rev/min. Both the suction and discharge pipes are 100 mm diameter, the former being 9 m long and not provided with an air vessel. The suction lift is 3·6 m. The discharge pipe is 475 m long, the outlet being 15 m above the pump and a large air vessel is fitted 15 m away from the pump. Assuming that the piston has simple harmonic motion, that $f = 0·01$ and that the area of cross-section of the piston rod can be neglected, sketch the indicator card for one end of the pump, stating the absolute suction and discharge heads at the ends and mid-points of both strokes. Take the barometer as 10·3 m of water and neglect losses at inlets and outlets of both pipes.

7. A reciprocating pump has two single-acting cylinders 80 mm diameter and 160 mm stroke, with the cranks set 180° apart. It supplies water to an accumulator having a ram 160 mm diameter, through a horizontal pipe 40 mm diameter and 60 m long. No air vessel is provided. The steady pressure on the accumulator when the pump is not working is 7 MN/m², and packing friction can be neglected in all cases.

Determine the pressure in one of the pump cylinders during a working stroke: (*a*) at commencement, (*b*) at mid-stroke, (*c*) at end of stroke, when running at 60 rev/min. Take the pipe friction $f = 0·01$, and assume s.h.m. for the pistons.

8. It is desired to compare the performance of the following three types of reciprocating pump:

(*a*) Single-cylinder single-acting pump without air vessel
(*b*) Similar pump with air vessel on delivery side.
(*c*) Three-throw or triplex pump without air vessel.

In each case the crankshaft speed is 75 rev/min, the diameter of the delivery is 150 mm and the mean discharge is 0·7 m³/min. Show by means of graphs, as nearly to scale as possible, the relationship between time, and the instantaneous velocity in the delivery pipe, for each of the three pumps.

From the shape of the graphs, what can you deduce concerning the relative water-hammer or inertia pressure in the delivery systems of the three pumps?

9. A three-throw reciprocating pump has single-acting cylinders 50 mm diameter with a stroke of 100 mm. Water is delivered into a pipe 75 mm diameter and 36 m long. There are no air vessels. Make a sketch on squared paper to show the form of the velocity variation of water in the delivery pipe during one revolution of the crank and determine the fluctuation in delivery

pressure due to inertia when the pump runs at 100 rev/min. Assume s.h.m. for the pistons.

10. A hydraulic machine delivering 34 kW with an overall efficiency of 75 per cent is operated by water supplied from a hydraulic accumulator through a 100 mm diameter pipe 90 m long. The accumulator has a 300 mm diameter ram with a stroke of 3 m and carries a load of 100 kN. The water for the system is supplied to the accumulator through a short pipe by a two cylinder single-acting pump having plungers of 200 mm diameter, a stroke of 300 mm and rotational speed of 40 rev/min. Assuming 5 per cent slip in the pump and using a friction coefficient $f = 0.007$ for the 100 mm diameter pipe, calculate the longest continuous period during which the machine can be operated at full power.

11. Describe a variable-speed hydraulic transmission system of the hydrostatic type, in which the input element has the form of a rotating-cylinder positive pump. If the shaft of this pump is driven at constant speed, explain how the speed of the corresponding output element or hydraulic motor can be varied from zero to a maximum. In a particular installation the pump has 11 cylinders, each 57 mm diameter, with a maximum stroke of 96.5 mm. It uses oil of relative density 0.95; it runs at 250 rev/min and it creates a pressure difference of 2.76 MN/m². The slip is 2 per cent. What maximum power would it deliver?

# Answers to exercises

## CHAPTER 1

(1) 1 m.  (2) 0·78 m.  (3) 7·4 m; 2·82 m.  (4) 0·267 m.  (5) $l < \dfrac{ds_o}{\sqrt{(8s)(s_o - s)}}$.
(6) 0·344 m; 0·909 m.  (7) unstable; GM $= -0·008$ m; 5 degrees.  (8) 374 mm; 16·3 mm.  (9) 1·66 m; 1·09 m.  (10) 70 mm.  (11) 39 m.  (12) 2 m.
(13) 1·29 m from base.  (14) 2·46 m.  (15) $\tau = \dfrac{2\pi k_G}{\sqrt{(g \times \mathrm{GM})}}$; 0·58 m; 3·88 m.
(16) 6 m; 4 degrees.

## CHAPTER 2

(1) (a) 1·41 MN; 4·75 m; (b) 1·11 MN; 4·56 m.  (2) 7·95 MN, 4·2 m up.
(3) 0·925 kN; 62·7 mm vertical.  (4) 4·25 m below water surface; 1 MN; 4·94 m.  (5) 88·26 kN; 197 kN m.  (6) 46·2 m.  (7) 30·8 kN; 31·8 kN m.
(8) 121 kW.  (9) 1 380 kN; 1 520 kN; 2 900 kN; 98·1 kN halfway down sluice.  (10) 17·85 kN; 0·835 m slant depth.  (11) 8·25 MN.  (12) 16·65 kN, 1·749 m below surface; 19·47 kN, 0·488 m from bottom flat.

## CHAPTER 3

(1) 104 kPa.  (2) 0·587 kPa; 60 mm; 4·4 mm.  (3) 20 mm.  (4) 11·6 kPa.
(5) 6·1 mm.

## CHAPTER 4

(1) 1·395 m/s; 17·12 litre/s.  (2) 9·07.  (3) (a) 127·8 l/min.; (b) 87 mm.
(4) 5·66 m³/min; 600 mm.  (5) 0·552 m³/mm; 3·21 m.  (6) Rising 29·3 mm/min; 185 m.  (7) 7·8 m/s.  (8) 167 kPa; 17 m.  (9) 0·945.  (10) 5·62 min.
(11) 14·5 m³/min; 0·63 m.  (12) 1 055 s²/m⁵.  (13) 1·38 kg/s; 830 m³/s.

## CHAPTER 5

(2) 1 h.  (3) 1·25 per cent; 1·41 per cent.  (4) 74 litre/s; 68·9 per cent.  (5) 280 mm.  (6) 570 mm.

## CHAPTER 6

(1) 980·7 m/s²; 25·9 m/s; 44·1 mm; 34 m.    (2) 82·6 mm; 211 rev/min.    (3) 600$\pi$ mm/s; 300$\pi$ mm/s; 182 mm; 318 mm.    (4) 726 mm; 101·5 × 10⁻³ m²; 466 mm.  (6) 32·5 rev/min.  (7) 35 cycle/min; 0·048 J.  (8) 28 rev/min.

## CHAPTER 7

(1) 4·67 kN; 15 kN.  (2) 24 m/s; 111 kW.  (3) 3·48 kN at 30°; (*a*) 1·51 kN; (*b*) 15 kW; (*c*) 75 per cent.  (4) 18 kN; 52·4 kN.  (5) 23°; 51 kW.  (6) 5·49 kN; 2·1 kN.    (7) 4; 134 mm; 1·24 m; 4 m³/s.    (8) 30 m/s; 97 per cent; 89 per cent; 50 mm.  (9) 24·5 knot; 93·2 kW; 14·3 per cent.  (11) 5·55 kN.

## CHAPTER 8

(1) 1·17 N/m².  (2) 600 × 10⁻⁶ m³/s; 8·13 mN.  (3) 2·3 Ns/m².  (4) 90 km/s per metre; 1·6 N/m².    (5) 521 mm/s.    (6) 0·925 Ns/m².    (7) 0·276 Nm. (8) 0·268 P.    (9) 11·6 kN/m²; 12·8 Nm; 1 kW.    (11) 600 × 10⁻⁶ m³/s. (14) 0·024 litre/h.    (15) 0·2 mm.    (16) 477 × 10⁻⁶ m³/s.    (18) 4·46 kN.
(19) $\dfrac{3\pi}{2} \eta \dfrac{R^4}{y^3} \dfrac{dy}{dt}$.

## CHAPTER 9

(1) (*a*) $\bar{u} = \dfrac{U}{2} - \dfrac{h^2}{12\eta} \left(\dfrac{dP}{dx}\right)$; (*d*) zero.    (2) $\dfrac{4}{3\pi} \dfrac{Wh^3}{\eta l d^3}$.    (3) $F = 6\pi\eta \left(\dfrac{R}{c}\right)^3 LU$.
(4) 13·8 mm/s.  (5) 1·415 P.  (6) 125 kN/m²; 45 kN/m².  (7) −5·6 MN/m²; 28 MN/m².  (8) $\eta U/mh$; 120 kPa.    (9) 795 kN/m; 280 N/m; 336 W/m.

## CHAPTER 10

(1) 736 N; 86 degrees.  (2) 380 kW.  (3) 6 630; 3·9 N.  (5) 11 m/s or 21·5 knot; 2·4.  (6) $24/Re = 24\eta/\rho u d$.  (7) 0·058 mm.  (8) 1·3 m/s; 18·8.  (9) (i) $P = \eta^3 \varphi(Re)/\rho^2 l$, i.e. $P \propto 1/l$ if $\eta$, $\rho$, $Re$ constant; (ii) 2·62.    (11) 21·6 kg/h.
(12) 9·75 m/s; 14 mm.    (17) 2·83 knot; 128 × 10³.    (18) $\dfrac{Q}{g^{1/2}H^{5/2}} =$

$\varphi \left(\dfrac{Q\rho}{H\eta}, \dfrac{Q^2\rho}{H^3\sigma}\right)$.    (19) $\dfrac{Q}{\omega D^3} = \varphi \left(\dfrac{\Delta p}{\rho\omega^2 D^2}, \dfrac{\eta}{\rho\omega D^2}\right)$; 611 rev/min; 2·2 m; 0·842 m³/min.

## CHAPTER 11

(1) 0·628 m/s; 1 350 m³/h.  (2) 4·75 m³/s; 1·6 m.  (3) depth 0·384 m; diameter 0·64 m.  (4) $f = 2g/c^2$; 1·33 m³/s.  (5) vertical diagonal 4·25 m.  (6) 615 mm.  (7) 0·27.  (8) 4·17 mm.  (9) 19·5 mm; 0·774 m³/h.  (11) 59·5 m³/s; 1/2 420.  (12) 3·18 m.  (13) (*a*) 1·57 m; (*b*) 16·9 m.  (14) 1·9 m bed-width and depth.

## CHAPTER 12

(1) 41 W; 6·19 m.  (2) 0·304 m³/s; 0·278 m³/s; 0.  (3) 6 min.  (4) (*a*) 0·315 m³/min; 1·716 m³/min; (*b*) 90 mm.  (5) 2 030 m.  (6) 36 kW; 4·5 m/s.  (7) New, overall; 1·48 m³/min; 1·37 m³/min.  (8) 5; 1·5 kW.  (9) 3·28 m³/min. (10) 0·198 m³/s; 0·396 m³/s.  (11) (*a*) 47·6 mm; (*b*) 386 kW.  (12) (*a*) 52·2 m/s; (*b*) 54·8 m; (*c*) 6 m.  (13) 0·412 m³/min.  (14) 76 per cent; 13·6 per cent; 1·5 per cent.  (15) (*a*) 55 m; (*b*) 36·4 m; (*c*) 19 mm; (*d*) 6·6 kW.  (16) 119 kW.  (17) 0·133 m³; 485 kN.  (18) $\delta P = \rho l \dfrac{du}{dt}$ or $\dfrac{\rho l}{a}\dfrac{Q}{t}$ ; 200 kN/m².

(19) (*a*) 50 kN/m²; (*b*) 3·04 MN/m².  (20) (*a*) 3·65 MN/m²; 1·435; (*b*) 0·5 MN/m²; 0·25 MN/m²; (*c*) $\delta P/(\text{MN/m}^2) = 2\cdot5/(t/s)$.

## CHAPTER 13

(1) $\dfrac{1}{\rho}\dfrac{dp}{dh} = \dfrac{-\gamma g}{a^2}$ ; $\dfrac{P_2}{P_1} = \epsilon^{-\gamma gh/a^2}$.  (2) 1·235; 7·38 km; 0·46; 0·383.  (3) 0·62. (4) (*a*) −5·8°C; (*b*) 0·91 kg/m³.  (5) 1 K per 100 m.  (6) 0·717; 22 kN/m². (8) $u = \sqrt{\left\{\dfrac{1}{6}\left(u_0^2 + 5a_0^2\right)\right\}}\cdot$  (9) 173 kg/min.  (10) $u = \left\{\dfrac{9}{89}\left(u_1^2 + 5a_1^2\right)\right\}^{1/2}$ ;

$\rho = \rho_1 \left\{\dfrac{16}{89}\left(\dfrac{u_1^2}{a_1^2}+5\right)\right\}^{1/2}$.  (11) 5·48 kg/s.  (12) (i) 44·1 kN/m²; (ii) 55·5 kN/m²;

(iii) 50·7 kN/m².  (13) $p_1^2 - p_2^2 = \dfrac{4fl}{d} RT \left(\dfrac{\dot{m}}{A}\right)^2$ ; 5·49 m³/min.  (14) 7 per cent.

(16) 3°C; 0·39  (17) $p = p_0 \left(\dfrac{5 + Ma_0^2}{6}\right)^{7/2}\cdot$  (18) 0·45 per cent; 8·3 per cent.

(19) 0·37 kg/m³.  (20) (i) 178 m/s; (ii) 172m/s; (iii) 3·5 per cent.  (21) 0·577; 101°C.  (22) 400 m/s; 825 kPa; 1·09; 0·927.

## CHAPTER 14

(1) 9·65 m³/s; 2·77 m.  (3) 24·5 min.  (5) 520 × 10⁻³ m³/s.  (6) 4·3 and 0·326; 460 mm; 434 mm; 1·8 kW.  (7) $d^2 = \dfrac{d_1}{2}\left\{\sqrt{\left(1 + \dfrac{8u_1^2}{gd_1}\right)} - 1\right\}\cdot$  (8) 18·65 m³/s.

(9) 4·48 m.  (10) (*a*) 375 mm; 338·4 mm; 11 per cent; (*b*) 1·98 m/s; (*c*) 1·616 m/s; (*d*) 1·31 m/s.

## CHAPTER 15

(1) 312 rev/min; 4·88 MW.  (2) 13·4 m/s; 60°; 6·9 m/s; 83 per cent.  (3) 9 degrees.  (4) 31·5°; 26·5°; 97·8 per cent.  (5) 84° 40'; 10° 42'.  (6) 11·9 m; 1·52 m; 3·9 m.  (7) 36·5 rev/min; 95·5 W.  (8) 278 Nm.  (9) 98; 147; 245; 343.  (10) 7 m/s; 84 kW.  (11) 100 mm water.  (12) 0·34 m³/min; 0·22 m³/min; 27·6 h.  (13) 558 rev/min.  (14) 10·5 m; 78 per cent.  (15) $Q_c$ = 1·6 m³/min; $Q_p$ = 1·9 m³/min; ratio 1·19.  (16) $100 \left(\dfrac{\text{rev}}{\text{min}}\dfrac{(\text{kW})^{1/2}}{(\text{m})^{5/4}}\right)$.

(17) 5·95 m; 65·5 rev/min; $735 \left( \dfrac{\text{rev}}{\text{min}} \dfrac{(\text{kW})^{1/2}}{(\text{m})^{5/4}} \right)$.    (18) 443 kW.    (19) 290 rev/min; 1 m. (20) (*a*) (i) 1960; (ii) 5; (*b*) (i) 2 260; (ii) 13·5 m³/min; (iii) 344 kW. (21) 2·9. (22) 286 mm; 11 stages. (23) 173 rev/min; 1 270 kW; 27·8. (24) 1·134 5; 43·8 kN; 2·77; 266 kPa. (25) (*b*) 6·25 × normal full-load torque.

## CHAPTER 16

(1) 63·7 m; 47 m; 28·3 m above atmospheric. (2) 76·5 rev/min; 6·6 W. (3) 73·5 rev/min. (4) 40 rev/min. (6) Suction 3·5 m; 7·9 m; 10·4 m. (7) (*a*) 7·56 MN/m²; (*b*) 7·12 MN/m²; (*c*) 6·24 MN/m². (9) 58·5 kN/m². (10) 10 s. (11) 31 kW.

# Index

# K

Kaplan, propeller turbine, 433, 455
Kármán, von, 292, 305

# L

laminar flow, 172, 180, 183, 185
lapse rate, 365
lift coefficient, 231, 258
lock gate, 37, 41
losses,
    at bends, 325
    at restrictions, 316
    contraction, 318
    effect of, 77
    expansion, 314

# M

Mach
    angle, 390, 411
    number, 241, 255, 385
    waves, oblique, 389, 391, 405
Magnus effect, 169
Manning's formula, 280, 301
manometers, 55, 58, 61, 64
meniscus, 51
metacentric
    height, 6, 11, 18
    radius, 14
meter, current, 92
Michell, 198, 216, 220, 269
momentum, 139
    force, 145, 151
    pressure, 153
Moody chart for pipe flow, 299

# N

nappe, 106
Newton's second law of motion,
    113, 115, 138
Nikuradse's curves, 261, 297
non-dimensional products, 246, 249,
    256, 265
notch,
    rectangular, 106
    vee, 105
notches,
    dimensional analysis of, 251
    dynamical similarity over, 251
nozzles, 117, 344
    choking of, 378, 384, 404
    convergent, 380
    convergent-divergent, 79, 403
    effect of back pressure on, 403

over-expanded, 404
supersonic, waves in, 390
under-expanded, 404
velocity and area changes in, 378

# O

oiliness, 176
open channels, varying flow in, 410
orifice,
    flow through an, 83
    metering, 87

# P

Papin, 428
parameters, non-dimensional, 246, 249,
    256, 259
Pelton wheel, 69, 140, 453, 459
periodic time of roll, 24
Petroff's equation, 175, 209
pi($\pi$) theorem, 247, 249, 265
pipe friction, gas flow with, 375
pipes,
    branched, 337
    smooth, resistance of, 289
    turbulent flow in, 292
Pitot tubes, 89, 90, 104, 386, 400, 402
Poiseuille, 171, 180
positive-displacement machines, 486
potential,
    energy of a fluid, 46
    flows, 168, 191
power, unit, 459, 465
Prandtl, 91, 188, 292, 399
Prandtl-Meyer expansion waves, 389,
    391, 404
pressure,
    absolute, 49
    at a point, 27
    atmospheric, 49, 54
    centre of, 36
    critical, 386
    difference, measurement of, 55
    gauges, industrial, 66
    head, 45, 46
    inertia, 355
    in moving fluid, 114
    momentum, 153
    saturation vapour, 50
    stagnation, 372, 385
    vapour, 134
    variation of, with depth, 28
propeller, 159, 273
propulsion, efficiency of, 156, 158, 161
pumps,
    axial-flow, 149, 434, 437
    axial-piston, 504
    boiler feed, 456

# U

units, xiv, xv, xvi
unit power, 457, 459, 465
unit speed, 457, 465
unsteady flows, 130

# V

valve,
  closure, 133, 352
  ported, 327
  seated, 486
velocity, temperature equivalent of, 369
velocity
  defect law, 303
  laws for pipe flow, universal, 304, 307
  of approach, 81, 87, 110
  of sound in fluid, 353
  of whirl, 435
  triangles for axial-flow machines, 433, 434
    for radial-flow machines, 438
vena contracta, 83, 85
venturi
  flume, 423
  meter, 79, 82
viscometer,
  Redwood, 178
  Ubbelohde, 178
viscosity,
  dynamic, 170, 172
  kinematic, 176

viscous
  action, 169
  films, 199, 218, 225
Viviani, 47
vortex,
  combined or Rankine, 126
  forced, 126
  free, 124

# W

wake, 188, 190
wall velocity laws, 304
water hammer, 134, 352, 355, 490
wavelet,
  gravity, 409
  surface, 409
  velocity of a, 384
wave-making
  forces, 241
  resistance, 243
wave, standing, in channel flow, 418, 422
waves,
  Mach, 389, 391, 405
  shock, 391, 394, 400
  stationary, in supersonic nozzle, 390
weirs,
  broad-crested, 253, 413, 415
  dimensional analysis of, 253
  dynamic similarity over, 251
  effect of, on channel flow, 418
  rounded, 413
whirl, velocity of, 435
wind tunnel, 75